The Truth About the Wunderwaffe

IGOR WITKOWSKI
THE TRUTH ABOUT THE WUNDERWAFFE

RVP Press
New York

RVP Publishers Inc.
95 Morton Street, Ground Floor
New York, NY 10014

RVP Press, New York

© 2013 Igor Witkowski / RVP Publishers Inc., New York

The Truth About the Wunderwaffe is an updated and extended edition of *Truth About the Wunderwaffe*, published in Poland in 2003 by European History' Press, Warsaw.
European History' Press is an imprint of the WIS-2 Publishing Company.

The original Polish edition was published in 2002 by WIS-2 Publishing Company, Warsaw.

Translated by Bruce Wenham
Cover designed by Tomasz Maros
Computer renderings by M. Ryś

All rights reserverd. No part of this book may be reproduced in any form or by any electronic or mechanical means, including information storage and retrieval systems, without persmission in writing from the publisher, except by a reviewer who may quote brief passages in a review.

For the illustrations used in the book every effort has been made to trace the holders of copyright and to acknowledge the permission of authors and publishers where necessary. If we have inadvertently failed in this aim, we will be pleased to correct any omissions in future editions of this book.

RVP Press™ is an imprint of RVP Publishers Inc., New York
The RVP Publishers logo is a registered trademark of RVP Publishers Inc., New York

Library of Congress Control Number: 2012946353
ISBN: 978 1 61861 338 7

www.rvppress.com
www.igorwitkowski.com

TABLE OF CONTENTS

ix Introduction

PART I: WEAPONS OF A TECHNOLOGICAL TURNING-POINT

3 **The Conception of Vengeance Weapons**
4 The V-1
10 The V-2
28 The V-3
33 The Rheinbote
35 Other Vengeance Weapons

45 **Luftwaffe: A Time of Quest**
45 The Me 262
51 The Me 163
53 The He 162
56 The Ho 229
60 Messerschmitt P-1101
62 Focke-Wulf Ta 183
68 The Jet Bombers

79 **Death Rays**
99 **The New Tools of Blitzkrieg**

113 **New Concepts for Conventional Weapons**
113 Energy Emitters
114 "Invisible" Aircraft and Vessels
116 Plastics
118 Underwater Warfare
123 Concrete Ships
123 Recoilless Weapons
128 Unusual Ideas
138 New Generation Small Arms
142 Infrared
152 Aircraft Carriers
154 Unusual Energy Sources

PART II: THE WEAPONS, WHICH COULD CHANGE THE COURSE OF WAR

161 **The Turbulent Development of Guided Weapons**
161 The "Feuerlilie"
163 The "Wasserfall" (C-2)
166 The "Taifun"
168 The Henschel Hs 117
170 The "Enzian"
174 The Rheintochter
177 The "Natter"
179 Air-to-Air Rockets
182 Air-to-Surface and Surface-Surface Rockets
184 Guided Bombs
186 Homing Warheads

195 **Ramjet Propelled Fighters**
205 **An Arsenal More Powerful Than American Nuclear Bombs**
221 **Biological Weapons**
224 **Nuclear Weapons**
227 **German Projects in Light of the American Drainage of Technology (Operations "Paperclip" and "Lusty")**

PART III: "KRIEGSENTSCHEIDEND"

239 **The "Bell"**
295 **More About the Physics of the "Bell"**
307 **How Hitler Planned to Win the War in 1944-1945**

324 Bibliography
329 Acknowledgments
330 About the Author

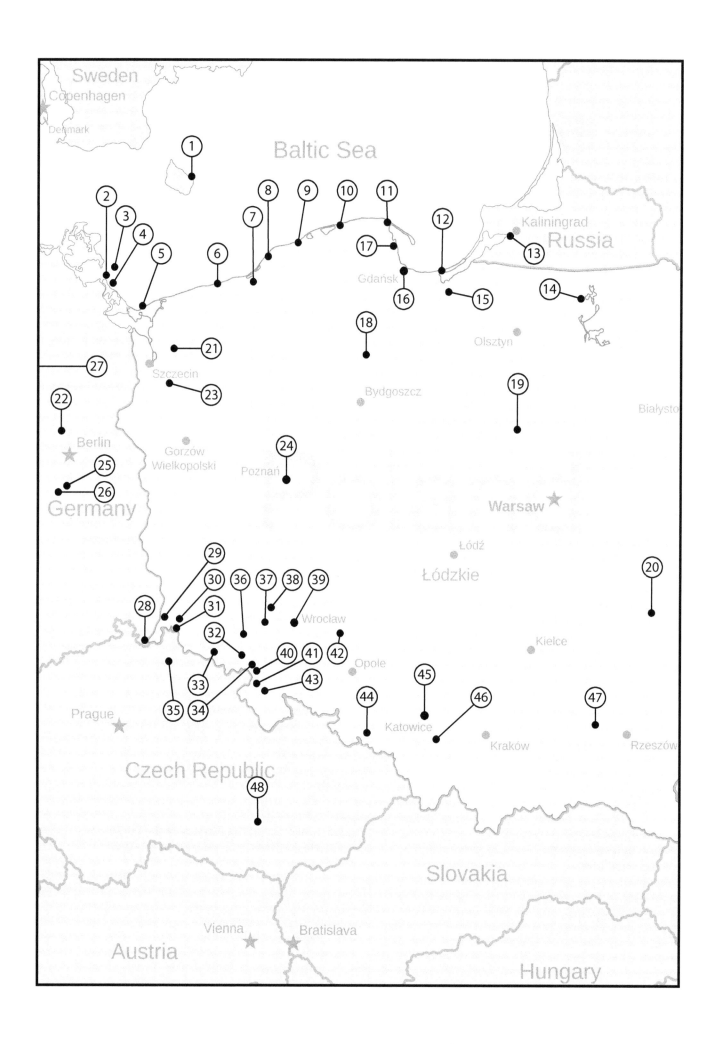

Selected sites related to research, development, and production of German special weapons during WW II, east of Berlin. Also the concentration camps without immediate extermination centres. Post-war borders.

1. Bornholm Island: "target" for the Rheinbote rockets
2. Peenemünde: V-1, V-2, etc.
3. Greifswalder Oie: tests of A-3 and A-5 rockets
4. Karlshagen: Elektro-Mechanische Werke, production of missiles
5. Międzyzdroje/Misdroy: V-3
6. Kołobrzeg/Kolberg: H. Coler
7. Koszalin/Köslin: long range missile schools
8. Darłowo/Rügenwalde: heaviest artillery test range, concrete ships
9. Ustka/Stolpmünde: firing range, also school for the crews of the new generation of submarines (T. XXI, XXIII)
10. Łeba/Leba: missile test range
11. Władysławowo/Grossendorf: experimental test range of the SS (detailed purpose unknown)
12. Stutthof concentration camp
13. Jesau: trials of the Hs-293 missiles
14. Kętrzyn/Rastenburg: Führer's main command post
15. Elbląg/Elbig: underwater silos for the V-2
16. Gdańsk/Danzig: stealth technology
17. Babie Doły, Oksywie/Hexengrund, Oxhöft: Kriegsmarine's evaluation centre. New types of torpedoes, midget submarines, propulsion systems
18. Bory Tucholskie/Tucholer Heide: V-1 and V-2 launch sites in the area near the Gacno village
19. The "Nord" test range: Schmetterling missiles
20. Majdanek concentration camp
21. Mosty/Speck: underground ammunition factory, also laboratory working on nuclear bomb
22. Oranienburg: nuclear laboratory (Auerwerke), also the Sachsenhausen concentration camp
23. Stargard, Miedwie Lake/Madüsee: tests of air-to-surface guided weapons
24. Pokrzywno/Nesselstadt: biological weapons
25. Kummersdorf: test range for tanks and artillery
26. Gottow: works on experimental nuclear reactor
27. Rechlin: weapon test centre of the Air Force
28. Zittau: Jägerstab
29. Zgorzelec/Görlitz, Łąki village: underground V-2 factory
30. Lubań/Lauban: GEMA-Werke
31. Leśna/Marklissa: V-2 engines factory (VDM)
32. Książ/Fürstenstein: Jägerstab's R&D dept., SS research
33. Kowary/Schmiedeberg: heavy water production plant, nuclear research facility, uranium mine
34. "Riesa" ("Riese"): underground complex, not finished
35. Zelezny Brod: command planning centre for the "guided, strategic weapons," not finished
36. Gross-Rosen concentration camp
37. Środa Śląska/Neumarkt: Wehrmacht's laboratories
38. Brzeg Dolny/Dyhernfurth: chemical weapons
39. Wrocław/Breslau: Rheinmetall plant and other objects
40. Ludwikowice/Ludwigsdorf: underground complex dedicated to weapons of mass destruction
41. Ścinawka Średnia/Mittelsteine: production of V-1 and V-2 components
42. Namysłów/Namslau: infrared technology
43. Kłodzko/Glatz: production of components for the V-1 (AEG)
44. Racibórz/Ratibor: graphite productions for nuclear research (Siemens)
45. "Udetfeld" (Mierzęcice): ME 163
46. Oświęcim/Auschwitz concentration camp
47. Blizna: V-1 & V-2 tests
48. Brno/Brünn: SS research and development

Above: Joseph Goebbels during a visit at Peenemünde
Left: Albert Speer, Erhard Milch, and Willy Messerschmitt

INTRODUCTION

The most secret and technically avant-garde weapons of the Third Reich are a complex issue, forcing reflections of a disparate character. They directly relate to both the management of the armament industry, the science as such, as well as the barbarous concept of a return to slavery and even to the role of the SS in the whole system of the war-time economy. This book is devoted only to technical issues, however it is worth making ourselves aware of the wider context of related phenomena. This is not only a historical problem—the conclusions resulting from these reflections have a timeless character and may also be useful in the future. If we concentrate on technical issues, it will become evident that a dominant feature of the whole scientific and economic system was its simply incredible efficiency. This is often presented and understood as some kind of "trump" of National Socialism, which is not only a false conclusion but also a convenient escape from a matter-of-fact analysis based on information.

Obviously it is not for me to justify these kinds of alleged relationships, I suspect that they are the result of superficial assessments. I myself, when analysing the functioning of science and the economy in the Third Reich, found no arguments or circumstances which would confirm this. I have a feeling that technological progress was accomplished not so much due to fascism, but against it. Hitler once said after all that: "I do not wish any intellectual upbringing whatsoever, knowledge may only demoralise youth." The only typically Nazi element, which appeared in this system and left its mark on the management of science and technology—all the time controlled by Hitler—was the party. It did not bring however any particular constructiveness, on the contrary, the blind terror and ignorance of the ruthless, incompetent official, blessed with excessive power. Not only did victims of the system share this view, but also many individuals from the management of the Ministry of Armament and War Production (Reichsministerium für Rüstung und Kriegsproduktion) with Minister Speer at the head. In the NARA archive I found among other things a report from the post-war interrogation of one of his officials, Kurt Weissenborn, who was Chief of the Weapons Office in Speer's Ministry. Here is how he described the influence of the "ideological element" on the war-time economy:

At Potsdam railway station a strange Mitropa train awaits under steam. Its third carriage is the restaurant car. It is the train Hubertus, belonging to a party member, Saur, Chief of the Technical Department in the Reich's Ministry of Armament and War Production. Engineers and industrialists of every kind, as well as civil servants from Speer's Ministry have already been sitting in the train for half an hour. Then a small man with a tense "ascetic" face, typical for pompous personages in brown shirts, hurries through the barrier, followed by members of his personal staff. The train departs. Like

Albert Speer and Field Marshal Erhard Milch

Karl Otto Saur

a storm he attacks the centres of the armament industry. Herr Saur's technical staff press on through the factory workshops, with him at the head brandishing his weapon and screaming in his booming, sometimes faltering voice. It takes him only a few minutes to fire the factory directors, replace the chief engineers, and hurl reprimands at members of his own staff in full view of everybody. On the long itinerary of the train's journey (having all possible right of way) far more engineers and industrialists wait for hours on platforms until they are taken onto the train for "interrogation," to be dismissed afterwards like schoolboys. As soon as the conversations and interrogations are over, the train driver receives an order by telephone to stop at the nearest station and all of a sudden the dismissed people are standing on a forgotten platform watching the departing "Hubertus" train. There was not an hour day or night in which people wouldn't be "dealt with" within a few minutes after hours of waiting. No technical intelligence, technical excellence or intellectual leadership whatsoever was allowed to speak here—only the brutal treatment of the individual. Saur introduced a caste division in the industry. But the industrial machine, in other cases very sensitive, fends off these blows, trained and learned in endurance at Gestapo interrogations and in everyday contacts with the party apparatus.

If however somebody tried to defend himself, he was ruthlessly silenced and removed from his post. If he was sufficiently young, he would be a soldier the next day. It wasn't the fear of a shortage of state contracts that made industry so conform to demands, but the fear of every man for himself and his family, which delivered obedience and the ability to endure the most humiliating treatment. Saur ruled his zoo in a lordly fashion with the help of abuse. I myself saw sixty year old industrial engineers crying in full view of everyone, for despite their toils day and night, they were treated like dogs. At the same time these difficulties were most often of such a nature, that they were unable to overcome them. It was easiest however for the incompetent High Technical Office to rid themselves of responsibility by casting the blame at an innocent man.

As I have mentioned, this was the only typically Nazi element in the whole system, which exerted its influence in every respect. Of course there was still the SS, but the influence of this organisation had a different character. The SS was responsible for the harnessing of slave labour in the gears of the economy. This undoubtedly had an influence on the phenomenon that despite ever greater supply limitations, the deepening deficit of strategic raw materials and infernally destructive Allied air raids, the production of this economy continually grew and at a rapid rate, altogether a three to four-fold increase was recorded with rapidly growing product modernity. The crisis and consequently also maximum of production fell in the summer of 1944. The contrast between the situation and the performance of the German economy is particularly evident if we take into account that carpet-bombing had commenced as early as the spring of 1942, whereas in 1943 almost twice as many aircraft were produced for the Luftwaffe as in the previous year, and in 1944 again almost twice as many as in 1943 (respectively 15,409, 24,807 and 40,593 aircraft). This paradox was best described (in his "Memoirs") by its chief author and unquestioned management genius, the Minister of the Third Reich's Armament Industry, Albert Speer. Here are his words[1]:

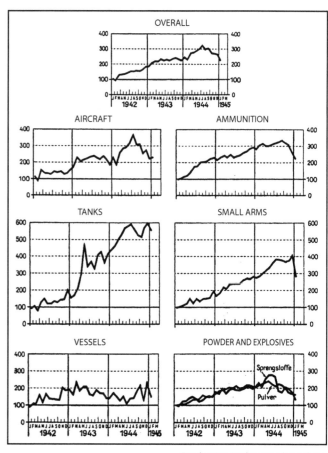

Production in the main branches of the German armament industry

Already in the half year after I had taken office we had achieved a significant increase in production in all domains turned over to us. Production in August 1942, in accordance with the "Coefficients of German Armament Production," had grown in comparison to the level from February by 27 percent in the case of small arms and 25 percent in the case of tanks, whereas the production of ammunition had almost doubled (a growth of 97 percent). All in all armament production had increased during this period by 59.6 percent. Obviously we had set in motion the reserves not used so far.

After two and half a years, despite Allied air raids beginning in earnest, we had increased total armament production: from a coefficient of 98 for 1941 to a peak coefficient of 322 in July 1944. At the same time employment had increased only by around 30 percent. It had been managed to decrease the workload by half. We had accomplished precisely that what Rathenau had prophesied in 1917 as an effect of rationalisation: "The doubling of production with the same equipment and the same labour costs."

Further on he also mentioned that:

The state of rapture of the first months, which the creation of a new apparatus managing industry had got me into, the successes, the expressions of acknowledgement, soon

Concentration camp prisoners working in an underground factory

passed; the period of the greatest concerns and intensifying difficulties had come. They resulted not only from labour force problems, the unsolved questions of material supply and courtly intrigues. The air raids of the British Air Force and the initial damage caused by them in the area of production made me temporarily forget about Bormann, Sauckel and the Planning Headquarters. However the prerequisites of an increase in my prestige were at the same time inherent in all of this. For despite gaps arising in production, we produced not less, but more.

The influence of the concentration camp system on the successes of the armament industry was considerably less than is commonly assumed. All in all 9 million people passed through

Semiconductor-based infrared detectors, manufactured in Germany during the war

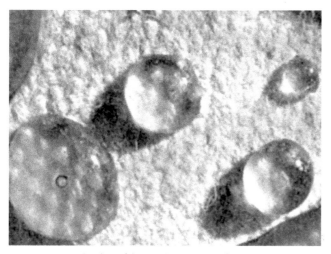

Synthetic fabric undergoing tests for water-permeability

these camps, obviously only a section of whom were exploited in industry. Apart from this, these people generally worked for a short time, since the tragic living conditions created a huge mortality rate. For these same reasons the effectiveness of such a worker was low.

Nevertheless, industry could only implement technical breakthroughs, which had already been accomplished. The key issue from the point of view of this book's subject matter is not therefore the management of the war machine as such. This problem can be narrowed down to matters related to the functioning of science. Since this was the source of the most important and greatest phenomenon from the point of view of the present day. Really unusual things happened there.

We must realise that this was a period of simply unimaginable acceleration in scientific and technical progress. The technology from the beginning of World War II did not differ much in principle from its state at the end of World War I. Let us take a look at aviation: aircraft with mostly wooden structures, covered in canvas still dominated. However a few years was enough for fully metal jet fighters to appear, equipped with radars and guided weapons. The production of supersonic fighters was prepared with propulsion of an even newer generation, for example ramjet and ram-rocket. The concept of a vertical take off and landing (VTOL) fighter was tested practically (the Triebflügel, Wespe…), as well as the concept of decreased radar detectability. Submarines were built able to remain submerged for weeks on end, and a whole group of guidance (homing guidance) systems existed, based on semi-conductor detectors. The materials presented in part three of this book prove that yet another step forward was made…

It was likewise for example in the domain of armoured vehicles. The beginning of the war passed away under the banner of tanks designed as a means of infantry support, with rather symbolic armament and armour so poor that it could be penetrated by a rifle armour-piercing projectile. Horses were still the core of most armies. At the end of war it was already just an issue of production capabilities to bring into service a tank able to fight day and night, with a gun stabilised during travel, powered by a gas turbine, with a hydro-kinetic power transmission system and hydrostatic turning mechanisms, and a defensive installation against chemical and biological weapons…

It was likewise in most other domains.

These were not only far greater changes than those which it had been possible to observe in the interwar period of the 1920s and 30s, but greater than those which have taken place since the end of World War II up until the present day (50 years!). Practically all modern trends in weapons development were initiated precisely at that time. It looks as if this was the greatest technological leap in the history of our civilisation, which is undisputedly a subject worth deeper reflection… The value of these achievements is clearly proven by the range of post-war drainage of German scientific and technical ideas by the USA and USSR (among others the number of captured patents reached 340,000). At the end of 2001, as one of the first researchers from the outside, I had the opportunity to thoroughly analyse historical documents from the US National Air Intelligence Center at Wright Patterson Air Force Base, where the technical intelligence service headquarters was located immediately after the war. From talks with some old airbase employees it clearly appeared that the

> Believe this development would be important for Pacific War. . . . The research directors and staff realise impossibility for continuation of rocket development in Germany. . . . They are anxious to carry on their research in whatever country will give them the opportunity, preferably United States, second England, third France.
>
> Excerpt from letter is as follows:
>
> Dr. von Karman estimates that here at this one place there is information immediately available that would take us at least two years of research in the U.S. to obtain. Also enough here to expedite our jet engine development program by six to nine months.
>
> Recommendations to the Commanding General, U. S. Strategic Air Forces in Europe (Lt. Gen. Carl Spaatz), from his Deputy (Maj. Gen. H. J. Knerr) included the comment, "Occupation of German scientific and industrial establishments has revealed the fact that we have been alarmingly backward in many fields of research. If we do not take this opportunity to seize the apparatus and the brains that developed it and put the combination back to work promptly, we will remain several years behind while we attempt to cover a field already exploited." In addition, it was suggested that immediate dependent families be allowed to accompany the scientists, a move considered essential in view of the political and economic factors involved in their general uprooting. As these and other communications indicate, it was believed urgent that immediate action be taken to transport scientists to the United States without delay. The motivating reason was to insure the employment of those top-ranking scientists who were without question the

A document from the archives of the National Air Intelligence Center

end of the war, when various German prototypes and plans were tested and examined, was for the Americans a period of "technical dazzle." In the airbase's files I came a cross among other things the following opinion of General Hugh Knerr, in 1945 deputy commander of US forces in Europe for administration:

Occupation of the German scientific and industrial establishment has revealed the fact that we are alarmingly backward in many fields of research. If we do not take this opportunity to harness the apparatus and minds that developed it, we will remain many years behind, simultaneously realising work that has already been accomplished.[2]

Afterwards, American president Eisenhower stated that "German technology outstripped Allied by a good 10 years. Fortunately the German leadership didn't exploit this superiority and realised too late what opportunities this had placed before them."[267]

This matter may deliver many valuable conclusions, even for the future, if we examine it in categories of development theory; why did this progress proceed so quickly, or, on the other hand, why was it relatively slow after the war?

The easy answer attempting to explain everything by the existence of total war, in reality concludes nothing, since the war afflicted a dozen or so countries, not exerting such an influence on them and besides there have already been many wars in history. Let us remember that what is involved is the leap of a few generations of equipment in the course of approx. 5 years. Today the average time to develop a new tank or fighter is the order of 15 years.

I have to admit that I have never encountered any deep analysis of this phenomenon, in connection with this I will naturally present my own opinion about the reasons which led to it.

There were of course many reasons and without doubt the strong pressure of state institutions was one of the fundamental. In the Third Reich however an additional factor was busy at work. Research and development work proved to be relatively profitable, both for small companies as well as large consortiums.

Karl Otto Saur, who was responsible for the organisation of industrial production in Speer's Ministry, stated during an interrogation on August 9, 1945, that the system of "fixed prices" imposed by the authorities greatly limited the profits accruing from mass production. According to him the consortiums "earned money (…) not through the amount [of produced goods], but through the constant development of new, complex types [of weapons]." For in this case profits were not so strictly limited as there was no way of unequivocally assessing the costs of work.

A crucial and rarely mentioned motor of technological progress in the Third Reich was also the necessity of the rationalisation of technological processes, caused by the deficit

A German electron microscope; the 1930s enabled, among other things, genetic research to be carried out, based on observations of the changes in chromosomes. Thanks to this, a cancer prevention programme was launched, that was about 30 years ahead of other countries; for instance strict norms appeared as early as before the war, regarding the maximum concentration of carcinogenes on the work-stand.

of raw materials and the labour force. In this way for example machine cutting was replaced by plastic tooling (forging, pressing, pressure welding and the like) to a greater degree than elsewhere, which was considerably less material and energy consuming. This led to such breakthroughs like the introduction to service of the MP-43 automatic carbine, almost entirely made with "plastic" technologies, or the miniaturisation of vacuum tubes to thimble size. In addition, it also promoted the production of plastics. However, even the aforementioned were not the most important causes. Another two rarely appreciated factors played an important role here:

1. It is worth reflecting on where progress in general comes from. In my opinion and certainly not only mine, it may be defined as the projection of a culture, understood of course not in the way of arranging cutlery on a table, but as a set of ideas, which were created by a given civilisation. In this case the most valuable achievement of Europe in the last few centuries comes into play, which is the tradition of intellectual criticism with its principle manifestation: the relativism of ideas. In moral categories it is often characterised negatively (after all, the technological achievements of the Third Reich also have a dubious moral dimension, if scientific and technological progress may be valued from a moral point of view). Relativism of ideas is however a necessary condition of progress, without it a given culture is as a general rule inclined towards self-preservation. This was not a sufficient condition but only a starting point to initiate a certain process, requiring the presence of one more factor:
2. When one analyses German war-time work, one is struck by the quite unusual way that science was controlled, which was harnessed to serve the armament industry. Research work was conducted simultaneously in many varied directions, with considerably limited "initial selection" on the part of academic science, which presently is very rigorous. Science was simply not controlled by the professors, at any rate not on the basis of full control. Thus the outright proverbial ostracism of academic science in the face of new ideas was suppressed or at least seriously restricted. Without this, even the V-2 rocket would probably have never been built. Initially British intelligence, based on the opinion of various professors, did not believe that it was possible to build such a large liquid propellant rocket. Its first fragments, delivered by Polish Home Army Intelligence (the AK, a Polish resistance organisation in occupied Poland) caused much consternation. Today these achievements seem to be quite plain to us, but then even the design of the swepted wing was an important psychological breakthrough. Let us remember how for a long time, up until the turn of the 1930s and 40s, the breaking character of armoured forces had not been realised in the majority of armies and tanks were treated as a means of infantry support, practically wasting their potential (France had more tanks than the attacking German armies). It was likewise with the concept of the diving bomber. It was likewise… it is possible to quote countless examples.

The more unconventional an idea, the greater and more irrational is the resistance to it being implemented. Einstein would probably have never convinced Roosevelt to produce a nuclear weapon, if not for the information about analogous German work. For even in 1923, Robert Millikan, a generally recognised authority and recipient of the Nobel Prize, categorically ruled out the possibility of an atomic nucleus ever being shattered. In 1932, the outstanding American astronomer, F.R. Mouton, denied the possibility of human space flight, on the same principle.

Whoever has spoken with professors about new ideas, knows that the resistance is very strong. The main criterion is whether they can be explained by already existing knowledge. In effect, science does not engage in what it does not know already, but especially in what it does not understand. This is presently the greatest barrier to development. As a result the exploitation of such phenomena, like e.g. the "separation of magnetic fields" described in part three, is currently almost impossible. Precisely because it is something authentically new. The rule dominating in these issues is that from whichever level we would not observe the world, we will not see more than is in ourselves anyway. Or otherwise: "There is no idea in the head—the eyes do not see facts." In the collective perception, breakthroughs appear suddenly, as a result of some kind of revelation, but in reality this is never the case. Information about a given aspect of reality is always present in the environment, but not always is noticed. From here it is only a step away from matching reality to existing theories.

The way in which the Third Reich was forced to change the way that it controlled science, is not therefore only a simple optimisation of what already exists, but can bring changes that are quite simply revolutionary. Without this, regardless of the size of available funds, we will turn in the same magic circle anyway. Obviously such processes may only exist under certain conditions, at a defined level of social consciousness. Something opposed in relation to present mass culture (which rather sanitises true information) is necessary, that tradition of intellectual criticism, "of the positive relativism of ideas." However despite appearances, war is not at all necessary for this, and certainly not Nazism!

PART ONE

WEAPONS OF A TECHNOLOGICAL TURNING-POINT

Niedersachswerfen, production of V-1 / V-2

THE CONCEPTION OF VENGEANCE WEAPONS

The weapons in question in this chapter's title were, contrary to all pretences, not just the V-1, V-2 and V-3. This conception has been developed and also concerns a series of projects, which one could call "the second generation of vengeance weapons." What was their specificity based on?

One can obviously answer that question by referring to the German classification—simply by indicating projects from the "V" ("Vergeltungswaffen") series. This question however is not so simple, for there existed a series of continuations already devoid of such a definition, in spite of having many features in common with the V-1 and V-2. So we must outline an agreed boundary, separating "vengeance" from "conventional" weapons. It seems that in this first case what was involved was first and foremost the creation of long-range weapons designed not for combating military targets, but clusters of the enemy's civilian population. Obviously this agreed boundary will never be very clear, which in connection with the adopted criteria, to a lesser or greater degree, will also comply with many of the weapons described in the following chapters, and first and foremost also strategic post-war missiles

For this reason one should view this programme in a way as a historic achievement.

To begin with it is worth observing where did the concept of "vengeance strikes" and "deterrence," so significant in the future post-war period, stem from. Perhaps this will surprise certain readers, but the author of these concepts was Hitler, although initially their significance wasn't clear-cut, as were the targets and principles of using "vengeance weapons." The first reason why the Führer became interested in them was simply because of their modernity. The army needed an injection of state-of-the-art technology, which with its qualitative superiority would compensate for any numerical shortages and its radically new combat possibilities would reverse the direction of the shifting fronts on the map of Europe. In short these were purely military reasons and motivated by the calculation of combat possibilities.

When with time it came to the hard technical facts, a second dominating factor emerged in the development of the "Vergeltungswaffen"—the terror factor. This was chiefly a result of the specific "optics" and features of Hitler's character, who preferred, often in defiance of plain military calculations—offensive action and had a tendency to prefer the psychological effects of weapons to rational and purely military effects. This was evident from the course of many operations. Perhaps the man best knowing Hitler and his inner circle, Minister of the Armament Production, Albert Speer wrote after the war that: *"…even as a leader he mainly took into consideration the psychological and not the military effect of a weapon's operation. An illustration of this was his idea of installing sirens on the bombs dropped by Stukas, whose demoralizing action was for him more important than the explosive force of the bombs themselves."*[1]

The Vergeltungswaffen question took more realistic shape when the Germans began to lose the Battle of Britain. Regardless of the psychological factors existing so far, the necessity arose of possessing a weapon capable of taking over the task of the air force bombers, that is which could inflict long-range blows not only without air superiority, but also without the cover of one's own air force fighters at all. Hitler insisted on lingering, tormenting attacks, leading to the creation of a psychosis of permanent, inevitable and unpredictable threat in the enemy.

Wernher von Braun in the early 1930s. Who would then have thought, that his rockets would take men to the Moon?

The V-1

The author of the conception of the "flying bomb" (Fernbombe), or as we would say nowadays—the first cruise missile, was Doctor Fritz Gosslau from the aerial company Argus. The idea itself arose shortly before the outbreak of war. From the start it met with the negative approval of the Luftwaffe, yet in spite of this Gosslau resolved to carry out research with the aim of constructing the first prototype by the end of 1941.

He planned to use a simple and cheap jet engine to propel his bomb, like the pulsejet engine which was designed in his company in 1939 at the order of the Air Ministry. This was an engine characterized by a large unit fuel consumption, but its simplicity and lack of having to use materials in short supply were crucial advantages under wartime conditions. The engine was built without any rotor, that is without a compressor and turbine. It was based on the piston engine with a compression, combustion and exhaust cycle, only that this took place in a completely different way. The combustion chamber was comprised of a long steel pipe around 0.5 m in diameter with an open rear section and air inlet with a system of valves at the front. So that the engine could operate it required a certain initial speed and operated in the following way: the air pressure at the front opened the valve inlet flaps mounted on springs and the combustion chamber was filled with the fuel composition. Ignition followed which gave a sudden increase in pressure in the pipe. This caused the air inlet flaps—valves to close. The combustion gases were expelled by the force of their own pressure to the rear giving a recoil. The large length of the pipe caused negative pressure to build in the combustion chamber under the influence of gas inertia, the inlet flaps opened and the whole cycle repeated again. The sparking plug was only necessary to start the engine, after which the cycle was executed automatically until the fuel was exhausted or its supply cut off. In the Argus As 109 engine the cycle was repeated 40-45 times a second, giving a similar sound effect as in the case of a loudly operating piston engine. On the other hand this was associated with a considerable load on the construction as a result of arisen vibrations. On the basis of this engine Gosslau designed the construction of the "flying bomb" with a one tonne warhead. The construction of the fuselage and wings was steel with wooden elements (mainly the wings). The skin plating was made of thin (<3 mm) steel sheet. Contrary to the V-2, the construction of the V-1 and especially of its propulsion, was maximally cheap and simple. The first design bearing the target name "Fieseler Fi 103" was presented for evaluation at the technical department of the Air Ministry on June 5, 1942. The worsening military situation and chiefly losing the "Battle of Britain" had resulted in a change in the previous military attitude in relation to this type of weapon. Hitler demanded vengeance weapons which could be produced and employed on a mass scale. The Fi 103 project was given the highest priority still in the same month. At the same time the missile was given the "deceiving" military designation Flak Zielgerät 76 (FZG-76, Flying target-76). Marshal

The V-1 in flight

Initially he intended to put this conception into effect with the aid of a new type of bomber. However this would have been a strange kind of aerial warfare, because in order to wage it—as the aerial designer Ernst Heinkel recalls—only 40-50 machines were to suffice, which flying at a ceiling of around 14,000 metres, and speed of 750-800 km/h would have been out of reach of Allied fighters (this became possible in any case relatively quickly, at the end of the war, along with the introduction of jet bombers).

In order to achieve greater "intimidation" they were to fly day and night (Hitler was particularly excited by the vision of millions of people leaping up from their beds just because of a few bombers), but each was to drop only one large bomb. This may seem incredible, but Hitler genuinely believed that thanks to a "terrorist effect" so minimal but modern, his forces would be able to reverse the course of the war. It is especially odd that this conviction persisted for so long, even when it was clear that the Allied air raids, waged on an incomparably greater scale against German cities, weren't able to break the German will to fight.

The first in the series of vengeance weapons is obviously the V-1, though work on the V-2 had a considerably longer history.

V-1 launch from a catapult

Milch, responsible for this project on behalf of the Luftwaffe placed great hope in it. In connection with the change in the military situation the previous serious shortcoming which was limited accuracy, decidedly too small with reference to "normal" targets ceased to play a significant role. Technological demands and the peculiarity of the predicted military application matched the "optics" of Hitler: limited resources and a large psychological effect.

Improvement of the Fi 103 design was charged to the Luftwaffe Research Station near Peenemünde (Erprobungsstelle der Luftwaffe, Karlshagen) appointing in August 1942 Major Otto Stams as its commanding officer. A launch catapult was erected for the missiles on the grounds of a local airport. In order to be a match for the military requirements it was necessary to redesign many components or design them from scratch. This work was included in a wider programme of developing guided weapons for the Luftwaffe by the name of Vulcan. Apart from improving the engine and airframe the gyroscopic autopilot and launch catapult were designed from scratch. Specialists from the companies Askania (the autopilot) and Rheinmetall-Borsig were employed in this. The Germans managed to complete the first prototype for in-flight trials at a rapid pace on September 1, 1942, despite the fact that the engine hadn't yet been completely examined, especially at high speeds, and caused numerous problems. Until then in-flight trials had been carried out only on the Messerschmitt Bf 109 and Bf 110. To this end a trial was carried out in the wind tunnel of the military research institute in Braunschweig before launching the engine along with the whole missile. The results were astonishing. It became evident that the engine's thrust fell rapidly along with an increase in speed. At a speed of 600 km/h, lower by approx. 100 km/h from the predicted cruising speed, zero thrust was measured! But it became evident that the cause of such results had been a test fault caused by a change in both the engine's operating characteristics and instrument readings under conditions of vibrational resonance in the confined space of the tunnel.

The first trial with the aim of verifying the missile's aerial properties was carried out on October 28, 1942 by dropping a trial specimen without engine from a Focke-Wulf Fw 200 aircraft. It confirmed the very good aerodynamic characteristics that had been predicted.

The first launch catapult was built. Two rails were installed on a slanting concrete platform 80 metres long, along which the Fi 103 "undercarriage" propelled by the rocket engine was to slide. The solid propellant rocket engine gained a short-lived thrust of 30 tonnes giving the missile an initial acceleration of around 15 G (360 km/h). In the final section of the rail launch runner the "rocket sledge" was brought to a stop and the missile commenced independent flight. The resojet engine was started while it was still situated on the launcher—before ignition of the rocket engine. After first checking of the catapult's operation on a concrete block similar in mass to the missile—approx. 2,3 tonnes and checking the operation of the fuselage with the resojet engine during launch, the turn came for a test "firing" of the missile with its own propulsion, which took place on December 24, 1942. A trial launch had taken place somewhat earlier from a Fw 200 aircraft. This equally had a practical significance, as the possibility had been already taken into consideration of carrying this new weapon by aircraft.

However, the results of trials didn't fulfil all expectations, among other things the speed of horizontal flight was too low—of the order of 500 km/h, but in general they were considered successful.

The missiles were launched in an easterly direction over the sea, so that the flight path would run at a distance of approx. 50 km from the coast. By May 1943, two new types of catapult

Trials of the Argus engine, on a special test stand

were had been constructed in Peenemünde. A piston driven by the gas (oxygen) from hydrogen peroxide (perhydrol) which decomposed under the influence of sodium permanganate was used in them to accelerate the missile and not the rocket engine. This same composition was also used to power the turbine that pumped propellant in the A-4 (V-2) rockets.

On May 26, 1943, a conference of the "Research Council for Long-range Weapons" was organised in Peenemünde, in which Generals from the Air Force and Ground Forces took part: Milch, Keitel, Olbricht and Fromm as well as Ad-

Test drops of the V-1 from an He 111 bomber

miral Karl Dönitz representing the Navy, and the Armament Minister Albert Speer. The objective of the conference was to determine the value and role of the two hitherto existing "rival" vengeance weapons: the Fi 103 and the A-4. After a long debate it was agreed with the chief of the rocket establishment, Dornberger, that both types of weapon had their own advantages and disadvantages which mutually compensated each other and should be treated not as mutually exclusive but complementary. In this connection both projects received top priority. This was a big success for the Luftwaffe, for in contrast to the A-4 both trials of the Fi 103 carried out in full view of the council's members had ended in embarrassment—one missile crashed just after launching and the other never managed to take off. It was clear that the "flying bomb" required further improvement and many months of research work. In the meantime the means and possibilities of constructing combat launchers were analysed. It wasn't clear if permanent concrete installations should be favoured, well defended against aerial attack, or rather cheaper but in return more numerous field launchers, easier to camouflage. Finally in June 1943, it was decided on a compromise to build 4 reinforced concrete launchers and 96 field launchers off the coast of the English Channel. The formation of the first combat units was commenced. Nothing meanwhile indicated that the technical problems would be quickly dealt with. Of the 68 missiles launched from ground launchers during the first two months since the conference in Peenemünde only 28 had had flights regarded as favourable (41%). Many of the missiles had crashed for unclear reasons just after launching. The Germans still hadn't managed to examine some of the target systems, including the crucial navigational system. The assumed deadline of commencing attacks on London December 15, 1943 looked therefore completely unfeasible. The production plans were equally unrealistic, assuming the delivery of 100 missiles in August 1943, 500 in September, 1,000 in October, 1,500 in November and later gradually from 2,000 to 5,000 monthly to the Luftwaffe. Simultaneously the scatter of tested missiles was so large that it didn't give a suitable probability of a direct hit on London. On September 10, Marshal Milch stated: *"I will be satisfied if the Fi 103 will work at all, so that we will be able to use them in combat."* This contrasted with his extremely optimistic expectations as to their military effect. He considered that London would not endure more than 2-3 days of massed attack. This was completely devoid of reason, since German cities had often functioned in conditions of at least equally damaging attacks.

The Volkswagen production plant in Fallersleben was designated, as the first one, to produce the Fi 103.

Contrary to plans, production was residual, in October 1943 only around two missiles came off the assembly line daily, which wasn't even sufficient to carry out training exercises. The bombing of certain key co-operating plants worsened the situation even more. At night on August 17/18, 1943 650 bombers from the Royal Air Force bombed the complex in Peenemünde. Under such conditions the fate of the V-1 programme was placed in doubt. The production plants of Fieseler in Kassel-Bettenhausen, which were to "assist" Volkswagen were also bombed by the RAF, at the end of October of that same year.

The necessity arose of an urgent, repeated analysis of the research, training and production plans of the V-1, and also the issue of counter-intelligence protection. As far as the latter was concerned, it was decided to abandon the use of foreign labourers. Due to the imperfection of the Fi 103's construction in November residual production was halted in Fallersleben. The training of soldiers on the rocket range in Zempin was also halted. At this time a much more urgent task was to increase production of fighters with respect to the intensifying Allied air raids.

The underground "Mittelwerk" structure near Nordhausen in the Harz mountains was adapted in the first place for the needs of producing the V-1 and V-2, construction of which had already begun as far back as 1933. In connection with this the newly created company Mittelwerk GmbH took over the facility. Two parallel tunnels were bored straight through the mountain of Kohnstein—main communication routes 10 m wide and 7.5 m high, which were connected to fifty traverse production halls each 140 m long, responsible for individual stages of produc-

The V-1 in the Imperial War Museum

tion. When the facility was taken over by Mittelwerk its underground cubature amounted to 875,000 m³ and the area of its tunnels and galleries—to 125,000 m². The production workforce was made up of prisoners from the Nordhausen concentration camp—chiefly Poles, French and Russians. In addition, the assembly of the "V" missiles and production of different sub-assemblies was also carried out in other underground complexes, among others near Helmstedt-Berndorf (the production plants of Askania Werke AG producing the control systems), electronic components being produced near Hadmersleben and Hersbruck in the underground factories of the AEG consortium.

The beginning of 1944 was a period of the next "swing" in Hitler's liking to the benefit of the V-1. Overcoming the technical shortcomings of this construction was only possible by way of intensive research. From now on the missiles were tested in large numbers, not only on the Baltic. The Udetfeld airfield in Upper Silesia (Mierzęcice) was also made available for this objective and in mid-March also the Blizna range in occupied Poland, from which so far only the V-2 had been launched. "Flying bombs" were launched in the region lying between Lublin and Chełm in Poland. In one of the German reports we read:

The V-1, close-up of the air inlet

In one incident, people as well as animals were killed —directly by the force of the explosion itself or shrapnel, in addition houses and other buildings had been burnt down. As a result of the 12th launching the village of Adampol was partially laid to ashes.

Finally the Germans managed to reduce the scatter to reasonable values. Between August 18 and November 26, 1944, almost 260 missiles were launched, at the same time only 17% hit the appointed target (30 km in diameter 225 km away, or 15 km in diameter 100 km away). In October however a rate of 32% had already been achieved, in November 46%, and the Germans finally managed to obtain the maximum airspeed planned earlier of 650 km/h, which made it difficult for enemy fighters to intercept the missile. Production finally got fully underway in the plants of "Mittelwerk" in Nordhausen, up till now "reserved" for the rocket industry. According to information from the Armament Ministry, as the programme of testing and improvements neared an end, the pace of production was increased (the period of greatest acceleration was October 1944) and even before the end of this year the Germans had managed to produce 23,748 specimens of the Fi 103 and in the period January-April 1945, 6,509 specimens (all in all 30,257). The search for a site to build the launch catapults was commenced as far back as the early summer of 1943. An "anti-aircraft regiment" was formed, Flak Regiment 155 (w) (the Fi 103s were officially "flying targets," therefore they had to be present in the equipment of an "anti-aircraft regiment") which was divided into four smaller tactical units, each composed of four field batteries. Four non-fortified launch catapults were provided for each battery which in connection with the field version were to make a total of 64 catapults.

The V-1 in the National Air and Space Museum in Washington

A V-1 falling over the centre of London

Apart from this, on the strength of Hitler's order from July 1, 1943 four fortified launch complexes were built, code-named "Wasserwerke." They were located in Desvres on the Pas-de-Calais, near Saint Pol—in this same region—in the vicinity of Dunkirk, as well as two on the Contentin peninsular—over 200 km further to the west—near the towns of Tamerville and Cherbourg. Theoretically each of them could launch daily up to around 100 missiles and were to a large extent self-sufficient, and could hold in their storerooms up to 1,000 specimens of the V-1 with the appropriate amount of fuel. The bunker ceilings, 5 metres thick, were impenetrable for the heaviest Allied bomb used so far. Their construction was begun in August 1943 and completed before the end of the year. In addition by May 1944 the construction of 24 combat position mock-ups had been commissioned. Despite this the location of the V-1 launch positions was established relatively quickly and easily. In principle this generally consisted of two reasons: the characteristic silhouette of the catapult and the fact of using numerous French companies during construction, which according to Gen. Erich Heinemann in command of operational use of the V-1, was simply a provocation of espionage. Air raids on these facilities had already begun at the beginning of November, i.e. long before many of them were completed. 57% of all positions had been bombed up to the end of December. Around 10% were completely destroyed and the another 10% suffered heavy losses. In spite of a successive rescheduling of the date of commencing attacks on London, to December 1, 1943, only around 70% of combat positions had been completed by the end of the year, at the same time many of them still didn't have a complete infrastructure. Such a quick and effective counteraction by the Allies was a big shock for the Germans.

As a result many changes were introduced to the plans, among others further work on launchers positioned inside bunkers was abandoned. The destroyed structures were replaced by new ones, many new annexes were built treating as top priority the issue of counter-intelligence protection, and dense natural vegetation was used for camouflage. Building work was carried out by German workers alone. In connection with the intensification of Allied attacks (but solely on "old" positions) in April 1944 the Luftwaffe General Staff stated that building work had only compensated for the losses suffered. Due to this 295 workers and soldiers had been killed and 401 injured by April 3. Simultaneously it became evident that the simple but very well camouflaged launchers were a considerably better solution than expensive bunkers. By June 1, 1944,

The Hungerford bridge in London after being hit by a V-1 missile

A crashed V-1 missile, found by US soldiers in France

80 out of 144 catapults were ready for use, two weeks later already 115 structures had been completed. Missile storerooms and repair workshops were situated in two large, suitably adapted caves near Nucourt. The preparations to commence military action slowly neared an end.

In connection with the start of the invasion in Normandy the order was given to "hurriedly" begin attacks on June 12, a deadline however which could not be kept. Combat units were still not suitably prepared, and there were problems with supply in connection with the opening of a new front and the bombing of many railway junctions. The Germans never managed to deliver concentrated perhydrol on time for the catapult's propulsion. As a result, on this day only 10 missiles were launched in the direction of London, at the same time only 4 reached England. Five (i.e. as many as 50%) had crashed just after launching.

Finally it was decided to "repeat" the commencement of operations three days later, shortly before midnight. 55 catapults opened fire, carrying this on until the afternoon of the next day (June 16, 1944). During these 24 hours almost 100 V-1 missiles were launched, by June 18 already 500. Depending on the launcher's position and other factors the time of flight to London amounted to approx. 25 minutes.

In the standard A-1 version the thin-walled warhead of the missile (weighing 830 kg) was filled with amatol—a composition composed of TNT and ammonium nitrate solidified from its liquid form. It was a few dozen percent weaker than cast TNT. On July 18, 1944, the B-2 version of the V-1 with the most effective warhead was introduced for use. It differed from the A-1 warhead solely in the type of explosive. This was trialen—composed of TNT, RDX and a small amount of aluminium dust, increasing the heat of the explosion. Its explosive force was almost two times greater than amatol. With regard to the large proportion of explosive in the total mass of the warhead, this version of the V-1 was particularly dangerous and with regard to the light character of London's buildings sometimes they destroyed entire districts of houses. Even Churchill could not fail to be astonished in "how can they do 8-10 times more damage than bombs similar in mass dropped on German cities." Although Hitler demanded the production of 250 trialen warheads monthly, carrying out his orders was not feasible, and towards the end of the war the warheads were even filled with a relatively weak mining explosive donarit. The V-2 rockets also possessed warheads filled with trialen (mass 1 ton). As far as the V-1 is concerned, other warhead types also existed, but they were used relatively seldom: from

A mysterious photograph taken over Antwerp, apparently showing one of the derivative versions of the V-1, probably with a different engine

Remnants of the V-1 in one of Mittelwerk's galleries

an 800-kilo aerial bomb with a decreased amount of explosive (min. approx. 500 kg) to a chemical warhead not introduced to production. The warhead of each version was equipped with three fuses: two contact (inertial and electrical contact) and a time fuse causing detonation independently of the others, after 30 minutes from launching. What this concerned was that in the event of crashing, or unintentional "landing" of the missile, its construction couldn't be examined by the enemy. However this caused serious complications in the event of an unsuccessful launch, after which the V-1 remained on the catapult. In such cases the only thing left to do was to wait in a safe place until the catapult had been destroyed or send in a volunteer who would attempt to remove the fuse.

After ten days of attacks the British had already recorded 370 direct hits on London. On June 28, a V-1 fell on the building of the Air Force Ministry, killing 198 people. Soon after the number of victims as well as number of destroyed and damaged houses began to number in the thousands. The production of London factories working for the war effort fell in effect by approx. 1/6.[3,4,5]

All in all the Germans ha managed to launch 6,046 missiles by the end of the year, at the same time 1,681 had crashed (27.8%), including 795 at a small distance from the launcher. The greatest intensity of operations was recorded in their initial phase: 1,000 V-1 missiles left their launch catapults within the first few days of operation, until June 21. 1,279 launches fell in the following, final year of the war, 986 missiles reaching Great Britain. The considerable number of unsuccessful flights revealed the frantic character of production preparations and use of this innovative weapon. This is evidence of the technical defects and shortcomings which could have been removed, but weren't with regard to limited time. A relatively large percentage of "losses" were caused by poor operation of the catapults, in spite of their simple construction. Within less than the first two months of operation, up until August 6, 34 missiles had exploded at launch.

After some time it was resolved to analyse the cause of failures. The 245 missiles which had crashed up until July 24, 1944 were added up and classified with respect to the reason for their occurrence. Most, 70 (28.6%), were caused by faults in the airframe's construction, 62 (25.3%) by incorrect catapult operation, 40 (16.3%) by engine problems, 34 (13.9%) by poor operation of the navigational and control systems (it occurred that missiles would fly "in circles" near the launcher), and 5 (2.0%) were a result of maintenance personnel errors. The cause of 34 incidents was not explained.

In the last months of the war, the Germans could no longer direct their Vergeltungswaffe against Great Britain, and Belgium also fell victim to them.

Summing up almost a year of their use in combat it is necessary to give the real effect of this. Overall all ground launchers launched 20,880 missiles, at the same time 18,435 actually reached their targets, according to available information. The following numbers fell on individual cities: 7,796 on London, 44 on Southampton, 7,687 on Antwerp, 2,775 on Liege, and 133 on Brussels. Apart from this around 1,600 V-1 missiles were dropped by KG 3 and KG 53 bomber groups in the direction of London, Southampton, Gloucester and Manchester.

Technical details of the V-1 (Fi 103 A1)

launch mass	2152 kg
warhead mass	830 kg
length	8.35 m
range	240 km
engine length	3.66 m
fuselage diameter	0.84 m

The V-2

From a purely technical point of view a considerably more interesting design was the V-2 rocket. Although it was one of the crowning achievements of technology at that time, this was the far-reaching development of entirely amateurish rockets from the beginning of the 1930s.[3]

Rudolph Nebel, Klaus Riedel and a group of rocketing fans constructed then, in 1930 the first test rocket called "Mirak," which was an abbreviation of the word "Minimumrakete." In this group was found the still unknown, young eighteen-year--old Baron Wernher von Braun (but also such a unknown person as the Polish Jew from Łódź, Ari Sternfeld, who, in 1926 patented the space suit, ended up as a veteran of Soviet space research). For trials the enthusiasts rented the old Reichswehr shooting-range in Reinickendorf in the suburbs of Berlin. It was September 1930. At the same time a simple (the first) liquid-propellant rocket was constructed. This construction consisted of two small fuel tanks for the propellant and oxidizer, as well as an engine placed above and between them. The verification of basic technical assumptions was involved.

The first results were not too encouraging, but after carrying out certain improvements a maximum flight altitude of the order of 1-1.5 km was achieved. The "Repulsor," since that was what the missile was called, entered history.

However at the same time work was being carried out on rocket missiles in the Armaments Department of the Reichswehr Ministry. They attempted to construct small solid-propellant rockets, of a range up to 8 km, which were designed to carry gas warfare agents. Through lack of any success after some time the research was suspended. In connection with this two officers responsible for a "frozen" until now liquid-propellant variant (Major Ritter von Horstig and Captain Walter Dornberger) were able to push "their" line of work, all the more easier that rockets of a greater range were essential for transporting poisonous substances—not creating too much a danger to their own armies, and such rockets would be very difficult to construct at this time on the basis of powder. However any kind of experience was lacking in this new field. Attention was turned therefore to the aforementioned group of amateurs, for nevertheless they had some knowledge and experience at their disposal. The chief of the Armament Department Col. Becker put the issue clearly: if these people are to work for the military, then this work must be kept secret, and this can only be achieved in seclusion and in the event of the civilians signing the appropriate obligations. Out of the civilians only Von Braun consented to fulfil these demands. At this time he was studying at Berlin university, at the faculty of military science, where he later gained a doctorate from the construction of a rocket powered by liquid-propellants. He was a model student and a certain kind of gem in the mind of Captain Dornberger. At only 20, he was appointed chief of Versuchsstelle West, one of the research posts on a rocket range in Kummersdorf. Gradually other specialists joined, famous later for accomplishments in the field of "V" weapons, among others the engineers Walter Riedel and Pöhlmann. Already back then it was possible to see in him at a glance a very talented and hard-working scientist, able to develop the methodology of problem solving and at the same time an excellent organizer, able to divide work. Nobody else brought such a contribution to the development of rocket technology as Von Braun.

His first "military" work was a relatively large engine powered by a 75% aqueous solution of ethanol and liquid oxygen. The computational thrust of 300 kg was however not achieved, as the engine exploded on the engine test bed. Luckily this occurred without casualties. After this it was resolved to construct a smaller model, of 20 kg thrust, which was made by the company Heylandt. There were no problems with this. Another scientist, Arthur Rudolph, after analysing the causes of the explosion of the first engine constructed an improved model, made of copper, which operated unreservedly for 60 seconds giving a large thrust of 330 kg. The group quickly gained new experience and created the basis for constructing the first, reputable rocket, and one should remember, that this all took place long before the outbreak of war and the military success of the Vergeltungswaffen. Conducting "civilian" rocket research was prohibited, which caused many of the previously undecided specialists to join the team, among others Nebel. In 1934 research on rockets was already treated very seriously by the military and shrouded in strict secrecy. The team of

An experimental A-3 rocket on the launcher, the test range near Kummersdorf

Von Braun numbered over 50 specialists. At this time, after the accession of Hitler to power he gave the order giving full responsibility and exclusive rights to the Heereswaffenamt as far as work in the field of rocket technology is concerned.

On the grounds of Versuchsstelle West two relatively large rockets were constructed: the Aggregat-1 (A-1) and soon after its improved version the A-2. It had a launch mass of 150 kg including 40 kg of propellant, length 1.4 m and was 30 cm in diameter. The engine thrust amounted to 300 kg. On Borkum Island in the North Sea, a modest rocket range was installed, from which two specimens of the A-2 were launched in December 1934. They operated continuously without fault, each achieving a maximum ceiling of 2,200 metres. These were the first rockets with a gyroscopic control system (directional stabilization). They constituted an important achievement. The "group" had proved the legitimacy of its funding. Von Braun's military superior at the time, Major von Horstig, wrote among other things: "He will undoubtedly accomplish great things in this area in future years." Successive research and construction successes produced the desired effect of a gradual increase in financial expenditures. In March 1936, analysing the predicted performances of future constructions and to an ever greater degree viewing rockets as a potentially crucial weapon, Becker and Dornberger resolved to formulate for the first time requirements in relation to an "important" combat rocket. They took as a point of reference the still record-breaking so-called counter-Paris gun—a super heavy gun from World War I with a range of approx. 125 km. Probably just then the "working" designation A-4 appeared for the first time (i.e. after later introduction to armament, the V-2) and the general technical requirements concerning this rocket were formulated. It was decided in the first instance that a smaller rocket should be developed, the A-3, in order to check the adopted assumptions on a smaller scale. A new, safer rocket range was designated. It was situated on Greifswalder Oie Island, to the north of Usedom Island on the Baltic. The A-3 had a mass of 750 kg and length of 7.7 m. A series of launches carried out in December 1937 revealed faults in the new, triaxial (tri-coordinate) con-

Take-off of the V-2

One of the V-2 prototypes during pre-launch preparations

V-2 missiles with military camouflage, positioned on mobile launch pads

to this objective. A year later the rocket experts had already moved to Peenemünde. Their possibilities grew immensely. Work was first of all concentrated on an installation lacking until now—a supersonic wind tunnel. The first only made research possible on small models—it had a cross section of only 10 by 10 cm, soon however a second one was constructed with a cross section of 40 by 40 cm. A flow speed up to around Mach 5 was achieved in it. In 1943, the number of those employed in the complex gradually rose from 50 to 15,000.

Progress of the rocket research carried out in Peenemünde gradually demonstrated the military significance of the rockets and the enormous developmental potential of this new means of warfare to the military. They could in the foreseeable future outstrip not only the Paris gun from World War I, but any kind of modern means of transportation, including aircraft. The arms race resulting from the intensive preparations of Germany for war enabled the scientists in Peenemünde to put into effect their bold intentions.

In November 1938, General von Brauchitsch ordered preparations for mass production of the A-4 to begin, so that it would be possible to cross over to this stage, as soon as research was complete. The management of this activity was entrusted to a special group assigned from the research and development department of the Heereswaffenamt, which was controlled by Colonel Dornberger.

This period was the final phase of work on the development of the revolutionary for those times ultimate engine for the A-4 rocket. The credit for this went to Dr. Walter Thiel, engineer Pöhlmann and countless other associates. As a result

trol gyroscope, but there weren't any problems with the propulsion itself. However many improvements were introduced, not modifying the propulsion unit. The new rocket, the A-5, was externally already similar to the A-4 (developed later), although it was obviously considerably smaller. Trials of it were carried out in the years 1938-1941. The final version had a mass of 1,300 kg and achieved a ceiling of 13 km, although still at subsonic speed.

In the meantime, at the end of 1938, the Wehrmacht General Staff had specified their first requirements in relation to the not yet designed A-4. It was to be operationally ready in units by the end of 1942. Intensive development of a research base was planned. In April 1936, Peenemünde was ultimately selected. General Becker (representing the Heereswaffenamt) and General Kesselring (the Luftwaffe) signed an agreement concerning the joint construction and use of a large research centre on Usedom Island. 20 million Marks were allocated

Test stand No. VII at Peenemünde. It was the test rig for the V-2's engines.

an engine with a thrust of 25 tonnes was obtained, but with very small dimensions (the length of the combustion chamber amounted to only 30 cm), although with a very high efficiency—around 95%. The first trials on this engine's test rig were carried out in the spring of 1939. This signified the achievement of a crucial stage on the way to making prototypes of the future V-2.

As it turns out, Wernher von Braun, christened as "the father of the V-2," had practically nothing in common with the development of the "A" rocket engines. The true author of the rocket's "heart" was the forgotten genius of German rocket technology, Dr. Walter Thiel, who also became famous for developing a powerful engine with a thrust of 200 tonnes for the A-9/A-10 unit as well as the first trial runs (theoretical analyses) with the objective of constructing a nuclear rocket engine—this plot was continued later in the USA and USSR. However the realization of this work wouldn't have been possible without the support that a team of scientists from Dresden Polytechnic gave to Thiel. They included first and foremost: Professor Beck, who until 1940 had been engaged in the analysis of alcohol combustion in pure oxygen at high pressure (he died in 1941) as well as engineer Hans Lindenberg, who from 1930 had been engaged in the research and development of fuel injectors for diesel engines.

Thiel himself was killed during an RAF air raid on Peenemünde on the night of August 17/18, 1943. This was probably the most serious loss inflicted on the Germans that night, as there was no more serious damage as far as research and production facilities were concerned (however many workers were killed).

Work on liquid-propellant engines, with oxygen as the oxidiser, had already begun before the war and the first success was the construction of an engine with a thrust of 1 tonne with a single injector, being at the same time the chamber of an initial composition of propellant and oxidiser. It became the foundation on which the V-2's engine was based. During the entire developmental path the key element, the injector, didn't undergo any significant changes, only the number of them was increased as well as the dimensions of the combustion chamber and nozzle. First an engine with three injectors and a thrust of 4 tonnes was made and finally the engine for the V-2 with 18 such injectors.

An increase in the engine's dimensions also resulted in the necessity of solving the problem of its cooling efficiency, since the larger the engine, the higher the temperature produced. In practise, for the first 5-7 seconds, several kilograms each of propellant and oxidizer flowed down—initially only under its own weight—into the engine of the combat rocket per second (this was the "warming up" of the engine). After the pump got going, within the course of 1 second up to 60 kg of spirit and 75 kg of oxygen reacted with each other. Paradoxically, solving this problem prompted precisely an enormous consumption of propellant—it was the best coolant for the nozzle. This also proved to be a "canon" in the case of future space rockets.[6,7] The rocket's nozzle was composed of two layers, between which flowed the spirit propellant. In this way the requirements (mechanical, durability and thermal)

Preparations for launching the V-2 from a railway platform

concerning the internal wall of the nozzle were considerably reduced—because it could now be resistant only to the difference in pressure—between the pressure in the combustion chamber and pressure of the propellant. In this way very serious problems were avoided and the engine could be made "only" of soft steel.

Work carried out almost simultaneously in the USA, by Goddard, which didn't strike on such a solution usually ended up with the engine being completely burnt out within only 5 seconds! The soft metal of the V-2 rocket engine worked however perfectly well in spite of the temperature in the combustion chamber reaching 3,000°C and falling to approx. 1650°C in the nozzle mouth. However trials carried out revealed that as a result of very efficient cooling the temperature of the internal walls didn't exceed 950°C.

One other factor was decisive in the aforementioned propulsion unit being highly evaluated after the war—namely it was relatively light—only 8% of the rocket's launch mass (for a com-

Inspection of the Peenemünde centre on May 26, 1943. First from the left (wearing a suit) is Dr. Thiel. This is one of the last photographs taken before his death.

parison: the simple it would appear V-1 propulsion unit took up 24% of the total mass). Thanks to this up to 69% of the mass of the V-2 was propellant, which for a long time was a difficult result to beat. On the subject of this engine's characteristics it is worth observing that work in this area was carried out in Germany on two tracks so to say, at the same time Dr. Walter was responsible for this second "track." Despite the independence of his work, his most famous engine, developed for the Me 163 rocket fighter was characterized by numerous similarities to the engine described above despite operating on a different propellant.

The next pioneering solution was the "thrust vectoring" control system developed for the V-2—by deflecting the exhaust gas streams from the nozzle. This ensured steerability also at very low airspeeds, i.e. directly after launching. Thanks to this it could be launched vertically from a small and simple launch platform (from a so-called "launch pad"). The lack of a large, stationary slide launcher constituted one of the most important trumps of the V-2 in comparison to the V-1. Besides, such a launcher would have to be so large that the point of producing such rockets would have been cast into doubt. The gas dynamic rudders were made of a heat-resistant material, such as graphite. These consisted of four plates, positioned just behind the nozzle mouth. They were coupled with the aid of simple chain transmissions to aerodynamic rudders in the fins.[7]

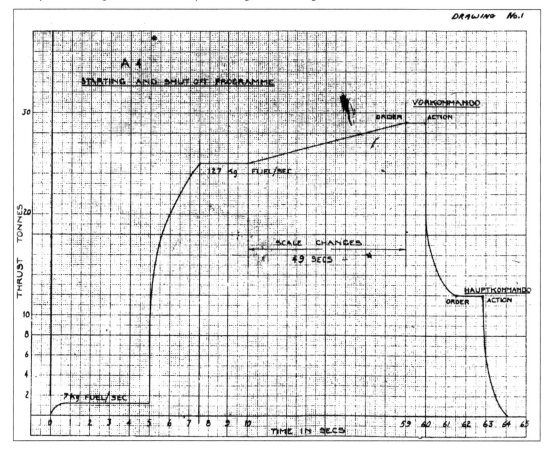

A graph showing the thrust of the V-2's engine (in tonnes, on the vertical axis) as a function of time in seconds

In connection with the development of the first large rockets, chiefly the V-2, a whole series of truly pioneering research programmes were put into effect. As a result the fundamental canons of the development of rocket technology were outlined, within which the rocket-nuclear arms race was to be realized for several decades after the war. The V-2 was among others the first rocket with an inertial navigation system (with a so-called inertial attitude reference system) and controlled by an electrical-mechanical flight sequence programming device. This principle was later exploited in all ballistic rockets with nuclear warheads. The flight sequence programming device didn't allow among other things the warhead to explode directly on the launcher or in its vicinity after a possible propulsion system failure, which at the beginning occurred quite often—the warhead was armed in flight only after the programme's completion. In practise, if the rocket crashed on the launch pad and possessed explosive, the warhead exploded anyway—in the aftermath of burning propellants and very high temperatures, the personnel however had at least 20 minutes for evacuation and could escape with their lives. Apart from this one could "theoretically" extinguish the fire. If not—if there existed the real danger of an unexpected explosion of the rocket along with the warhead, the entire infrastructure of the given unit could be destroyed (under conditions of use in combat)—the rocket's warhead had such a large explosive force that "normally" it destroyed entire housing quarters. This was the first serious problem which had to be solved during research and development work.

The next problem of this importance was the necessity of determining the conditions which would accompany a re-entering of the atmosphere (**at an almost vertical flight trajectory a maximum ceiling of 172 km was achieved, whereas 100 km is considered the agreed boundary of the Earth's atmosphere,** and the maximum airspeed exceeded 5,000 km/h). It was already clear at first glance that these conditions were incomparable to anything previously known.

First and foremost they created the danger of the skin plating melting and softening of the hot steel from which it was made, and even its oxidation by fast, "frictional" hot air. Much effort therefore was put into determining the temperature distribution in the nose of the "falling" rocket. First of all it was endeavored to calculate, or rather estimate the maximum temperature, based on the results of measurements carried out in a supersonic wind tunnel (in Peenemünde and Kochel). To this end a special model was made with bimetallic sensors sunken just under its surface. On this basis it was estimated that the skin plating of a real rocket shouldn't heat up to a temperature higher than approx. 600°C. In further succession special sensors were constructed which were installed in a measurement version of the V-2's warhead. There were very many of these sensors, miniature disks fixed to the skin plating, each of which was characterized by a somewhat different melting temperature and connected to an electrical sensor, sending an electrical impulse in the event of the disk melting. These were in turn connected to a telemetric device on board the

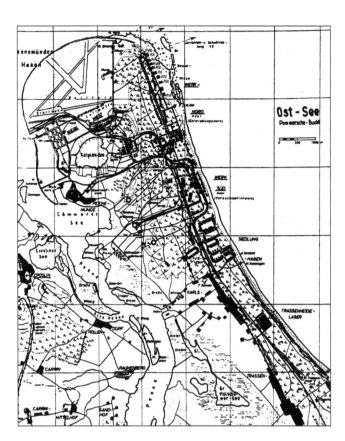

An original German plan of the centre near Peenemünde

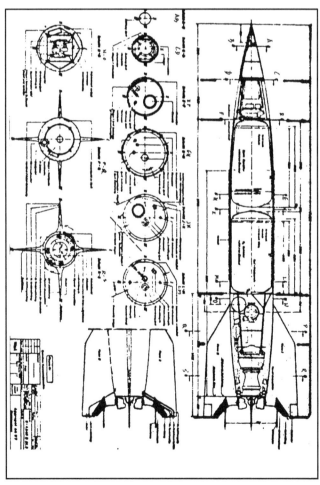

Original German technical drawings of the V-2

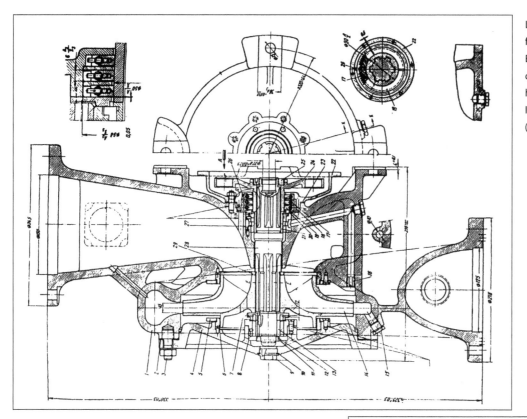

Left: A cross section of the V-2's fuel pump (original drawing)
Below: The reactor, in which catalytic decomposition of the hydrogen peroxide took place. Its diameter was just 12 cm. (original drawing)

rocket, which sent the appropriate parameters without delay by radio to a ground position. Parameters specified in this way revealed that in reality the temperature reached approx. 650°C (and so this indicated that the rocket would have been visible at night, while entering the dense layers of the atmosphere, when its airspeed reached a maximum—it would have shone with an orange light). In connection with this it was judged that no significant modifications in the construction were necessary, although the later affirmed fractures of the fuel tanks and leaks in the installations evoked by the "pressure" of the shock wave showed that nevertheless the significance of these problems had been underestimated. As a result the rocket's development was blocked for several crucial months—just when the pressure to bring it into production was the greatest.

No less a cosmic challenge was the problem of designing a suitable fuel system, including of course chiefly the pump, fulfilling such extreme requirements, that British and American experts considered in advance this problem to be insolvable—they changed their minds only when handed components of the V-2 (supplied by Polish Home Army Intelligence). This concerned a device which was to be small, light and reliable, and able to force into the combustion chamber (in which a gigantic pressure prevailed anyway) propellant at a speed of 150 kilograms per second(!!!), **in the course of about two seconds from the moment of activation.** In spite of this a pump fulfilling the above requirements was constructed, with overall dimensions so marginal in the scale of an entire rocket reaching as high as five storeys, that it could be lifted by one man alone. It was a relatively flat turbine 47 cm in diameter and with a power of 500-600 HP (the "Walter turbine," developed by the company HWW from Kiel). Its extremely efficient source of

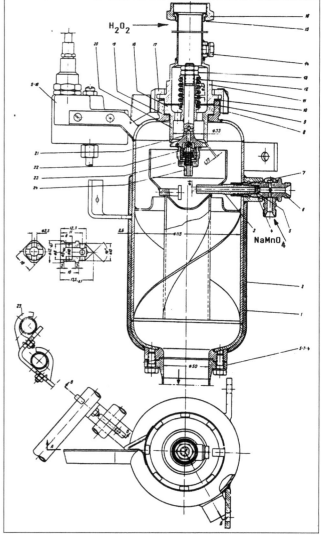

energy was composed of hydrogen peroxide concentrated to 80%, decomposed catalytically (with the aid of an aqueous solution of calcium permanganate). The reaction occurred at a temperature of 460°C, giving a composition of steam and oxygen at a pressure of 24 atmospheres.[6,7]

The outbreak of war in September 1939 coincided with the achievement of significant progress in research work, on account of which the project was given high priority. This permitted the assurance of suitable funding and the supply of regulated strategic raw materials, although on the other hand the time "allotted" for completion of research was shortened about twice over. From this time a deadline of September 1941 was in force for the scientists, at which time the A-4 rocket was to be ready to bring into mass production. A multitude of researchers and civilian designers were to this end strengthened by around four thousand technically educated soldiers, who after the completion of work were to form the core of the Wehrmacht's combat rocket units. Soon an even greater acceleration of work was formally demanded, referring additional workers to Peenemünde. The research and development programme was given the highest of all priorities, although at the same time as a result of Hitler's intervention preparations for mass production were suspended.[6]

Putting these plans into effect however also required an extension of the research base. Regulations were relaxed concerning the rocket programme's secrecy protection, many outstanding scientists from the institutes and universities of the whole Reich were invited to Peenemünde and asked for their co-operation in solving the most key technical problems. Despite achieving a significant advancement of work, much remained for them to do.

There was no shortage of those willing to co-operate. Working for the "flourishing" military arms complex was a very favourable alternative to the prospect of serving on one of the new fronts being prepared in Europe.

The first months of 1940 were however a poor period in the history of the German rocket programme. Serious restrictions appeared in the delivery of raw materials which was influenced by Hitler's dislike towards all long-term research programmes of new weapons. He extrapolated too easily the experience from the first "lightning war" (Blitzkrieg) expecting relatively quick settlements in the development of the military situation in Europe. On this basis he considered similar programmes in a way as "devourers" of resources, which could prove to be decisive for the course of future campaigns. Many programmes fell victim to this short-sighted policy, among others also the programme to develop an atomic weapon, in which up until 1941 the Germans still had been several years ahead in realizing, in relation to the Americans.

The second unfavourable event was namely the armaments minister Fritz Todt, a hostile man with regard to projects to construct long-range rockets. He considered (otherwise rightly, though on the basis of different assumptions than Hitler), that it was a gigantic waste of materials and scientific-technical potential, which held no promise of gaining valuable

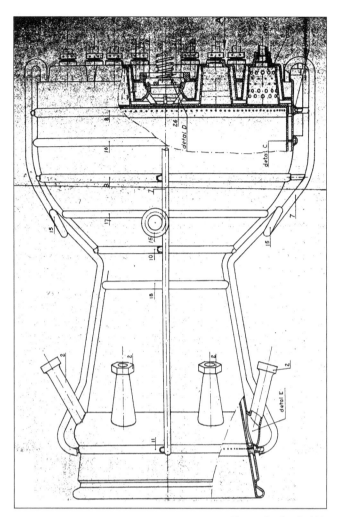

The engine for the V-2

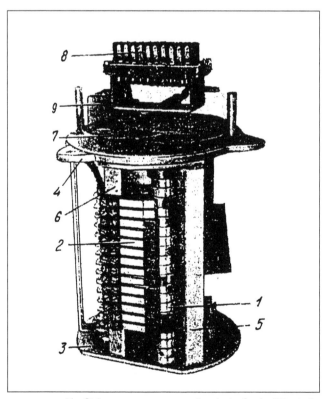

The flight sequence programming device for the V-2 missile

A fragment of the inertial navigational system for the V-2 missile

effects from a military point of view. One should remember that in its peak period around 200,000 people worked in Germany in order to carry into effect the programme of guided rockets. As early as the spring of 1940, work going on at that time concentrated on overcoming two fundamental problems: ensuring the right accuracy of the rocket and improving the technology of making the engine, since instances of its disruption during operation at maximum thrust turned out to be relatively frequent.

As far as the problem of the rocket's accuracy was concerned, analyses in Peenemünde revealed that errors in the inertial navigational system arose to a large extent in the initial flight stage —during acceleration to cruise speed and gaining of the assigned trajectory. This could be avoided by applying in this section of the flight path an additional radio guiding beam ("Viktoria Leitstrahl") enabling the correction of the navigation equipment's instructions by the ballistic rocket's flight path computer.

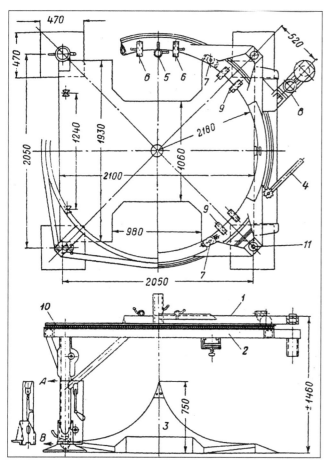

A plan of the V-2's launch pad

The first instances of the military use of this version (1944) revealed the significant susceptibility of "Viktoria" to jamming, although the conception itself of correcting the flight path in its initial and relatively unstable stage was recognized as very apt.

The Germans planned in this case to obtain an improvement through a significant increase in the beam's frequency to approx. 600 MHz. This work was conducted by an experimental military establishment in Peenemünde under the "aegis" of a specially appointed "Plenipotentiary for Research in the Field of High Frequency" (Bevollmächtiger für Hochfrequenzforschung—BHF). This was a temporarily appointed department, whose principle task was to speed up the closing of the gap, which had arisen between British and German research on high resolution radar— operating on centimetric waves. BHF also effectively "promoted" progress in derivative areas—e.g. concerning materials absorbing radar waves, beam weapons "paralysing" electronic systems and engine ignition systems and the "guiding beams" themselves.

On the strength of an order from BHF, probably in the autumn of 1944, the design of two new guiding beam systems, code-named "Libelle" and "Gloria," was commenced. Work was conducted in a small team under the direction of Dr. Faulstisch and engineer Battac. The code-names given above in a way referred to one entirety—"Gloria" was the modified equipment of the rocket, whereas "Libelle" was a ground transmitter. As much as the second component was completely new, this "on board" component was only modified in relation to its predecessor. Even the antenna system and receiver were left unmodified, adding only a high frequency converter (the previous amplifier served now as a preliminary amplifier (pre-amp). This resulted from the tight schedule as well as limited possibilities of "manoeuvre" in the tightly packed rocket's fuselage, which after all hadn't been designed with this equipment in mind.

The ground component on the other hand was designed practically from scratch. At the moment of evacuation of the centre in Peenemünde almost all of the plans were complete, but the Germans never managed to complete any prototype.[8] (The plans and possibly some hardware, has been captured by the Soviets and used in connection with their derivative of the V-2: the R-2 missile, that has been introduced into service in 1950). For the time being let us return however to the interrupted description of the initial stage of work...

In the summer of 1940, the position of the rocket programme fell further on the list of priorities. Now Hitler ceased to display any interest in it at all, considering the rocket not to be competitive with respect to the long-range bomber, which could carry far greater "combat load."[8] Irrespective of this the escalation of counter-intelligence operations forced the leadership of Peenemünde to dismiss around a thousand Polish forced workers. Even though the development of the A-4 rocket was quickly nearing an end, the centre on Usedom Island came very close to being shut down.

Only the intervention of Marshal von Brauchitsch changed the situation. At the turn of July and August he gave the order to classify the rocket programme (by the code-name "Rauch-Spur-Gerät") as among the most important programmes in the development of new weapons and unexpectedly restored it to the

The tail section of the V-2, with visible graphite steering elements

highest priority. Dismissed workers once again began to be assembled. The decision was made to extend further the research and production base in Peenemünde.

The "Reich's chief architect," Albert Speer, a rocket enthusiast, became responsible for this development (later, after the mysterious death of Fritz Todt in an air crash in February 1942, Speer was appointed armament industry minister and chief of the Todt Organisation). As he himself admitted, he developed the infrastructure in Peenemünde in spite of Hitler's negative stance and was impressed by the achievements of the local scientists, who he understood well in contrast to Todt. A few years later he became the key person in the development and production of "special weapons," since it was he who directed the construction of the gigantic underground installations for them.

The construction of a production line for the A-4 rocket was however suspended, until its development was completed. After the next "war of priorities," which was fought as early as the autumn of 1940 Hitler finally "came to like" the rocket programme, first reinstating it with priority 1A, and later giving it the highest priority SS. The reason for this of course was the outcome of the "Battle of Britain." For this same reason the argument of the rocket's uncompetitive nature with respect to the bomber ceased to have any significance. Its military role had been changed.

At the beginning of March 1942, the Germans managed to assemble the first prototypes of the A-4 rocket—the future main vengeance weapon of the Third Reich. Despite work on the engine not being carried through, cases of the combustion chamber disrupting were still very frequent, at the end of March it was planned to carry out the first launch. But the rocket exploded during a prior static trial on the test rig. Competition with the Luftwaffe and the Fi 103 as well as the necessity of quickly "proving" that the rocket's development had been carried through, for without this preparations for mass production couldn't be started, was what also caused indirectly many future catastrophes. In spite of Hitler's or-

ders, the chief of the Heereswaffenamt's rocket department, Dornberger, formed a substantial work group, which saw to the drawing up of production plans and began preparations, mainly concerning a "piloting line" in Peenemünde. Apart from fifteen prototypes the Germans planned to produce 600 rockets for the research centre only in order to perfect their construction. After a certain development it was estimated that the "piloting line" would be able to turn out around 5,000 rockets a year. This estimation was however based solely on the availability of the work force and raw materials, and not ensuing from realistic technical possibilities. The construction of two installations for the production of liquid oxygen on the grounds of the centre was also planned with a joint productivity of 3,000 tonnes a month. Although these were chiefly theoretical preparations, they allowed the embryo of the future structure managing production to be created. The production plans were in any case soon verified as a result of protracted problems with the engine and in part with the control system, and those concerning the "piloting line" were still up-to-date only for several months. Soon there was an event which very

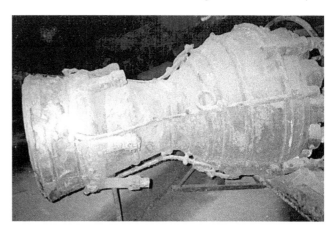

On both photographs: a V-2 engine, found in the underground Mittelwerk complex

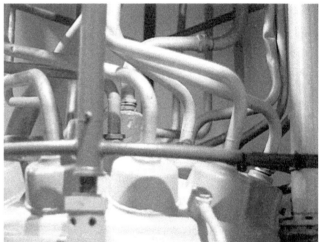

The main part of the V-2's engine with details:
the injectors with pipes supplying them with oxygen plus the fuel pump

nearly stopped the entire research programme. This was a massed RAF air raid on the centre near Peenemünde, carried out on the night of August 17/18, 1943.

For four hours (from 23.23 to 3.13) approx. 600 B-17 bombers dropped on a relatively small area one and a half thousand tonnes of high-explosive and incendiary bombs.[9]

A fact worth emphasizing is that carrying out this air raid had been possible thanks to reports of Polish Home Army Intelligence.[10,11,13,14] Besides, this wasn't the first or the last case in which the Poles supplied the Allies with intelligence information about German technology crucial from the point of view of the war's course. One can give many such examples, among others the cracking of the German desiphering machine, "Enigma" by Marian Rejewski and two of his friends, later the delivery of V-2 components, the delivery of accurate technical data and plans of the FMG-37 radar as well as the Heinkel He 177 bomber...[10,11,12,13] As far as the V-2 is concerned, the first information had already been captured by the Vienna outpost of Polish Home Army Intelligence shortly before Christmas 1942, but was made light of. In the summer of 1943, before the air raid, the Polish Home Army "Lombard" network operating in Pomerania made a thorough study of the centre on Usedom Island and as a result an accurate plan of it reached London. It was merely updated by RAF aerial reconnaissance. In spite of such good preparations the air raid turned out to be a tragic mistake, and chiefly those who had risked their lives delivering information to Polish Home Army Intelligence fell victim to it.[3]

Work was seriously slowed down as a result of the August bombing, but certainly not as the Allied commanders had imagined. In spite of the death of several important scientists (among others Dr. Thiel), the damage to the research infrastructure caused by the air raid was not large. The workers' "small town" mainly fell victim to it, and first and foremost the barracks of forced workers. Work was quickly resumed. Launch of the fourth A-4 prototype took place on October 3, 1942. This was the first (and for a long time the only) trial flight that ended in complete success. The rocket achieved a maximum ceiling of 60 km, and range of almost 200 km. The engine operated according to plan, for 61 seconds. The airspeed in its final stage amounted to over 1,200 m/s, i.e. around Mach 3.75. It seemed that the technical difficulties had at last been overcome, however in reality this was still a long way off. This favourable start proved however that overcoming these problems was possible.

On December 22, 1942, Hitler whole-heartedly signed the order to commence mass production. Soon afterwards in Speer's ministry a special committee was formed to engage in the supervision of research, production and use of "vengeance weapons" (Entwicklungskommission für Fernschiessen), in which apart from ministerial officials were also included representatives of the Heereswaffenamt, Air Ministry, and companies involved in production. The first production schedule based on realistic possibilities was established, which in 1943 was mainly to be assigned to research and training purposes. Its start was fixed for April (5 rockets), increasing this gradually to 10, 20, 40, 60, 105, 200, 400 and 700 specimens a month. It was anticipated that a significant quantitative increase could take place only at the beginning of the following year. These relatively small quantities were distributed to three production plants: the piloting line in Peenemünde, where the technical personnel of private companies were to gain experience, the Zeppelin plants in Friedrichshafen as well as the Henschel plants in Wiener Neustadt. The company "Mittelwerk GmbH," which took over the underground facility "Mittelwerk" (previously the central store of the Reich's fuel and lubricants) was not formed until September 24, 1943. It was to ensure the primary portion of the V-2's production.

The enormous range of resources which this rocket's programme of development and production devoured once again exposed the conflicts of its potential effectiveness and military value, and the intensifying aerial attacks on Germany again steered the discussion to anti-aircraft rockets, all the more that realistically the V-2 was still not ready for production. Speer, i.e. the person directing the entire armament production of the Third Reich, sent his special plenipotentiary to Peenemünde, Professor Carl Krauch, so that he could carry out an evaluation of the situation. He became acquainted as to the range of existing problems and familiarized himself with other designs being worked on at a similar or smaller intensity (the Fi 103, the Me 163 rocket-propelled fighter, the surface-to-air "Wasserfall" missile). As a result, just as Todt had done earlier, he recommended the complete cessation of work on long-range rockets. He recognized that production of the V-2 would be senseless, nor would there be the conditions for it, if he permitted that Allied bombers continually ruined at such speed the German industry and infrastructure. Krauch recommended in connection with this a re-orientation of the research and development potential committed to the "vengeance weapons" programme to the benefit of anti-aircraft rockets, and in particular the "Wasserfall" rocket, on which work had begun several months earlier. They could however be completed by the beginning of 1944, since many solutions had already been tested in the V-2.

The argumentation of the special plenipotentiary was logical and completely justified, but was based on the "defeatist" assumption that inevitable defeat was possible, which was for Hitler completely unacceptable. In 1943, he no longer accepted this type of comment as fact, as they didn't suit his plan for the course of the war (if such a plan ever existed).

One of the entrances to Mittelwerk

The facts indicate after all that he had a rather vague idea of the problems involved with the development of this class of rocket weapons—for example when Von Braun described the destructive effect of the V-2's one tonne warhead, he demanded the development of a rocket with a ten tonne warhead. Hitler's megalomania probably had a significant influence on this kind of attitude, manifesting itself in a fascination for several hundred tonne cannons (with an operating life the order of a hundred firings), or in demanding the construction of a thousand tonne tank. The fact remains that it was precisely Hitler's fascination that decided the fate of the V-2. After watching a colour film portraying the launch and successful (the best so

The galleries of Mittelwerk were high enough, so that the main fuselage sections of the V-2 could be positioned vertically

An unfinished V-2 missile in Mittelwerk's assembly gallery

Mittelwerk: the main transport tunnel. Railway tracks can be seen on the left.

far) flight of one of the next prototypes he appointed General Dornberger to Major and conferred on Wernher von Braun the title of Professor. This flight took place on May 26 and in it the rocket covered a record distance of 265 km, falling at a distance of only 5 km from the designated target.

Hitler demanded the doubling of production plans to 2,000 rockets a month, which however proved to be impossible to put into effect in the near future with regard to the bombing of the Zeppelin and Henschel production plants. The situation was rescued only by the gigantic "Mittelwerk," where already by the end of September 1943, 3,000 prisoners from the concentration camp in Nordhausen were working for the production needs of "vengeance weapons," and a year later this number had already risen to over 13 thousand. In addition production was started in two underground facilities in the region of Niedersachswerfen near the towns of Lehesten (here worked prisoners of the camp in Buchenwald) and Dernau (prisoners supplied by the camp in Natzweiler).

During this time intensive training of the Wehrmacht units was conducted. A rocket school was formed in Köslin (Fernraketen Schule) conducting training of both officers as well as soldiers, lasting around six weeks. This was mainly theoretical preparation—for the aim of conducting combat training an SS military training ground situated on Polish territory was taken over, being an adapted artillery training ground of the Polish Army, code-named Heidelager. It was situated in the region of Blizna, at the branches of the rivers Wisła (the Vistula) and San, around 150 km to the north-east of Kraków (Cracow). By mid-November 1943 two artillery regiments had been formed there, numbering in total six field batteries, each of which had three mobile launch pads at its command. The training ground's location, lying beyond the area of the Allied Air Force's interest seemed to guarantee safety. The Germans however had underestimated the fact that the rockets fell on terrain to a large degree controlled by the partisans, i.e. hostile. It was precisely from here that British Intelligence, thanks to Polish Home Army Intelligence captured the first moderately detailed information concerning the construction of the V-2. The earlier, amateurishly planned air raid on Peenemünde had in reality only proved how weakly the Allies were informed of German rocket research (around 80% of bombs had fallen beyond the grounds of the centre).

The conducting of trials in inhabited territory presented the next opportunity for Polish Home Army Intelligence. Missiles launched from the Blizna rocket range travelled in a northerly direction and fell in an area near the town of Sarnaki near Platerowo. Both Blizna and Sarnaki had been thoroughly penetrated by Polish Home Army (AK) Intelligence. The first specimen of the V-2 fell in the region of Sarnaki (between the villages of Mężenin and Ogrodniki) on the anniversary of Hitler's birthday, in April 1944. The Germans however quickly reached the debris and scrupulously removed it.

The resistance movement had been expecting this for a long time, among other things it had had at its disposal an accurate map of the Blizna rocket range as far back as the end of 1943, bought for 2,000 RM from a German member of staff. They planned to capture a complete missile, even considering ambushing a train transporting rockets and "abducting" one of them. Soon however an easier opportunity occurred. During the first days of May 1944 a V-2 fell into the river Bug and didn't explode. It didn't even sustain serious damage, and experts from Warsaw under the leadership of Prof. Groszkowski from Warsaw Polytechnic could even inspect in this operation the electronic circuits (the valves hadn't been shattered!). It was resolved to transport the extracted and disassembled rocket to England. On July 25, at a forest airfield near Tarnów, a DC-3 aircraft landed and snatched the precious consignment from "under the noses" of German units billeted a kilometre further on. This was already relatively late, but still before use of the rocket in combat.[11,14,15,16,17] Let us return however to the turn of 1943/1944...

In Peenemünde work was still carried on, mainly with the objective of increasing the rocket's technological susceptibility, i.e. introducing changes to the construction with the aim of making production easier and cheaper. The updated plan assumed the production of 200 specimens in December and in the first quarter of 1944 300, 600, and 900 rockets a month respectively. But this plan was also not kept to, as only 56 rockets had been produced by the end of January. However the specimens leaving Mittelwerk were not complete. Their electrical equipment was missing, which was later installed in the DEMAG plants in Falkensee near Berlin. They also didn't have warheads, which were only installed shortly before launching.

The first production batch revealed the existence of serious technological imperfections. The pressure systems were full of leaks and poorly made welds, damaged or destroyed parts had been installed and wires had been wrongly connected. The rockets of the first batch were practically not fit for use at all. This problem was later considerably reduced, but the Germans never managed to completely eliminate this, first and foremost with regard to the high susceptibility of the rocket's complex, precision systems to sabotage by prisoners.

The next "surprise" was brought in time by training flights from the rocket range in Blizna. Apart from the fact that of the first eight launches only one had ended in success, it later became evident that the majority of rockets reaching the planned target area as had earlier appeared, had in fact exploded at an altitude of a few kilometres above the ground. This was clearly some fault which hadn't been detected during trials carried out on the Baltic. After all there had been no such possibility. The area of impact had at that time been determined on the basis of defined markers (coloured sheets floating on the surface) being present in the water. Their presence in the expected area was treated as tantamount to the flight's successful course, to say nothing of the fact that no trials had been carried out with warheads, which was entirely understandable taking into consideration the number of rockets falling in the area, or in the immediate vicinity of the centre in Peenemünde. After studying the problem it became evident that the cause was too high thermal stresses in the skin of the fuselage's central section, leading to it tearing apart and then to heating of the liquid oxygen and alcohol tanks, which in turn led to an explosion. Since the rocket heated up in the dense layers of the atmosphere, the explosions took place relatively low above the ground. In order to reduce stress in the construction, a layer of glass wool was placed between the skin and propellant tanks, which gave the desired effect. The next, still unsolved problem were cases of the sudden and unexplained ceasing of the engine's operation in various stages of flight (in general shortly after launching). Here it was possible to take advantage of the results of static research, which revealed that the operation of the rocket's propulsion unit was accompanied by strong vibrations, originating chiefly from the engine. It was surmised that they could cause the fracture and disconnection of pipes in the propellant system (including nozzle cooling) and in connection with this were reinforced.

Only now could one ascertain that the V-2 rocket showed a state of technology making its military use possible. It was already the early spring of 1944. Trials of the new solutions were no longer carried out at Blizna rocket range, as the Soviet summer offensive had made this impossible. A new rocket range was created at the end of July code-named Heidekraut in the Tuchola Forests, several dozen kilometres to the east of Tuchola. There the research programme was completed. As a matter of interest I add that the first piece of information which aroused the curiosity of the AK Intelligence in this place, and led to its disclosure was the observation by the local population of soldiers who sometimes put on sheepskin coats and thick fur gloves (even in Summer). It later proved to be those who were employed in the filling of liquid oxygen tanks.

In Germany 1944, was a period of the systematic increase in dominance of the SS, which gradually took control of an ever greater number of institutions. In March, the SS made its first attempt to take control of the V-2 rocket programme. Himmler summoned Wernher von Braun to his headquarters in Possessern, in the East Prussia and proposed a significant widening of co-operation (Von Braun had been a member of the SS as far back as May 1940 and had the rank of SS-Sturmbannführer, the equivalent of Major), proposing the position of a member of staff, support of his work and better "access" to Hitler, and this counted most at that time in the Reich. Von Braun didn't feel qualified for this type of "bureaucratic" task and didn't accept the offer. Soon by strange coincidence he was, with several other scientists, arrested by the Gestapo. He was accused of sabotage. As it turned out (and this was the truth) they repeatedly and critically expressed their opinion about the military use of the rockets, stating that they had only done it out of compulsion, and during the period of discouraging "fluctuations of priorities" had engaged themselves completely in the design of space rockets, entrusting work on the A-4 to personnel of a lower grade. The position of the SS was

Mittelwerk: a present-day entrance

The remnants of a narrow-gauge locomotive, which provided transportation within the underground factory. On the right is a penal bunker for the prisoners.

so strong that any attempt to free them conducted "through official channels" by General Dornberger was denied any chance. Marshal Keitel was completely afraid to take the floor in this matter. Only the intercession of Speer, a person closely connected with Hitler, led to the "temporary release" of Von Braun, Riedel and Gröttrup. Himmler repeated an attempt to take control of the rocket programme after the assassination attempt on Hitler in July 1944, when a further significant increase in the influence of the SS was recorded. This time the actions had the approval of Hitler and in connection with this were crowned with success. SS-Gruppenführer (the equivalent of Lieutenant General) Hans Kammler was appointed as the officer responsible for all matters concerning the A-4 rocket, and he performed this function almost to the end of the war, when like many other higher SS officers he "vanished" in mysterious circumstances.

Despite overcoming construction problems and developing the production base the rocket programme was still plagued by difficulties.

The V-2 in a way lost the competition to the V-1. Hitler, impressed by the first military attacks with the participation of the latter demanded an increase in the speed of production of the V-1 at the cost of the V-2. The speed of production of the rocket was reduced (transitionally) to 150 a month, despite that now it had already been managed to achieve a high technical standard—out of the rockets launched in the Tuchola Forests around 75% landed within the limits of the designated area.

Modifications introduced into the construction (reinforcements) had an impact on the mass of the warhead, which was reduced from 1,000 to 976 kg. Modifications were also introduced giving a significant increase in range, initially up to 320 km, and later after enlarging the propellant tanks up to 470 km, but the Germans never managed to introduce these modifications to production specimens on time. By September 1944 over 600 rockets were leaving the production line in the plants of Mittelwerk each month and at maximum operational speed it was possible to produce around 30 daily. Liquid oxygen was condensed in sufficient amounts for about 30 launches daily. As Von Braun summed up after the war over 200,000 people worked for or served (in combat units) the needs of the German rocket programme. As Speer in turn judged in his memoirs, this was just the same potential which would have been necessary to develop and produce by approx. 1945 a nuclear weapon.

Data concerning the production of the V-2 is incomplete. It is estimated that on the grounds of Peenemünde over 300 rockets were produced and in the Mittelwerk plants in the consecutive months of 1944 the following numbers: January: 50, February: 86, March: 170, April: 260, May: 440, June: 132, July: 86, August: 375, September: 629, October: 668, November: 662, and December: 613. In each of the first two months of 1945 600-700 rockets were produced and in March—the final month of production probably around 400.

An entrance to the underground factory in Leśna (Marklissa). This was one of many places where components were manufactured for the V-2.

	_		Lager				
Beschr.	Hesdin	Aumale	Rouen	Berg.	Cassel	Mont-didier	Summe
Erdaushub m 3	12 000	5 000	1 500	8 000	3 100	--	29 600
Erdbewegung m 3	1 300	7 000	6 000	27 000	--	400	17 400
Planierung m 2	2 500	5 000	1 500	1 500	2 400	1000	13 900
Stampfbeton m 3	300	1 000	200	3 000	--	100	4 600
Fussbodenbeton m 3	50	700	200	--	--	--	950
Strassenbeton m 3	--	800	700	--	--	--	1 500
Eisenbeton m 3	1 400	2 000	100	--	--	--	3 500
Kiestransp. m 3	2 000	9 000	9 000	--	--	--	20 000
Steintransp. m 3	1 800	300	--	--	--	--	2 100
Eisentransp. to	160	--	--	40	--	--	200
Zementtransp. Sack	17 000	30 000	10 000	--	--	--	57 000
Bomben ausgegraben	3	12	36	1	--	58	120
Bomben gesprengt	3	8	4	--	--	5	20
Bomben transp.	--	4	32	9	--	95	140
Maschinen m 2	3 500	--	--	--	500	--	4 000

A German list of the costs of constructing stationary launchers on the coast of France

At first the Germans planned to wage military operations solely from large bunkers, assuming that such a complicated weapon system like the V-2 required a complex infrastructure of good workshops and installations essential during preparations for use, and that the entirety should be assured good protection against aerial attacks. Waging military operations from field positions couldn't really be imagined (wrongly). At Hitler's order the construction of four enormous bunkers was commenced in August 1943 on the English Channel. Work was begun in Watten—the "Kraftwerk Nordwest" facility, in Wizernes—the "Schotterwerk Nordwest" facility (both in the region of Calais), as well as near Sotterast—the "Reservelager West" facility and Hainneville—the "Ölkeller Cherbourg" facility (both in Normandy, the latter was finally converted into a V-1 launcher).

These were gigantic bunkers, completely self-sufficient and therefore easy to detect while they were still being built. These fears proved to be justified for at the turn of the summer and autumn of 1943 the Allied Air Force carried out around 100 air raids on them, dropping tens of thousands of tonnes of bombs. Under conditions of complete Allied air supremacy, holding these bunkers, in spite of five-metre thick ceilings was recognized as unrealistic. It was resolved to launch all of the rockets from field launchers, relying rather on the effectiveness of camouflage and not resistance to bombs' explosions.

By August 1944, when SS-Gruppenführer Kammler took command, 45 camouflaged combat field positions, 20 different types of field warehouses as well as a series of installations for the production and storage of propellants had been prepared. The Normandy landings soon forced the Germans to transfer them further to the north even before the start of combat launches.

Hitler gave the order to commence launching on September 15, 1944. In contrast to the V-1, whose launch equipment

One of the V-2 rockets captured by the Americans, on display at the National Air and Space Museum in Washington

The crater that was made after a V-2 exploded on the launcher, at the White Sands range after the war

was large, immobile and easy to detect, the V-2 used small truss launch platforms, which were mobile and practically undetectable. Launches were carried out not only from well prepared positions, but also from stretches of road, forest glades and all other places, if only they assured the possibility of access and protection against detection of the columns of military vehicles.

As opposed to the V-1, the V-2 was not only indestructible in the final stage of its flight (with regard to the speed of the order of 3,500 km/h) but was also immensely difficult to detect directly before launching—due to the high mobility of the rocket batteries. The typical flight time of a rocket from western Holland (from where the majority were launched) to

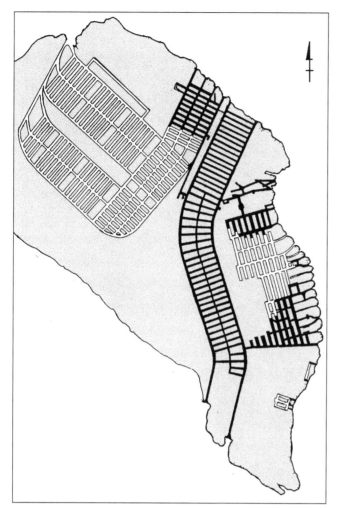

A plan showing the system of underground factories under the Kohnstein mountain. The galleries that were actually constructed are marked black, the ones that were planned: white.

London amounted to around 5 minutes. The engine turned off after 60-70 seconds from the moment of launching, at an altitude of the order of 35 km. Then through ballistic flight the rocket ascended further, gaining a maximum ceiling of the order of 100 km. Hence the V-2 was the first space rocket.

When using a simple version of a navigational system (only gyros) the rocket's scatter, i.e. the magnitude of deviation from the designated trajectory reached up to around 20 km. The application of a guiding beam or inertial navigational system (gyros, plus accelerometers, plus flight path calculator) allowed the scatter to be reduced 5-10 times and was originally designed to combat standard military targets. This last possibility was taken advantage of mainly in the later stage of operations, combating targets situated in Belgium and France.

The V-2 warhead detonated, as opposed to the V-1, after driving into the ground. In the case of open ground this didn't involve as a rule large damage, but a direct hit on a specific structure (e.g. a building) almost always resulted in its complete destruction, like what happened for example with a direct hit on a station of the London underground (used as a shelter), as a result of which over a thousand people were killed outright. An explosion inside a building literally tore it apart, also causing loss to its surroundings. The first two rockets were launched in the direction of London on September 9, 1944, gradually increasing the number of attacks. Gradually the number of targets was also increased, from the end of September also attacking the cities of Belgium, France and Holland.

By October 3, 156 V-2 rockets in total had been launched, out of which 52 were in the direction of Great Britain (London 30, Norwich 22), Belgium 42 (mainly at Liege 17 and Hasselt 10), France 45 (mainly at Lille 15 and Paris 10) and 17 at the Dutch city of Maastricht. On October 12, Hitler gave the order to direct all rocket attacks against London and Antwerp.

Up until the end of military operations by the rocket units—at the end of March 1945, no less than 3,170 V-2 rockets, as is estimated, had been launched, of which the majority, 1610—had been launched against Antwerp.

Second on the list of targets was London—1,359 rockets. All in all 1,664 had been launched against cities in Belgium, Great Britain around 1,400, France 73 and Holland around 20. The final target of the V-2 was the German town of Remagen, occupied by the Allies, in the direction of which 11 rockets were launched.

The retreating rocket units reached the town of Celle near Hanover a month before the end of the war, where on April 7, they destroyed their equipment so that it wouldn't fall into enemy hands. On the basis of conservative estimates one may assume that around 70% of launched rockets hit their designated targets.

However the arsenal of "V" weapons didn't end at this.

Basic tactical and technical details of the V-2 rocket (version B)

launch mass	12,700 kg
mass without propellants	4,008 kg
warhead mass	1,000 kg
length	14.04 m
fuselage diameter	1.65 m
maximum engine thrust	25,200 kg
range	approx. 300 km

The V-3

The V-3, the next "vengeance weapon" of the Third Reich, represented the materialization of a completely different conception than in the case of the V-1 and V-2.

This was a long-range gun, only that of a very peculiar kind—a pioneering construction from a technical point of view. As opposed to the other "vengeance weapons," this was a great hope of Hitler from the very beginning, up until almost the very end of its short life. The chief author of the V-3 was engineer Coender, technical director of the Röchling Eisen- und Stahlwerke works.[3,9]

In general the Germans planned to achieve an increased range due to an increase in the projectile's muzzle velocity, gained in turn by maintaining a very high pressure in a long barrel not only in the initial stage, but for the entire period of the projectile's motion in the barrel. This could be achieved only by bringing about the strongly progressive combustion of a powder charge (i.e. the speed of combustion grows strongly with time), or by applying many powder charges, initiated in turn, in due measure of the missile's motion in the barrel. In the V-3 it was precisely this second principle that was exploited. The powder charges were placed in special side chambers, for the whole of the barrel's length. Only the first charge was placed conventionally behind the projectile, in the barrel's axis.

This was the so-called multichamber gun, "christened" by the Germans with the official code-name "Hochdruckpumpe," a high pressure pump (abbreviated HDP), unofficially called the "centipede" due to its shape (Tausendfüssler). This 150-mm calibre gun was designed to shell solely one target: London. For this reason, and with regard to its great length and generally complicated construction, it was to be mounted on a permanent, concrete-steel base with the barrel at a permanent angle of elevation. Although the idea itself of such a construction was born at the end of World War I in France (in response to the so-called counter-Paris gun), it was put into effect for the first time a quarter of a century later in Germany.

In May 1943, Speer initiated a meeting with Hitler in order to present the design of the new "vengeance weapon," at which he arrived together with the owner of Röchling—Hermann Röchling. Hitler at once liked the idea and demanded the construction of prototypes. It was ultimately planned to use 25-30 guns in combat with a combined rate of fire of the order of 300-600 shots per hour.

The V-3 was a smooth-bore gun and the first for which subcalibre projectiles fin-stabilized on the flight path were developed (presently constituting the basic type of tank ammunition). The projectile's diameter amounted to 100 mm, length 2.5 m and mass 140 kg, of which 25 kg was explosive charge. The projectile was aligned in the barrel with the aid of fins in its rear section and more or less at one third of the length from the front with the aid of elements discarded after leaving the barrel (the so-called sabot). A short sealing cylinder was placed behind the projectile, performing the function of a piston. It was planned to achieve a range of the order of 160 km. Prototypes were to be installed at two special firing ranges, near Hillersleben, approximately 20 km to the northwest of Magdeburg and near Misdroy on Wolin Island.

Hitler's interest rose further in due measure of the growing problems with the V-1 and V-2. In August 1943, at a meeting with Speer he demanded, despite that prototypes hadn't even been tested, the selection of a site for the V-3 combat positions. To this end the Germans decided to build a large bunker near the town of Mimoyecques in north-western France, housing underground ten combat batteries, each of which was to number five guns. The distance from this site to the centre of London amounted to 153 km.

In the autumn of 1943, the first segments of the "centi-

The first version of the V-3

pede" were assembled in Hillersleben and in October the first firing trials were commenced, but were unable to give reliable information concerning the future combat version. Trials of a 20-mm miniaturized version of the gun were also conducted, with equally modest results. Not until the turn of October and November was the first full length prototype completed in Hillersleben and in January 1944 in Misdroy. Firing was immediately commenced with target subcalibre projectiles. It was affirmed that they constituted a successful design, which bore fruit with the order to increase the speed of production to 10,000 a month. However the construction of the gun itself hadn't been fully examined, with regard to it being fired using reduced powder charges. Despite this, the mood was very optimistic and Hitler was convinced that the hopes pinned on the V-3 were completely justified.

Not until March 1944 did it become evident that these hopes had been somewhat premature. The "fully-sized" gun, consisting of 32 segments with a total length of 130 metres, achieved a projectile muzzle velocity when firing (still not full) powder charges placed in all of the chambers, of the order of 1,100 m/s. Achieving the appropriate range required a projectile muzzle velocity at a level of 1,500 m/s, which was expected to be achieved using full charges. However when they began to be used serious problems appeared.

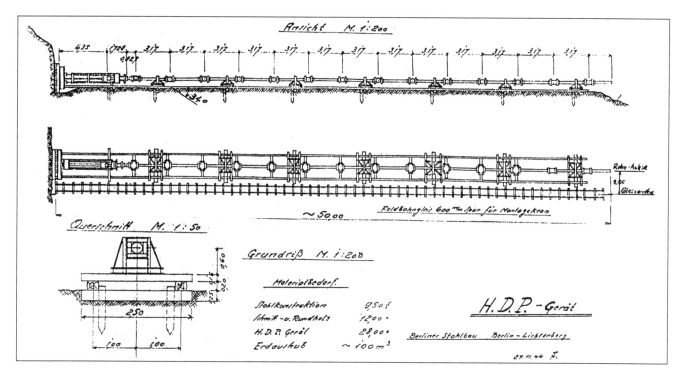

A German plan of the V-3 from November 1944

It became evident that the durability of the barrel segments was simply too small, many segments being torn apart during firing. From a present-day perspective, one may judge that the applied single block barrel had no chance of withstanding the existing pressure. This would have required the application of a layered barrel, connected straight away using the so-called auto-frettage method, which would have ensured a different load distribution and even greater durability than was required in the case of the centipede. During World War II this technology was still in its infancy. The next problem proved to be the lack of accuracy. Namely it became evident that at this speed the projectile wasn't stable in flight and up till then the Germans had managed to produce about 20,000 of them. At this moment Hitler's docility took its revenge against Röchling, who had demanded from the start that the Heerswaffenamt not be informed about this project's existence until trials had been completed. He simply feared that the military would be hostile to such an unconventional construction and would reject it after simple calculations. In this way a project containing obvious faults had been accepted for realization without an expert theoretical verification. This was at least the conclusion of a group of military experts from the field of artillery who with General Leeb from the Heerswaffenamt, finally took part in trial firings. It was March 1944. The military was for cancelling the project, which would have surely happened, if not for the deep commitment of Hitler.

So work was continued. Despite the design of a new projectile (significantly lighter with a mass of up to 80 kg) being commissioned to the Institute of Aerodynamic Research in Göttingen, production of the former was continued, although in reduced numbers. For the decision had been taken not to inform Hitler of the scale of problems which had appeared.

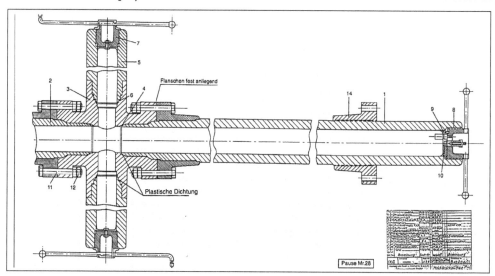

Original technical drawing of one of the V-3's sections

An American trophy—a shell for the V-3

A close-up of the V-3 prototype

In the meantime essential corrections had been introduced into the construction. Side chambers were employed, that were positioned not perpendicular in relation to the barrel but inclined backwards at a certain angle, and the projectiles were improved. In spite of this, firings resumed in July near Misdroy brought further failure. About one third of the entire barrel was torn apart. Work was still continued, although it was obvious that achieving the planned projectile muzzle velocity was unrealistic. This became evident when the gigantic combat bunker in Mimoyecques had already been completed.

Simultaneously a special Wehrmacht regiment on Wolin Island, assigned to operate the V-3, was in the final stage of training. It numbered approx. 1,000 soldiers and was commanded by Lieutenant Colonel Bortt-Scheller.[18] Contrary to pretenses, the problems reported in achieving the target range didn't at all signify that this project was at once doomed to failure. Simultaneously, although probably independently, a new type of long-range ammunition was developed and tested among other places in Hillersleben, which could have put the whole undertaking in a completely different light. It concerned an artillery projectile with additional ramjet propulsion (Staustrahlantrieb), i.e. the so-called Trommsdorff-Geschoss.[20] Several modifications of it existed, among others 105 and 150-mm calibre. In the latter case this was also the calibre of the V-3 "super gun." I do not know if combining the virtues of both weapons was considered, but one way or another this was possible. Tromms-dorff-Geschoss is described in one of the following chapters.

Irrespective of the fact that work on the V-3 itself was very interesting, a very curious and little known issue is presented by the story of the construction of the gigantic underground system in Mimoyecques. This construction was kept so top secret, that even General Leeb, Chief of the Heerswaffenamt found out about the whole undertaking by accident, while inspecting fortifications on the coast of France (at the end of 1943).

The Germans initially planned to place 50 guns underground, finally deciding on 25. But the huge system of underground fortifications still came into play, 430 miners and approx. 5,000 skilled workers from the Ruhr coalfields being assembled for its construction in the autumn of 1943. The core of the complex was made up of five large adits each 150 metres long, entering the mountain at an angle of approx. 45º. Five guns were to be situated in each of them arranged side by side. The Krupp consortium supplied the armoured covers sealing the adit mouths so that only the tips of the barrels protruded. In addition it was decided to cover the whole surface of the mountain with a layer of reinforced concrete 6 metres thick. Together with the hard rock itself this was to protect the complex against any weapon of that time.

The main complex of horizontal tunnels, including storerooms and a railway line, was situated at a depth of 30 metres, approx. 10 floors down. Lift shafts ran down from this level, through which it was planned to supply the ammunition. Successive levels of tunnels were situated at a depths of approx. 80

Shocking are the small dimensions of the weapon that was supposed to "bring the British to their knees"

and 110 metres (the deepest). Directly above only the muzzles of the 25 guns, as well as narrow ventilation shafts betrayed the existence of the complex. Even two high voltage lines, running from the interior of France, were situated underground. This construction swallowed up a million tonnes of cement, steel and gravel etc. In spite of this it proved to be possible to destroy.

Here also Polish Intelligence provided the Allies with invaluable services, and specifically the groups of Major Grabowski ("Lille") and Wł. Ważny (pseudonym "Tiger"). Grabowski received an order from London to cut the power line leading to Mimoyecques. A group of commandos were dropped in to assist (Raszka, Bronicki-Łoziński, Fijak, Kral and others). Reference was made to a section where the line ran on the surface. Since after every operation the Germans repaired the electric feeder line, it was cut all in all 16 times. These were however only temporary operations. In England a decisive strike was planned.

On August 12, 1944, i.e. already after the landings in Normandy, an unusual "Liberator" took off from a base in Norfolk—loaded with ten tonnes of high-explosive. It was commanded by Lieutenant Joseph Kennedy, brother of the future president of the United States. Before reaching the English coast the crew were to jump out with parachutes, and escorting aircraft were to then take control—by radio. Finally the "Liberator" was to strike the V-3 complex.

But this never happened. 28 minutes after taking off, still over Great Britain, the flash of a huge explosion lit up the sky. The "Liberator" had ceased to exist. It has never been unravelled if this was an accident, or the result of German intelligence activity.

Immediately an alternative plan was put into effect—it was resolved to bomb the gun complex with the heaviest bombs of that time, the five tonne "Tall Boys." This was the first time that this bomb was to be used in combat. How much the British feared the V-3 is evidenced by the fact that only a couple of weeks later the whole complex was seized. One of the German witnesses of the air raid, Col. Walter, described as follows the first use of "the earthquake bomb": "It seemed that the whole mountain was shaking and at any moment would collapse. A shower of large and small stones fell from the ceiling, everything creaked. Even people with strong nerves couldn't stand it for long underground."

When Churchill later saw the construction in Mimoyecques, he said: "London could have expected from this place the most decisive blow of all." Let us remind ourselves that a projectile was to fall on London every 12 seconds. Later Churchill's son-in-law, Minister Sandys, wrote to him in a report referring to the V-3 gun: "It could be completed and used to shell London. As long as it exists, it poses a potential threat to England." He recommended "destruction of the bunker as long as our units are still in France."

It becomes evident that just as the French didn't trust the British, the British didn't trust the French (e.g. the armies in the French colonies generally, despite being under no compulsion, declared to be on the side of the Vichy government, during a landing in North Africa the British soldiers were dressed in American uniforms). Heedless of possible protests from de Gaulle the British resolved to destroy the complex in Mimoyecques. On May 9, 1945, British sappers detonated explosive charges placed at various locations of the underground complex's highest level—probably on the same principle the Russians had attempted to destroy the "Mittelwerk" complex. Several days later both entries to the railway tunnel were blown up, using up as much as 25 tonnes of high-explosive for this objective.

In spite of this, most of the underground cubature is in all probability still intact. Perhaps one day it will be possible to reach inside and open this "locked museum"?[3,9,18,19]

The issue of various derivative versions of the "V" weapons as well as numerous alternative constructions constitutes a relatively little-known problem. Now a short summary.

The Rheinbote

Despite that the "Rheinbote" missile presented below was not officially numbered among the "V" series of weapons, it should in fact be treated along with them, since it constituted an attempt to create a rival for the V-1 and V-2 in the form of a long-range solid propellant (powder) rocket.

It was an initiative of the designers from Rheinmetall, Borsig, accepted for realisation in June 1941 by the artillery department of the Heereswaffenamt. In a way this project arose due to previous experience resulting from the construction of a series of powder rocket engines in the latter half of the 1930s, among other things the launch engines for gliders.

In the summer of 1941, Rheinmetall presented three long-range rocket designs (all four-stage) to the Heereswaffenamt for assessment and selection. The lightest of these rockets was to have a mass of 1,750 kg (of which 625 kg was propellant), warhead mass of 200 kg and predicted range of about 100 km. The "medium" rocket was to have a launch mass double in size—3,500 kg, 1,220 kg of propellant and a 500 kg warhead. The predicted range was 110 km. The heaviest variant was for those times a true giant in this rocket class and was to be comparable in terms of size to the A-4 rocket. The Germans planned it to have a launch mass of up to eight tonnes, 2,800 kg of propellant, a 1,250 kg warhead and a range of the order of 120 km.

But officers from the Wehrmacht were rather sceptical with regard to the whole project (chiefly Dornberger), first and foremost due to the high consumption of powder in short supply and predicted low accuracy of the rocket, putting its military usefulness into doubt. Which was why only the first "light" variant was accepted for further realisation, and this probably wouldn't have come about if not for problems at the time with developing the A-4 rockets. It was decided however to considerably reduce the warhead mass (to only 40 kg), since as calculations showed, this was to double the rocket's range. Clearly therefore the "Rheinbote" was classified in advance as yet another "psychological-vengeance" weapon, and not purely military for in the face of no navigational and guidance system, this variant would be solely fit for combating large superficial targets.

The range, approximate to the expected range of the A-4 rocket also suggests that a similar use was taken into account. In order to design prototypes of the missile a forty-man re-

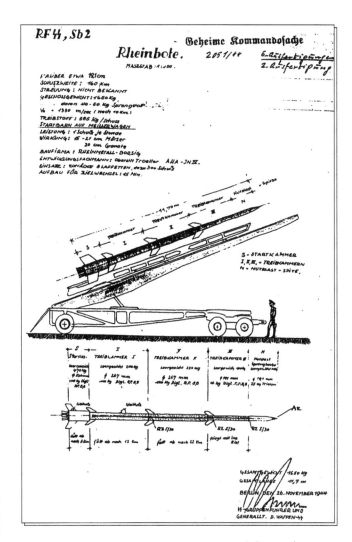

A German document, containing the first draft of the Rheinbote rocket

search group was formed in the Berlin plants of Rheinmetall, and the missile was given the working designation Rh Z 61. The first specimens were ready for rocket range trials as early as November 1941. They were commenced on a range at Leba on the Baltic. This was a Luftwaffe rocket range (also called at this time the "small Peenemünde"), which Rheinmetall already regularly used, testing different kinds of aerial armaments. The rockets were launched in the direction of Bornholm Island occupied by the Germans, 170 km away, on which a suitable research apparatus had been positioned. At this time however only individual

The Rheinbote on the launcher

stages were launched independently. Material difficulties (the project was not encompassed by any priority) resulted in that complete rockets were not launched until April 1943. The trials were recognised as successful, and the final stage of one of the rockets even fell at a small distance from the observation post on Bornholm Island, and could later be examined.

In connection with this the programme's realisation was continued. Despite the "Rheinbote" still not being classified in any broader armament program and so not being entitled to any priority and allocations of raw materials, the research group still managed to unofficially "economize" a supply of materials from the long-range weapons programme to build the next 30 prototypes.

But for unclear reasons the rocket parts were not delivered until the beginning of 1944, and of the wrong dimensions. In connection with the significant delays that had arisen it was resolved to seriously accelerate work. The production

A close-up showing the six nozzles of the Rheinbote's main rocket engine

of the first batch of 200 rockets was ordered. A special artillery unit was formed, whose training was commenced at Leba. In the meantime rocket range trials were conducted, but which revealed the existence of fundamental faults in all of the launched missiles. Problems were caused by the irregular burning of powder, which in several cases led to the missiles exploding. There were also cases of the fins dropping off when the sound barrier was exceeded.

It wasn't until the end of 1944 that the research group managed to overcome these problems. As in the case of the V-2, after the unsuccessful assassination attempt on Hitler at Rastenburg (July 20, 1944), the SS also took control of the "Rheinbote" project. Unexpectedly Kammler as well as other SS officers, in defiance of the "sceptical" stance of the Heereswaffenamt officers, became enthusiastic supporters of the new rocket. The site of carrying out in-flight trials was moved from Leba to the Tuchola Forests (in Poland). By mid-December they had managed to produce not much more than 100 rockets. An additional 220 were to be supplied by the end of January 1945. During this time attempts were continued to improve the rocket, but the results were still far from those expected. Out of twelve rockets launched in the first half of December as many as five failed to a significant degree (mainly as a result of explosions), the others were characterised by significant scatter, ranging from approx. 50 to 160 km.

Despite that the "Rheinbotes" were not mature, some of them (the test batch) saw combat on the western front. The only confirmed case is that during the Christmas Eve of 1944 several dozens of them has been fired in the direction of Antwerp, from a distance of 165 km. Error in calculations caused however, that they reached a distance of about 220 km! Being a controversial and not very successful project, in addition with no defined use, it met a similar fate as the unfortunate multichamber gun.

On February 6, 1945, SS-Gruppenführer Kammler made the decision to cease work.[3]

The Rheinbote in combat position

A unique photograph, showing the Rheinbote rocket disassembled for transportation

Basic tactical and technical details of the "Rheinbote" missile

total mass	1,656 kg
mass of individual stages	I - 710 kg, II - 380 kg, III - 360kg, IV - 166kg
warhead mass	40 kg
total length	12.9 m
maximum flight speed	approx. 6,000 km/h
duration of thrust (in total)	approx. 15 s
range	200 - 230 km
ceiling	70 km

Other Vengeance Weapons

The long-range weapon projects presented above testify to the enormous scientific and productive potential of the Third Reich. These weapons, and particularly the V-2, astonished the world with their innovation and modernity and in many cases were developed and perfected after the war in other countries (e.g. the V-2 in the USA, USSR and in France). However the German programme encompassing them didn't quite end at this. Many other projects existed (the majority of which lived to see only the prototype stage) that were much more revolutionary, forming quite simply the most interesting group of weapons developed during World War II.

In what directions was work carried out? Realising the limited effectiveness of the "vengeance weapons" constructed according to the existing conception, not only was an increase in range strove for, but great emphasis was also placed on increasing accuracy. When analysing these more advanced projects it also becomes evident (which is not seen in the history of the V-1 and V-2's use in combat) that the Germans did not plan to limit the conception of "vengeance strikes" just to the territory of Europe.

Work in the aforementioned directions was put into effect in the first instance with reference to the weapons that had already been produced. The first manifestation of this was the modernization of the V-1 missile and a version with a greater range and airspeed being brought into production. In 1944, the order was given to produce a new turbojet engine for the V-1. BMW and Porsche presented their designs. With this engine the missile would have gained a range of 500 km, and with a cruising speed amounting to 800 km/h would have been an extremely difficult target for British fighters. Moreover the Germans planned to equip it with a remote guidance system (similar to that already employed in guided aerial bombs) with a television camera fitted in the nose section transmitting a picture of the target and receiver-transmitter radio system transmitting in one direction the picture from the camera and in the other the guidance commands. This version would have been a "genuine" cruise missile, capable of the precision striking of small targets, which wouldn't have put many armies to shame long after the war. In the spring of 1945, intensive work was also carried out on the application of a guiding beam in the

Fragments of the Rheinbote rocket which exploded shortly after launch, found at the test range near Leba

A suicidal version of the V-1 (the Fi 103 Re-4, without nose section)

The A-4b missile at the beginning of 1945

initial flight stage, analogous to that in one version of the V-2. The V-1 was also to be the first "vengeance weapon" that the Germans planned to use against the United States.

A plan existed to install launchers on the most state-of-the-art Type XXI submarines, but the Germans also never managed to put this into effect. However an unknown number of V-1 launchers were installed on older U-Boats and within the confines of operation "Elster" it was attempted in 1945 to use them against the United States. The operation ended in a fiasco. According to some intelligence reports, American counteraction was so swift and effective, just because they suspected, that these missiles are carrying biological warheads. The last attempt to turn into reality Hitler's "great" dream, or rather a nightmare—to destroy New York—"the capital of Jewishdom" has failed. It was April 1945. The unguided "Ursel" rockets developed in 1942 in Peenemünde were to serve the same purpose, but it is not known if they were ever used in combat.[3,22,23]

At the end of the war in an act of desperation the dying Third Reich planned to adapt (by installing a cockpit) a portion of the V-1 rockets to the role of suicidal aircraft, designed to destroy particularly important facilities. This version of the V-1, known as the "Reichenberg Fi 103 Re-4" met however with the opposition of the Luftwaffe command, who called this conception "a suicidal plan for the Luftwaffe." Even Hitler treated it with great reserve. In spite of this, several dozen pilots were trained and 175 "suicidal" V-1's were produced, but were never used in combat. The training was used mainly to obtain additional information concerning the flight properties of the missile.

A rich developmental programme was also put into effect in relation to the V-2. The Germans also planned to use this weapon to attack American cities. Code-named "life jacket," floating "underwater silos" were developed with a displacement of 500 tonnes. Each housed one rocket, essential apparatus and a cabin for personnel. After commencing pre-launch procedures the crew were to abandon the silo shortly before automatic launching of the rocket. It was determined that a new Type XXI submarine could simultaneously tow up to three silos. But these plans were abandoned, as there was no possibility of carrying out suitably accurate topographical control of such a floating launcher (however, ascertaining exact coordinates of the targets was still possible—in the last months of the war American counterintelligence has arrested a group of German agents equipped with beacon transmitters). By the end of the war the shipyard in Elbig, to which this task had been commissioned, had managed to produce only one silo.

Based on the A-4 a new, far more revolutionary rocket was developed with a far greater range. This was a completely new design, although it was designated the A-4b, suggesting "only" a derivative version of the A-4. This was arranged however only because resources wouldn't otherwise have been available to carry out its trials. The A-4b was to attain a greater range mainly due to the application of ... wings. After the ballistic flight stage, gliding (side slip) was to occur, permitting a range of 600 km to be attained. In order to improve control during gliding the tail fin was enlarged, adding aerodynamic control surfaces (during engine operation the A-4b, like the A-4 was controlled with the aid of graphite elements deflecting gas streams from the nozzle). In December 1944, the decision was taken to build 20 prototypes of this rocket. Although the first two launch trials ended in failure, work was continued and finally a launch took place on January 24. Everything looked promising, however while crossing into a glide one of the wings fractured and the planned range wasn't achieved. A manned version of the A-4b with a small cockpit was also developed, with retractable landing gear and an additional small jet engine permitting a further increase in range. However this version was only carried through to the drawing stage.

Although in-flight trials of the A-4b rocket were not carried out until 1945, plans of its production had existed much earlier and this conception had been developed almost in parallel with work on the A-4. Based on this project a more revolutionary rocket was developed as early as 1941, with a tailless "delta" aerodynamic system, designated the A-9. Even then the Germans had planned to use it as the second stage of the most revolutionary rocket ever designed in Peenemünde, the intercontinental A-9/A-10 ballistic missile, also known as the "Amerika-Rakete," as it was designed to destroy the main cities of North America. However this version of the A-9 was to transport not a tonne of high explosive, but a powerful nuclear charge. General Dornberger wrote in his "Memoirs" the following about work on the A-9:

> *Hundreds of calculations were made to determine the trajectory which would give the greatest range. Finally it was determined that the missile would achieve a maximum speed of 4,500 km/h at a ceiling of 19 km and then pass into a glide on a slightly curved path with the peak at a ceiling of almost 29 km. After reaching the target area, at an altitude of about 5 km it was to go into a dive, like the Fi 103 (V-1).*
> *We were only a step away from advancing from an unmanned A-9 rocket with a fully automatic guidance system to a piloted version. This extremely fast aircraft, with wings only around 13.5 m² in area had no military significance. Special wing flaps would enable it to land, after travelling around 640 km in 17 minutes, at a speed of only 160 km/h. The development of the A-9 didn't however satisfy our ambitions. We wanted to cover a range of thousands of kilometres. Our own private and exclusive sphere of activity only began beyond the range of the heaviest aircraft.*
> *Only by abandoning a single-stage system to the benefit of a multi-stage rocket, i.e. discarding the "dead" mass, which had already fulfilled its task and so improving the rocket's mass ratio could we hope for such an incredible increase in range.*
> *Such was the source of the A-9/A-10 project's origin. The goal in this case was to lead to the situation in which the engine of the second stage (A-9) would commence operation only when the missile had achieved a high enough speed due to the first stage, thereby acting as a means of auxiliary thrust."*

An alternative was catapulting in order to give a high initial speed to the A-9. On the basis of calculations and experience with the V-1 launchers a long, inclined catapult was designed, capable of giving the A-9 rocket an initial speed of 1,290 km/h. This speed would have been sufficient for the propellant-filled rocket to smoothly commence flight.

However a better plan, which increased the range con-

Photographs of the A-4b missiles from January 1945

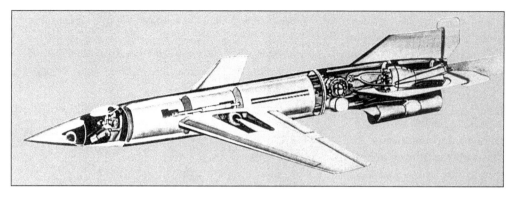

A manned version of the A-4b with additional jet propulsion. The manned version of the A-9 rocket was developed on its basis. (original drawing)

siderably, was the construction of the A-10, the first stage for the A-9/A-10 system, which would have weighed 87 tonnes, the propellants having a total mass of 62 tonnes. The A-9 was positioned on top of the A-10. The latter, with a thrust of 200 tonnes maintained for 50-60 seconds would have given the A-9 stage an initial speed of 4,350 km/h. After the propellants of the first stage had been exhausted the A-9 engine was to be started and separation to take place. Soon afterwards the A-9's angle of climb was to be increased, which was to attain a maximum ceiling of 56 km. Then a long, supersonic glide was to commence.

The Germans planned to achieve with the basic version a range of the order of 5,500 km. For understandable reasons the sole variant of armament was to be a nuclear warhead. The A-9 rocket programme was the first case in which a long-range means of transportation had been directly combined with a nuclear weapon (work on its construction in the Third Reich was almost completed). The effective use of this rocket required a new approach to the main problem up until now, accuracy, occurring this time on a proportionally larger scale, which was too large, even taking into consideration the warhead's destructive radius and large size of the New York or Washington agglomerations. The Germans planned to solve this problem by employing a manned version of the A-9, similar to the manned version of the A-4b. The rocket was to reach the eastern seaboard of the United States by travelling over arctic territory and approaching from the north-east. The pilot, all the time flying at a relatively high altitude, was to aim the rocket on target, drop the warhead over it (or rather nuclear bomb), and assisted by auxiliary propulsion alone continue his flight to the south, so as to finally land on the plains of Argentina, traditionally friendly with the Germans.[24,25]

Formally this was a neutral country, but was in fact connected to the Third Reich by ties of strong co-operation, among others due to an individual from the military attaché at that time in Berlin, who from 1940 was the future President of Argentina, Juan Domingo Peron (as it in any case became evident after the war, appearing on a list of SD agents). Thanks to this individual Argentina still managed to be of service to the German programme of "special weapons," for on his order in the years 1945-1947 diplomatic posts in Austria and Italy issued around 2,000 passports, enabling the evacuation of many hunted individuals, including scientists and the precious documentation of new weapons. The problem of accuracy wasn't however as big as it may appear from the present perspective, the correction of the final stretch of the trajectory

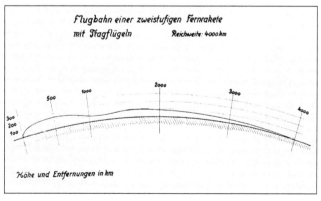

An original sketch showing the trajectory of one of the A-9/A-10's versions

Von Braun's military I.D.

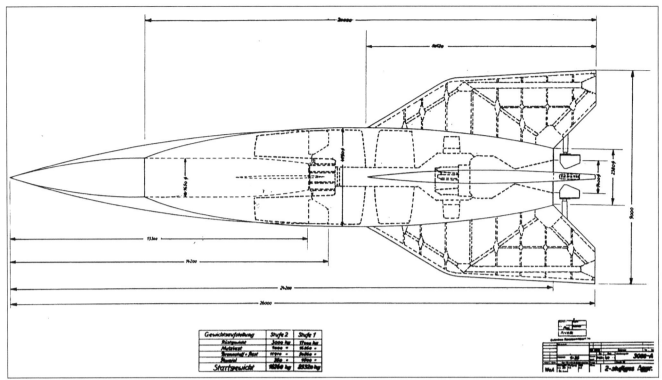

A cross section of the A-9/A-10 missile.
When positioned vertically on the launch pad
it would have been as high as a ten-storey building. (original drawing)

could be achieved without any pilot, for instance: on November 30, 1944, the U-1230 has delivered to the USA a group of agents equipped with beacon transmitters (in connection with the operation "Elster"?).

Let us however return to the A-9/A-10 project.

The Germans never managed to complete any prototype by the end of the war, but all theoretical work and the complete documentation, in the form of engineering drawings, had been made, commenced as early as 1941. Dr. Thiel, later killed as the result of an air raid, and Dr. Walter had suggested using six perfected engines from the A-4 for the propulsion of the A-10, with a combined thrust of 180 tonnes. Later however it was decided to develop a new, single engine with a thrust of 200 tonnes. In the final version the rocket was to attain a ceiling of 180 km in the course of 1 minute.

For the needs of the A-9/A-10 project's research and production, the construction of a gigantic, multi-level underground complex was commenced under the direction of the SS at the end of 1943. It was code-named "Zement" (Cement). This facility, with a total area of galleries and workshops of around 65,000 m² was situated under a mountain massif on lake Traun, near the town of Gmunden in north-west Austria. Around 3,000 skilled workers were to be employed there. In a neighbouring valley engine test rigs were to be constructed for the new engines and launch pads for the rockets.

The "Amerika-Rakete" would have been the first "vengeance weapon" which actually had a chance to influence the course of the war to a significant degree, obviously if it had been employed correspondingly early.[21,24,25,7]

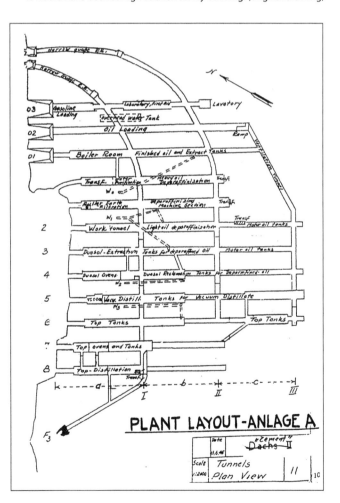

"Zement," plant layout, Anlage A (BIOS)

Ending this discussion of German long-range weapon projects one cannot help venturing an assessment of their role in the history of the arms race and their significance on the course of World War II.

Seen from a present-day perspective these projects had a very large influence on the development of military technology in general. Paradoxically this issue looks completely different, as far as their significance on the course of World War II is concerned.

The wrong strategy and the potential of these weapons being in fact limited by large scatter contributed to this. Of the types that were used in combat only the "precision" version of the V-2 (with an inertial navigational system and additional

One of the entrances to the "Zement" complex

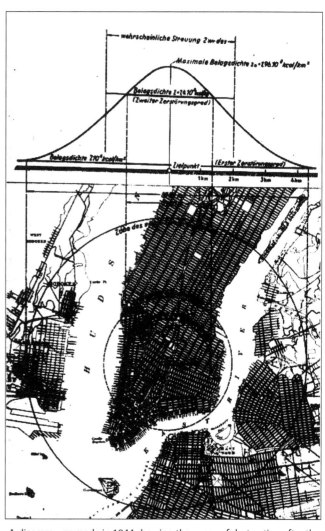

A diagram was made in 1944 showing the zones of destruction after the explosion of an A-9/A-10's nuclear warhead. In the centre is Manhattan Island. (original drawing)

guidance by a guiding beam method) could have actually threatened many important facilities, but its potential was not used correctly.

Only certain types, which the Germans never managed to bring into mass production on time commanded a real and significant military potential (the version of the V-1 with a new propulsion and television guidance system, the A-9/A-10 rocket). In reality therefore a huge disproportion had existed between the scale of resources used and their influence on developing the military situation. The maxim promoted by Hitler of "little resources, large effect" in reality had turned into the principle of "huge resources, small effect."

Albert Speer, the Third Reich's Minister for Armament and War Production, attempted to portray the inner history of these operations in his "Memoirs":

We were again two years late. The Soviet winter offensive forced our units to retreat. The situation became critical. As often happened in moments of crisis, Hitler, guided by an exceptional short-sightedness, declared to me at the end of February that he had ordered Meister's Corps to

"Zement," one of the galleries

The Me 264 "Amerika Bomber" was meant to be an alternative to the "Amerika Rakete"

destroy all railway lines, so as to paralyse the Russians' supply lines. All of my reservations came to nothing—that the ground in the Soviet Union was frozen solid, that bombs could inflict only superficial damage, and at the same time, as it followed from our experience, we had managed to repair the more susceptible German railway lines in a relatively short time: everything was in vain. Meister's Corps was brought into a senseless operation, not able, needless to say, to interfere with Russian operations.

Hitler's further interest in the strategy of selected points on the agenda vanished in the face of his idiotic vengeful intentions against England. At the moment that the Meister Corps was being crushed we still commanded a sufficient number of bombers to execute these plans. Hitler however speculated the unrealistic hope that several massed air raids on London may persuade the English to abandon their aerial war offensive against the Germans. And precisely for this reason he demanded even in 1943 an improvement in the production of new heavy bombers.

The fact that far more lucrative targets could be found in the east made no impression on him even in the summer of 1944, when from time to time he would still agree with my argumentation. [key branches of Soviet industry were concentrated in huge, monopolistic behemoths, rem. by the author] Both Hitler and our Luftwaffe General Staff were not capable of waging an aerial war in accordance with the technological requirements, instead stubbornly residing in obsolete military conceptions; as with the opposed side. (…) Hitler's manoeuvre, despite the tactical errors of the Allies, helped them to achieve success in the aerial offensive of 1944; not only did he bring to a standstill the development of a jet fighter, and then ordered it to be converted into a fighter bomber—but he also wanted to take revenge against England with the assistance of a new, large rocket. On his order from the end of July 1943 enormous production resources were allocated to a long-range rocket known by the code-name V-2, 14 metres long and weighing over 13 tonnes. He demanded that 900 rockets be produced monthly. It was absurd in 1944 to oppose enemy bomber fleets—which for several months had been dropping around 3,000 tonnes of bombs everyday on Germany, using 4,100 four-engined aircraft for this objective—with rockets, which would be able to transport daily to England only 24 tonnes of high-explosive; this was equal to the weight of bombs dropped during one attack of six flying fortresses.

This was probably one of my biggest mistakes in directing the German armament industry, in that not only did I agree with this decision of Hitler, but I even supported it. Our efforts should rather have been directed towards the production of defensive surface-to-air rockets.

Tactical and technical details of the "A-4b" missile

take-off mass	13,000 kg
warhead mass	975 kg
engine thrust	27,500 kg
overall length	14.06 m
fuselage diameter	1.65 m
wingspan	6.2 m
maximum flight speed	5,500 km/h
range	595 km

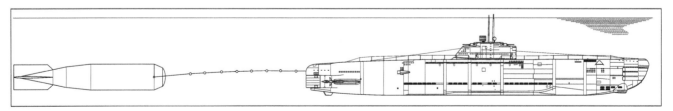

The Type XXI submarine towing a silo with the V-2 missile

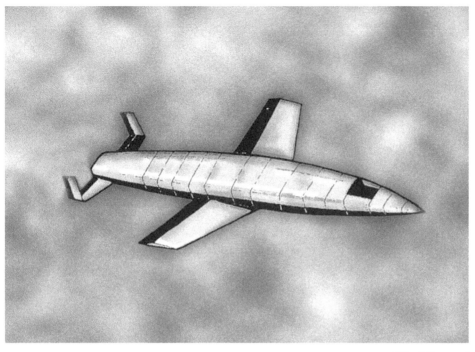

"Thor's Hammer"

designed for aerodynamic trials. This design was initially intended for civilian use, but in 1939, under pressure from the military, Sänger transformed its use and from this point on it was a kind of space bomber with an intercontinental range. It was to be propelled by liquid propellants, and Sänger planned to employ one enormous rocket engine with a 100 tonne thrust as well as two considerably smaller engines, fixed to the sides.

It may astonish one that this was a single-stage spacecraft (as opposed e.g. to the A-9/A-10 rocket and American Space Shuttle, which possesses two additional discarded solid propellant engines). However this wasn't a fault, but rather evidence of the superiority of Sänger's conception. For it was predicted that the launch would not require the use of components lost irretrievably. A launch ramp was designed for this objective —a kind of catapult, or rail launcher three kilometres long. Though it may seem that this would have been a gigantic construction, it still would have been shorter than the standard runways at modern airports (3,600 m). The spaceship was to be accelerated on the platform by a "launch module" considerably larger than the "bomber" itself—an assisted take-off unit equipped with rocket engines and combined thrust of 600 tonnes. They were to operate for 11 seconds, giving the primary object an initial speed of 1,850 km/h. The final section of the ramp, inclined to the horizontal at an angle of 30° would have ensured a climbing flight already in its initial stage (one should suppose that it would have

Tactical and technical details of the "A-9/A-10" missile

	version I	version II
take-off mass	85,320 kg	100,000 kg
I stage propellant mass	51,700 kg	62,000 kg
II stage propellant mass	11,850 kg	8,800 kg
I stage thrust	200 tons	200 tons
II stage thrust	28.1 tons	28.1 tons
overall length	26.00 m	-
fuselage diameter	4.15 m	3.5 m
span of fins	9.3 m	-
range (pilotless version)	ap. 8,000 km	ap. 8,000 km
maximum flight speed	ap. 11,900 km/h	-
warhead mass	ap. 1,000 kg	ap. 1,000 kg

The "Amerika-Rakete" had, at least in theory, an interesting alternative, designed under the code-name "Thor's Hammer."[26]

This was to be a so-called rocket-propelled plane, i.e. more popularly—a space shuttle. Work was carried out on it under the direction of Dr. Eugen Sänger as early as 1936 (!) in the Institute of Rocket Research created by him. It was the first "specific" design of a spaceship that would have been able to carry a human crew beyond the Earth's atmosphere.

"Thor's Hammer" was to have, like the present-day Space Shuttle, an unusually flattened fuselage, ensuring additional aerodynamic lift and facilitating breaking on re-entering the atmosphere (as well as during "gliding" after this—through which it was to have a greater range). The fuselage was to have the shape of a flattened spindle with dimensions 28 x 3.60 x 2.10 m. The crew's cockpit was to be positioned at the front, but completely hidden in the fuselage. In all probability a full-scale prototype was never built, but it known that as early as 1938 the assembly had begun of a model at a scale of 1:20,

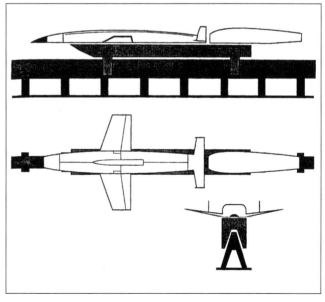

"Thor's Hammer" in take-off configuration

been most convenient to build such a rail launcher on the slope of a mountain with the appropriate profile). The cruise engines would not have been activated immediately after leaving the rail launcher, but only after several seconds, after attaining an altitude of 1,200 m. They would have operated continuously for 8 minutes, leading "Thor's Hammer" with a speed of 22,100 km/h to the point at which ballistic flight would have commenced—145 km over the surface of the Earth. At this moment all of the propellant would have been used up. After covering several thousand kilometres the space bomber would again begin to approach the thin atmospheric layer of our planet and only there could have made use of its fuselage's flattened surface and small 15 m wingspan for the first time. It would have then bounced off the layer of air and again commenced a stretch of ballistic flight (such a possibility in fact exists—in the 1960s when the "Apollo" spacecraft were designed under the direction of von Braun, such a calculation of flight trajectory posed a serious worry—if the landing module entered the atmosphere at too sharp an angle—it would have been threatened with burning up, and on the other hand been deflected off it).

The A-4b missile, one of the derivatives

Such "leaps" would have been repeated many times. This would have permitted several objectives to be accomplished simultaneously. First: it would have extended the range. Second: it would have solved to a large degree the problem of the skin plating overheating, since in the upper stages of flight it would have been cooled every now and then, and in the final stage the rocket-propelled plane would already have had a sufficiently reduced speed for this problem not to exceed the technical possibilities of that time. Third: every "low" stretch of flight created the convenient opportunity to drop a possible bomb (the possibility of a nuclear bomb however only really began to be considered in 1944).

"Thor's Hammer" was to be characterised by a simply unimaginable range for those times, amounting to 23,500 km. The question arises: why such a large range?

Since it posed no secret at all that the target for such a weapon would have been first and foremost the United States, New York, which was an obsession of Hitler, or Washington, one can without any problem determine the approximate region of landing. The space bomber would have reached the eastern seaboard of the USA from the north-east, flown across the territory of this country and then over the Pacific, flying all the time roughly in the same direction, flown across Australia and then would have turned again in a north-westerly direction, so as to finally commence descent over Africa. The former Union of South Africa would not have come into play for the same reason as Australia—this was enemy territory, however Namibia, i.e. former "German South West Africa" appeared considerably more favourable. It was perpetually inhabited by a population to a large degree friendly, and the always sunny Kalahari desert could have appeared a likewise convenient landing ground as the desert base "Edwards", which the Americans selected considerably later. With a small change in course "Thor's Hammer" would in turn have landed in North Africa.

Protruding undercarriage was predicted for its construction so that it would have been able to land in exactly the same way as the "Space Shuttle", familiar to us from numerous television reports.

After carrying out a preliminary design Dr. Sänger concentrated on researching the phenomenon of the construction's heating as a result of drag. It is known that this research was at an advanced stage as early as June 1939. He also worked on an engine with a thrust of 100 tonnes, but without much progress. This work was discontinued in 1942, because another team from Peenemünde, designed an analogous engine for the A-9/A-10 rocket.

Wind tunnel in the most important and most state-of-the-art German research facility working on aerodynamics: the Luftfahrtforschungsanstalt Hermann Göring (LFA), Völkenrode, near Braunschweig. It was estimated that it was responsible for around 20% of the activity in this field in the Third Reich. It was hidden inside a forest and perfectly camouflaged (never detected by the Allies), in this respect being the only known analogy to the complex in Ludwikowice, described in the last chapter.

Ernst Heinkel (left) next to wind tunnel

THE LUFTWAFFE: A TIME OF QUEST

Out of all the advanced aerial constructions, which arose in the Third Reich, the Messerschmitt 262 probably exerted the greatest influence on the course of World War II. Out of all of these often quite unconventional aircraft it was one of the few to live to see mass production and use in combat.

The Me 262

The Me 262 became to some extent a symbol of the huge technical leap which took place during the war in aviation and technology in general, although this wasn't the first jet aircraft. The name of the truly pioneering construction belongs to the Heinkel 178—also a German construction, the design of which was completed well before the start of the war, which in Europe is considered as the invasion of Poland (de facto the war began with aggression on China).

Like many pioneering works both itself as well as the conception which it represented didn't gain the understanding of the military from the start. For at this time demands were shaped chiefly on the not so distant after all World War I (the period of truly fast development in science and technology was only about to come). The airspeed of the order of 400 km/h was still considered sufficient (in the course of several years of warfare this value was to be doubled).

In fact the first working turbo-jet engine in history was constructed by Joachim Pabst von Ohain, a professor at the University of Göttingen. This was in September 1937. Soon afterwards von Ohain accepted a permanent position in the Ernst Heinkel aerial concern and the local plant was named in his honour. In a relatively short time a prototype of the aforementioned first jet fighter was constructed, on the basis of his engine. This was an exceptionally simple construction, built chiefly for a practical verification of the usefulness of the new engine itself to propel an aircraft. The first He 178 prototype soared into the air five days before the start of war with Poland.

At this time however advanced work was already being carried out on two different types of jet engine in BMW. One of them, after development, was to be used later in the propulsion for the Messerschmitt 262. This was the first engine in which the centrifugal compressor was replaced by an axial compressor—considerably more efficient. Currently only such a solution is employed. At the same time in the plants of Messerschmitt various conceptions of constructing a new fighter were being studied. It was to be characterised by the dimensions of a standard propeller-driven fighter, and since it was expected that the new engine would be characterised by a thrust of the order of 3 kN

The He 178

(approx. 300 kg) it had already been decided at the very beginning on a twin-engined variant. On this basis it was assumed rather that the performance would not be very promising, among other a combat flight duration of the order of 30 minutes. However very many modifications were yet to occur before the final solution was achieved…

The development of the engines themselves was an absolutely crucial issue, since it was them that determined to a significant degree the performance and construction of the aircraft.

The starting point for further work was the necessity of a significant increase in thrust with a simultaneous restriction in fuel consumption rate. The designers from BMW aimed to achieve this mainly by an increase in engine diameter. For this reason it became impossible, as had been earlier suggested, to "build the engines into" the wings' construction. They would have to be suspended under the wings. A certain accidental benefit was the simplification of the construction of the wings themselves. However, an almost twofold increase in thrust to a value of approx. 600 kg was achieved. The engines were designated the BMW 003.

All in all five types of turbo-jet engines were constructed in the Third Reich, four of which during the war—the aforementioned BMW 003, which in the years 1941-1943 underwent several modifications, the analogous in relation to it Junkers Jumo 004, as well as the somewhat larger HeS 011 from Heinkel-Hirth. This had a thrust increased from 8 to 9 kN (in comparison to the Jumo 004) and was made in 1944 as a development of the not very successful, earlier "test" HeS 08 engine. In 1945, another one was to enter production, but didn't, being considerably larger and more state-of-the-art, than the previous, the BMW-018 engine. A thrust of 34 kN would have made it possible to implement the conception of large jet bombers.

On July 27, 1941, the first BMW 003 engines were delivered. They were installed on one of the Me 262 prototypes, but both failed in the first flight, and this just a few seconds after the aircraft had taken off. The experienced pilot, test pilot, managed to execute a quick turn and land safely on the same airstrip from which he had taken off. Cursory inspections of the engines already revealed a serious structural defect of the compressors, which had led to their destruction. The whole programme would surely have been suspended, if not for a "rescue" from the Junkers plants, also working at this time on jet propulsion. They made available their latest Jumo 004 engines, similar to the BMW engines, although somewhat larger. Their installation required a certain redesign of the engine nacelles and wings, but these actions turned out to be right. The new engines didn't present so many problems as the BMW engines and finally "eliminated" the latter.

The target configuration of the aircraft had thus been determined, though the results of further flight tests demanded further modifications to be carried out. Among others the undercarriage was modified, as a result of the aircraft's bad steerability on the runway, when the control surfaces were situated in a zone of turbulence behind the wing. The wheel under the rear

The BMW 003 engine on a test rig

The HeS 011 engine without casing

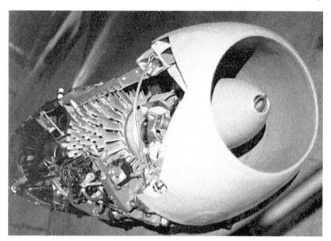

The Jumo 004 jet engine with partially dismantled casing, a specimen examined after the war at Wright Field Air Force Base

section of the fuselage, forcing the aircraft to lean backwards during the take-off run, was replaced by the undercarriage that had the third leg placed under the nose, all were retracted during flight. Disturbances of laminar flow around the body, revealed during flights at high speed forced certain modifications of the wings to be carried out. Installing the undercarriage leg under the nose admittedly solved one problem, but created another additional aerodynamic drag during acceleration. Finally it was managed to eliminate this problem as well, by installing two solid propellant rocket engines (each of thrust 450-1,000 kg) under the fuselage of each aircraft before flight. Thanks to them the Messerschmitt 262 not weighed down furthermore, took off from the runway after an approximately four hundred metre take-off run. At the same time the main engines were upgraded, which reduced their weight together by 180 kg and increased the total thrust to a value of 1,800 kg. At this configu-

The Me 262

ration the aircraft exceeded a speed of 800 km/h with ease, the V-9 test version with a smaller cockpit windshield (this variant was not continued later with regard to limited visibility) even exceeded during diving a speed of 1,000 km/h.

The time had come to make a decision concerning mass production…

It was decided to carry out an official demonstration, to which Hitler, Göring and many high-ranking officers of the Luftwaffe were invited. Its date was set for November 26, 1943. After an impressive demonstration of the new fighter's performances Hitler became its fervent enthusiast and ordered preparation for mass production to commence, though de facto this had already begun earlier.

Apart from the plants of Messerschmitt a whole series of subcontractors from various lines of industry were involved. Two underground complexes were designated to assemble the aircraft: near Weimar and Nordhausen.[26,27,28,29,30]

That was how it was to be in 1943, but a year later, when the Me 262 was encompassed by the so-called "Jägerprogramm", increasing the priorities of new aerial weapons, there were already more of these factories. For devastating Allied air raids had at the same time increased the significance of underground factories. Several large underground complexes were to serve for the production of the Me 262 alone: "Bergkristall" near Linz[31] (the only one completed produced 987 aircraft, mostly within one month) was considered one of the most state-of-the-art factories in the world, "Lachs" in Thuringia,[32] whose tunnels were supposed to have a total length of 26 km, approx. 40% completed, as well as at least one factory of the "Weingut" series in Bavaria, in the form of tremendous semi-underground bunkers[33] ("Weingut II"). Even a launch catapult for the aircraft would have fitted under a ceiling 362 m long, 97 m wide and min. 5 m thick.

Furthermore a series of components for the Me 262 were produced or were to be produced in other underground factories—among others in the following facilities: "Salamander"[34] (Przyłęk/Poland), "Mittelwerk"[35] (1463 engine specimens), "Zechstein"[36] (Rabstein/Czech Republic) and "Flugzeugwerke Eger" (Cheb/Czech Republic).[34] As it was in the case of other types of aircraft, or rockets, the secret weapons of the Third Reich and its underground economy was one and the same issue, which however I have described in a separate book.[37]

Let us return to the year 1943 and work on the Me 262.

When it already seemed that finally a successful weapon system had been made, actually able to tip the balance of aerial victories somewhat in favour of the Third Reich, whose industry was being more and more ruined by enemy carpet bombing, problems then appeared… political.

Hitler, in spite of the situation contradicting this, thought

Ground support, anti-tank version of the Me 262, with MK 214 50-mm cannon

all the time in terms of attack and not defence, and began to demand that the Me 262 be "transformed" into a bomber. This trend intensified particularly after the invasion of Normandy.

He imagined that thanks to bombers of this type being too fast and "elusive" for the enemy it would be possible to easily halt the onslaught of armies spreading deeper into France. In reality however it was completely unfit for the role of a bomber and the unforeseen weight considerably deteriorated its performance.

Minister Albert Speer, a person from the highest circles of German power, probably best described this whole affair in his "Memoirs":

The Me 262A-1a/Jabo

Along with the visible deterioration of the situation Hitler accepted with more and more difficulty each argument opposing his decisions and grew even more despotic than up till now. His obstinacy also exerted a decisive influence on the field of technology; he was about to make worthless the most valuable of all our "wonder weapons": the Me 262, our ultramodern fighter jet equipped with two engines with a speed of over 800 kilometres per hour and ceiling considerably outstripping all enemy aircraft.

While still an architect, spending time at the aerial works of Heinkel in Rostock in 1941, I experienced the deafening roar of the first jet engine to be put to test. The designer, Ernst Heinkel then insisted on this amazing invention being used in the air force. During an arms conference at the aerial weapons evaluation centre in Rechlin in September 1943 Milch passed on to me with no comment a telegram which had been delivered to him; it contained Hitler's order to cease preparations for mass production of the Me 262. Admittedly we resolved to evade in some way this order, but work could not be carried out at such a speed as before, which was essential.

Three months later, on January 7, 1944, we were suddenly summoned with Milch to headquarters. Information from the English press about British tests of a jet aircraft drawing to a close caused a very change in Hitler's stance. Now he impatiently demanded a larger number of this type of aircraft as soon as possible. Since all preparations had been neglected, we were able to promise a monthly production of 60 pieces only by July 1944; from January 1945 we were to produce 210 machines monthly.

Already at this conference Hitler mentioned that he intended to use the aircraft that was constructed to be a fighter as a fast bomber. The experts were disappointed; then they obviously still believed that with the help of irrefutable arguments they would finally be able to convince Hitler. But it came about differently: he obstinately ordered the removal of the entire on-board weapons, at the same time increasing the bomb load. He believed that jet aircraft had no need of defence, for commanding superiority in speed they cannot be attacked by enemy fighters. Full of distrust of the new invention he decided that in order to ensure the safety of the cockpit and engine it should be initially used first and foremost in a straight flight at a high altitude, and to reduce the risks associated with this still untested type of aircraft a reduced speed should be assumed.

The effectiveness of this small bomber, with a bomb load amounting to around 500 kg and primitive aiming devices, was ridiculously low. But as a fighter every jet aircraft, distinguished by superior characteristics, could have shot down many four-engined American bombers, which were carrying out systematic air raids on German cities, dropping on them thousands of tons of high explosive.

At the end of June 1944 along with Göring we made a repeated attempt to convince Hitler, but again ineffective. At this time the pilots of the fighter air force had tried

The Me 262V-1a/U1: a two-seater night fighter, equipped with radar

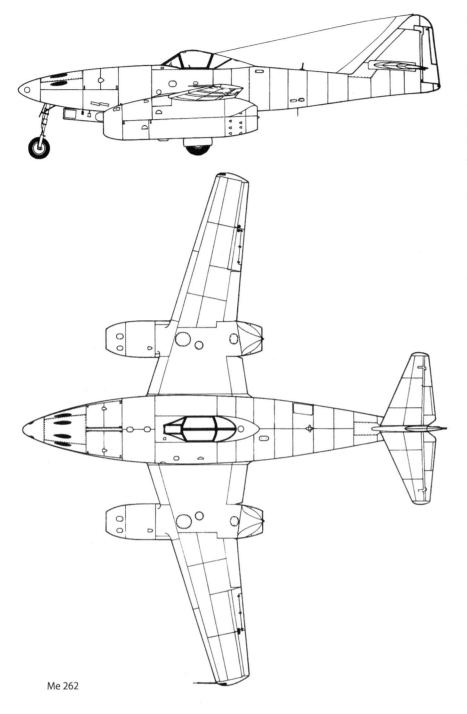

Me 262

ing speed could attack the motionless formations of American bombers. The more insistently we tried to dissuade him from his decisions, the more stubbornly he persisted in his opinion. He consoled us that in the distant future he would certainly consent to using these machines as fighters.

There existed only several aircraft prototypes, as to whose ways of use we objected in June; despite this Hitler's order had to be echoed in far-reaching military planning. Since it was from this fighter aircraft that general headquarters expected decisive changes in the aerial war. In the face of our desperate situation everyone who could give evidence of a certain knowledge of these problems tried to affect a change in his decision: Jodl, Guderian, Model, Sepp Dietrich and naturally the air force Generals, persistently opposing the ignorant decisions of Hitler. However this met with his disapproval, since he scented that all these attempts in a way called in question his expert military knowledge and orientation in technical matters. In the autumn of 1944 he at long last in his characteristic way freed himself from the whole dispute and growing doubts, simply forbidding this subject from being brought up.

out the new machines and demanded that they be used against American bomber fleets. Hitler shifted his ground: quoting various not well-thought-out arguments, he said that the fighter pilots would be subject to considerably higher G-forces, caused by a violent change in ceiling and flight direction at high speed; while the machines, gaining a higher speed in air combat, would find themselves in an inconvenient situation in relation to the slower and hence more responsive enemy fighters. Hitler, who had already once made up his mind in such a way, now wouldn't be convinced by any arguments; he didn't want to understand that these aircraft flew higher than the American escort fighters and that thanks to overwhelm-

Although initially Hitler demanded that using the designation "fighter" in relation to the Me 262 be ceased altogether, finally it was managed to achieve a certain compromise, the fighter bomber modification (Me 262-1a/Jabo) was treated as the base version, made by installing two locks under the fuselage for the suspending of 500 kg bombs. Apart from this there were no differences in relation to the "normal" A-1a fighter version.

In spite of this, work was carried out on specialised bomber versions, for carrying out dive-bombing attacks (in this instance the pilot was to be the bomb-aimer) and horizontal attacks. This version acquired an altered nose section, in which a small cockpit for the bombardier had been "placed" with a new bomb sight called the Lofte 7H. It was given respectively the designations: Me 262-2a/U1 and Me 262-2a/U2 (later changed to Me 262A-4).

A ground attack anti-tank version was also constructed

with a single anti-tank 50 or 55-mm cannon (BK-5, MK-114 or MK-214) protruding in a characteristic way from the aircraft's nose. The anti-tank Me 262 wasn't however mass produced. At this stage the fighter version was also modified, initially modifying chiefly the armament—the aircraft were also armed with a different type of cannon, and unguided rocket launchers were added under the wings, very effective against enemy bombers—usually a single rocket was sufficient to destroy the target.

However in the autumn of 1944, a completely new version was constructed, a night fighter. It was equipped with an on-board radar, with aerials installed on the surface of the nose section.

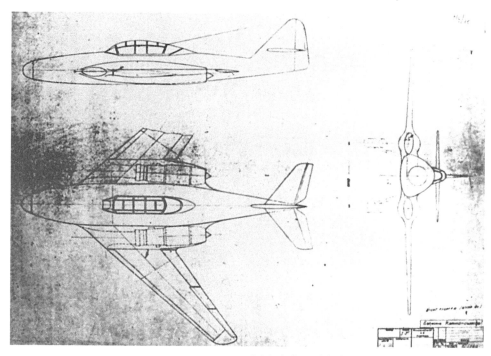

Original plans of the high-speed Me 262 HG III version

The night fighter was developed on the basis of a two-seater combat training version, in that now the seat behind the pilot was occupied by the radar operator and other equipment, including a "friend or foe" radio identification system. The fuselage length was slightly increased in order to enlarge the capacity of the fuel tanks, which in the training version was up to 1,650 litres less than in the original version. It was also possible to suspend additional fuel tanks. The conversion of the combat training aircraft into night fighters was commissioned to the Lufthansa repair shops in Berlin, Staaken. The modifications were made in the following way: the B-1a/U1 and B-2 acquired the FuG-218 radar as well as an unusually useful system which located the emissions of British bomber H2S radar sights (night attacks were carried out mainly by British aircraft). They bore the designation GuG-350 Naxos.

Out of the aforementioned night fighter versions only the first managed to be used in combat. Several dozen specimens supplied during the period January-April 1945 mainly took part in the defence of Berlin which could boast of many spectacular victories. The commander alone of the unit, which acquired them, Lieutenant Walter shot down 29 aircraft, including two four-engined bombers.[38]

The Messerschmitt 262 Schwalbe versions described above

Specimen of the Me 262 captured by the Americans after the war

The Heinkel He 280 was the Me 262's rival, but was characterised by a significantly worse flight performance

were physically constructed, but several very interesting, more avant-garde developmental versions were found on designers' drawing boards or in prototypes at different stages of development. First and foremost aerodynamics was worked on… Three versions were designed with new wings designated HG I, HG II and HG III (from Hoche Geschwindigkeit—high speed). If the HG 1 version was "restricted" to the employment of a new tail plane with a greater sweep of the leading edges (from the "high-speed" Me 262V test version) and large fuselage-wing connections before the centre-wing sections, which reduced the aerodynamic drag at an increase in aerodynamic lift, then in the HG II version completely new wings were to be employed. They were characterised by a greater sweep and lifting surface similar to that in the HG I version. The Me 262 HG III was the continuation of this developmental line—here the sweep of the wings leading edges reached as much as 49°, simultaneously the engines were built inside the wing-fuselage streamlined transitions. This was one of many German solutions that were executed after the war by many designers from other countries. These versions in principle remained on paper, although model tests had been commenced. A modification by Professor Lippisch, known for his unconventional constructions also didn't live to see realisation, in which he proposed to "shift" the pilot's cockpit to the rear section of the aircraft, housing it in a large triangular vertical fin.

Among other things alternative propulsions were also worked on. The Me 262 C3 was to be a fighter in which apart from the engines used so far it was intended to employ additional rocket propulsion. A container was to be suspended under the fuselage, discarded after attaining the appropriate altitude, part of which was to be a liquid-propellant rocket engine as well as fuel tanks and oxidiser. After being discarded this container was to descend to the ground by parachute, after which it was to be prepared for repeated use.

The Me 262 Lorin project presented a much more valuable conception. Apart from the standard Jumo engines it was anticipated to install above them two large but light ramjet engines (a type of jet engine without turbine and compressor), which were to be started after attaining a suitable airspeed. This was the variant which was probably most ahead if its time. This aircraft (like the HG III version) was to attain an airspeed approaching the speed of sound.

In the end it is worth mentioning yet one more propulsion variant—after the delayed improvement of the BMW 003 "rival" jet engines an aircraft with a combined propulsion system was referred for testing, composed of the aforementioned engines and the BMW 718 rocket engines (placed in a single housing). The latter were propelled by a composition of concentrated nitric and sulphuric acid (oxidiser) and aniline solution. Their working time—approx. 3 minutes made possible in practise ascend to a ceiling of 7,500 metres in approx. 1.5 minutes! A ceiling of 12,000 metres was attained in a time of less than 4 minutes! All in all approx. 1,500 specimens of all versions of the Me 262 were produced.

Summing up one should observe that the Me 262 and individual solutions employed in it became a model for several post-war constructions, among others the Soviet Su-9 fighter was modelled on it. German jet engines were also closely analysed in various countries.

Tactical and technical details of selected versions of the Me 262

	A1a	B2A	HG III
wingspan [m]	12.65	12.65	12.65
length [m]	10.60	10.75	10.60
height [m]	3.85	3.85	3.85
wing area [m2]	21.70	21.70	28.50
empty mass [kg]	4,000	4,764	4,323
take off mass [kg]	6,775	7,700	6,697
max. speed [km/h]*	870	841	1,100
range [km]	845	–	–

*at horizontal flight at alt. 6,000 m

The Me 163

World War II was not only a period which saw the debut of jet aircraft, but also rocket aircraft. The only German type which was employed in combat was the Me 163.

The Me 163, the fastest fighter of World War II and the first tailless aircraft used in combat was, like the Me 262, a result of work started before the outbreak of World War II.

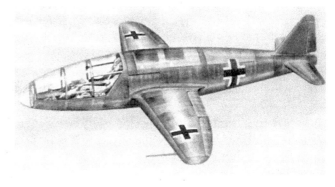

The He 176, a rocket plane design from 1938

This work absorbed enormous potential, completely incommensurable to the gained military effect (the shooting down of a dozen or so enemy aircraft was substantiated, whereas 364 specimens of the Komets were delivered to the Luftwaffe). Although it is said that the chief designer of this aircraft was Alexander Lippisch, work on it was commenced at the end of 1938 under the direction of Dr. Rohrbach, initially it wasn't supposed to be a fighter aircraft at all, but… liaison.

The Me 163 was to become an intercepting rocket fighter only at the end of the thirties after using Lippisch's previous work on aircraft without tail planes, characterised by the employment of rocket propulsion. The rocket engine was designed by a chemist from Kilonia, Hellmuth Walter. During work it was gradually improved, so that its thrust grew from an initial 135 kg to 1,500 kg. The direct predecessor of the Me 163 was a test aircraft designed by Lippisch at the end of the thirties, which acquired the designation DFS 194. Owing to problems with the propulsion, initially a piston engine was installed on it. Only in 1940 did the DFS 194 perform a successful test flight with the "Walter RI-203" rocket engine and became an object of interest for the Air Ministry, which commissioned the development of the aircraft later called the Me 163A. It acquired the new, improved RI-203 engine with a thrust of 17.5 kN. In the spring of 1941 the first glide was carried out in the area of Lechfeld airfield (the prototype was towed by an Me 110), and in the summer flight tests of four aircraft already with their own propulsion were commenced. During one of them, on October 2, H. Dittmar beat the world horizontal airspeed record achieving a velocity of 1,004 km/h at an altitude of 3,600 m.

The test results persuaded the Air Ministry to order a prototype of the Me 163 B fighter. It acquired the Walter 109-509A (R II-211) engine with a regulated thrust of 3-15 kN fed with a composition of "T-Stoff" and "C-Stoff" (hydrazine, methanol and water). Permanent slats were added to the wings protecting against falling into a spin at high speeds, and the appearance of the fuselage also changed. In April 1942, the first prototype of the Me 163 B V1 was made. Soon afterwards Messerschmitt A.G. started the construction of 70 pre-mass production Me 163 B-0 aircraft. From February 1943, exploiting the growing number of new machines, the EK-16 (Erprobungskommando-16) test unit commenced training of pilots and developed an optimal warfare technique.

The mass produced fighters acquired the designation Me 163 B-1a. Their armament consisted of two MK-108 30-mm cannons with a supply of 120 ammunition pieces (version B-0 had two 20-mm cannons). The 170 kg R II-211 A-2 engine had a regulated thrust from 1kN to 17kN. Like in previous versions, the B-1a undercarriage consisted of an undercarriage bogie jettisoned after taking off, and landing took place on a hydraulically lowered skid.

In 1944, Japan purchased the licence to produce the Me 163 B. Until the end of the war a couple of prototypes were built there in the plants of Mitsubishi called the J8M1 Shusui. Just before the end of the war production of the Me 163 C was started. It was equipped with an additional rocket engine with a 3 kN

The Me 163 awaiting combat alert

The shockingly simple cockpit of the Me 163

A prototype of the Me 263 during assembly and at an airfield, partially covered

The Me 263

thrust. Flight endurance was increased to approx. 12 minutes, the practical ceiling also grew (to 16 km). The wingspan grew by 0.5 m, and take off mass by 1,000 kg. This version was never managed to be used in combat. D version was also prepared. The design was passed on to the plants of Junkers, where it was produced under the name of Ju 248, which was changed later to Me 263. In December 1944 the decision of mass production was taken, but before the end of the war no mass produced aircraft was built. The Me 263 had a longer fuselage, redesigned wings, rocket engine with a greater thrust (20 kN) as well as retractable undercarriage.[27,29]

It is a little-known fact that units located on present Polish land played a large role in the training of Me 163 pilots. In Rudniki to the north of Częstochowa preliminary training was carried out on gliders, whereas at the current military airbase of Mierzęcice near Katowice the target aircraft were already used (then it was called "Udetfeld" in honour of a dead Luftwaffe ace). In Mierzęcice were tested among others the Me 163's armed with multiple rocket launchers.

Technical details of the Me 163 B

Empty mass	1,905 kg
Take off mass	4,110 kg
Length	5.69 m
Wingspan	9.32 m
Lifting surface	19.62 m^2
Maximum thrust of engine	17 kN (1,700 kG)
Engine working time	approx. 8 min.
Maximum speed	
at sea level	835 km/h
at an altitude of 3,000 m	960 km/h
Rate of climb	approx. 60 m/s
Time-to-climb	
to an altitude of 2 km	1 min. 46 s
to an altitude of 6 km	2 min. 26 s
to an altitude of 12 km	3 min. 45 s
Practical ceiling	12,000 m
Range	approx. 100 km

The He 162

The Me 262 and Me 163 were not the only innovative fighters (jet, or rocket), which were brought in to arm the Luftwaffe. There existed one more …

The rapid loss of control of German skies by the Luftwaffe with thousands of Allied bombers devastating the German armament industry made it essential that the Germans carry out fast, immediate countermeasures.

In March of 1944, the framework of such an immediate "fighter programme" (Jäger-Programm) was outlined, with the aim of delivering to the Luftwaffe fighters, which would be relatively easy to produce and service, not requiring the consumption of too many strategic raw materials in short supply, but in return produced in their thousands (among others in numerous underground complexes). These were to be relatively small single-seater aircraft, propelled by single jet engines that were available at the time. With regard to larger and larger problems with the replenishment of staff shortages in the Luftwaffe even the recruitment and accelerated pilotage training of teenage boys from the Hitlerjugend was anticipated. In view of the consequences of similar undertakings in other countries (the USSR and Japan) this would have been a tragic move.

At the end of August 1944, the fundamental technical guidelines were specified and presented as a list of demands to the companies: Heinkel, Arado, Blohm und Voss, Focke-Wulf and Junkers. The aircraft was to be propelled by a BMW 003 en-

The He 162

gine and gain a maximum speed of the order of 750 km/h, the lifting surface load should not be higher than 200 kg/m². At a take-off mass not exceeding 2,000 kg the length of the take-off run was established at around 500 metres. Since the "people's fighter," as it was christened, was to operate from simple, densely scattered airfields, 20-30 minutes was considered as a sufficient flight time. It was expected to be armed with only two MK-108 cannons.

In a way a reflection of the structural simplicity was the very short time that remained for presenting the designs for acceptation, which was to take place on September 20, 1944. Mass production was to commence on January 1 of the next year. Practically only Heinkel matched these severe demands. A mock-up of its aircraft and preliminary design was presented on September 23 to the inspector general of the Luftwaffe. The result of this was Heinkel being awarded a contract for production (even before the prototypes and individual solutions had been examined!), which followed on the same day, as well as the aircraft being given the designation He 162. However this was a construction characterised by an exceptionally small technical risk—it was just very simple. In order to simplify the fuselage's construction and lessen the danger of foreign bodies being sucked into the engine from the area of the runway the engine was placed externally above the fuselage. This also lessened the risk of the wooden fuselage catching fire.

Design work was finally completed at the end of October, beginning the production of four prototypes. The first of them took off from Schwechat airfield in Vienna on December 6. Although in general the flights proceeded favourably, in any case taking advantage of the amazing, to put simply, agility of this simple construction, the first prototype crashed at the beginning of December 1944. Most of the other prototypes also crashed. One of the main reasons was surely sabotage by forced workers. I have at my disposal a report from a relative who participated in the assembly of tested aircraft in Schwechat, and who recalls that e.g. commonly the drills with a somewhat larger diameter were used to make the openings for bolts, which wasn't detected by the Germans.

In the aftermath of the test flights certain modifications were carried out: the wing construction was reinforced, a "more powerful" version of the BMW 003 engine was employed with a thrust of around 800 kg and the fuselage's construction was modified. Companies were contracted for mass production.

The He 162 at Vienna's Schwechat airfield

The fuselages were to be produced in the Heinkel plants in Barth in Pomerania, in the Junkers plant at the "Seegrotte" facility in Aschersleben near Vienna, and in the complex of Mittelwerk (Dora) near Nordhausen—both facilities were located underground. Wooden components—first and foremost the wings and control surfaces were to be produced by a series of small, often purely craftsman establishments. This was caused by the significant contribution of manual work. These establishments were grouped into two regions: around Erfurt and Stuttgart. Flexible wooden components were produced by Behr in Wendlingen. Final assembly was also to be carried out by various establishments—around 1,000 aircraft monthly were to leave the Heinkel factory in Rostock, the same in the plants of Junkers (various), and around 2,000 were to leave the Mittelwerk complex each month. These were truly astronomical numbers, albeit completely unreal. Many of the guidelines and expectations concerning this fighter were in any case unreal, which was repeatedly stressed by Luftwaffe officers as well as numerous experts from industry. The chief accusation which was made were the too poor combat performances of the Volksjäger, not enabling it to compete with Allied aircraft as well as the at least questionable usefulness of very poorly trained and completely inexperienced pilots from the Hitlerjugend.

The Generals of the Luftwaffe demanded an increase in production of the Me 262 instead of the He 162, as a tried and tested aircraft enjoying a very good opinion. Professor Willi Messerschmitt took the same stand in his memorandum to the Air Ministry, which stated that "the He 162 constitutes from a

A He 162 captured by the Russians

The Henschel Hs 132. Despite similarity to the He 162, this plane was to be a diving bomber, replacing the Ju 87

technical point of view a step backwards," and that "by now the imposition itself of tasks that the people's fighter is to meet has been based on false grounds, since other fighter aircraft already existing today can better meet all of these tasks."

The turn of 1944 and 1945 was however a period in which rational arguments seldom met with Hitler's understanding and a larger and larger influence on directing the war machine of the Third Reich was gained by the SS and incompetent ideologists from its inner circle, uncritically supporting even the most absurd conceptions in competition for their leader's favours.

The programme to construct the "people's fighter" was therefore continued. However the tragic situation of the German economy let itself be more and more known, and only in February 1945 could one speak of commencing production, and therefore definitely too late for the He 162 to play any meaningful role,

Unfinished fuselages of the He 162 in a gallery of an underground factory, probably "Schildkroete"

obviously there was no question of realising the ambitious plans of mass production on an enormous scale. Production plans in 1945 of 1,000-5,000 BMW 003 engines monthly remained only on paper. In March, a month crucial for production of the He 162, only 100 pieces were produced, of which 60 were allocated for Ar 234 bombers. All in all by the end of the war only approx. 250-270 "People Fighters" had been produced, at the same time the first of the few aerial battles in which they took part only took place on the second of May.

Many developmental versions of this fighter remained on paper only, including those that had more powerful jet engines (among others the HeS 011), Argus resojet engines from the V-1, and wings with a negative sweep.[26,27,29,30]

Tactical and technical details of the He 162 A2

length	9.05 m
wingspan	7.20 m
take-off mass	2,805 kg
maximum speed	approx. 840 km/h
range	approx. 600 km
armament	two 20-mm cannons

The Ho 229

Out of the many types of German jet fighters, including those not mass produced, one deserves special attention.

It was the "flying wing" of the Horten brothers: the Ho 229. This aircraft constituted the crowning achievement of the rapid, though rich career of the young, in their twenties, designers from Bonn.

The first already described construction from this group was the Me 163 rocket fighter and its Me 263 derivative, however this was only one of many designs of this kind.

The famous brothers Reimar and Walter Horten from Bonn were pioneers in this field, not only in Germany but all over the world.

They built as early as the beginning of the thirties (in their parents' garage!) their first aeroplane—glider, characterised by a tailless configuration. In 1934, it won a prize at one of the aviation competitions that year. It was the Ho I. This was still a completely amateur construction, however five years of strenuous work still remained until the outbreak of war. In the years 1935-1937 three other gliders were built in the "flying wing" configuration (Ho II-Ho IV), at the same time an eighty horsepower Hirth engine was installed on the first of them (Ho II M) during testing. All of them were one-man gliders, at the same time the Ho III—a development of the Ho II, was massed produced, and this before the war had ended.

The Ho V in flight

The Ho 229

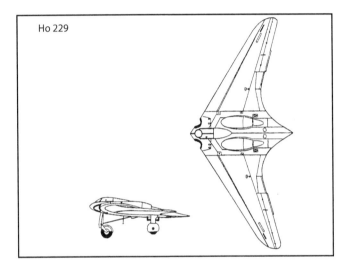

Ho 229

Horten's plane finished in Argentina: the I.Ae.37

The most famous glider of the Horten brothers, built in 1941 in Königsberg's Aerial Production Plants, was the Ho IV with its highly swept wings. One of its developmental versions, known as the Ho VI competed successfully at aviation shows with the best glider in the world at that time—the Cirrus, constructed in Darmstadt. With relatively large dimensions (wingspan: 24.20 m), the Ho VI was characterised by a rate of descent (glider sinking speed) equal to only 0.43 m/s. After them other gliders were made, among others the Ho XI and Ho XIV, but obviously aircraft with their own form of propulsion were the most significant. Many of them were made, in any case on the basis of very different sources of funding (despite the fact that in 1939 Reimar became chief designer in the tailless aircraft department of the Heinkel concern, none of the developed aircraft were built in this producer's production plants).

The first aircraft with its own propulsion was the Ho V, whose origins reach as far back as 1936. This was an experimental construction, thanks to which many new solutions were tested, among others steering in the horizontal, with the aid of the wingtips being turned in their plane—at that very moment this didn't turn out to be useful—causing the prototype to crash. Apart from an unconventional aerodynamic system, the Ho V was also a pioneering construction in another respect—it was constructed largely of plastics, and in the remaining sections mainly of wood. A celluloid skin plating combined with a small amount of metal in the construction would have (in the event of its possible use in the future war) made this aircraft very difficult to detect. If the radar wave absorbing coating (in the form of plastics and paint, see the relevant chapter), later invented by the Germans, had been added to this we would have had to deal with the first "genuine" aircraft of the "stealth" class. But the Germans didn't go along this path. The Ho V was a two-seater aircraft, propelled by two Hirth HM-60R, 80 HP piston engines. It was built thanks to the financial help of Dynamit A.G.

At the beginning of the forties information began to reach Germany concerning work being carried out in the USA by an outstanding designer, Northrop, on a "flying wing" aerodynamic system. This resulted in the Air Ministry awarding in 1942 permanent research funding to the Horten brothers. Thanks to these resources an aircraft was made being a development of the Ho V, the Ho VII. It was also characterised by a two-man crew and twin-engined propulsion, however considerably more powerful engines were applied (2 x 240 HP) as well as a reinforced airframe with a somewhat larger lifting surface. The Ho VII was to be a training aircraft, but from the start it wasn't known for what specific training it would be designed for. Only at the beginning of 1945 was the decision taken to use it for the training of pilots who were to fly the new Ho 229 jet fighters (Ho IX). In connection with this it was given the designation Ho 226, 20 specimens being ordered.

The Ho IX on the other hand was a further-reaching development of the conception so far, but with regard to the application of jet propulsion and predicted military use the airframe was completely redesigned. Its structural strength was increased and a profile employed that was adapted to high airspeeds. In this area the previous research results and calculations of Professor Busemann were made use of, who in the latter half of the thirties had been engaged with this problem. The centre section of the wing was widened, acquiring a trailing edge with a sweep gradually passing into the negative, so that in this section it would be possible, as a "substitute" to the fuselage, to employ a relatively "thick" profile, enabling the housing of both jet engines, and in the frontal section air intakes as well as the pilot's cockpit, relatively heavy armament (four MK-108 30-mm cannons plus ammunition) and in the night fighter version also an on-board radar. The aircraft's propulsion was to consist of two Jumo 004 engines. The take-off mass amounted to 7.5 tons. It was referred for production in the plants of Gothaer Waggonfabrik at the personal intervention of Göring, on whom its air display in January 1945 had made a big impression (the prototype achieved a speed of 800 km/h). The aircraft acquired the designation Ho 229. This was decidedly the most mature construction of the discussed group, however by the end of the war only a short prototype batch had been managed to be produced. Therefore the Ho 229 found no use in combat, which is a pity, since its comparison with classic constructions would have been very interesting. In 1945, on the basis of the version described above, a fighter-bomber version was constructed with a bomb load of 2,000 kg as well as a night fighter. Both these versions were characterised by two-man cockpits, instead of one-man as in the basic fighter. After the war's end one of the Ho 229 prototypes was transported to the USA, where it was given a thorough examination. The Ho IX was probably the first prototype in which the technology of reducing "radar visibility" was tested. The wings (whose skin plating was initially made of plywood) were coated from the inside with a composition of wooden dust, charcoal and glue, which was to absorb the radar beam, to some extent shielding in this way the metal supporting structure. However the results of these trials are unknown to me. But they could have been promising, since the shape of the aircraft itself was from this point of view very favourable.

At the beginning of 1945, two companies, Gotha and Klemm, received the first mass production orders for the Ho 229 (fighter): the first for 53 specimens, and the second for 40. A night fighter, equipped with the FuG-244 "Bremen" radar was to be produced in order of second importance. The execution of these plans had no sooner begun when the end of the war interrupted them.[40,41,42]

At the same time, in the years 1944-1945, the Horten brothers had also been developing a series of different and interesting aircraft, but all of which remained at the model testing stage, or were just plans and calculations.

They were:
- The Ho VIII intercontinental passenger aircraft designed with the post-war period in mind, propelled by six Jumo 222 piston engines (each 3,000 HP) placed at the wings' trailing edge. In retrospect these were to be turbo-propeller engines. The passenger cabins were to be situated in the bulge of the wing's centre section, making up a residual fuselage. A take-off mass of 120 tons was predicted, very large for those times. In 1945 a flying model at a scale of 1:2 was made.
- In response to the programme to build the "people's fighter" the Ho X light, one-man jest fighter was constructed, obviously built in the configuration of a "flying wing", like all the other constructions of the Horten brothers. With a wingspan equal to 9.2 metres it was much smaller than the Ho 229 (correspondingly: 16 m), somewhat resembling the Me 163 rocket fighter. The Ho X was to be ready for production only by 1946.
- The Ho XII light training aircraft being a development of the older Ho IV glider, taking into consideration the latest achievements in aerodynamics, among others airframe tests of the American Mustang. The wingspan was to amount to 16 metres.
- The supersonic Ho XIII fighter, on which work had begun at the beginning of 1944, which was to develop a speed of up to 1,800 km/h at high altitude. This was a very interesting construction with respect to a wing construction different to that employed so far. A very large sweep of the leading edges, amounting to 60°, very thin wing sections (camber of the order of 10% of wing width), an acute camber of the frontal sections, as well as the jet nozzle being positioned precisely in the wings' plane were all designed to reduce the aerodynamic drag of supersonic flow around the body. The aircraft possessed a fin (also with a large sweep) with rudder. In the fighter's first version a narrow cockpit was to be situated in its central section. Later it was intended to place the cockpit and one-track undercarriage (with additional wheels on the wingtips) in an underwing pod. As much as it was intended from the start to house a single, large jet engine with an afterburner in the centre section of the wings, finally it was decided on using two "hybrid" (turbo-jet rocket) BMW 003R engines, suspended under the central sections of the wings. Such an engine was made up of a standard turbo-jet unit, ensuring a thrust of 1,000

The Ho 229 in flight

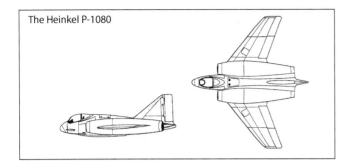
The Heinkel P-1080

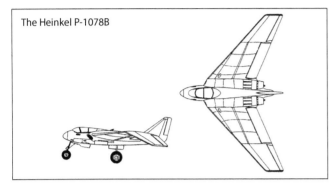
The Heinkel P-1078B

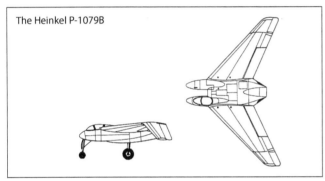
The Heinkel P-1079B

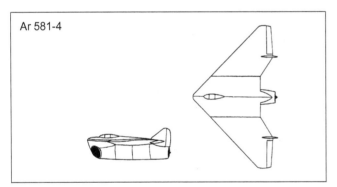
Ar 581-4

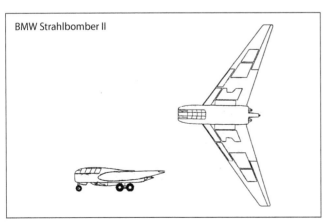
BMW Strahlbomber II

kg and liquid fuel rocket motor with a thrust of 400 kg. At the war's end the first Ho XIII prototype was found at an initial stage of construction, and was to be ready for flight tests in the middle of 1946. A wingspan was planned approaching 12.5 m.

- The Ho XVIII long-range fast bomber. It was to be propelled by six modernised Jumo 004H jet engines. It was devoid of fuselage—crew, bomb load and much of the on-board equipment has been situated in the wing's centre section— and the engines were to be placed under the wings. The armament was to be made up of four tons of bombs as well as two anti-aircraft gun turrets; one on the nose and the other on the end of the fin. A maximum range of the order of 8,000 km at a cruising speed of approx. 800 km/h was planned. Part of the bomb armament and the main undercarriage legs were to be retracted into the central sections of the wings. The Ho VIII was to be characterised by the following parameters: wingspan: 30 metres, lifting surface 156 m², take-off mass of the order of 34,000 kg (somewhat less in a future variant propelled by four HeS-011 engines). Work on this aircraft was commenced at the beginning of 1945 and was to be completed two years later.

The Horten brothers were not the only people in Germany carrying out work on the described aerodynamic system. They had very strong competition in the form of Professor Dr. Alexander Lippisch from Munich, at that time one of the most outstanding scientists in the world engaged in aerodynamics and aviation engineering. Lippisch's work is described in the second part of this book. It is worth to mention, by the way, other German designs of "flying wings," of which there were many more, contrary to commonly prevailing opinion.[43]

The P-1111 aircraft was among others designed in the plants of Messerschmitt. It was to be propelled by a single HeS-011 jet engine. It was characterised by an exceptionally large sweep of approx. 45° of the wings' leading edges, as well as of the fin. It was expected that this would enable it to achieve a speed of approx. 1,000 km/h.

The design office of Heinkel, after the Horten brothers and institute directed by Professor Lippisch, was the third significant source of interesting designs from the discussed group of "flying wings," able to boast of a more modest in number but equally ambitious output.

In the final ten months of the war three types of such aircraft were built there. The first of them was the P-1078. Two modifications of it were made. The first was classic, (P-1078A) propelled by one jet engine, and characterised by a similar construction lay-out to Kurt Tank's Ta 183 and the Messerschmitt P-1101. The second modification (P-1078B) was characterised on the other hand by a flying wing design (devoid of tail plane with a significantly reduced fuselage) and propulsion in the form of two jet engines. The P-1708B was to be a single-seater fighter with a take-off mass of 3,900 kg, developing a speed of over 1,000 km/h. The large area wings (approx. 20.5 m² at a wingspan of 9.4 m) would have enabled a large cruise altitude to be achieved of the order of 13,500 m.

A further development of this project was the P-1079 design, two modifications of which were made in parallel, classic and tailless, designated as A and B. The second differed from the P-1078B with a completely eliminated fuselage and fin. The pilot was placed in a cockpit situated in front of the left section of the wing's central section, and a housing situated on the right hand side that matched its shape was to house a radar and gun. This was to be a single-seater night fighter.

The P-1080 design on the other hand presented a somewhat different developmental route. In terms of the general construction layout it differed less from the P-1078A than the P-1078B, but two new ramjet engines with a large diameter (90 cm) were employed in it, developed by Dr. Sänger. They were to be fed by conventional fuel. After the aircraft had been accelerated with the aid of four powder propellant rocket motors the main engines were to be started, giving a total thrust of 1,170 kg at a speed of 500 km/h and 4,370 kg at a speed of 1,000 km/h. A serious problem which was never managed to be solved before the end of the war was the very high temperature produced in the combustion chambers, reaching 2,500°C. The P-1080 was to be a fighter, equipped with an on-board radar hidden under the nose fairing. No prototype was managed to be built before the end of the war.

To end it is worth mentioning several designs of this kind that arose in other design offices: Junkers, Arado and BMW. The first of them was the Junkers EF-130, constructed in 1944. This was a classic "flying wing", devoid of fuselage as well as conventional control surfaces. A two/three-man crew was predicted and propulsion in the form of four BMW 003 jet engines placed on the rear part of the wing's central section. As in the case of most constructions described in this chapter, the end of the war interrupted design work.

Arado also attempted to construct its own tailless aircraft (this time a fighter). It acquired the designation Ar-581-4. It was to be propelled by one HeS-011 engine. It was to possess triangular wings, with an outline similar to an isosceles rectangle triangle, characterised by a wingspan of the order of 10 m.

A somewhat closer equivalent to the Junkers construction was a BMW design, known as the BMW Strahlbomber II (type II jet bomber). This was to be an aircraft with a take-off mass of 31,500 kg propelled by two BMW 018 turbo-jet engines placed centrically at the rear, developing a thrust each of 3,450 kg. A very large bomb load for those times was predicted, significantly exceeding 10 tons. This was one of several types of bombers, studied within the confines of a programme to build a new, fast long-range bomber. To end these descriptions of jet fighters of the Third Reich one should mention work on their second generation, in relation to the Me 262. This concerns the aircraft: Lippisch's P-13B, the Ta 283, the "Triebflügel" (described in the second part of this book), as well as the Messerschmitt P-1101 and P-1110 and Kurt Tank's Ta 183. Out of the last three probably only the P-1101 and Ta 183 lived to see the prototype stage.

Messerschmitt P-1101

The P-1101 was developed from July 1944 until the end of the war on the basis of an order from the Air Ministry which came in the middle of 1944. On the basis of the results of this construction's testing (after the introduction of possible modifications) a new tactical fighter was to be made, which could constitute a complement to the Me 262. It was to be better suited for dogfights with enemy fighters at close quarters, more agile and most importantly was to develop in horizontal flight a speed of up to approx. 1,000 km/h. It was however considerably simpler in construction than the Me 262 and less "demanding" with respect to the consumption of numerous raw materials being more and more in short supply.

The P-1101, already with mounted engine (probably mounted after the war), with mock-ups of the cannons, but without the engine enclosure, fuselage-wing transitions and at part of the fuselage

The P-1101: front and rear views

It was to be propelled by a single HeS-011 jet engine, but due to production problems the BMW 003 engine was used temporarily. The engine was placed in the fuselage, at the bottom of its central section, with the air intake in the nose. The fuel tank was situated above the engine (sufficient for approx. half an hour's flight), and in front of it, the cockpit. Additional fuel tanks were placed in the wings.

The main key to the P-1101's success was to be the use of the latest achievements in aerodynamics, including first and foremost highly swept wings as well as control surfaces. On one of the prototypes built in the first few months of 1945 the wings were installed on adjustable clampings, enabling the sweep of the edge of attack to be changed in a range of 35-45°. However this procedure could only be carried out on the ground.

At the moment of the war's end the aircraft's first prototype was 80% completed (waiting for the engine to be installed) and after being seized by the Americans was referred to the laboratories of Bell—so as in the case of the Ta 183 the capitulation of Germany didn't mean the end of this project's life. The Americans however installed the wings on bearings, obtaining an aircraft which could change its geometry during flight.

After the entire aircraft had been constructed it was given the designation X-5. During flight tests it became evident that it was possible to realise most of the intentions of Messerschmitt, but not without certain faults. The first of them was a relatively large instability, resulting from the engine axis being located "to one side" in relation to the aerodynamic centre. The aircraft was also very difficult to fly out of a spin, which led to it crashing and the programme's continuation being put to an end.

Low endurance was also a problem. It had been attempted to remedy this even in Germany, constructing the P-1106 developmental version, in which the cockpit had been shifted significantly to the rear, gaining in this way space at the front of the fuselage for an additional fuel tank, which could have extended the flight duration to around an hour.

In spite of certain shortcomings the P-1101 remains one of the most crucial constructions from the period of World War II.[29,43]

Tactical and technical details of the P-1101

length	9.17 m
wingspan	approx. 8 m (see: description)
take-off mass	4,070 kg
maximum speed	approx. 980 km/h
range	approx. 500 km
armament	four 30-mm cannons

However this aircraft wasn't the "last word" of Messerschmitt. The P-1110 is considered, like the P-1101, one of the most momentous aviation constructions which arose in the described period. In terms of construction layout it is no different from many types of jet fighters created long after the war, it became their prototype. It exerted a significant influence on such constructions as: the Saab J-32 Lansen, Folland Gnat, Dassault Etendard, Hawker Hunter, Supermarine Swift or Grumman F-11 Tiger.

Originally it was to be an aircraft developing at high altitude in horizontal flight a speed close to the speed of sound (the order of 1,000-1,100 km/h). Along with high manoeuvrability this would have given it a decisive superiority over enemy fighters. The first was realised mainly by a significant reduction in the fuselage's cross-section, giving it a streamlined shape as well as by using a high thrust engine (1,300 kg), the HeS-011, which was placed in the fuselage, in its rear section, like in the case of contemporary fighters. Employing wings and control surfaces of a high sweep facilitated the achievement of high manoeuvrability, all the more that these wings were adopted from the P-1101. However with regard to the limitation of the fuselage's cross-section it wasn't possible as in the P-1101 to move the air intake into the fuselage's axis, since the pressurised cockpit, fuel and armament were placed at the front. This problem was solved in two ways, constructing in this way two versions of this fighter. In the first the air intakes were placed on the sides of the fuselage, above the rear section of the wings. However this involved disturbances of airflow at high angles of attack.

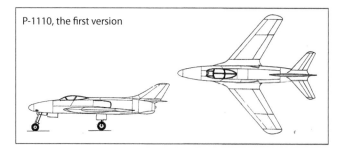

P-1110, the first version

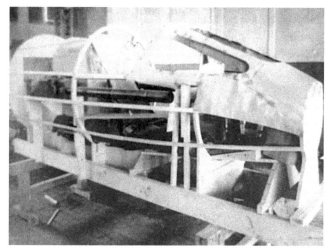

P-1110, the last version

In connection with this an unconventional ring-shaped air intake was employed in the second version encompassing the whole circumference of the fuselage's central section, at the same time the control surfaces has been changed, from classic into "V-shaped."

Tactical and technical details of the P-1110 (refers to the I version):

length	10.36 m
wingspan	8.25 m
take-off mass	4,290 kg
maximum speed	1,015 km/h
range	approx. 1500 km
armament	three 30-mm cannons

Focke-Wulf Ta 183

A very interesting design, competitive in relation to the Messerschmitt P-1101, arose in the works of Focke-Wulf, under the direction of their chief designer Kurt Tank. It was the Ta 183, one of many designs of the Third Reich, which however played a significant role in the development of jet aviation directly after the war.[27,43,44,45] Its plans, captured by the Soviets in 1945, were used among others in the design of the MiG-15. The Ta 183 constituted a significant step forward in the development not only of aviation technology, but also technology which like in many other cases was forced on the Germans by savings and the need to simplify the technological processes. It was estimated that production of this aircraft would be even four times less labour-consuming than in the case of the Me 262

Work was begun on it at the beginning of 1942, when the Air Ministry expressed interest in a conception promoted by Tank of a single-engine jet propulsion fighter, though at this time the general construction layout and aerodynamic assumptions hadn't yet been determined. The type of engine also hadn't been chosen, although from the start the Jumo 004 was favoured—it was the most technologically advanced at that time.

The design office under the direction of engineer Mittelhuber was appointed, which began designing many different versions of the aircraft—since this was to be a pioneering construction it was possible to make use of previous experience only to a limited degree, and a developed and verified theory was also lacking. This was heightened by the fact that as many as nine different versions of the Ta 183 emerged, before the final configuration was chosen. The first design (P.Ia) assumed the construction of an aircraft with wings of a negative sweep and control surfaces in the shape of the letter "V" (combining the functions of the vertical tail unit and tail plane, positive sweep). This was connected with the fact that the engine was to be placed over the fuselage and the described control surfaces not placed in the engine's gas streams. But this was an "inconvenient" conception, especially with respect to stability.

In the P.II design a more conventional layout was employed —classic, with simple wings and control surfaces, and the engine placed under the fuselage. Placing the engine above was returned to in the P.III design, but continuing with the classic aerodynamic system, with the fins and vertical tail units on the tips of the tail plane. The air intakes to the engine were situated on the sides of the cockpit.

The development of this was the P.IV, in which mainly the rear section of the fuselage was modified.

The P.V, P.VII and P.VIII designs represented a different developmental line. The engine was placed in the rear section of the fuselage, to which trapezial wings had been fixed in a mid-wing monoplane configuration. The control surfaces were connected directly not to the fuselage, but to the wings —through two tail booms. The fin and tail plane were placed

Mock-up of the Messerschmitt P-1110 cockpit. Striking are the cannons mounted just by the pilot's windshield

Test flight of a model (?) of the Me 1110

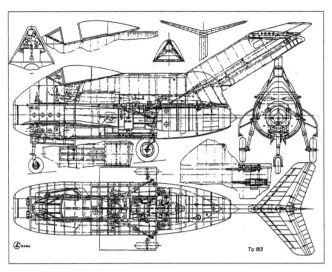

Original plans of the Ta 183 V1

above the engine's gas streams. This was the same conception which led to the independent construction in Great Britain of the Vampire. But although a single prototype of the P.VII was built, this line was not further realised in Germany.

Modifications between the individual versions of this developmental line were minor—in the P.VII the air intakes were moved from the sides of the cockpit to the wing roots, and an additional, auxiliary liquid fuel rocket motor was installed, which was to be used during take-off and dogfights. Use of the Heinkel HeS-021 engine was also anticipated.

However the design which was to lead to the final version arising, approved for introduction to armaments, took on a different shape.

The P.VI became its prototype, an aircraft built in mid-wing monoplane configuration, with the engine built onto the fuselage. The most critical feature was the application of a very large sweep to the wings as well as the control surfaces. The fuselage was short in this variant, however the control surfaces themselves were lengthened (by giving them a large sweep), which acquired an outline (inclined) of the letter "T." Such a construction lay-out was later launched in the West as well as the East, among others in the MiG-15. Only slanting aerodynamic surfaces made it possible to make full use of the advantages of jet propulsion. After introducing certain modifications—mainly within the control surfaces, and shift-

Mock-up of the Ta 183 in a wind tunnel

ing the cockpit further back, the final configuration of the Ta 183 arose. Propelled by a HeS 011 engine it was to develop a speed of the order of 1,000 km/h. It had wings with a sweep of the edge of attack equal to 32° and its armament was to be comprised of two MK-108 30-mm cannons. Construction of prototypes was commenced in January 1945 (but due to technical reasons with the Jumo 004 engines). On February 23, 1945, a contract was even signed for mass production of the Ta 183 (in Bad Eilsen), obviously this remained only in theory.

However work was continued after the war's end … in Argentina, where Professor Tank, his team of engineers and documentation had been evacuated to in 1947. On the basis of the results of work on the Ta 183 an aircraft was constructed there: the Pulqui-II, which was presented in flight during a military parade in 1952. Although it was then regarded as one of the best jet fighters, it didn't enter mass production. In any case work was also continued in Argentina on the Horten brothers' "flying wings."

Events taking place in Buenos Aires were the start of Kurt Tank's design being continued. Their chief hero was a Major from Argentinean military intelligence and previously also research worker of the Caltech institute in California, Gallardo Valdez. Shortly before Christmas 1947 he was awaiting a scientific trip to Moscow, but soon this was placed in doubt, when the order came "from the front line" to take up the position of Air Attaché in Stockholm. Before the contradiction of orders had been explained, he had to cancel all plans.

For in late autumn of that year (spring in Argentina) the intelligence establishment in Madrid had received a report that a group of outstanding German scientists and designers "waits" in Norway, and should be immediately transferred to Argentina. The group made use of false documents issued at the end of the war and could be at any moment exposed.

The mission of their transfer was entrusted to Major Valdez himself. In the final days of 1947 he made his way to Sweden, not however to take up his new position but to await more precise information concerning a meeting with the Germans, after which he was to make his way to Oslo. This followed very quickly, for the meeting had already been arranged by Muret, the Argentinean consul in Norway. Directly after the meeting, so as not to give the enemy time to react, the Germans as well as Valdez made their way to the airport, where a specially prepared plane was already waiting.

It took off on a forty-hour flight to Buenos Aires (almost 13,000 km), interrupted by short stopovers so as to replenish fuel. Almost all of this time the Major didn't talk to the Germans, with the exception of a few trite words in English. He remembered only that the surname Matias or Matthies was spoken. It should have been: Pedro Matthies—this was the false name of the Prof.-Ing. Kurt Tank—one of the most outstanding aerial designers of the 20th century. He had brought with him two of his most important associates and a suitcase filled with microfilmed engineering drawings. Initially the entourage was to be larger, but several of the Professor's associates were arrested in Denmark and the aircraft had to take off without them. But

The Pulqui II, a post-war derivative of the Ta 183

thanks to the help of the Argentinians Tank quickly co-opted new associates and "reconstructed" the research team. The aircraft factory near Cordoba was placed at his command. There he was to complete work on his most promising design, the Ta 183.

The no less outstanding Reimar Horten also arrived here (in 1948), co-designer (along with his brother Walter) of the famous German "flying wings," including the ultramodern Ho IX/Ho 229 jet fighter, which in 1945 developed a speed of over 800 km/h.

In the plant near Cordoba an outstanding aerial designer of the Henschel concern also turned up, Julius Henrici. There was no shortage also of experts from other famous companies, such as Fieseler, Messerschmitt and Focke-Achgelis. The latter was a pioneer in the production of helicopters.

Tank's arrival became an event for Peron himself, soon he presented Tank with a memorandum detailing the technical possibilities of the country and general assumptions concerning the development of the air force. In return Tank proposed the development of four types of aircraft: light training, reconnaissance, medium bomber (these three types were to have airscrew propulsion) as well as a new jet propelled fighter. This last one met with the greatest interest and gained the favour of the Argentinean president right away. It was to be based on the design of the Ta 183 aircraft. In this case it concerned the version with a highly placed tail plane (control surfaces in the shape of the letter "T"), on which the well-known aerodynamics expert Hans Multhopp had worked on. This design had been led to the model testing stage in a wind tunnel in Germany, but as opposed to the Messerschmitt P-1101 a full scale prototype was never built. A rival version with a twin-boom layout (similar to the British "Vampire" fighter) never left this stage, but it was precisely its authors, Ludwig Mittelhuber and engineer Ulrich Stampa, who were to assist Tank in Cordoba. Paradoxically, Multhopp was employed by the British and developed their design further.

However the Argentinean fighter I.Ae.33 "Pulqui II" was not only to be the conclusion of works on the Ta 183. Significant modifications were introduced to it and as a result it was characterised by a better performance than had been assumed in Bad Eilsen. By the force of events it was necessary in any case to rely on completely different components. The Ta 183 was to be propelled by a no longer existing engine (after all not completed up to the very end of the war) the Heinkel HeS-011, which according to the most optimistic estimations was to ensure a speed close to 1,000 km/h in horizontal flight. In Argentina the Rolls-Royce Nene-2 engine was employed, characterised by similar dimensions and regarded at that time as very modern (despite the centrifugal compressor not being a very far-reaching solution). This was a very similar version to that which the Russians copied and applied in the propulsion of their MiG-15, being one of two equivalents of the Pulqui-II, beside the American F-86 Sabre. Incidentally this comparison alone exposes a very high testimony to Kurt Tank's team, which to a large extent created their design as early as during the war i.e. from a technical point of view in a completely different era and by using considerably simpler resources than those which were later put to the disposal of design offices in the USSR and USA.

The British assessed Tank likewise, who interrogated him right after the war, but paradoxically never induced him to co-operate at all. As Tom Bower[48], one of the best experts in the Allied "hunt" for German scientists, described:

Tank along with his team of aerial designers was proud of his achievements and annoyed that the British, instead of immediately rewarding his talents with an offer of permanent employment, unceremoniously sent him back to Germany. Handel Davis [British aerial designer—endnote I.W.], one of many who had spoken with Tank for hours on end, wasn't surprised at all that Tank hadn't been employed by the British: "He was such an important person, so great, that it would have been very difficult to include him in and adapt him to any design team.

Let us return however to the issue of the design itself, or rather redesign of the aircraft in Argentina. In all probability the Nene-2 engine was generally better than the Heinkel engine, but the passage from the axial to the centrifugal (radial) compressor signified primarily an increase in diameter of the engine compartment and fuselage (see: the bulging fuselage of the MiG-15) and threatened a clear deterioration in performance. So as to avoid this the entire construction had to undergo numerous modifications. But on the other hand a higher powered engine with an improvement in aerodynamics gave an increased chance of approaching the speed of sound. Otto Pabst, one of Tank's associates and another outstanding figure appearing in this little-known episode in the development of aviation, described this problem as follows.[45]

> *So as to retain a similar fineness ratio of the fuselage as in the Ta 183 it was necessary to significantly extend it in accordance with its increased diameter. Only in this way was it possible to reduce the compression wave on the nose, retain small aerodynamic drag and enable a speed of 1,000 km/h to be exceeded.*

Also, apart from essential corrections, efforts were made to preserve Multhopp's fundamental aerodynamic lay-out. This is why the sweep of the wings (large: 40°) is just the same as in the Ta 183. The same applies to the leading edge of the fin, which is still more slanting in relation to the fuselage's axis than in the case of the wings.

A very serious design problem was the lack of a wind tunnel in Cordoba for testing the behaviour of the airframe under conditions of high speed and angles of attack. The only data which was at disposal were the results of tunnel tests of wooden models carried out in Bad Eilsen and of course the feeling of the designers as a result of their huge experience. This was the fundamental reason why efforts were made to keep modifications to the aerodynamic lay-out to a complete minimum. Efforts were also made not to reduce the stability and to the greatest possible degree comply with the area rule—the area of fuselage and wing cross-sections measured in turn from the nose to the tail were to remain in a strictly defined, optimal relationship. This was the simplest formula in order to avoid surprises during test flights.

However in spite of this, efforts were made before designing was completed to check in some way the aerodynamic properties of the aircraft before the pilot was placed in it. For the aerodynamic lay-out itself with such large sweeps of the leading edges was something new and of a revolutionary nature. Nobody from either Tank's original or future team had yet designed an aircraft, which could be comparable in this respect.

The lack of a suitable wind tunnel was finally compensated for in this way that a very faithful wooden model of the Pulqui II was constructed to a scale of 1:10 balanced exactly like the target aircraft, equipped with movable control surfaces and a radio control system so as to carry out feasible flight tests. This peculiar, remote-controlled glider was then dropped from a bomber flying at medium altitude, from which it could be controlled. These were very similar to the actual conditions. This flying model not much more than one metre in length was built under the direction of ... Reimar Horten, who was briefly assigned to Tank's team before he formed his own team at the end of 1948, realising even more avant-garde conceptions. The choice of his personage was not accidental, since Horten had much experience in constructing completely wooden aircraft (for a change: the Pulqui II was one of the first aircraft with a completely metal construction). The first test flight of the model took place on October 20, 1948, and was piloted personally by Tank himself. Pabst, being the unofficial chronicler of this work described the course of trials as follows:

> *It became evident (...), that the "Planeador" [in Spanish: glider], which still possessed a tail designed for the Ta 183, demonstrated at large angles of attack a by far insufficient operation of the rudder which first and foremost responded only at very large inclinations and then very violently. These statements of the "pilot" led to a very rapid redesign of the fin's trailing edge and the rudder's axis was shifted to the rear, the surface of the fin was also increased and took on a more trapezial shape. The control surfaces themselves were also enlarged. As further flights demonstrated, these modifications removed the source of problems so far. Equipped with an engine the construction of the Pulqui II could be carried through.*

Up to this time a large section of this aircraft's fuselage had already been made—almost complete fuselage halves (of a semi-monocoque duralumin construction). The design work was carried out not by Tank himself, but by a team of German engineers under his direction, supervised in turn by many Argentinean engineers, who once again checked the results of calculations. Argentinean draughtsmen also made most of the engineering drawings.

These prototypes were made using almost craftsman's methods. As a result of frequent faults of obsolete presses, even many skin plating components were modelled by hand. This was quite simply the first Argentinean aircraft with a completely metal construction, in connection with this it became obvious that any kind of production on a large scale would not be possible under present conditions and in the event of any interest in this substantial investment would be essential. Such interest was far from evident—for almost at the same time—in the years 1947-1949 the Argentinean air force purchased the British Gloster Meteor jet fighter. This was an older type and as it was soon to become evident, significantly inferior to the Pulqui II, but at the time in hand there was no urgent need to bring a new fighter into armament. At the turn of May and June 1950 two prototypes were constructed: one with no engine, for static trials and the other complete, with the Nene 2 engine.

On June 16 of that year, it underwent its first, test flight. Captain Weiss of the Argentinean air force settled behind the controls, who had previously flown many hours on the Meteors. The take-off took place on a grassy airfield—the undercarriage after all had been made in accordance with the demands of the Luftwaffe from the time of the war, though this was perhaps not justified under the new conditions. For a long time preceding this event Tank had pressured the Argentineans that he should pilot the aircraft in its first flight, supporting this with quite obvious arguments. These suggestions were strongly rejected by the Argentineans. They in turn argued that they could not allow the exploitation of this event for national and propaganda purposes to stress the possibilities of their country to be given up. The first take-off was to be only a formality, the true flight tests in which it was intended not to interfere with the Germans were to take place later. On June 16, therefore the aircraft was only tested in a short horizontal flight. Apart from light and minor damage to the landing flap of one of the undercarriage legs there were no surprises. In all stages of the flight the aircraft behaved in accordance to plan and after landing Weiss could present a positive opinion.

The first true test flight was achieved as early as the next day. This time the honour of settling into the cockpit fell to Otto Behrens—until recently an SS-Oberscharführer and one of the best pilots of the Third Reich. The time had come to test the properties of the aircraft in a large range of parameters. At low and medium speeds it behaved pretty well, but it became obvious that above a speed of around 700 km/h problems with stability appeared, and the aircraft became more and more difficult to control. As soon as these problems appeared, Behrens stopped the flight. During landing the right undercarriage leg snapped and so it was a very hard landing. The pilot didn't suffer injury, but this flight in general revealed significant defects and imperfections. This is how Pabst described this situation:

> *This led to detailed testing of the undercarriage legs. At the same time it became clear that the strength of the shock absorbers had been underestimated to a significant degree and the oil pressure hadn't been taken into consideration at all. In the new construction the air shock absorber was separated from the oil shock absorber [the aircraft possessed hydro-pneumatic suspension—endnote I.W.], thanks to which the strength became sufficient. The fuselage undercarriage leg was also significantly modified, although in this landing it hadn't played any role. Apart from incorrect functioning of the suspension, it was proposed to examine the aircraft in a wind tunnel—with regard to an observed loss of longitudinal stability during landing. During renewed assembly of the machine modifications were introduced with the objective of correcting flight properties, which led to a sharpening in profile of the fuselage's nose section.*

As a result a significantly better construction was actually obtained. From October 1950 to May 1951, it underwent 28 flight tests, but now Tank piloted it himself. Certain minor modifications were introduced, mainly in the construction of the control surfaces. But it was already possible to say that this was now a mature construction. During one of these flights Tank achieved a speed of 1,040 km/h, i.e. very close to the speed of sound. Attempting to check the properties of the aerodynamic lay-out with its highly swept wings at a very large angle of attack he caused it to stall, in which the aircraft literally stood vertically in the air. Nowadays this phenomenon is already well known, defined by the English term "superstall," but at that time was completely new. Not only did Tank have problems with it, but it surprised designers in other countries to an equal degree (where German conceptions had been absorbed).

After this incident, in which it had very nearly led to a crash, Kurt Tank resolved however that the aircraft needed to be tested in extreme conditions once again. These are his words:[45]

> *I must repeat the whole spectacle once again, only in this way can I arrive at the essence of the problem. To what was the cause, or will this not lead to the aircraft falling at its given position?*
> *I pull the control stick towards myself and ascend to 9,000 metres. Again I pass into a flight under conditions of stalling and soon with the breakaway of the air streams [from the wing's surface—endnote I.W.] I fall vertically like a stone. But this time I immediately pump [fuel] with the throttle lever and in a short time the machine was again under control. (…) When I land it becomes clear to me that for the first time I had constructed an aircraft which may endanger the pilot's life.*

After these tests another modification was introduced, namely a ballast was installed in the nose section of the fuselage. Shifting the centre of gravity to the front turned out to be effective in reducing the risk of stalling.

When it appeared that the Pulqui II, alias the second incarnation of the Ta 183, was finally a tested and mature construction it was decided on the first public display. On February 8, 1951 a prototype piloted by Tank landed at the civilian airport Aeroparque in view of thousands of cheering people, almost in the centre of Buenos Aires. This event received an appropriate propaganda setting, emphasising the scientific and technical advance which Argentina had experienced on behalf of the war—yet it concerned not only that the knowledge and skills of scientists of the Third Reich been exploited, but also, and rather first and foremost that a record export of Argentinean products and raw materials during the war had given money for the realisation of this and other breakthrough conceptions. The air show at Aeroparque airport didn't take place without a characteristic, joint toast of Tank with Peron (in view of foreign diplomats) and words glorifying Argentinean—German co-operation. Taking into consideration the realities prevailing in the world, at that time this was an outrageous phenomenon … It is no wonder that incidents appeared (mainly in the USA) in which Argentina was described as the "Fourth Reich"— a couple of books appeared on this subject, among others *Secret Agenda*, in which mention is made of Juan and Eva Peron's Argentinean *Fourth Reich* as well as *The United*

States, the German-Argentines and the Myth of the Fourth Reich, published in 1984.

Despite that, after the show in Buenos Aires and Tank's earlier flights it seemed that the Pulqui II was almost a complete construction, further tests revealed the existence of certain shortcomings. Excessive self-confidence took its revenge on the design and research team resulting in loss of life. At the end of May 1951, the Argentine pilot Manuval in unclear circumstances used his ejection seat and died during ejection (the Pulqui II was one of the first aircraft that possessed a discarded windshield and ejection seat). As it became evident, the seat had been mounted incorrectly.

After this accident a second improved prototype was ready for testing, which among others acquired additional fuel tanks in the form of internal wing torque boxes. This allowed the amount of carried fuel to be increased from 1,875 to 2,600 kg (approx. 3,700 l), thanks to which the range rose by over 1,000 km to 3,090 km. The maximum flight duration rose by over an hour to almost three hours. Despite an increase in mass by keeping the engine used so far, thanks to modification of the construction it was possible to increase the maximum developed speed in horizontal flight to 1,057 km/h. Taking these parameters into consideration one should observe that they were almost identical as in the case of "rivals" from the USA and USSR: the F-86 Sabre and MiG-15. These aircraft developed a speed up to approx. 1,070 km/h and had a similar range. All were developed at practically the same time.

The modified, second prototype of the Pulqui II was also destroyed in a crash. Otto Behrens died in it. This time however no technical reasons were affirmed. Behrens was testing the aircraft in gliding conditions at high speed with the engine turned off. At a critical moment he didn't manage to ascend and hit the ground.

At the beginning of the fifties testing of the aircraft formally neared an end, but despite this still no decision had been made to develop (or rather build) a suitable production base. It seemed that a state of peculiar suspension persisted, awaiting settlement. Finally it came, but was different from that which had been expected. In 1955, Peron, the chief henchman of the Germans, was removed from power as a result of a coup and work on continuation of the Ta 183 was finally stopped. However in the meantime several other types of aircraft were constructed …

With the considerable help of German designers among others arose:

- the twin-engined, one-man I.Ae.30 Nancu fighter, modelled on the British Mosquito;
- the twin-engined multi-role I.Ae.35 Huanquero (light bomber or transport aircraft). It was developed under the direction of engineer Paul

The Ar 234B

Klage, working within the confines of Tank's group. It was propelled by two relatively poor Argentine 750 HP in-line engines. The maximum bomb load amounted to approx. 1.5 tons and maximum range, approx. 1,500 km. At the turn of the forties and fifties a small number of 44 specimens were produced. Soon afterwards they were replaced by new machines:

- The I.Ae.50 Guarani, also developed under the direction of engineer Klage. It possessed two turbine turbo-prop engines from the French company Turbomeca, allowing a speed of 500 km/h to be developed. Soon a modification appeared:
- The Guarani II with a modernised pressure cabin, of which 31 specimens were produced.
- With the participation of Tank's group the Pentaturbino passenger aircraft was also designed, which could carry 35 passengers and was propelled by five turbo-prop engines. However work never went beyond testing a model on a scale of 1:1.

A prototype equipped with four BMW 003 engines

The Ar 234 after the war

Unquestionably the most interesting was however a construction created by the team of Reimar Horten. This was to be an unusually avant-garde, fast jet fighter, obviously constructed in the lay-out of a "flying wing" (though with a fin and vertical tail unit). The aircraft had the outline of a long isosceles triangle of surface area 48 m², at the front of which were situated air intakes to the engine and between them an unusually innovative, completely transparent (i.e. also on the sides and bottom) cockpit. This aircraft acquired the designation I.Ae.37. Work was executed on it in three stages. First, at the beginning of the fifties, a version without propulsion was constructed, which was intensively tested in flight; the aircraft had such a large lifting surface that it showed good characteristics also without an engine—the landing speed on approach was only the order of 100 km/h.

In the second stage the I.Ae.37 acquired the Derwent-5 Rolls-Royce engine. At this stage it was tested at medium airspeeds.

In the third and final stage it was to acquire an engine with afterburner, enabling supersonic speeds to be achieved. Work on this aircraft was never completed and as far as I have been able to determine, it never exceeded the speed of sound.[44,45,46,47,48]

One way or another this was however a momentous construction and as in the case of Tank's Pulqui II, constituted the crowning achievement in the impressive post-war career of this designer and visionary, whose work was even quoted publicly by the designer of the American B-2 stealth bomber.

In passing it is worth observing that the "Argentine continuation" took place also in this way, that many famous pilots found their way there—Luftwaffe aces, including Hans Ulrich Rudel (then: ODESSA resident for Bariloche), Inspector General of the fighter air force, Adolf Galland, or Inspector General of the bomber air force Werner Baumbach. This last individual died in Argentina during tests of a developmental version of the Hs 293 guided missile.

The Jet Bombers

In the Third Reich a whole series of jet bomber designs also arose, although only one type was brought into armament (not counting the bomber version of the Me 262). This was the Arado Ar 234 "Blitz."[26,27,29,39]

The Ar 234 was described at the beginning of this section of the book, contrary to alphabetical or chronological order, since out of the most advanced German aerial constructions

The Ju 287 was the Third Reich's heaviest jet bomber

only these portrayed types were managed to be used in combat on a significant scale.

This aircraft was initially designed as a fast medium-range reconnaissance machine, in time however its main purpose became to carry out bombing flights. This was the first jet bomber brought into service. Overall from the summer of 1944 to April 1945 little more than 200 specimens of several different versions were produced.

Work on this aircraft was commenced at the turn of 1940/1941. It was to be propelled by the pair of engines already known from the description of the Me 262: the BMW 003 or Jumo 004 (depending on the version). Finally however, since the BMW engines ensured a smaller thrust and simultaneously were lighter, it was decided on installing four of them—each pair in a shared engine nacelle under each wing. In this way arose the future versions: B (2 x Jumo) and C (4 x BMW). The latter was characterised by as much as 20% greater airspeed at a similar range. In exchange the range of the C-5 was even increased to over 1,000 km. The main differences between these two aircraft resulted precisely from the different types and number of engines. A P version was also constructed propelled by the Heinkel-Hirth HeS 011 engines designed by Joachim Pabst von Ohain, interesting from a structural point of view, but incomplete in technological terms. With regard to prolonged problems in removing various faults, only 28 aircraft were constructed with these engines. The application of two twin-flow Daimler-Benz engines was also considered (target designation DB-007), but work on them was not completed in the required time. However these were in all probability the most state-of-the-art jet engines ever designed in the Third Reich. A complement to the aforementioned groups of jet engines were additional rocket motors used in assisting the take-off of aircraft with the Jumo 004 engines.

The standard armament variant of the Arado 234 consisted of two 500 kg bombs suspended under the wings or a single one-ton bomb.

Up until the end of the war the possibility was studied of this aircraft carrying guided weapons, among others the Fritz X guided bomb, Hs 293 air-to-surface missile and V-1 missile.

The cannon armament (defensive) was modest. It was made up of two underslung MG-151/20 20-mm cannons. This followed simply from the fact that the best defensive measure against enemy fighters was to exploit the very high airspeed of the jet bomber, about 100 kilometres an hour greater than in the case of airscrew-propelled enemy aircraft. In the event of

The Ju-287 during flight

encountering fighters while carrying out the combat mission the Blitz could as a general rule escape safely. The execution of reconnaissance tasks was made possible by two aerial cameras: the Rb 50/30 or Rb 75/30 mounted in the fuselage.

Tactical and technical details

	Ar 234 B2	Ar 234 C5	Ar 234 P3
length [m]	12.62	12.90	13.30
wingspan [m]	14.41	14.41	14.41
height [m]	4.28	4.28	4.28
wing area [m²]	27.0	27.0	27.0
empty mass [kg]	4,900	6,570	5,995
take-off mass [kg]	ap. 10,000	11,150	10,675
max. speed at altitude 6,000 m [km/h]	735	870	820
range [km]	770	1,020	-

Two variants of the Ar 555
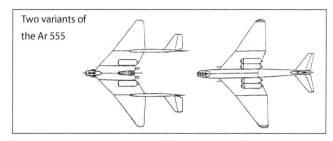

Blohm und Voss P-188
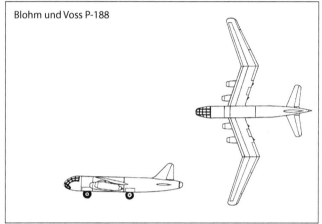

Bv-170
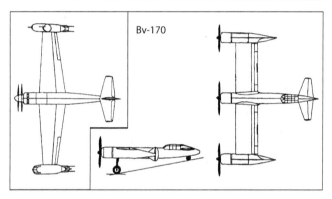

Blohm und Voss P-178
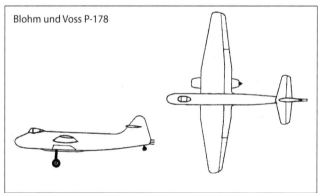

Blohm und Voss P-194
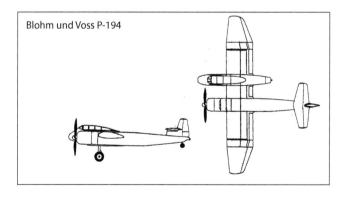

Blohm und Voss P-192
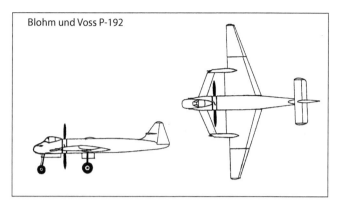

The Messerschmitt P-1109 with wings twisted to achieve maximal sweep
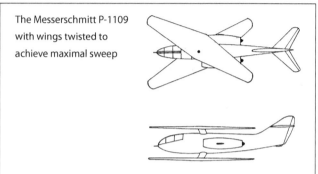

Blohm und Voss P-208
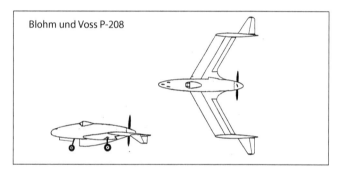

Blohm und Voss P-212
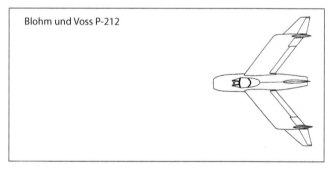

Blohm und Voss P-215
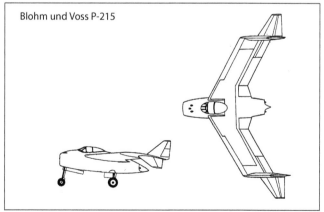

A serious re-valuation in the conception of waging the aerial war was caused by the intensifying Allied carpet bombing air raids, especially from the beginning of 1944. A particular problem was combating formations of British bombers operating at night, with regard to the small number of night fighters available. The ignorance of Hitler alone contributed to this to a large extent, who unfavourably viewed the use of the Me 262 as a fighter and delayed work on, for example, the He 219 Uhu intercepting night fighter.

Only in October 1944 (when it was already too late) was conversion of the training-combat Me 262 into a night fighter commenced. This same requirement also applied to the Ar 234. But since this was a one-man aircraft it was necessary to "economise" room for a second crew member, the radar operator. Such a possibility existed after removing the camera situated just behind the cockpit.

However the Ar 234 didn't fit into this new role very well, mainly due to the lack of an armoured cockpit. Finally a completely new version was designed with an armoured cockpit (the aforementioned P version) with room for three crew members. But the choice of this final configuration only followed in February 1945, in connection with this it never ended up being mass produced. Many alternatives to the Ar 234 arose, work on two of which was crowned by the construction of prototypes and their test flights. This concerns the Heinkel He 343, an aircraft very similar to the Ar 234, as well as an entirely different, unconventional construction: the Junkers Ju 287. It was significantly heavier and had a greater bomb load than the Arado.[27,29,30]

However this aircraft never ended up being mass produced, up until the end of the war only several prototypes were built (in 1944). This occured because it was developed later than the Ar 234—work was begun at the beginning of 1943. In order to speed up work many components and sub-assemblies that had already been tested in other aircraft were used in its construction—among others a modified (later altered further) fuselage from the Heinkel He 177 Greif heavy bomber, the main undercarriage legs from the Junkers Ju 352 and front undercarriage leg from a captured American B-24. The propulsion of the first prototype was made up of four Heinkel-Hirth 011 turbo-jet engines each with a thrust of 12.75 kN (1,300 kg), but due to the fact that they were still unfinished six BMW 003 A-1 engines with a smaller thrust were used on the remaining prototypes.

All of these solutions had already been tested, but what constituted a truly new feature of the Junkers 287 was the aerodynamic lay-out, which had never been applied before. Namely the aircraft acquired wings of a negative sweep. They were supposed to improve the machine's flying qualities, mainly by maintaining manoeuvrability at an increased critical velocity of up to Mach 0.85 and also by maintaining good manoeuvrability at low airspeeds.

The cost of this was problems with wing rigidity, but during prototype flights the innovative aerodynamic conception confirmed its virtues. Being able as a bomber to develop a

The Fa 233 represented one of many types of helicopters constructed in the Third Reich

The innovation of a conception meant great risk, although thanks to this sometimes constructions arose that defeated their rivals. An example of such a situation was constituted by the Dornier Do 335 "Pfeil" with two propellers: one on the nose and the other on the tail.

The FL-282 helicopter

speed of 650 km/h in horizontal flight the Ju 287 would have been particularly dangerous to the Allies. Carrying 2-4 tons of bombs it could have effectively attacked targets located e.g. in Great Britain constituting an alternative to the V-1 missiles and V-2 rockets.

The final period of the war was marked by the design of simple and cheap aircraft. In this photo: a mock-up of the Me 328 in a wind tunnel. It was meant to be propelled by two resojet engines.

Tactical and technical details (V-1 prototype)

length	18.28 m
wingspan	20.10 m
empty mass	12,510 kg
take-off mass	20,000 kg
max. speed	650 km/h
bomb load	2-4 tons
range	up to approx. 4,500 km

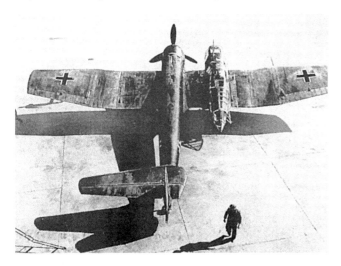

The Blohm und Voss company, leading in the promotion of unconventional aerial concepts, produced among others a short series of asymmetrical BV-141 aircraft

The introduction of jet propulsion was obviously a very important inspiration for developing innovative aerial constructions, not only bombers. If in the initial phase only the propulsion was in principle new, the aerodynamic concepts remaining traditional, then soon a second generation of jet aircraft began to arise, already fully utilising the advantages of the new propulsion.

It is worth paying attention to some of these incomplete designs, even as a contrast to aerial technology from the period directly preceding the outbreak of war.[27,43]

An example of this generation were e.g. the designs of jet bombers from Arado, which in the future were to replace the Ar 234.

The E.560 arose in the first instance. A modification in relation to the Ar 234 consisted mainly in the application of new swept wings, with a sweep of the leading edge of around 20°. At the beginning of 1945 a model of this aircraft was tested in a wind tunnel.

In addition to this at least three more advanced designs arose: the Ar 555 (two designs) and the Arado II. The latter was a further modification of the E.560, at the same time the sweep of the wings was further increased, a tail plane of a similar sweep was applied and the fuselage construction redesigned, giving it a more streamlined shape. In the nose section it was intended to mount four MK-108 30-mm cannons, which would have also enabled fighter tasks to be carried out. The propulsion unit was limited to two HeS 011 engines.

The Me 328 (a specimen without engines) in its first flight

THE LUFTWAFFE: A TIME OF QUEST | 73

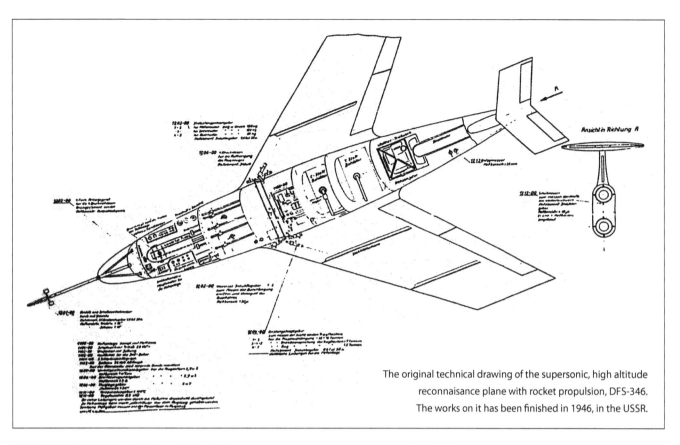

The original technical drawing of the supersonic, high altitude reconnaisance plane with rocket propulsion, DFS-346. The works on it has been finished in 1946, in the USSR.

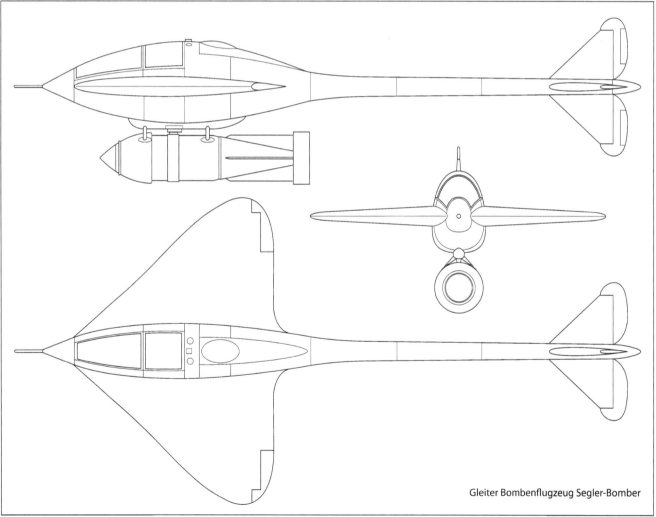

Gleiter Bombenflugzeug Segler-Bomber

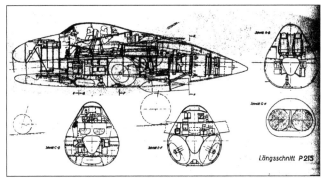

The original cross-section of the P-215's fuselage

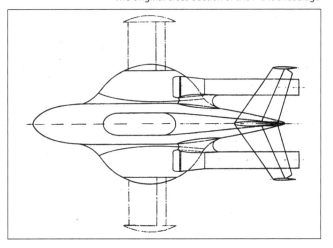

An original drawing of one of the Me 328's planned derivatives

The next stage of work was represented by the Ar 555 design, two fundamental modifications of which arose. Their common feature was a wing with a deflection of the leading and trailing edges; their sweep was greater near the fuselage and smaller in the case of the outer section. The application of 2-4 jet engines of an unspecified type was predicted, suspended under the wings in direct proximity of the fuselage. If the first version acquired a fuselage and control surfaces like in the Arado II, then its derivative presented a completely different conception. The actual fuselage was eliminated, increasing only the thickness of the wing's centre section. A small cockpit was to be situated in its frontal section, in the middle the bomb bay and at the rear—on the trailing edge—an on-board gun, turret gunner, was to be also the bombardier. The continued rear section of the fuselage with control surfaces was "divided" into two narrow tail booms, which were attached to the central sections of the wings. The advantage of such a conception was the possibility of shifting the gun turret "responsible" for defending the rear hemisphere of the aircraft to the central section, which made possible a significant reduction in the mass of the ("divided") fuselage.

If the subject is about innovative bomber designs, one can't possibly omit the conception which Daimler-Benz attempted to realise in 1944.[21,27,43]

This was one example of a way in which it was intended to solve the problem of transporting weapons of mass destruc-

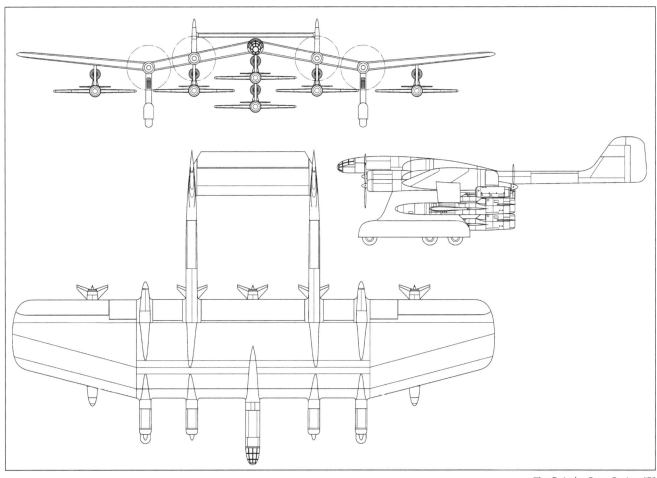

The Daimler-Benz Project "F"

tion (among others chemical) to distant targets, i.e. American.

In 1944 Daimler-Benz A.G. joined work in this area. It concentrated on plans to construct enormous strategic aircraft of a peculiar type, carriers of jet bombers or guided (remotely or by suicide pilots) flying bomb aircraft, constructed according to a conception similar to that which the Mistel designs were based on.

The designs of two carrier aircraft arose:

Design "A" assumed the construction of an aircraft propelled by six HeS 021 jet engines each with a thrust of 3,300 kg, which were to be mounted in nacelles above the wings. The undercarriage problem was solved in an interesting way, in consideration of the plan to transport the bomber under the fuselage. The undercarriage was made up of only two legs —tall pylons placed under the central sections of the wings. It is conceivable that fuel tanks were intended to be placed in them. In their lower sections, three wheels were to be placed in a row inside long housings.

For understandable reasons the range of the carrier according to design "A" was not strictly defined, but it was to enable the bombing of targets located on the eastern seaboard of the USA. It was not armed since detachment of the bomber was to take place while still over the relatively safe Atlantic.

The bomber itself was to be a simple and relatively classic construction, if the control surfaces in the shape of the letter "V" are not counted as well as missing … undercarriage. This followed from the specific conception of this system's application. The carrier's task was to "deliver" the bomber to the limit of its range, detach it and … return fast to base. After turning on its engines the bomber was to commence flight in the direction of the target, which under conditions of descent was to enable a speed close to the speed of sound to be attained in the region of the coast—thanks to which it was possible to resign from defensive armament. After dropping the bomb or bombs it was to land on its "belly." The crew was then to be evacuated by submarine. The predicted lifting capacity is unknown, but it is common knowledge that it was to be propelled by two BMW-018 jet engines each with a thrust of 3,450 kg. This conception assumed an enormous waste of resources—the predicted military potential was totally incommensurate in relation to the complexity and costs of the whole system. The issue would have appeared different in the event of a nuclear bomb being carried, but in 1944 this was only theoretical.

A certain modification of the aforementioned conception was the "B" design. In this case the carrier was to have slightly redesigned wings and a more economic piston propulsion, 6 DB-603 1750 HP engines driving 4 airscrews situated in front of the wings and two pusher airscrews. It was to carry a bomber like that in the "A" variant or up to six smaller jet aircraft, either suicidal or remote-controlled from on-board the carrier.

Both designs were presented as early as 1944 to the Luftwaffe high command, but were not accepted for realisation.

A modern and similarly unconventional jet propelled bomber was also developed in the plants of Blohm und Voss, famous until now chiefly for the production of hydroplanes.[27,43]

The P 188 was the design of a bomber that was supposed to be an equivalent to the Ar 234, but it never entered mass production. Two versions arose—the first, propelled by four jet engines underslung separately (the P 188.01) and the P 188.04 version, in which the engines were housed in two nacelles, two in each. The most characteristic feature of this bomber were the wings with a positive sweep in the central section and at the same time with a negative sweep of the outer sections. It also had non-standard undercarriage—both main undercarriage legs in a lowered position were located precisely under the fuselage's axis, and two additional undercarriage legs with smaller wheels were to fulfil the function of stabilising the aircraft in the horizontal, lowered from the outer sections of the wings. The application was predicted of four Jumo 004 C engines already known from the Me 262 and Ar 234. Apart from the bomb armament carried in the bomb bay, the installation of four 20-mm cannons was predicted, fixed in the fuselage as well as 2-4 13-mm machine-guns in one or two turrets located in the rear section of the fuselage.

Tactical and technical details of the P-188.01

length	17.5 m
wingspan	27.0 m
take-off mass	23,800 kg
max. speed	approx. 820 km/h
range	up to 1,500 km

Above: the Messerschmitt P 1109
Below and right: Lippisch's "trapezial" version of the P-13b

The Ho 229

The Focke-Wulf Ta 183 armed with X-4 missiles

DEATH RAYS

The aforementioned issue in this chapter's title is, contrary to pretences, very extensive and has a rich, well documented history. As early as the beginning of the war Allied special services received numerous reports on this subject.

Proof of this is even a summary from British Intelligence, in which these weapons were mentioned as early as November 11, 1939. Its author, Dr. R.V. Jones, a physicist from Oxford and Chief of Scientific Intelligence in the British Air Ministry, mentioned in his report among others:[5]

> bacteriological weapons, new combat gases, flame-throwers, gliding bombs, aerial torpedoes, unmanned aircraft, long-range guns and rockets, new torpedoes, submarines and mini-submarines, **death rays, rays stalling engines,** magnetic mines.

As it was to become clear, this was only the beginning of a range of research programmes, conceived on a wide scale. I have to admit that in my archival searches I never concentrated on attempts to find materials confirming that this work had been carried out. In spite of this, to some extent when other reports presented themselves, I found myself in the possession of a large number of documents on this subject. By the force of events this gives the impression of the large scale of this work, today practically forgotten.

A considerable number of reports from American special services were presented to me in the form of copies by Mr. Henry Stevens from the USA (unfortunately without signatures). I have based the entire first half of this chapter on them.

Let the starting point for describing this issue be the first "important" report, which reached Western special services officers and forced them to take a closer interest in this phenomenon. The event in question took place in the middle of November 1944. Its first moderately accurate account was included in an appendix to a special report on this subject, bearing the date 16.XI.1944 (DOCUMENT 42). It relates to a P-38 American reconnaissance aircraft.

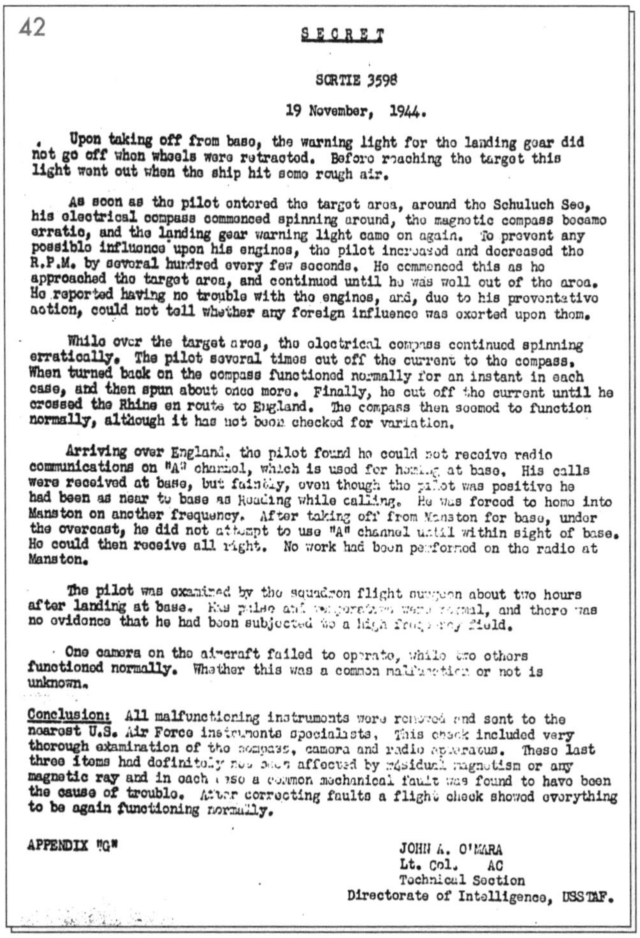

SECRET
SORTIE 3598
19 November, 1944

Upon taking off from base, the warning light for the landing gear did not go off when wheels were retracted. Before reaching the target this light went out when the ship hit some rough air.

As soon as the pilot entered the target area, around the Schuluch Lake, his electrical compass commenced spinning around, the magnetic compass became erratic, and the landing gear warning light came on again. To prevent any possible influence upon his engines, the pilot increased and decreased the R.P.M. by several hundred every few seconds. He commenced this as he approached the target area, and continued until he was well out of the area. He reported having no trouble with the engines, and, due to his preventative action, could not tell whether any foreign influence was exerted upon them.

While over the target area, the electrical compass continued spinning erratically. The pilot several times cut off the current to the compass. When turned back on the compass functioned normally for an instant in each case, and then spun about once more. Finally, he cut off the current until he crossed the Rhine on route to England. The compass then seemed to function normally, although it has not been checked for variation.

Arriving over England, the pilot found he could not receive radio communications on "A" channel, which is used for homing at base. His calls were received at base, but faintly, even though the pilot was positive he had been as near to base as Reading while calling. He was forced to home into Manston on another frequency. After taking off from Manston for base, under the overcast, he did not attempt to use "A" channel until within sight of base. He could then receive all right. No work had been performed on the radio at Manston.

The pilot was examined by the squadron flight surgeon about two hours after landing at base. His pulse and temperature were normal, and there was no evidence that he had been subjected to a high frequency field. [The Americans probably knew what they were looking for!—note I.W.]

One camera on the aircraft failed to operate, while two others functioned normally. Whether this was a common malfunction or not is unknown.

Conclusion: All malfunctioning instruments were removed and sent to the nearest U.S. Air Force instruments specialists. This check included very through examination of the compass, camera and radio apparatus. These last three items had definitely not been affected by residual magnetism or any magnetic ray and in each case a common mechanical fault was found to have been the cause of trouble. After correcting faults a flight check showed everything to be again functioning normally.

APPENDIX "G"
JOHN A. O'MARA
Lt. Col. AC
Technical Section
Directorate of Intelligence, USSTAF

At exactly the same time in which the aforementioned report arose, on November 16, 1944, the American War Department for the first time recapitulated information concerning the possibilities of new German weapons, which could have had a connection with the "adventures" of the aforementioned reconnaissance plane. Only then did it become clear that this wasn't the first report on this subject and that in fact the Americans were aware in which direction German work was heading. Here is the summary (DOCUMENT 1):

COPY
SECRET
WAR DEPARTMENT
MILITARY INTELLIGENCE SERVICE
WASHINGTON

16 November 1944.
SUBJECT: Evaluation of Reports on Rays or Charges to Neutralise Aircraft Motors.
TO: Major F. J. Smith, Post Office Box 2610, Washington, D. C.

1. The Military Intelligence Service is in receipt of various reports dealing with rays for stopping or neutralising aircraft motors such as:
a) An ultra-violet ray for stalling airplane motors is being worked upon hard, but as yet without practical success.
b) Experiments were being made on another electrical apparatus whereby vehicles coming within its field could be stopped.
c) The famous "Death Ray" is spoken of again and it is said that several airplanes have been made to fall in the neighbourhood of this installation.
d) The use of radio ray emits wave that slows down the motors of Allied planes flying over region.
e) In February it was reported that the Germans were working on some mechanism to stop aircraft motors.
f) A source who states that he has done much work for the P. T. T. at Tempelhof in Berlin and for the Reichspostforschungsanstalt claims that 80% of their laboratory work is on aviation. A high official and an engineer there have informed him that they have completed a new A/A weapon which will be turned out in mass production in two or three months.
g) Experiments with "Death Rays" were conducted by AEG Siemensstadt Berlin at Tempelhof in 1939. Guinea pigs were killed at a distance of 200 metres.
h) An individual employed "on electrical matters," not named, or otherwise described, told the prisoner that the Germans had for years been experimenting with these death rays.
i) It is believed that it is a sort of magnetic beam, capable of stopping the motors of planes at a great distance.
j) Experiments of which he knew, (1) destroying the functions of aircraft motors with induced magnetic fields, (2) exploding aircraft in the air by direct ultra-violet ray.
k) This weapon which emits waves or rays based on piezoelectricity is a development of the death ray. It is known that tests were made in 1938 and that at a distance of eight hundred meters an automobile motor was successfully stopped.
l) The Germans will use within the next three weeks a

new generator which is capable of stopping motors.
2. An hypothesis has been received stating that by use of rays or charges the atmosphere surrounding the airplane engine could be ionised, causing the ignition system to spark at joints and also where the insulation has been cracked or broken, thereby causing a short circuit of the spark plug.
3. Any evaluation or opinion you may be able to render on the existence or practicability of any rays or charges, to include probable type of emission, range, size of installation required, would be highly appreciated.
For the Chief, Military Intelligence Service:
S/ Merillat Moses,
Lt. Colonel, F. A.
Chief, Scientific Branch, MIS.

American military intelligence analysed and collected information concerning German "death rays" until the very end of the war, and even after its end. The instance alone of the aforementioned P-38 reconnaissance aircraft (whose pilot was a certain Lieutenant Hitt) bore fruit to detailed analyses as far back as January 1945. (see: DOCUMENT 30). The most important part of these actions was the formation at the end of 1944 of a particular research project, whose task was to deliver possible explanations of the observed effects on the grounds of physics (DOCUMENT 51).

Although from the start practically all possible explanations were taken into consideration, even giving thought to possible substances, which sprayed by aircraft could have given such an effect, in the end a single leading hypothesis was relatively quickly put forward. It was recognised that from a practical point of view the only solution would have been to construct an electromagnetic wave generator (radio frequency), which would induce current in all of the aircraft's circuits.

Calculations carried out revealed that at an aircraft skin plating thickness amounting to 1 mm (duralumin) the greatest beam "penetration" would have been obtained for waves at a frequency of 75 kHz, in other words at long wavelengths. But in order to emit energy which could destroy aircraft circuits over an area of 5 square miles (approx. 16 km^2) thousands of tons of wires and power consumption at a level of several hundred MW would have been needed. This is how it must have been in the case of the (long wave) variant analysed by the Americans, when there was practically no possibility of focusing or concentrating this energy on selected targets. The local analysts didn't take into consideration another possible way—at a significant increase in beam frequency—to "radar" frequency the aircraft's skin plating begins to reflect the electromagnetic wave to an ever greater degree, but simultaneously it is possible to easily focus this wave and despite worse "penetration" it is possible to obtain a greater effect with a smaller power consumption. The Americans still didn't know that the Germans had gone precisely down this path ... However in the aforementioned analysis it was announced that even at long wavelengths, in order to invoke sparking in aircraft circuits (instead of their destruction) a beam one hundred times less powerful than that which followed from initial calculations would have been sufficient.

On the basis of the aforementioned report, soon afterwards work was commenced on possible counter-measures in relation to the new German weapon. As a result on February 6, 1945, the Director of Technical Services of the American Air Force, Col. Bunker, prepared an extensive document containing recommendations for air force units, in order to reduce the vulnerability of aircraft to the effects of the described energy (DOCUMENT 48). It was addressed to the Director of Air Technical Service Command, residing at Wright Field air base, near Dayton in the state of Ohio.

In this letter however the view was presented that the most effective strategy under existing conditions would be to gather information on the location of the German installations and ... to avoid them, until a greater amount of information had been gathered. It was predicted that jet-propelled aircraft would be unaffected by the effect of the aforementioned energy, which suggested that in the conceivable future they would be able to operate over dangerous areas. In the meantime therefore attempts were made to gather a greater amount of intelligence information, which would allow the arisen danger to be better evaluated. But only residual information continued to come flooding in short reports, which until the end of the war gave no grounds for a comparatively complicated explanation of this phenomenon. An example of such a report is DOCUMENT 5, dated January 25, 1945. This is the summary of a German POW's testimony, in which he states that work in this field was carried out as early as 1934, and with success.

Then it succeeded in disabling a combustion engine situated 150 metres away. Somewhat later, but still before the outbreak of war, in 1938, two German aircraft factories were to receive the task to develop an analogous weapon, effective in relation to the engines of enemy aircraft. This work was carried out somewhere "between Augsburg and Munich."

Therefore one can see that in spite of this research being carried out for several years, information describing the work's entirety never reached the enemy, at least that which was situated in the West …

Therefore it was only possible to unravel the mystery of German "death rays" long after the war. Information currently available indicates that work was carried out in two fundamental directions. The first (but not the only) was connected precisely with the effect on engines and other systems of enemy aircraft. We know that this work was inherently connected with work on radar and in a way constituted their side effect. We know this since the laboratories accountable for developing just such an effective system constructed first and foremost radar stations. These were the so-called GEMA-Werke, situated in the town of Lubań (then Lauban) in Lower Silesia (Niederschlesien). As Leszek Adamczewski wrote—a journalist engaged in this subject for a long time[49]:

> *When I announced an appeal in the* Luban Review *that people come forward who had stayed in Luban during the war or shortly after and had seen or heard about some kind of secret phenomena, our readers responded. They associated a well discovered in a park with the top secret activity of GEMA-Werke.*
> *In 1943 the GEMA arms factory was transferred from Berlin—bombed by the Allies—to the modern for those times Lubań works of Gustav Winkler, in which—according to information available to me—a top secret program from the field of radar was developed. Nearby operated the largest labour camp in town. It was called Wohnheimlager GEMA and mainly Russian and Polish women were held there.*
> *The recently deceased Stanisław Siorek, explorer and former Security Service officer of the Polish People's Republic, confirmed shortly before his death that the well accidentally discovered in the park near Esperantists street was none other than a former ventilation or evacuation shaft of an underground section of GEMA-Werke, probably flooded after the war.*
> *One of those who answered Skowroński's appeal was Józef Bujak from Lubań. He found himself in this town shortly after the war. He worked in an engine rewinding workshop, where an old German by the name of Glaubich was also employed. And it was he who informed Bujak about secret experiments carried out in GEMA Werke in the final period of the war.*
> *They relied—recalled Bujak—on the creation of some kind of magnetic field, because cars travelling near GEMA Werke came to a halt! It was possible to observe*

this on a stretch of road around 300 metres long. A section of the Lauban-Goerlitz road had been closed earlier to traffic. When the experiment was paused the vehicles continued on their way as if nothing had happened. Glaubich vowed that he saw how not only vehicles with spark ignition stopped but also diesel! And I am no longer in any way able to explain this ...

Bujak ventured several times into the ruins of GEMA-Werke, set alight by the retreating Germans. Metal structures around 15 metres high with a cabin at the bottom still stood on the factory site. These were revolving structures, and the cabins were stuffed with electronics.

Of course—adds Bujak—this wasn't like modern electronics, but it looked like some kind of complicated radio installation.

Everything points to the fact that a large underground research-production facility was associated with this work, which to this day remains undiscovered. Regardless of this it is very likely that this wasn't the only facility engaged in this work.

The aforementioned work, i.e. the construction of devices emitting radio wave beams, which with regard to their power and frequency stopped the operation or caused damage to the electrical systems of aircraft flying overhead is however not all that the Germans worked on in the field of so-called electromagnetic weapons.

After all it is common knowledge that among other things work was also carried out on some kind of "X-ray laser"—a source of coherent X-ray or gamma radiation—which is well-known to be lethal to living organisms at high intensity. Thanks to searches carried out in German archives it has been possible to establish that in 1944 a special Luftwaffe research establishment received the task to develop such a weapon, situated in the town of Gross Ostheim. Materials relating to this work are currently located in a civilian establishment, the Karlsruhe research centre (Forschungszentrum Karlsruhe), and were disclosed several years ago. An extensive study-report on this subject was revealed among others, prepared for the Luftwaffe high command and dated July 12, 1944. Unfortunately the available copy of this report is in such poor condition, that a large part of it is illegible. In spite of this one can make out that three different lethal radiation emitters were worked on and that the construction of such a device, effective as an anti-aircraft weapon lay within the Third Reich's range of possibilities and its military use was possible in a relatively short time—that is to say still before the end of the war. The third and most mature version of the weapon assumed the irradiation of a target 5 kilometres away at a rate of 7 rads a second for 30 seconds which, as affirmed in the report, was completely sufficient to totally paralyse the aircraft's crew. In the event of the aircraft flying at a different altitude an appropriately shorter or longer time of irradiation was required. A certain curiosity among others resulting from this report, as was determined, is that the aluminium skin plating of the target aircraft rather increased and not decreased the effectiveness of the new weapon, as far as its effect on living organisms is concerned. Under the described conditions the skin plating would have acted like some kind of microwave oven. The issue of applying X-ray generators on the battlefield remains to this day unclear, however if this weapon had been used at all it would surely have been on a small scale. As in the case of the other revolutionary conceptions described in this book, all the time the topical question remains if such a weapon is a part of contemporary arsenals and if so, on what kind of scale?

In spite of its futuristic character it turns out however that the scientists of the Third Reich had worked on an even more avant-garde weapon—the so-called particle beam weapon—an emitter of high-energy particles or ions, which would paralyse living targets with a kind of invisible "micro-projectile," converting its kinetic energy into lethal radiation only at the moment of impact. In the case of the atoms of heavy elements probably even a single, high energy ion would have been sufficient to kill the "victim." As far as I am aware calculations for such a variant were not carried out, however it is no secret that even a single quantum of high-energy gamma radiation (a single photon!) could kill a human being and regardless of secondary radiation, with a "direct hit" on the head could even raise the temperature of the brain by several degrees. Electromagnetic radiation of such a gigantic energy can only be generated at an enormous cost and on a small scale in the largest accelerators.

A heavy metal ion however could fulfil this task considerably better and would be one of the best candidates to be designated the "perfect weapon of the future." Obviously heavy ions accelerated to a speed comparable to the speed of light would come into play. By the way, some time ago information reached me concerning the death of two people, a death quite mysterious (in modern times, but not in Poland), which was accompanied by microscopic holes in some window panels (unusual in as much as 0.5-mm in diameter with partial melting of the edges, however shattering of the glass was not observed around the holes). Perhaps therefore someone has already mastered this lethal technology to a considerably greater degree than the Germans did during World War II …

As far as their work is concerned, documents from the US Military Intelligence supply information on this subject. The first to a degree exhaustive report is presented by the testimony of a German POW, a certain Karl Schnettler, who was taken prisoner on December 1, 1944. His testimony (the official report reproduced in this book as DOCUMENT 47) concerns therefore by the force of events an earlier period. This prisoner declared that particle beam weapon experiments were carried out in an underground laboratory near Ludwigshafen, belonging to the I.G. Farben concern. In September 1944, it was to be moved to a location either in Heidelberg or Freiburg. However the prisoner knew nothing on the subject of experiments carried out there.

In return he comprehensively described the laboratory near Ludwigshafen. Experiments were to be carried out in an underground shelter 25 metres wide, 50 metres long and 8-10 metres high. The shelter walls were up to one metre thick. The floor as well as a long strip 4-5 metres wide, running along the inside walls approx. 2 metres above floor level were covered by a 3-5-cm thick Igelit plastic layer. At one end of the shelter was situated a recess where "electronic tubes" (ion guns) rested on special trolleys, referred to as "Fangpole" and "Spruehpole." Before commencing tests these "tubes" were moved into the focusing area and shielded by semi-circular quartz plates. In a recess at the other end of the shelter were also electrical control devices. Similar plates shielded an observation and control station. Nearby was a target stand for the "tubes," probably acting as ion or particle guns. This consisted of a cuboidal 1.25-1.5 metre high column, surrounded by massive 5 cm thick quartz plates. It follows from the text that they formed some kind of "aquarium" and were as such connected to a vacuum pump. All components of this stand were covered with a thin layer of Igelit.

The testimony also includes the account of an experiment which was carried out on rats in April 1944. They were killed instantly, at the same time for the duration that they were subjected to the effect of radiation their bodies emitted a phosphorescent glow, lasting for a split second.

Schnettler stated that their bodies immediately decomposed and as a gas (literally: Sodium vapours) was sucked out by the vacuum pump. This statement most explicitly met with the interrogating officers' disbelief, who at the end of this sentence placed a question mark in brackets.

The POW also gave a number of names of scientists taking part in this work. On behalf of the Kaiser Wilhelm Institute they

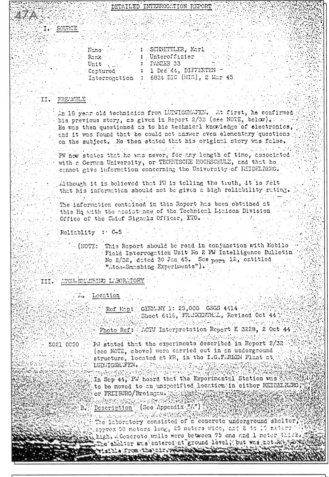

```
47C
    SECRET            - 5 -         6824 DIC (MIS)M.1075
                    APPENDIX "A"

                ATOM - SMASHING LABORATORY
                   I.G.FARBEN/Ludwigshafen

    LEGEND
    1. Recess for
       Electronic Tubes.
    2. Quartz shield.
    3. Focusing Area.
    4. Power Contact.
    5. POLYTRON(?).
    6. Power Cable.
    7. Target Stand.
    8. Observation and
       Control Station.
    9. Power Station.
   10. Switch Station.
   11. Transformer
       Station.
```

were engineers: Kalb, Meissner, Falke, as well as trainee Haeringer, whereas on behalf of I.G. Farben in Ludwigshafen engineers Wendt, Raithel (or Raitrel) and Edlefsen.

I have enclosed a diagram of the laboratory in Ludwigshafen, which was enclosed in an American report on the subject of the discussed work.

In pursuit of new, unconventional anti-aircraft measures and guided by the hope that precisely these unusual methods will turn out to be effective in halting the relentless waves of American and British bombers, the leadership of the Third Reich even reached for such an odd "sword" such as clouds causing enemy aircraft to ice up. Obviously this refers to artificially generated clouds.

As is common knowledge, the icing of aircraft is a very significant problem and the effectiveness of de-icing systems (electric heating) is a fundamental condition of flight safety. Even in summer at a ceiling of a few thousand metres sub-zero temperatures always prevail and regardless of clouds, a certain amount of water in the form of vapour is always present in the air. Under certain conditions, depending on humidity, pressure and temperature it undergoes condensation. However, it is possible to influence these conditions to a large degree, introducing additional water into the atmosphere, or (considerably more efficient) introducing a substance facilitating its condensation. Various types of free radicals are most effective in this role, forming condensation nuclei, or the crystals of defined substances or gas substances crystallising in the air forming microscopic "grains," around which water vapour condenses, giving water droplets and then ice crystals. Even smoke, at certain altitudes, becomes a very effective means of condensing water vapour. If meteorological knowledge and favourable atmospheric conditions are exploited for this objective, then at a surprisingly low cost (if we compare this to complicated weapon systems) it is possible to create a serious danger to enemy aircraft. These kinds of measures were developed after the war in various countries although, as follows from available data, have never been applied as a weapon.

In the Soviet Union light unguided rockets were developed in the 1960s (very simple and cheap), whose warheads were filled with powdered silver iodide (AgJ). With their use fields of delicate cotton were protected very effectively against thunderstorms and hailstorms. It was done in this way that a "barrier" of condensation nuclei was placed before an approaching atmospheric front, thanks to which the clouds "disposed of" their ballast before they reached the protected crops.

Similarly these phenomena are currently exploited in South Africa, but with the use of considerably simpler resources. Thick clouds passing over dry and under normal conditions failing crops are "forced" to irrigate them, with the use of … ordinary, cardboard smoke candles. Several to a few dozen such candles are suspended under the wings of a light aircraft, which then flies into the clouds, where the pilot electrically ignites them. This method is so effective that usually the use of 2-4 candles is sufficient in one flight to induce rainfall.

But what did the Germans work on?

The answer to this question we again find in declassified documents of the US Military Intelligence. In one of the reports concerning new German weapons we also find information on this subject (DOCUMENT 39). It included the following account:

In their efforts to cope with Allied bombing attacks, it is reported the Germans have introduced their newest weapon—"Icing Gas." The new weapon, it is stated, operates on the principles of accelerated icing induced by an extremely low temperature zone, including crystallisation and condensation through a temporary cloud causing the immediate icing of objects passing through it. The equipment is said to consist of a cylindrical reservoir tube secured under each wing of a fighter aircraft. These tubes are filled with "Azote," a liquified nitrogen with a temperature of minus 250°C, combined with liquid air, and a gas outlet is affixed.

Proposed tactics consist of cutting across the bomber path perpendicularly and releasing the tube facing the bomber, or, flying over the bomber and diving across its path releasing the gas at close range, the higher the altitude the better the results.

According to German sources, extremely satisfying tests have been made on robot target planes at altitudes of 7-8,000 metres. It is alleged that these targets were brought down at once and that on reaching the earth they were found to have a coating of ice about two inches thick. It is further alleged that German fighters have brought down some isolated heavy bombers by employing this method. Hopes are also being built on this weapon as a good defensive measure against a pursuing plane.

Other American documents contain information describing attempts to exploit substances of a different kind, which after being sprayed into the air were supposed to damage the engines of enemy bombers. DOCUMENT 52 describes these substances in the following way:

Two types of gas are known to have an application in the case of aircraft. One is to cause premature ignition, which would lead to the cylinder heads being torn off, the other is supposed to decrease the viscosity of the oils used in lubricating the engines. Under laboratory conditions, when operational arrangements are not taken into consideration, these gases give the impression that their application is realistic. However it is doubtful if with adequate fighter escort it would be possible to use any of these concentrated agents against the enemy giving any serious effect. Similarly, if anti-aircraft cannon projectiles were to be used against this target, the possible concentration gained probably wouldn't be more dangerous than the accurate firepower of conventional anti-aircraft artillery.

However the list of unusual German conceptions associated with anti-aircraft weapons doesn't end at all at this. The last of the presented documents (DOCUMENT 44) reveals even stranger ideas, whose sense and exact use we can only guess at. Here is its short excerpt:

(…) I have recently been looking into the evidence for a reported enemy radiation capable of interfering with aircraft ignition up to 3,000 metres and operating in certain high-priority target areas. (…) Not so long ago several pilots reported flying through thousands of transparent glasslike bubbles which, although they had no adverse effect, were thought to a be new weapon. A considerable number of ground and air observers saw new-weapon possibilities in an unusual pink cloud phenomenon which persisted for about an hour over the front line, but without any noticeably adverse effects.

It looks as if "icing gas" was even used on the battlefield. Otto Skorzeny described this in his memoirs, who during this period, at the turn of 1941-1942, fought within the confines of the SS "Das Reich" division.[50]

(…) to our left is situated Khimki, Moscow's port. From here it is only 8 kilometres to Moscow. On 30 November, without a single shot, the 62nd reconnaissance regiment belonging to Hoepner's Armoured Corps moves in here. It is not known why this opportunity was not exploited. Our motorcyclists retreated.

Here begins the next mysterious episode in the battle for Moscow, which has escaped the attention of many historians. In order to oppose the horrifying rockets of "Stalin's organs" we applied a new type of rocket missile filled with liquid air. These were similar to enormous bombs and as far as my competence allows me to estimate—their effectiveness had no equal. Their use immediately had an impact on the enemy's defensive forces. The enemy used huge loudspeakers for propaganda purposes (more than banal and crude). By means of them several days after first using our missiles the Russians threatened to respond with gas attacks if we continued to use rockets filled with liquid air. From that moment, at least in our sector, they were never used again. I don't think they were used on other stretches of the front as well.

However the "ice weapon" also had its opposite alternative in relation to it, the "fire weapon." At the beginning of the 1970s the Americans used a "new" weapon in Vietnam, unknown up until then. The effect of its first use was horrifying but at the same time exceptionally unusual: the Vietnamese inspection team which arrived at the scene of attack saw disorderedly scattered bodies, frozen in strange positions … and bearing no traces of any kind of external injury. One could only see that blood had flowed out earlier from their mouths. With regard to the traces' unusual nature accusations appeared in which mention was made of the use of banned biological weapons. The Americans rejected these allegations, although the whole truth about the "new" weapon came to light much later, only after many years. It turned out that this unusual means of warfare had indeed been used on a large scale for the first time, unusual, but however rated among conventional weapons. These were the so-called fuel-air charges. The aerial bomb credited for such a weapon little resembles a classic bomb. It has no thick, steel shell. It is rather a squat, thin-walled container, made of thick sheet metal plate, cylindrical in shape. It contains no explosive whatsoever, if the fuse is not counted. Namely it contains compressed methane, or other volatile hydrocarbons (ethane, hexane, a special composition…). The essence of the "new" weapon is that the explosive composition is created only after the charge is dropped by the proper (a suitable ratio) mixing of the container's contents with air.

A fundamental technical breakthrough is on the other hand that the arisen composition doesn't burn—like e.g. in a combustion engine's cylinder, but detonates. This difference requires a certain explanation:

Combustion, deflagration (the combustion of black powder or smokeless powder in a barrel, explosion of a squib…) differs in this way that the principle factor in propagating the reaction is heat, or in other words flame, i.e. an expanding medium, heated as a result of the reaction. The speed of combustion is usually relatively low, rarely exceeding 2,000 m/s. This last parameter is directly connected to a principle feature of explosive combustion, the small destructive ability.

The main factor in spreading the detonation is on the other hand something completely different, namely the so-called shock wave. This is not directly related to the propagation of the medium—the shock wave and blast are two different matters. It constitutes rather an analogy to a sound wave (whose transition is accompanied only by oscillation of the air density, and not its displacement), however by definition the shock wave always moves at a speed greater than that of sound, and weakening then changes into a sound wave. These are speeds reaching up to approx. 9,000 m/s, in connection with this it is accompanied by a huge oscillation in density—if matter had no atomic structure, it would be an infinite density.

The standard fuel-air charge creates a powerful shock wave as well as a very powerful blast and heat wave. The chamber containing half a ton of methane allows an explosion to be obtained which in terms of its destructive force is comparable to the explosion of approx. 1.5 tons of TNT (high explosive). Therefore it is a difference of around 3 times. In practise however it is considerably larger, since in a classic bomb as a general rule most of the mass is made up of shell. So if we compare the armament alone, the fuel-air charge may be even 5-10 times more effective than a "standard" aerial bomb. The difference therefore is simply colossal, at the same time one should bear in mind that it was achieved at relatively low cost and in a relatively simple way (achieving an analogous increase in effectiveness e.g. by an increase in accuracy would surely have been incomparably more expensive).

The fuel-air charge possesses one more "interesting" feature: since the explosion takes place at a large volume, a phenomenon arises, which has no visible significance in the event of some explosive exploding. As it expands the expanding gas bubble produces a zone of negative pressure in the centre—after expansion some kind of implosion follows. It was precisely this phenomenon which was responsible for trickles of blood around the mouths of the dead Vietnamese—blood from perforated alveoli had been sucked out as a result of the "implosion."

In short what is involved is an extremely dangerous weapon. An excellent example of its possibilities may be presented by a certain unusual incident from 1991, and the First Gulf War. One of the British reconnaissance units (SAS) scattered over the territory of Iraq observed at a certain distance the explosion of a fuel-air charge and without hesitation interpreted it as … the explosion of a small nuclear charge, an analogous report was sent to headquarters, creating much confusion. This report was treated seriously, since SAS units are very well trained from the point of view of identifying the effects of using different types of weapons and nobody expected them to commit such a "blunder."

The above fact says however much about this weapon's possibilities. One may without greater exaggeration define the fuel-air charge as a "transient link" between classic ammunition and battlefield (tactical) nuclear weapons.

It turns out, that this weapon had also been worked on … by the Germans during World War II …

This work was carried out under the code-name "Hexenkessel" ("Witches' cauldron"). In it participated among others a ballistic research laboratory situated on the grounds of the district of Berlin, Gatow, the laboratory of a certain Dr. Zippermayr and the company Dynamit A.G. Krümel. However the Germans never used volatile hydrocarbons for this purpose, but … fine coal dust. This is a material making higher requirements, but characterised by a similar effectiveness. Certain data on this subject is provided by contemporary research on coal dust explosions in mines. It turns out that an explosive composition is formed by grains 1 mm in size, when their "concentration" exceeds 12% of the mass of air. The approximate limits of explosiveness amount from 45 g to 1,000 g of dust, suspended in 1 m³ of air. The most powerful explosion occurs at a content of 300-500 g in 1m³ of air.[51] It was managed to obtain certain data concerning the "Hexenkessel" project thanks to access to a file of the aforementioned Dr. Zippermayr. In August 1945, he was interrogated by an American counter-intelligence establishment in Austria and basic information on the subject of the work which he participated in was enclosed in an official report from the interrogation (DOCUMENT 54). Here is an excerpt of it referring to this work:

In 1923 and 1924 Subject (Mario Zippermayr) was an assistant professor in the Technical High School in Karlsruhe. During the last half of 1924 Subject went to Vienna to set up his own private laboratory in which he undertook scientific experiments for various industrial enterprises until August 1939. During this period he also received his doctor's degree in Engineering. In August 1939 Subject was drafted into the Luftwaffe as a private. In May 1942 he was included in a group of technically skilled soldiers who were remain in the Luftwaffe but who were to be utilised according to their technical skill. Zippermayr was sent to Vienna and told to set up his own laboratory again and that he would receive supplementary equipment to further experiments which he was to conduct for the Luftwaffe. His laboratory was located at Wien, Bezirk 19, Weimarerstrasse No. 87. He was furnished a staff of 35 people who assisted him in the work. Experiments were conducted on three main projects; development on the L-40 Torpedo; two anti-aircraft rockets known as Enzian and Schmetterling; and a jet propelled high speed plane. This work was financed by the Reichsluftfahrtministerium and was under the direction of the Chef der Technischen Luftruestung.

The L-40 Torpedo was one that could be dropped at high speed from a high altitude, constructed so that its mechanism would not be damaged upon contact with the water. It was slow in its descent and had a self directing mechanism. The L-40 Torpedo was successful but was not put into production because the type of plane necessary to its launching was not being produced.

The Enzian and Schmetterling were anti-aircraft rockets that were charged with a coal dust explosive strong enough so that upon explosion the concussion could break the wings of a bomber. This item also was proved to be successful by August 1943 but orders for its production were not issued until 9 March 1945.

The jet-propelled high speed plane was an outgrowth of technical knowledge obtained in the development of the torpedo L-40 and was only in the early stages of development. By the end of the war it was to fly at a speed of one thousand miles per hour and was of a radical design with only one wing which ran parallel to the plane.

In January 1945 the laboratory and staff was moved to Lofer but Subject stayed in Vienna where he continued ex-

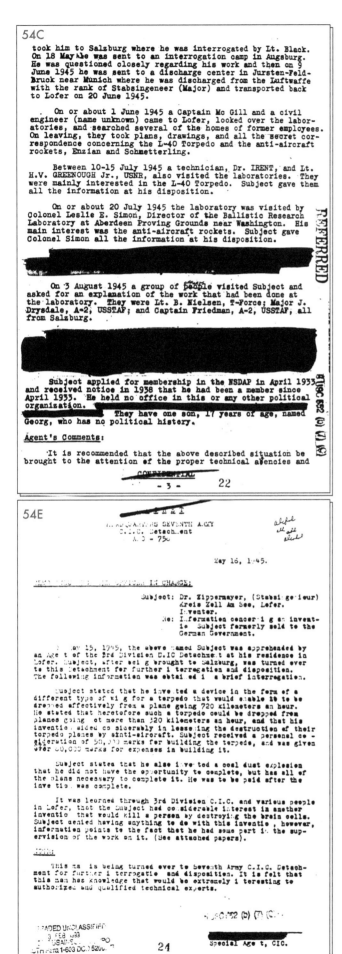
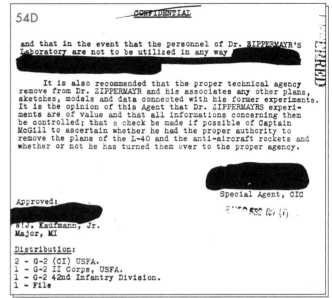

periments in coal dust explosives. On 1 April 1945 Subject came to Lofer to continue his work, at which time he had a staff of approximately eighty workers and technicians. The Lofer laboratory is dispersed, the main group of buildings being located in a place just outside of Lofer called Hochtal. Hochtal is enclosed by a circle of mountains with a dirt road as the entrance. Two other shops are located in Lofer itself.

Subject conducted his experiments until 8 May 1945. When the American troops came he reported the presence of 2,000 kilos of explosives at the Hochtal laboratory to the CO of the occupying troops. At the time all buildings and equipment were in excellent condition. At the present time, however, buildings and equipment are in a condition which indicates wholesale looting and vandalism. On 15 May 1945 Subject was arrested by a CIC Agent who took him to Salzburg where he was interrogated by Lt. Black.

The American documents indicate that the work of the aforementioned individual was not restricted in the very least to the above issues, although he himself wasn't willing to admit this (one cannot exclude that the reason for this was experiments carried out on people). Here is an excerpt from DOCUMENT 54E, including information on this subject:

It was learned through 3rd Division C.I.C. and various people in Lofer, that Zippermayr had considerable interest in another invention that would kill a person by destroying the brain cells. Zippermayr denied having anything to do with this invention, however, information points to the fact that he had some part in the supervision of the work on it.

So yet another mystery still remains, which has so far escaped the attention of researchers engaged in the history of World War II. Perhaps sometime it will also be possible to unravel it.

I discovered many documents referring to the "death rays" between 2000-2001 in American archives, first and foremost in materials remaining after the so-called Alsos mission, the thorough study, directly after the war, of the German nuclear programme and the issues associated with it.[52] The "death rays" are also mentioned in an analysis from British Intelligence, devoted to particle accelerators and specifically the so-called betatrons.[53] They permitted a directed high-energy X-ray beam to be achieved. It was written in the British report that:[53] "The Luftwaffe authorised work in the hope of obtaining death rays to combat enemy aircraft."

The Alsos mission documents prove in turn that the betatron was treated as a starting point for constructing a weapon system. Many research groups were employed to put this task into effect, however the most crucial appears to be a group focused on Dr. Wideröe, and the company BBC from Mannheim. Wideröe gave the impression of being a very active and creative scientist. Only up until September 1943, at least ten of his patent applications had accompanied the research.

The last available report from Dr. Wideröe on the subject of the work's state dates from September 15, 1944, which suggests that approx. four months was still needed to make the "weapon." It was to be an accelerator using 200 MV (million Volts) electrical discharges.

A report from the chief of the Luftwaffe research laboratories working on this problem—Prof. E. Schiebold, also indicates a similarly advanced state of work—dated from May 1944.[52] It mentions among others the use of the Luftwaffe's resources, the construction of a "huge workshop" for research purposes on the grounds of the laboratories in Gross-Ostheim, as well as the assembly of experts released from the Army. It may seem that all of this clearly indicates a laboratory stage of work, but the data is far from complete; it ends at a certain moment and probably doesn't encompass everything. For example, there is nothing about the activity of the aforementioned company Gema. Despite all of this the large number of employed individuals and institutions is intriguing, making omission of this subject in contemporary studies completely unjustified.

This is even more striking if we make ourselves aware of the fact that in the files of the SS Reichsführer's Personal Staff existed a special folder, in which correspondence was found referring to a "weapon" emitting directed energy with the objective of combating enemy aircraft (I found these materials in 2001).[54] We also find here traces of the high inner circles who were involved in the whole affair. In the folder are among others reports of the companies ELEMAG, AEG and the Reich's Scientific Research Council. From them it also doesn't follow that any kind of practical use was reached, among others, the chief of the Planning Office in the Reich's Scientific Research Council (*Leiter des Planungsamtes des Reichsforschungsrates*), Prof. Werner Osenberg stated on February 7, 1945, that in spite of work being carried out "for several decades" it had given no specific results whatsoever. In light of this the position of the office of Plenipotentiary for Research in the field of High Frequencies (*Bevollmächtiger für Hochfrequenzforschung*, BHF) could have been interesting, which should be an authoritative instance in this affair.

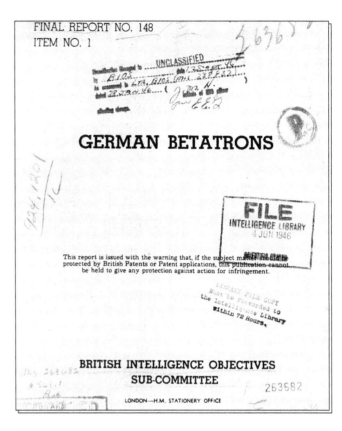

A German document from the files of the "Alsos" mission, containing a brief description of the project regarding "death rays," related to Gross-Ostheim

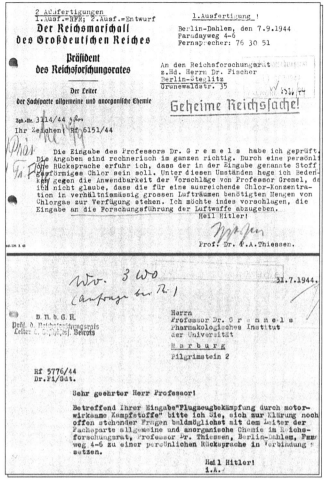

Two documents from the Reich's Scientific Research Council, regarding chemical agents, supposed to stop the engines of Allied bombers
(via Alsos)

Such an opinion was included in one of the documents from January 1945.[54] It followed from it that with regard to the "high burden" of the research establishments subordinated to the Plenipotentiary, they are not able to become involved in this area.

I did not find therefore in German source materials explicit confirmation of the military application of a directed energy beam, although the number of institutions alone appearing in this context is surprising and doesn't very well suit a project devoid of any practical aspect.

If the subject is about "electromagnetic weapons," then one should mention one more project which was after all also directed against Allied bombers.

Out of the hundreds not to say thousands of Allied special service reports concerning "new weapons" of the Third Reich, many described revolutionary or simply unusual conceptions in the field of artillery weapons. Sometimes errors or exaggerated assessments appeared in these reports, which made the new German solutions more amazing than they were in reality. Here is an example, an account included in document 4 from December 29, 1943:

According to reports from France, based on information originating from industrial circles, the new German weapon is a heavy, long-range cannon. It fires projectiles containing phosphorus and other chemical substances absorbing oxygen from the atmosphere within a radius of hundreds of metres from the site of explosion, which creates lethal conditions for all living organisms.

Of course the Germans had in fact worked on a heavy, long-range cannon, but for a fact the projectile fired from it wouldn't have worked in the described way. A flood of similar and very diverse reports resulted in that a certain group of it was never devoted the proper attention, which in actual fact contained equally unusual details, but which referred to a weapon which only after the war could one ascertain without a shade of doubt, that had been intensively worked on by the Germans (this fact is not universally known till this day).

It refers to the so-called electromagnetic cannon, a weapon in which the projectile is not accelerated by gases created by a burning propellant charge, but by an extremely powerful magnetic field, maintained for a fraction of a second. The principle of operation of the electromagnetic cannon is relatively simple (in this respect it constitutes an analogy to an electric linear induction motor), but the technological challenges which accompany attempts at constructing the valuable device result in that the weapon is still presently regarded as of the future. This is so also because it offers the possibility to overcome many technical barriers, impassable in the case of the classic cannon. The so fundamental barrier which engineers and the military dream of overcoming is the projectile's muzzle velocity—a crucial parameter determining the penetrative ability of the basic type of antitank ammunition—armour-piercing projectiles with hard cores, penetrating the armour only thanks to the kinetic energy. No modern tank fires projectiles characterised by an initial muzzle velocity greater than 2,000 m/s. Exceeding this "magical" limit would be extremely difficult on the basis of classic solutions and would entail costs totally incommensurate to the effect gained. This would entail first and foremost a sudden decrease in the barrel's life—probably to a level of the order of 100-200 shots (the barrel working life of 120-125 mm modern tank cannons is the order of 500-1000 shots, at the same time one should bear in mind that such a barrel costs on average tens of thousands of dollars).

The problem of the propellant itself would also require new solutions; the gunpowder would probably have to be replaced by something completely different, with an appropriately greater speed of combustion and greater energy, which would involve the risk of such a propellant detonating (this would most certainly result in the destruction of not only the cannon itself, but also the entire tank). Also difficult to overcome is the endurance barrier of the barrel to bursting—this is strictly defined. It is no longer possible to significantly increase the endurance of steel and "thickening" the barrel gives very little—its internal layers would fracture anyway. Tactical requirements also pose limits on the barrel's length.

One can in any case summarise the above problem in a considerably simpler way—it is a generally well-known fact that since the end of World War II classic firearms haven't undergone great change. The basic weapon of the modern infantry, the automatic carbine (a semi-automatic/automatic weapon for the so-called assault round), differs little from the first construction of this kind, the German MP-43 from 1942. This constitutes an excellent illustration of the problem, as does the fact that the basic machine gun of the Wehrmacht—the MG-42 (currently the MG-3) is produced till this very day, with no large modifications, for the needs of the Bundeswehr and of at least ten other armies.

With the passage of time and progress made in other types of armament, the need of abandoning classic barrel weapons becomes therefore ever stronger, which as can be seen, in principle achieved the limit of its possibilities as far back as half a century ago. This need is accompanied by ever greater efforts in work on analogous, but in terms of quality completely new solutions.

Despite that such searches have by no means been going on since the present day, I am convinced that the decided majority of experts currently questioned about the most far-reaching "successor" to the classic cannon, would unambiguously point to the electromagnetic cannon.

Paradoxically, the passage of 58 years which has passed since the described war-time work was interrupted, has not in the least deprived it of its topical interest. On the contrary, it makes us look at it with particular attention, since it is continually a source of inspiration.

Let us return therefore to the source materials from the time of the war …

Available data clearly indicates that the Germans never intended to exploit the virtues of the new weapon with the objective of re-arming their tanks—here the lack of a suitably powered energy source which could be installed on a tank stood in the way. Therefore stationary weapons came exclusively into play. The attractiveness of the electromagnetic cannon for the leaders of the Third Reich was something else; it was the only throwing weapon which could pose an effective alternative to the guided V-1 missiles and (particularly expensive) V-2 rockets (especially in the face of problems with the unrealistic conception of the V-3 cannon). One should remember that a similar potential was employed in the development, production and use of the V-1 and V-2 as was in the United States in the "Manhattan" project, and work on a nuclear weapon, mythical in terms of its scale. The creation of competition in this field opened up therefore access to enormous resources, considerably greater after all than may be obtained at present…

Irrespective of the aforementioned virtues of the "electromagnetic cannon" one should bear in mind that the high projectile speed may not only bear fruit to a high range, but also a high accuracy, since the shortening of the projectile's flight duration turns out to be crucial, particularly in the case of moving targets. It was in all probability precisely these factors which decided that anti-aircraft defence became the second area in which the new German solution was applied. However this became evident only after the war.

The next intelligence document, this time from March 30, 1944, includes information exclusively on the subject of the electromagnetic "super cannon":

The new throwing device:
Consists of a solenoid [electromagnetic coil—endnote I.W.] 900-1,000 metres long, buried at a small angle at the edges of the Cotentin district. It is connected to a high voltage 10,000 kilowatt (10 MW) industrial strength cable, installed by C.C.M.

A somewhat earlier report, from March 5, 1944, but based on data from February (DOCUMENT 10) includes considerably more information. It states among other things that these projectiles possess additional rocket propulsion (whose mass makes up a small part of the projectile's mass). This information is interesting and finds an analogy in contemporary work, but also demands certain comment. In the case of projectiles with a very high airspeed, such a propulsion may seriously influence the range, even if (paradoxically!) it doesn't give a significant thrust. It involves a physical effect relying on the fact that in the described conditions a very large part of the aerodynamic drag arises due to a vacuum formed directly behind the flying projectile. Additional rocket propulsion (and even a greater tracer) enables, apart from its principle function, this vacuum to be filled by combustion gases.

Here is this extremely interesting document:

Precisions on the characteristics of Guns and Shells.

3 Feb 44
We were able to show the outline of the "secret weapon" gun to an officer of the "Preparation of Terrain," who participated in the installation operations for these guns. He confirmed the exactitude of the outline, but as to the calibre of the rocket grenades, he thinks that they are 700 millimetres. The grenade: man's height.

7 Feb 44
The rocket shell placed on a car crosses a tunnel about 3 kilometres long (in concrete). The attraction is effected by an electromagnetic system. The power for the launching of the shell is that of 7,000 kilowatts raised to 10,000 kilowatts. The tunnel exit is inclined at a 50° angle.
Contrary to everything we have previously said, the container is propelled with the shell—the latter detaches itself only after it has been ejected.
Many accidents and breakdowns are caused by magnetic fields formed in the tunnel.
Diameter of the shell: 650[mm]; weight: 13 tons of which 8 tons are explosives. (During the construction, the excavations were camouflaged and painted in green) (…)

10A

```
TRANSLATION          SF 7196

                            (Feb. 44
                Date of    (5 Mar 44
                            (Diffusion: 8/3/44

        GERMANY   ARMY
                  FRANCE
        Various Information - Secret Weapon
```

Precisions on the characteristics of Guns and Shells.

3 Feb 44
We were able to show the outline of the "secret weapon" gun to an officer of the "Preparation of Terrain", who participated in the installation operations for these guns. He confirmed the exactitude of the outline, but as to the claiber of the rocket grenades, he thinks that they are 700 millimeters.

The grenade: man's height.

7 Feb 44
The rocket shell placed on a car crosses a tunnel about 3 kilometers long. (in concrete). The attraction is effected by an electromagnetic system. The power for the launching of the shell is that of 7,000 kilowatts, raised to 10,000 kilowats. The tunnel exist is inclined at a 50° angle.

Contrary to everything we have previously said, the container is propelled with the shell; the latter detaches itself only after it has been ejected.

Many accidents and breakdowns are caused by magnetic fields formed in the tunnel.

Diameter of the shell: 650 - wieght; 13 tons of which 8 tons are explosives.

(During the construction, the excavations were camouflaged and painted in green.)

9 Feb 44
We are concerned with a gun, firing shells which functions not by flashes or explosion, but by emission of a cold wave - reaching 160°.

About 100 guns have already been set and numerous emplacements are being arranged.

(These guns will only be used in case of debarkation).

Intelligence on New Weapon Emplacements in the Hilly North Section.

CONFIDENTIAL

10B

Gleaned from a train conversation with a workman, employed by the Germans, 3 large operations (secret weapon) are in progress, in the Cherbourg vicinity: At Valognes, Couville and Martinvaast.

11 Feb 44
From a conversation held at Moissac with the workers, it appears that all along the coast from the cape of the Hague to Frouville there are being constructed platforms of reinforced concrete destined to received pieces of long range artillery. According to one of the workers who seemed to be chief of the crew, or a foreman, these platforms have a surface of 100 square meters, the lower portion is arranged in strong rooms (3 or 4), intended to receive the munitions, a command post and, in the case of the largest, probably an electro-guns group, since all this has to do with an electric cannon capable of bombing the English coast and a considerable distance inland.

Allied airplanes come frequently to bomb and destroy some of these works, which must be reconstructed at top speed. To accomplish this the occupying authorities offer 19 francs; 90 centimes an hour to the workers, who are volunteers for the city and port of Cherbourg.

It is estimated that there is a layer of 6 meters 50 of reinforced concrete on top of each strong room.

24 Feb 44
We are concerned with a provisional gun, electrically charged and fired. The barrel alone is 48 meters long. It launches a projectile having a diameter of from 30 to 35 centimeters, and a length of 4 meters and a weight of 450 kilograms. Range: 240 kilometers. Firing cadence: 1 shot every 30 or 45 seconds. The bore of the barrel which serves at the same time for the propulsion, is changed during a 3 hour delay.

These guns are in shelters, 120 to 450 meters, having a pyramidal or ovoidal roofing - very thick. The gun placed at ground level having only the battlement embrasure uncovered.

In the interior are found:

    Central Electric Power Station and
    Munitions Depot.

There is no detonation, only a mewing announces the departure.

Many guns are set up. German industry is making a considerable effort to multiply their numbers.

In Calais and its vicinity, the Germans blow up the fortifications they had constructed. They have prepared for the flooding of this territory in the case of an invasion.

- 2 -

CONFIDENTIAL

10C

28 Feb 44
An informer who helped build many emplacements for shell launching rockets, claims that there are 38 of these between Dieppe and Dunkerque.

These apparatuses comprise of a hole more or less inclined of a diameter and depth X, in which a coating of concrete is poured, in this shell is installed a guide (not of the gun) which brings about the firing of the rocket shell.

This rocket shell, weighing 5 tons at departure, comprises of 4 tons of explosives on arrival at target (1 ton serving towards the engine propulsion). Its range is 400 kilometers approximately, the levelling deveiation course is that of 10 kilometers - launching rockets which cannot be oreinted.

An underground casemate - containing projectiles, is constructed under each rocket thrower.

O-2 Comment:

Report consists of conglomerate facts poorly expressed and arranged and for the most part a probabl

Special attention however is invited to following extracts:

1. Under date of 7 Feb.: Tunnel - it is inclined at a to degree angle. Weight of shell 13 tons on which 8 tons are explosives.

2. Under date of 8 Feb.: large site at Martinvast and supply site at Valognes are noted. Couville is often quoted with reference to Martinvast as it is the nearest town of fair size.

3. Under date of 24 Feb.: Firing cadence: 1 shot every 30 or 45 seconds—replacable bores.

The extract data listed above are not to be accepted as true, but represent the more important statements in the report. They should be weighed and compared with more credible information.

---

11 Feb 44
From a conversation held at Moissac with the workers, it appears that all along the coast from the cape of the Hague to Frouville there are being constructed platforms of reinforced concrete destined to receive pieces of long range artillery. According to one of the workers who seemed to be chief of the crew, or a foreman, these platforms have a surface of 100 square meters, the lower portion is arranged in strong rooms (3 or 4), intended to receive the munitions, a command post and, in the case of the largest, probably an electro-guns group, since all this has to do with an electric cannon capable of bombing the English coast and a considerable distance inland (…). It is estimated that there is a layer of 6 metres 50 cm of reinforced concrete on top of each strong room.

24 Feb 44
We are concerned with a provisional gun, electrically charged and fired. The barrel alone is 48 meters long. It launches a projectile having a diameter from 30 to 35 centimetres, and a length of 4 metres and a weight of 450 kilograms. Range: 240 kilometres. Firing cadence: l shot every 30 or 45 seconds. The bore of the barrel which serves at the same time for the propulsion, is changed during a 3 hour delay (…)
These guns are in shelters, 120 to 450 metres, having a pyramidal or ovoidal roofing—very thick. The gun is placed at ground level having only the battlement embrasure uncovered.
In the interior are found:
Central Electric Power Station and Munitions Depot,
There is no detonation, only a mewing announces the departure.

Many guns are set up. German industry is making a considerable effort to multiply their numbers (...)

The document on the opposite page creates more questions than answers. If the information presented can be treated at all as fact, then what happened to these gigantic cannons when the Germans were forced to abandon this territory? Could such cannons ever have existed?

Their construction would have been an unquestionably very difficult task, yet one should affirm decisively that they could have existed. This challenge, despite its spectacular character, didn't exceed however the possibilities of the German economy.

If the underground facilities (shelters) accompanying these cannons had been blown up and then camouflaged by the retreating German army (and this would surely have happened), then it is not at all obvious that later they would have been uncovered and studied. It is enough to look at the history of these German underground industrial facilities, the entrances (the mouths and openings of communication and installation tunnels) to which were blown up and camouflaged, and that it was possible to professionally carry out such tasks. As a result many such facilities to this day have not been discovered, and often not even located. These are currently gigantic tombs, as in such cases the people (prisoners) working in them were never set free. Such principles of conduct were the result of a special order given by the Reichsführer SS. Departures from these principles were very rare.

So could the Allies have got to know only part of the truth about German electromagnetic cannons?

Other data also indicates this, as a special report, referring to work on the electromagnetic cannon, prepared at the beginning of 1946 by a special committee from the US Military Intelligence[55] also contains no information about which we now know from other sources only in the present day. In actual fact this is an account of the role which was fulfilled by a certain establishment—one of many—engaged in this work, the Berlin laboratory of the Heereswaffenamt (the Army Weapons Office). We currently know that a significant role in the aforementioned project (by the way, even its code-name remains a secret) was fulfilled by the establishment in Peenemünde—after the war projectiles for the experimental electromagnetic cannon were found there. The work described in this chapter therefore constitutes one of many examples of an issue which we know existed, but for a fact we only know part of the truth. What then does the aforementioned report specifically include? It can be divided into two basic sections—a section including analyses of a general character, as well as an account of work on an anti-aircraft electromagnetic cannon (relatively small) with a very high rate of fire.

Here are extensive excerpts of the American report:[55]

Attempts to replace powder by electricity are not new. Many schemes were tried by this laboratory, and the method developed during the last war by Fauchon—Villiplee was finally adopted. The great difficulty is, of course, the developing of enough power to launch a projectile. It has been possible here to accelerate a body weighing 12 g to 1,100 m/s in a 2-meter tube, which corresponds to an acceleration of about 30,000 times gravity. Coupling two such tubes was not very successful, 1,200 m/s being obtained.

The projectile is accelerated by a "linear motor," which consists in its simplest form of two conductors (parallel rails) across which the projectile completes the circuit by means of fins attached to its rear. When current passes through the circuit, the projectile travels forward. The conventional electromagnetic equations apply to the process.

The principle was demonstrated to me in a 50 cm. tube with the projectiles submitted with this report. After firing, the edges of the copper fins are melted. The muzzle velocity in the demonstration was low.

The difficulties in the way of a suitable power source are formidable. Lead-sulphuric acid storage batteries with very thin plates are used for experimental work; 9,000 kW is available from this source. Condensers to give 2,000 volts with 24,000 microfarads capacity are also installed. The great advantages of the electrical scheme would be:

1. gun barrel unnecessary;
2. higher velocity than is attainable with powder;
3. higher efficiency than with powder;
4. lower energy cost than with powder.

Work was about to begin on a projectile 1 cm. in diameter, weighing 60 or 70 grams. An A.A. gun was planned (see the attached report), which would have required a maximum current of 1,500,000 amperes at 1,300 volts to launch a 4 cm. projectile [a power somewhat less than two gigawatts!—note I.W.]. The power was to be obtained from three unipolar generators weighing 150 tons each.

Projectiles electrically launched would have to be fin-stabilised since they could not be rotated. Fin-stabilisation presupposes wind-tunnel work, which explains the close co-operation between this target and the Peenemünde wind tunnel personnel at Kochel.

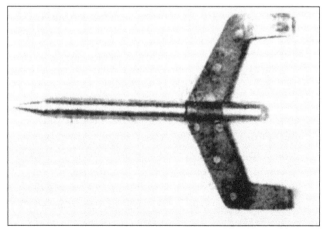

A projectile for the electromagnetic cannon, found in Peenemünde
(via Alsos)

OUTLINE, 4 CM A.A. ELECTRIC GUN
I PREFACE
Speeds of powder propelled shells are limited. Theoretically they approach, according to van Langweiler, 2,810 m/s. This figure was substantiated by him experimentally when he attained 2,790 m/s. This velocity, however, is only realisable for a certain mass. In the case of heavier bodies, the ultimate velocity is less—in any case under 2,000 m/sec. Nonetheless, a great many practical factors prevent the attainment of these theoretically possible (with powder propelled shell) muzzle velocities. The only known way of reaching higher velocities, e.g., accelerations, is by electrical means. Quite a few electrical systems have been proposed and published, particularly in the German patent literature. The most promising system is that of the Frenchman, Fauchon-Villiplee, and models built and experiments made of them by Gesellschaft für Gerätebau. Experiments were still being conducted just before the end of the war, and preliminary pertinent results were published in this company's reports, the latest of which was issued 18 January 1945.

Electrically there are no means of storing energy in such a concentration as is offered with powder. For this reason, within the velocity range of the powder gun, the electrical gun cannot compete with the conventional gun. However, for shell velocities above the conventional A.A. gun range the electrical gun has its field. This field has been considered to embrace the following: anti-aircraft guns, long distance artillery, or launching devices for large rockets. Initially, consideration of the electric gun was directed towards the achievement of a long distance gun wherein a shell would be fired with a muzzle velocity of 2,000 m/sec. This idea was dropped, however, with the advent of the V-1 and V-2 weapons, or laid aside, at least until such time as much higher velocities are attained.

Due to the recent large advances in aircraft ceilings and speeds and the inability of conventional anti-aircraft weapons to counter these improvements, the electric gun becomes a possibility for anti-aircraft. By increasing the muzzle velocity, the probability of hitting the target is improved and the range is increased. If the guns have fixed emplacements, this cannot be considered as a ruling disadvantage inasmuch as the majority of "Home A.A." is of this type.

Launching ramps for large rockets are a possible use for the electric gun principle. These rockets are unable to withstand large accelerations because of their delicate control mechanisms and, therefore, the length of their take-off ramp is relatively long in order to obtain required take-off velocity and still stay within maximum acceleration limits. Also, it can be stated that the more controllable the acceleration can be made, the shorter the launching ramp, that is, the flatter the acceleration-time or distance curve can be made, the more efficient the cycle from the point of view of length of launching ramp. Thus the electric

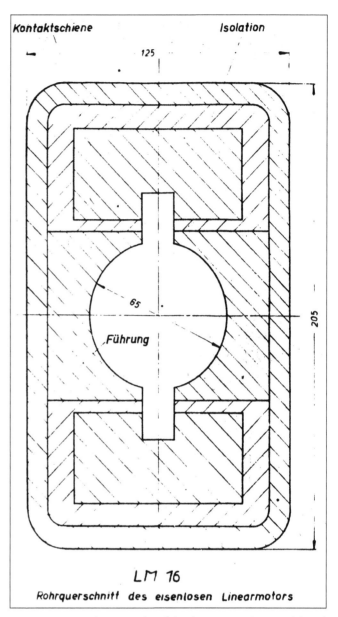

A cross section of the electromagnetic cannon's barrel

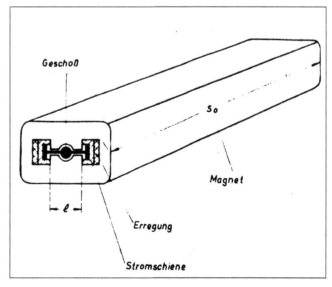

A general diagram of the German electromagnetic cannon's construction
(via Alsos)

gun can be considered in competition with other methods for the launching of large rockets which have maximum acceleration limits.

II RECOGNITION AND REQUIREMENTS OF PROBLEM
The rapid progress of science made possible the practical consideration of the electric gun. The Gesellschaft have presented a published preliminary plan, dated 9/10/44, for an electric anti-aircraft gun with a muzzle velocity of 2,000 m/sec and an average rate of fire of 6,000 rounds per minute. On the basis of this preliminary plan, discussions with the Chief of OKL (Oberkommando der Luftwaffe, The Air Force High Command), TLR and A.A. Defences have been held, and the following requirements were laid down:
A) Muzzle velocity: 2,000 m/sec.
B) Pay load of shell: 500 grams [this refers to the mass of explosive—in the next section of the report the projectile mass is given, which amounts to 6.5 kg—note I.W.],
C) There should not be a rapid firing of six rounds per barrel, as proposed in original plan, for it is feared barrel sway may be excessive. Instead, six guns firing simultaneously should be linked to one power unit.
D) The battery should have a burst of fire every five seconds, which gives the battery a firing rate of 6 x 12 or 72 rounds per minute.
E) The gun barrels, according to calculations, should have a length of 10 meters and for reasons of time be built into a standard carriage. A suggested one is the 12.8 cm A.A. Gun Carriage.
F) The Chief of OKL-TLR requests that an experimental plant be erected in the shortest possible time consisting of a complete power unit and an experimental gun with three barrels. (...)

V THOUGHTS FOR THE TACTICAL USE AND FOR FURTHER DEVELOPMENT OF THE GUN
From the following considerations it seems to be unsuitable to install the six guns belonging to a battery on the circumference of a circle with max. 40 m diameter, and to connect these guns only to one set of machines:
1) If the machines fail, the whole battery forcibly goes out of action.
2) Installing these guns on such narrow space, they are—and particularly also the connections—not protected enough against air raids.
3) The heretofore required connections of 20 m length need too much raw material.
4) The development must consider the transportability of the whole installation. With connection to one set, a change of location is impossible; so it should be attempted right from the start to furnish each gun with its own set of machines. In addition, it would reduce the weight of the machine installation.
It had been mentioned previously that the decrease of

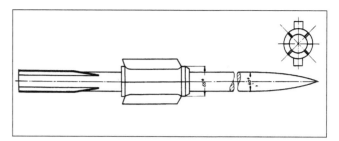

The final projectile for the electromagnetic anti-aircraft gun

rpm of the generators after six rounds amounts only to 4.8%. It could easily be doubled, i.e., 10%. This would mean an energy economization of 50% and, therefore, also a reduction of weight of 50%. If each gun gets its own machine set, this later would require only 1/6 of the so far calculated energy-content, so that totally the weight for the generators of the installation would be reduced to 37.5 tons. [a fundamental restriction on the system's mobility—note I.W.]
The new machines planned further on run only with 200 m/s angular speed [on the circumference?—I.W.]. This angular speed should be easily increased to 300 m/s—especially with unipolar machines, which would correspond to an increase of the energy-content per kg rotor-weight of more than double. This would mean a further decrease of the weight of 50%.
The testing of these ideas still requires intensive work for several weeks. One could, in our opinion, achieve, under circumstances, that the first employed guns are transportable.

So much, as far as the American report from 1946 is concerned ... A supplement to the information included in it is comprised by that referring to tests carried out in Peenemünde (according to some sources also near the town of Schlosskranzbach). It follows from it that a modified version of the described anti-aircraft cannon was also constructed, with a shortened rail, or "barrel" (to 8 metres) and firing a lighter projectile, with a mass of 2.88 kg. However a considerably greater muzzle veloc-

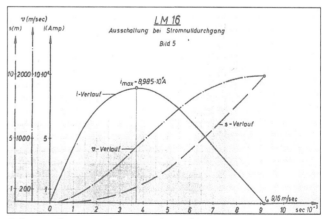

A diagram illustrating the course of one of the experiments. From this it follows that the projectile reached a maximum velocity of 2,000 m/s

(via Alsos)

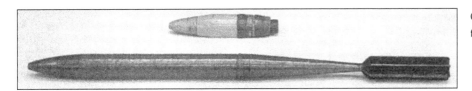

One of the projectiles designed for the 40 mm electromagnetic cannon

ity was gained, exceeding 2,500 m/s. Therefore the estimations formulated in 1944 turned out to be true. This cannon was supplied by a set of capacitors, giving a short impulse of current with a gigantic, to put simply, intensity of 3 million Amps. The capacitors were in turn supplied by a generator, driven by two turbine engines. I have recently found in the documents of the Alsos mission[52] an original German report referring to this work. As far as its contents are concerned, it agrees with the American study (or rather translation). It contains however original engineering drawings, depicting the final 40 mm calibre ammunition. These plans have been reproduced in this book. A glance at the drawing of the final projectile allows one to guess at why it was worked on in Peenemünde. It is a miniaturised version of the projectile for the V-3 cannon, tested after all on a neighbouring island. Or perhaps the electromagnetic cannon was supposed to be one of the alternatives to the multichamber cannon?

THE NEW TOOLS OF BLITZKRIEG

The armoured vehicles of the Third Reich are rarely described as a domain in which any significant breakthrough was accomplished. This was the case however, though it rather referred to the final period of the war. The designs created at that time sometimes bring to mind the state from the 1960s or even 1970s, and not the 1940s. They considerably outdistanced the analogous accomplishments of other countries, although these are very little known facts.

At the beginning of this chapter I will take the liberty to quote one of the analyses of American intelligence from 1945, as it contains some very interesting remarks about the trends in the field of tank construction that appeared in Germany in the final months of the war. Any kind of information on the trends from this admittedly decadent, yet remarkably important period very seldom appears, since it simply foreshadows what was yet to take place in the field of armoured vehicles after the war, often several decades after the war! Here is the quotation:[56]

The increase in firepower of German tanks (by installing higher calibre or longer guns) did not come to an end along with the arrival of 1945. Plans were being elaborated to employ more powerful armament in all tanks and self-propelled guns.

From year to year an ever greater percentage of armoured vehicles were equipped with guns of limited movement in the horizontal. These activities were supported by infantry and artillery divisions, but encountered the opposition of the armoured forces themselves.

Close to being introduced to armament was a fully tracked, light base vehicle, on which various types of guns and howitzers would have been mounted. It was to be used by self-propelled field artillery units.

The trend to build bigger and heavier tanks was practically stopped at the end of the war. The development of 150 and 200 tonne vehicles, commenced in the years

The Pz.Kpfw IV was for a long time the main tank of the German Army

The Ferdinand/Elefant self-propelled gun, the first mass-produced combat vehicle with electromagnetic power transmission

1942-1943, proceeded very slowly due to a lack of interest at the highest level. Many people in the ground forces and armoured vehicles industry considered that the "Tiger II" ("Königstiger") tank, due to its size and weight, demanded a greater production effort than would have been justified by its value as a weapon.

The developmental period of a German tank, from the start of its development to the commencement of production, lasted around 2.5 years before the war and during the war was shortened to approx. 15 months.

Hulls

A review of the hulls of modern German tanks creates the impression that the ballistic factors were treated with high priority during development. Flat plate [rolled] is almost exclusively used, and all surfaces are sloping to the greatest possible degree. Plates of all sizes are connected in such way that they have dovetailed with each other, in addition the joints have been reinforced with special profiles.

This must have led, in many cases, to a considerable increase in the production costs. As yet there is lack of data, on how much these methods improved armour protection. [The American officer perhaps didn't know, but this resulted from very specific circumstances. Along with the introduction by the Soviets of the I.S. (Joseph Stalin) series of tanks armed with a 122-mm guns, an abrupt increase in the projectile weight ensued. Even if it did not penetrate the armour, the force of the explosion was so great that it often caused cracking of the welds—note I.W.].

There is nothing to indicate that armour significantly thicker than that used in the "Tiger II" tank was considered. The super heavy tanks, whose construction was being developed, had armour plates approx. 30 % thicker than in the "King Tiger."

Power transmission system

There existed a very clearly defined trend to replace petrol engines with diesel air-cooled engines. Such engines were to enter production in 1945, for 15-20 tonne vehicles, however they were developed for larger vehicles and were far from being mass produced.

The Tiger I tank

The application of coolers located outside the watertight section of the hull (engine compartment) in the "Tiger" and "Panther" tanks has many advantages and is worth consideration in the development of new vehicles.

Fully automatic gearboxes and [hydrokinetic] torque converters were to be found under development. The German engineers considered them promising.

Hydraulic control systems transmitting power to particular tracks through the intervention of hydraulic pumps [so-called hydrostatic turning mechanisms, enabling turns to be carried out at any given radius, without the need for uncoupling the kinematic link with the engine—note I.W.] were built and tested. The results were considered highly satisfying. (...)

Driving system

Large diameter wheels are used in all modern vehicles. Suspension with the wheels overlapping each other, employed in the "Panther" and "Tiger I" tanks and half-track vehicles, caused many problems due to additional resistance appearing when the wheels sank into mud, or when they iced over. The creation of driving systems, in which the possibility of the wheels rubbing against each other would be eliminated, was considered essential.

Wheels with rubber bands on their circumference, enclosed in turn with steel rings, were very positively evaluated. This simultaneously ensured protection of the rubber layer and significantly extended the wheel lifetime.

Bakelite and other composites used in the suspension [as linings for frictional shock absorbers, still used in this role today—note I.W.], proved to be correct in many applications and require further examination.

The use of suspension in the form of torsion bars has not become widespread in the last few years [this relies on a damper spring in the form of a rod (rod spring) performing the role of the shock absorber and the torsional shaft being subject to turning. It passes through the hull near its bottom and on one side is rigidly connected to the vehicle's side, and on the other passes through a bearing to the outside, where it ends in a rocker arm, on the end of which is located the wheel. This solution is currently standard and is slowly being superseded by hydro-pneumatic suspension (as in modern buses)—note I.W.]. The Germans aimed to create a cheap suspension, but which would not take up space inside the hull.

Turret

The turret design of the "Tiger II" vehicle was considered better than any other type and according to the same criteria the turret of the "Panther" tank was redesigned. (...)

Stabilisers for gun-sight devices and guns were under development [enabling effective firing to be carried out on the move—note I.W.]. (...)

Miscellaneous

In the "Tiger II" tank all devices, including the gun suspension, were fixed only to the floor or roof. This practise seems worth copying.

The tanks were made ready to overcome water obstacles at a depth of 6 metres. However this was never employed in combat operations, since it was not that imperative, so as to justify additional costs and labour-consuming preparations.

The development of armoured vehicles was realised simultaneously in four directions:

1. Already produced tanks were modernised, first and foremost the Pz. Kpfw. V "Panther" and Pz. Kpfw VI "Tiger."
2. The conception of super heavy tanks (the prototype of the "Maus" tank and study design of the "Ratte" tank) was verified.
3. Tanks were designed being a continuation of a previous developmental line: the Pz. Kpfw. IX and Pz.Kpfw. X.[57]
4. Prototypes of the "E" series of vehicles were constructed, separate and alternative in relation to that in point 3.[58]

It could appear that the most avant-garde conceptions would come about with the participation of designs from points 3 and 4, however this is not entirely true. For example, it was planned to significantly modernise the "Panther" tank to the extent that it would not be much inferior to its likely successors (the E-50, Pz. Kpfw. IX). It was planned to modify the armour, employ a system enabling effective firing to be carried out both at night and on the move (a gun stabiliser), replace the conventional petrol engine with an entirely new generation engine with the power increased almost by half (manoeuvrability was the weakest side of tanks from that period), as well as equip the "Panther" with a radically new power transmission system and turning mechanism. Despite appearances this work was well advanced and came close to being fully realised. The main obstacle was the collapse of the economy at the turn of 1944-45, caused by air raids, and not technological problems. One of the few components that would have been left unmodified was to be the gun. Despite having a lower calibre than for example in the "Tiger" tank, it was considered sufficient for a vehicle of this weight category. In this respect the "Panther" had nothing to be ashamed of even in confrontation with the heavy Soviet IS-2 tank (the Joseph Stalin) armed with a much larger cannon. I will take the liberty to quote some excerpts from an article touching upon this problem, written by an expert on Soviet armoured vehicles. It was dedicated to the IS-2 tank:[59]

Paradoxically the greatest weakness of this tank was its armament. One of the requirements made to the vehicle designers was precisely armament that permitted all currently employed and future enemy tanks to be engaged in combat. As we know from articles in Nowa Technika Wojskowa from 2 and 3/01, various types of guns were

The Tiger II/Königstiger tank, the version with Porsche's turret

tested on the IS, preferring initially however the 122-mm D-25 gun. In reality it was characterised by considerably better armour piercing parameters than the 76 and 85-mm guns available at the time, all the same compared to the enemy's gun it wasn't at all a revelation. One should remember that the D-25 had the ballistics of an A-19 corps gun, designed above all for combating targets with indirect fire, when shooting at fortifications the initial velocity of the projectile didn't play such a great role. Combating armoured targets was an altogether different affair. Here the initial velocity has a decisive significance on the kinetic energy which the projectile has at the moment of hitting the target. A 25 kg projectile from the D-25 gun fired with an initial velocity of 781 m/s had a similar ability to penetrate armour as a 4.75 kg sub-calibre projectile fired with an initial velocity of 1,120 m/s from the 75-mm gun of a "Panther," not mentioning the 88-mm long-barrelled "King Tiger" and "Jagdpanther." Thus the IS-2 gun didn't at all have such fantastic capabilities—undoubtedly not at the typical distances at which armoured encounters took place—that is to say up to 1,000 m—penetration right through the "Panther" after hitting the frontal hull plate was unrealistic. I do not deny however that even without armour penetration the impact of a 25 kg projectile (also high-explosive, H.E.) eliminated a tank and above all its crew, from combat for some time. The situation changed along with the introduction of the hollow-charge projectile, but this occurred after the war. (…)

The next matter associated with the armament of the IS was the ammunition reserve and rate of fire. The ammunition reserve amounted to only 28 pieces, i.e. almost three times less than in the "Panther" and "King Tiger"! Additionally the ammunition was divided, which caused the gun to be reloaded in two cycles, which was in any case unavoidable as the complete round weighed over 40 kg. This in turn limited the rate of fire to 2-3 shots per minute, while the analogous rate for German tanks, whose guns were loaded with complete rounds, was two to three times better. And this cannot be overestimated on the battlefield. Taking into account that at a distance of 1,000 m or less, the armour of the IS did not protect it against 88-mm KwK 43, 75-mm KwK 44 as well as PaK 40 projectiles, this caused that the tactical advantage was on the enemy's side."

All of this does not mean however that the fire power of the "Panther" and other already produced tanks was to remain at the same level. As was mentioned in the previously quoted intelligence analysis, tank gun stabilisers had been worked

The "Panther" medium tanks

on. Moreover the aforementioned tanks were equipped with night vision gun-sights on a small scale (operating in the infrared band, the so-called active). The two types of devices mentioned below were designed for this out of the dozen or so types that arose in the Third Reich.

Both the F.G. 12/50 and F.G. 12/52 were initially mounted only on armoured personnel carriers (equipped as such they were code-named "Falke"), soon however a modification was made designed for the "Panther" tank, which was code-named "Puma." The sight which until then had been mounted on 7.92-mm machine-guns was adjusted for the 75-mm tank gun. A small number of these systems were used in combat, to very good effect, also decided undoubtedly by the surprise factor. In spite of this, probably for purely irrational motives, this equipment also had many opponents in the Wehrmacht.

The first serious clash of opinions occurred in August 1944, during one of the staff briefings of Ground Forces Command (OKH), in which the future counteroffensive in the Ardennes was planned. Already on the basis of initial engagements on the Western Front the majority of generals concluded that effective operation of armoured units would only be possible at night, obviously on condition of being equipped with the appropriate gun-sights and observation systems. It mainly concerned the F.G. 12/50 and F.G. 12/52 sets. For the facts themselves were shocking; only during one week of July that year (23-31 VII) the Allied air force destroyed approx. 400 German tanks. In connection with this not only were night vision systems prepared, but even special uniforms masking soldiers in the infrared band (they turned out to be unnecessary, since the enemy did not have such devices). Without doubt this would have been the most befitting of possible moves, potentially a classic example of gaining the upper hand through the astonishing application of a new means of warfare at the opportune time; the counteroffensive in the Ardennes was after all to be fought in winter, when the conditions for observation at night were at their best.

One of the high ranking general staff officers present at the aforementioned briefing stated, however, to the surprise of the others that:[63] "Gentlemen, I cannot understand what all your modern stuff is about, the front is happy with what we have done so far."

In response several generals left the briefing room and the costly night-vision devices, which had already been managed to be mounted on the vehicles were dismounted and stored in

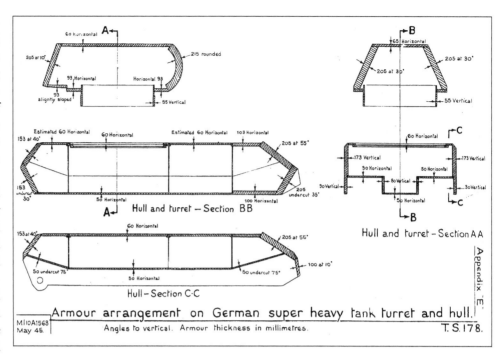

Cross sections of the Maus tank

The Maus in late 1944 at the Kummersdorf firing range

The "Maus," a portrait photo

one of the inactive mines in Austria. Only a symbolic number were left, which could not have a noticeable influence on the situation's development. As I have already mentioned, from available information it appears that this equipment was used in combat only once in the West and no earlier than on April 9, 1945, during fighting near the towns of Wietersheim on the Wezer, when several Panthers decimated a group of British tanks, defending a bridge head. The new invention could no longer reverse the course of the war… From American intelligence materials it appears that on a slightly larger scale and in conjunction with the "Uhu" system described later, it was used again on the Eastern Front. It is not known exactly how many "Panthers" were upgraded in this way, in all probability it was one or two companies. The results of their operations exceeded the boldest expectations, among others during just one night 67 "blind" Soviet tanks were destroyed.[64] It is not difficult to imagine what would have happened if in accordance to plan not a few companies but 2-3 armoured divisions had been rearmed in the described way. The course for example of the Soviet January offensive would not have been so obvious.

Some sources state that a certain number of tanks and MP-43 carbines were equipped with night-vision sights and used in action as early as March 1945, during the last German counteroffensive in Hungary (operation "Frühlings-erwachen"). This was in all probability at Lake Balaton. All in all over 1,100 pieces of the F.G. 12/50 had been produced up until this time.

One of the little known fields associated with the development of armoured vehicles in the Third Reich is the extensive (though seemingly not very fascinating) issue of hydraulic power transmission systems. This refers both to some kind of hydrokinetic clutch (non-rigid coupling), transmitting the torque from the engine to the gearbox, as well as analogous device, ensuring a smooth alteration in the vehicle's turning radius—almost without loss of power as opposed to solutions to date, in which the driver disconnected the power of one of the caterpillar tracks in order to turn. This was aimed on the one hand at reducing power loss. For example, the vehicle's acceleration increases significantly, when the kinematic connection with the engine is not uncoupled during gear changing. On the other hand this was simply a result of seeking the most sensible way of replacing the conventional clutch, which has its durability limitations and cannot operate at any given load—e.g. as a rule there is no rigid coupling in locomotives—if it was connected at high engine revolution speeds it would simply burn out, and the train would not move. In hydrokinetic equivalents on the other hand there is no problem with friction, since there is no rigid contact surface. The simplest variant of a torque converter of this kind is a type of dual turbine. In an approximately cylindrical hermetic casing filled with oil arc situated two rotors equipped with suitable rings of blades (not connected to each other, although located on one axis). When one of them starts to rotate it causes movement of the hydraulic liquid, which conversely forces the other rotor to move. Contrary to appearances the losses are not high and with the exception of low revolution speeds do not exceed 2-4%.

The unfinished hull of a E-100 tank, on a special transportation platform

This type of device was not a German invention—the Americans and British had worked on it before the war—but the German company Voith from Heidenheim was the first to commence the development of a whole series of models, destined for combat vehicles. Hydrokinetic power transmission was to be acquired by the "Panther," "Tiger II"/ "Königstiger" (both versions) as well as the E-25. This work was not fully carried through –being discontinued on the strength of an administrative decision from August 8, 1944. If however we take into consideration that work was simultaneously carried out on electric power transmission (a current generator /elec-

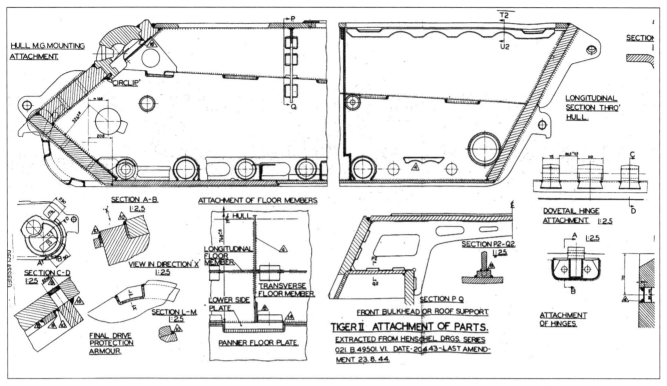

Cross sections of the hull of the Tiger II tank

tric engine, used in the "Ferdinand"/"Elefant" self-propelled gun and "Maus" super heavy tank), hydrostatic turning mechanisms were developed and it was also intended to employ diesel engines relatively modern at that time, then this would be rather suggestive of the 1970s of the 20th century.

If on the other hand World War II is concerned, then the only hydro-kinetic power transmission produced by Voith which found wider application, was a system designed for locomotives, carrying a power of up to 1,800 HP.[56,60]

Apart from hydrokinetic and electric power transmission, there arose yet another alternate, allowing the engine's growing power to be more fully exploited. It was an automatic gearbox with magnetic (electromagnetic) clutches. It was developed by Z.F.-Zahnradfabrik in Friedrichshafen and designated the G/EV/75. It had six gears. Its use was considered in new versions of the "Panther" and modified versions were to be installed in the E-10 and E-25 vehicles.[62]

As far as the engines themselves were concerned, all tanks brought into armament were propelled by petrol units. In the nearest future it was intended to replace them with Diesel engines.

At the same time Daimler-Benz, BMW and Heinkel-Hirth hurried along a completely different path, constructing propulsion units that constituted the next generation in relation to Diesel engines, namely turbine. This was a step, which combined with others would bear fruit to a true revolution on the armoured battlefield. For example an engine was developed that generated 1,000 HP, whose turbine was only 32 cm in diameter(!). The "Königstiger" (Pz Kpfw. VI) tank was among others to acquire such a propulsion unit, designated the GT-102.[61] The Germans had managed to overcome several fundamental

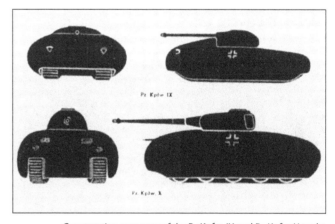

Construction concepts of the Pz.Kpfw. IX and Pz.Kpfw. X tanks

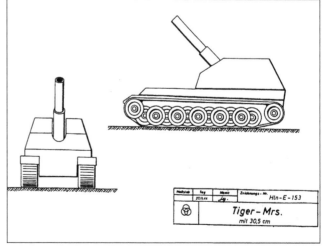

A draft of the 305-mm self-propelled mortar, mounted on the Tiger's hull

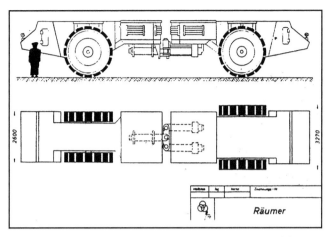

The "Räumer-S" (original drawing)

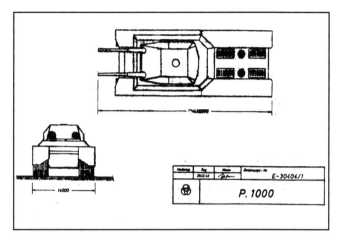

A draft of the "Ratte" tank (original drawing)

problems—among other things they mastered the production of turbine blades "empty inside"—for the purpose of cooling, and developed effective ceramic protective coatings for these blades, which increased their lifetime by approx. ten times.

The only shortcoming of such an engine was the on average two-fold fuel consumption than in the case of a conventional equivalent, however the advantage over the petrol piston engines used at that time was colossal. First and foremost it was possible to detect a marked increase in power with a simultaneous decrease in the engine's mass and capacity. In addition the cooling system was eliminated and there was a lack of engine vibration—the latter influenced after all aim-

ing. This overcame the chief limitation of World War II tanks, giving the impression that the three main "coefficients": firepower, armour protection and manoeuvrability were incompatible with each other. With the exception of light tanks, as a general rule the latter fell victim to compromise, for example the "Tiger" propelled by a 600-700 HP engine (depending on the version) had a road speed of only 30-40 km/h.

At the end of the 1960s of the 20th century a restricted military circle experienced a presentation of the "revolutionary" KPz-70/MBT-70 German-American tank (not introduced to mass production). This "super tank," half developed after all by the Krauss-Maffei plants dominating during the

The "Räumer-S," captured by the Americans

war, possessed among other things a high calibre stabilised gun, turbine propulsion, hydrokinetic power transmission and hydrostatic turning mechanism as well as a defence system against weapons of mass destruction. All of this seemed entirely new—in reality this technology was already quarter of a century old and wasn't the only such case.

Simultaneously with the modernisation of already existing tanks, yet other conceptions were developed:

Two new types were designed (the Pz. Kpfw. IX and Pz. Kpfw. X), which were to be successors to the "Panther" and "Tiger." They were to be characterised by fully cast hulls and turrets, in the form of monolithic elements. Little is known about them, only their approximate plans.[57]

A whole series of "E" fighting vehicles was developed, in addition:

The "Maus" and the "Ratte" were designed, the so-called super heavy vehicles and enlarged developmental versions of the "Tiger" and E-100 tanks.

The "E" series represented a new generation of combat vehicles consisting of five types:[58]
1. The light E-10 with a mass of 10-15 tonnes.
2. The E-25 vehicle with a mass of 25-30 tonnes, approximate to the "Jagdpanzer 38(d) Hetzer" tank destroyer.
3. The E-50 tank with a mass of approx. 50 tons, which was to replace the "Panther" tank.
4. The E-75 –the successor to the "Tiger." This would have been the first mass-produced tank with a hydromechanical power transmission system.
5. The super heavy E-100 vehicle, with a mass of 130-140 tonnes.

The last was admittedly the least necessary from a military point of view, but work on it was the most advanced.

Production of the E-100 was abandoned shortly before completion of the first prototype, the remainder were scheduled (in 1943) to be introduced to armament at the turn of 1944 and 1945. The E-100, which was named the Adler ("Eagle") was 40 tonnes lighter than the "Mouse," though almost the same turret and thick frontal armour was to be used. The hull, though somewhat lower, was even wider—despite having still heavier armament! The main reason for this was a greater concentration of heavy armour at the front. Its decidedly strongest feature (though of doubtful usefulness on the actual battlefield) was its armament, which was to be a 150-mm gun even heavier than in the "Mouse." A contrast to this was the propulsion, the engine from a twice lighter though still not very mobile 700 HP Königstiger, which would have enabled a road speed not much exceeding 20 km/h to be developed. In June 1944, at the proposal of Hitler himself, it was ordered to discontinue work on assembly of the Adler prototype, though it was continued at a very slow pace (the company Henschel in which assembly took place assigning only three people to this task, a lack of spare parts) right up until January 1945. A direct competitor to the E-100 was the super tank "Maus" with a combat mass close to 200 tonnes, ten trial specimens of which

An unfinished prototype of the 170-mm self-propelled gun, using the enlarged hull of a Tiger II tank. Note the two-level arrangement of the interior.

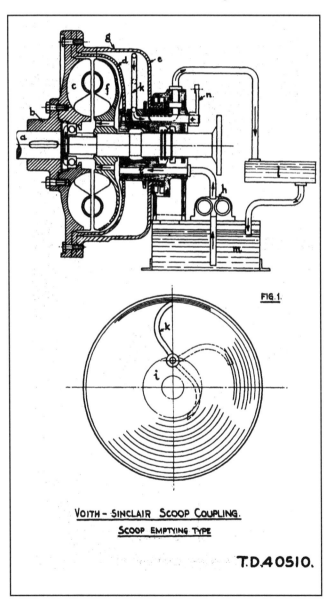

A cross section of the hydrokinetic power transmission system, developed during the war for tanks by the company J. M. Voith

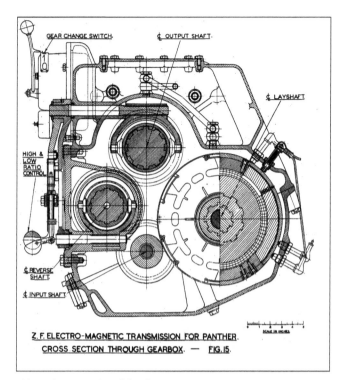

A lateral cross section of the electromagnetic power transmission system for the Panther tank

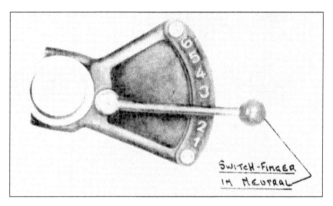

A fragment of the above system, the gear-changing switch

and the Berlin company Altmarkische Kettenfabrik was chosen to assemble completed vehicles.

At the same time it is worth observing that both the Maus as well as the remaining super heavy tanks were not to be conventional "front breaking" (assault) vehicles," though with respect to their construction lay-out and appearance they did not differ much from standard tanks. Due to their forecasted low manoeuvrability (both the traction parameters themselves, fuel consumption as well as cross-country capabilities and bridge and road passing capabilities) they were simply not able to carry out the predicted tasks for "normal" tanks. Instead of forming standard combat formations they were to rather constitute mobile fortified lines, continually on the move and able to redeploy strongly armed and armoured bunkers in a given direction. In some sense, as in the case of the giant and expensive, but of doubtful effectiveness railway guns, this was an attempt to return to the positional World War I, or perhaps rather an expression of certain sentiments and personal weaknesses.

In any case here the mistake also let itself be known of belittling its purely defensive capabilities—the first version of the "mouse" was not even equipped with a single machine gun! It was a characteristic reflection of Hitler's perception of warfare.

However many original and interesting solutions were used in this tank, the most important being without fail the power transmission system. It was an electromechanical system. This relied on the power from the engine to the track sprockets be-

were found at various stages of construction. The "Mouse" was very likely the most odd combat vehicle of World War II.

It was designed by Professor Ferdinand Porsche. The decision to commence production of this unusual construction fell on November 29, 1941, taken by Hitler (who as is common knowledge was a megalomaniac, and this wasn't to be his last word in this field) directly after an "inspirational" talk with the professor. He presented the first vehicle design to Hitler as early as June 1942, at that time consideration of its possible armament was also commenced. Two variants were taken into consideration—in both two guns were to be mounted in the turret—a 75-mm gun and 150-mm gun or a long-barrelled (with a barrel length of almost 7.5 m.) 105-mm gun. Finally however the "medium" 128-mm KwK 40 L/55 gun became its basic armament. The contractors were also chosen. The production of the turret and hulls was turned over to the Krupp plants, the construction of the electrical equipment to Siemens-Schuckert, the suspension was to be made by Skoda

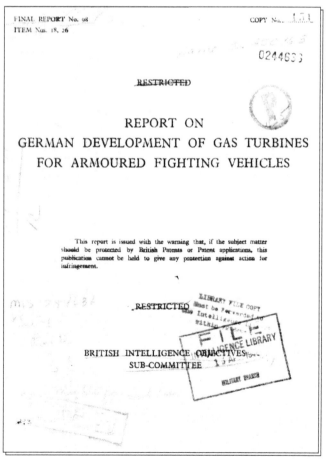

The front page of the report mentioned in the text

ing transmitted not directly, through the gears and clutches but through the intervention of an electric generator and two side electric engines. This was not a completely new solution, as early as 1940 Porsche had designed an electromechanical power transmission system for the VK-3001(P) prototype tank, which never entered production, losing to the Tiger. After redesign however it became the self-propelled Elefant gun, introduced to armament on a limited scale (88 pieces), better known by its unofficial name Ferdinand. It was the first vehicle with such a transfer of propulsion, which was used in battle.

The construction of the Maus tank and the preparations alone for its production consumed a huge amount of resources and were associated with big problems, totally disproportionate in relation to its real combat effectiveness. These problems resulted from the unusual size of the vehicle itself and the overall dimensions of its armament, for example the almost 1 metre recoil of the main gun and 1.52 m long rounds forced huge dimensions on the turret, whose mass exceeded 50 tonnes, comparable to the mass of the Tiger tank, the main gun itself weighing 7 tonnes. It housed four crew members: the commander, gun-layer operator and two loaders. They had after all to perform exceptionally "arduous" functions, since the round (fixed) for the 128-mm gun weighed as much as 56 kg.

It was one of the most heavily armoured tanks. The hull front was formed by armour plates 205 mm thick, inclined at the angle of 35° and 55°, the turret front was made of 215 mm thick profiled plate. In addition the remaining armoured components were unusually resistant to piercing and were over 150 mm thick (which best testified to the role of this vehicle conforming more to a "travelling bunker"), only the turret roof was made of 65 mm thick plate.

A very interesting outcome of the application of an electromechanical power transmission system was the way of overcoming water obstacles. Namely that the Maus could overcome them up to a depth of 6 m, i.e. completely submerged (of course assuming that it would not "dig into" the slimy bottom). A second tank of this type would have sufficed for this with its engine switched on, from which cables would have led to power the electrical engines.

The first specimens of the tank were ordered as early as the middle of 1942 and the first serial batch of 180 in March 1943. Yet despite the fact that the program received high priority and was enthusiastically supported by Hitler, the practice was far removed from the intentions, Allied air raids in addition to structural difficulties contributing to this. The first prototype was not ready until the end of 1943, and was still incomplete, lacking the turret. The second and only complete prototype was not delivered until November 1944. At the end of war several others were found at various stages of assembly. The "Mouse" remained only a curiosity…

However the gigantic and from a military point of view useless Maus and Adler tanks were not at all the peak of Hitler's aspirations in the field of armoured vehicles. They were indeed midgets compared to the 1,000 tonne tank, whose design he ordered during one of the staff briefings devoted to the

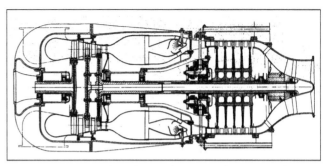

Selected types of German turbine engines for tanks

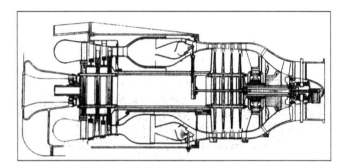

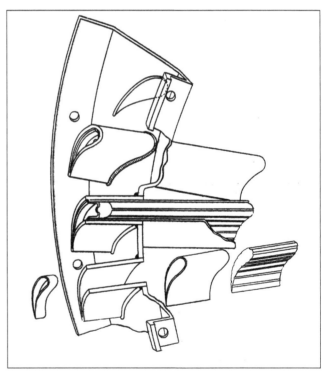

A detail regarding one of the above engines; the way of making hollowed blades

future of tank production. It was June 23, 1942, and engineers Grote and Haker were assigned the task to develop it. This "super giant," which was managed to be given the name Ratte ("rat") likewise contrary as in the case of the Maus, was to be built in a conventional lay-out, if not taking into account the arming of the turret with two (identical) high calibre guns, the order of 200-mm. Being around 32 m in length and almost 10 m. high, the Ratte would have been some kind of "land cruiser," however let us not delude ourselves; the structural problems associated with its possible building would have

The way the GT-102 turbine engine was built into the hull of the Königstiger tank

been totally insurmountable for the German war-time economy. Hitler's idea constituted some kind of collision of utopia with reality, and so work on this tank was halted even before the completion of its design.

The Germans were also credited however with many much more sensible innovations in the field of armoured vehicles—it is enough to mention the introduction of hollow-charge and sub-calibre projectiles to the role of tank armament (among others for the very successful 45/55-mm conical barrel gun employed on PzKpfw. IV tanks). An equally crucial step was the introduction to production of core armour piercing ammunition, with a uranium core. It was intended to use for this purpose several hundred tons of surplus uranium from the delayed nuclear program (the range, where Germans tested this ammunition, near Mielec in Poland, is to this day the most radioactively contaminated place in the country).

Also worthy of attention are the little known achievements of the Third Reich in the field of artillery armament. Except for the development of railway guns as enormous as the "super tanks," a whole series of pioneering fin-stabilised sub-calibre rounds with discarded sabots were developed (mainly in Peenemünde). They were designed for several new smooth barrel guns—initially for the long range multi-chamber V-3 gun, the famous "centipede." However a new 800-mm (!) gun was also designed, for which a most technologically advanced projectile was especially constructed. This gun, with a mass of the order of 1,500 tonnes, was to use the modified chassis of a Ratte tank. The chassis of the Tiger II tank was also to be used in the construction of a long-range gun, this time 170-mm calibre. The prototype of this gun was under construction in 1945, and was designated the 17 cm K44 Sf/Gw-IV. It was also intended to install 210-mm and 305-mm mortars on the very same chassis (Bär).[65]

So as one can see, the "Maus" was not the only "super heavy" vehicle, with a mass of over 100 tonnes. There existed one more, although it wasn't a tank, but… a mine-clearing vehicle. Apart from the mass itself, 130 tonnes, its construction lay-out was highly unconventional. The vehicle consisted of two hulls, connected through an articulated joint. Each hull had only one pair of wheels, but almost 3 m in diameter. The wheels were steel, equipped with very thick rubber overlays, resistant to mine explosions. The "Räumer-S," as it was called, also existed in the form of only a few prototypes.[65]

THE NEW TOOLS OF BLITZKRIEG | 111

Above: The "Tiger"
Below: The "Panther"

Propaganda poster featuring an image of the Thor mortar

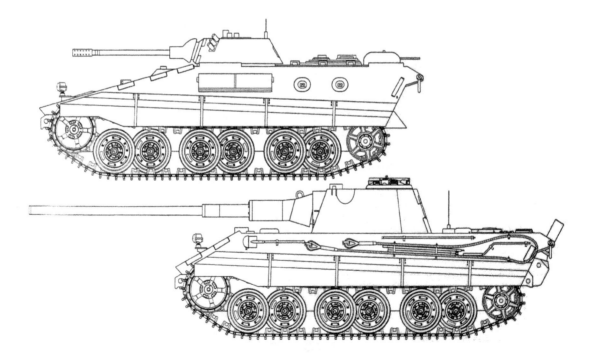

The pair of combat vehicles that was to transform Wehrmacht into the army of the new era. The E-50 tank equipped with a new, stabilized 88-mm cannon and an 1150 HP turbine engine with hydrokinetic transmission. The other drawing shows the second member of the family—the "Büffel" infantry fighting vehicle (which, as such, is something altogether different than an armoured personnel carrier!). It was armed with an automatic 30-mm gun. The main components of these vehicles were tested by February 1945. The main source of the interest in turbine engines was that they could use very low-grade synthetic fuels, much easier to produce from coal than regular gasoline.

NEW CONCEPTS FOR CONVENTIONAL WEAPONS

The present chapter is by no means devoted to all innovative conceptions from the domain of conventional weapons, which were promoted or simply verified in the Third Reich, but only to the most interesting and less familiar.

Energy Emitters

I will begin with the truly unusual inventions of the "sound cannon" and "wind cannon" (Windkanone), which generated a directed blast wave.[65]

The sound cannon was an incomparably simpler construction. It was a large, massive paraboloidal reflector 2.3 metres in diameter, in whose focus was detonated a small explosive charge (it is common knowledge that among others 19 kg nitroglycerin charges were applied). Somewhat later a composition of methane and oxygen was decided on, detonated in a chamber placed in the paraboloid's axis, whose length made up 1/4 of the wavelength of the generated sound. The advantage of this solution was the possibility of attaining a high frequency of explosions (800 or 1,500 per second) and continuous operation of the device. Its effect on a human being consisted mainly in deafening and paralysis of the nervous system (among others, as was affirmed, point light sources were seen as lines). Work was carried out in a research establishment subordinate to Speer's Ministry near Lofer in Austria, and the person responsible was Dr. Richard Wallauscheck. This invention's operation was based on the fact that a shock wave behaves just like every other plane wave, that is to say it can be focused into a point, is subject to interference, diffraction etc, in short from a geometrical point of view is completely predictable. One should at the same time to bear in

Detonation chambers,
to be placed in the "sound cannon's" axis

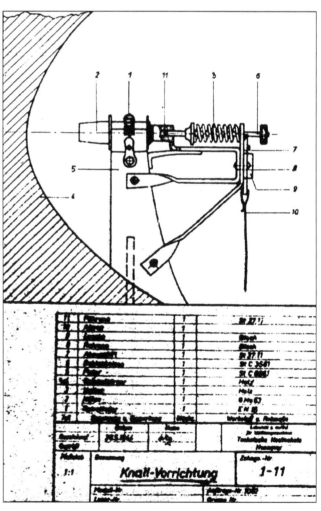

An original technical drawing of the "sound cannon"

The "Windkanone," the experimental "gun," generating and accelerating high-speed toroidal vortices. A smaller version was tested as a portable thrower of "tabun," a kind of personal chemical weapon, externally resembling a flame thrower. In the case of the latter device a "donut" (soliton) was spinning very fast and it almost didn't disperse up to a distance of about 50 m, which made it behaving more like a projectile. The vortex itself was obviously invisible. Generally these weapons illustrate the German innovativeness, for they probably would be quite hard to imagine by conventional scientists.

A development of the sound "cannon" was in turn a tubular device generating a powerful, relatively directed blast wave. It was developed by one of the companies from Stuttgart and tested at a research testing ground in Hillersleben. As American intelligence determined, penetration of a 25 mm thick board at a distance of 200 metres was achieved, however the effectiveness further quickly fell along with distance. For this reason the plan to use its prototype for the anti-aircraft defence of one of the bridges over the Elbe river, at the end of the war was abandoned. Not all areas of scientific and technical quest turned out to be "dead ends" after the war…

"Invisible" Aircraft and Vessels

Some were the beginning of promising and to put simply, futuristic trends in armament technology, although they remain practically unknown to this day. An excellent example is constituted by a German research programme connected with the development of materials shielding against detection by radar, sonar and the like—in short an area currently defined by the term "stealth."

Only recently I was able to enter into the possession of source materials; documents, describing this work. It concerns the aforementioned analysis of the activity of institutes subject to the Reich's plenipotentiary for research in the area of high frequency (BHF).[8] In it was written that the chief executor of this work was the company I.G. Farben, co-operating in this field among others with Gdańsk Polytechnic (Danzige Technische Hochschule).

Two models of the "sound cannon," a generator of a directed shock wave

Work was co-ordinated on behalf of this institution by the director of its chemical laboratories, Prof. Klemm. The chief authority on "purely radar" issues was engineer Karl Roewer, from an unspecified institution.

It was written that two code-names functioned in relation to the described project, of which at least the first sounds quite modern, the "Schwarzes Flugzeug" ("Black aircraft"), the second being the "Schornsteinfeger" ("chimney-sweep"). Well, another blank spot in the history of the Second World War…

Unfortunately, it will also be possible for me to fill this in only partially…

Information on this subject was gathered through the interrogation of the aforementioned people. Officers from the

mind, that a shock wave is decidedly the most effective carrier of energy apart from the area of nuclear physics. As opposed to a sound wave, the shock compression at the front of the shock wave has a completely abrupt (nonlinear) character, and in the case of powerful explosives this may be a change of density (e.g. of the air) the order of a million times. This signifies a very high density of destructive energy.

The effective range of this "cannon" was the order of several hundred metres. It was to be used against living forces—and in all probability was, but never under combat conditions. It was intended on a similar basis to construct the "light gun," but it was probably never completed.

U.S. military intelligence also found their way to Klemm's laboratory, where he worked in the final months of the war, in the town of Schmalkalden (Thuringia). In the town of Travemünde near Lübeck a system was found for researching the properties of new materials, where samples of them were also found – these were panels, made by a pressing method from powders of an closely not determined (at the time of the report's preparation) composition. The constituent substances themselves were produced in small amounts in the laboratories of I.G. Farben in the town of Hoechst.

The general rule, according to which work was carried out consisted in aiming at obtaining of materials characterised by magnetic conductivity and dielectric constant, that would be possibly close to the respective parametres of the air. During the war materials absorbing mainly medium frequency waves, up to approx. 100 kHz, were tested and produced on a small scale in Travemünde. In the development phase were the materials protecting against detection by radars working on higher frequencies (newer)—shortly before the end of war it has been completed, in Travemünde, an outfit to examine their properties in this respect.

This work, although unusually crucial, was never treated with top priority in the Third Reich. The described materials were used almost exclusively in an experimental field. The only exception known to me is represented by Type XXI submarines, which were mass equipped with snorkels shielded with this kind of material. The end of the war signified the arrival of a transitional period of a fall in interest in "stealth" technology. Only in the second half of the 1950s did the Americans recall German work, when they were developing "antiradar" paints for the supersonic SR-71 "Blackbird" reconnaissance aircraft, which carried out its first flight in 1962. This field is presently experiencing dynamic development, radar detectability being considered one of the most important parameters of the modern military aircraft.

When I already wrote this book, I found further Allied intelligence papers on this subject—and these are not just ordinary reports. It became evident that apart from those mentioned earlier, there still existed at least four such summaries![66,67,68,69] This obviously testifies to the significance of this issue for the "winners." Additional information, in extensive form, had been placed in a report from the American intelligence operation "Lusty," extensively described in part two of this book. All in all it follows from it, that the I.G. Farben institute near Frankfurt (Hoechst) was not the only establishment working in this field. The following were also mentioned:

1. The Institute of Inorganic Chemistry in Danzig (Gdańsk). (Prof. Klemm).
2. The company Osram—"Studiengesellschaft für Elektrische Beleuchtung"—Berlin (Dr. Friederich).
3. The laboratory of the Degussa consortium, 8 km from the town of Konstanz, near Bodensee (Prof. Fuchs and others).
4. The ceramic laboratory of the company Lutz und Co in the

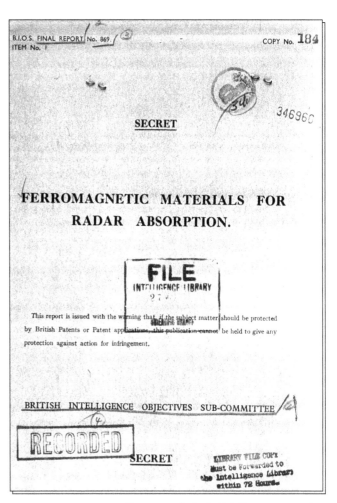

The covers of selected intelligence summaries pertaining to German "stealth" technology

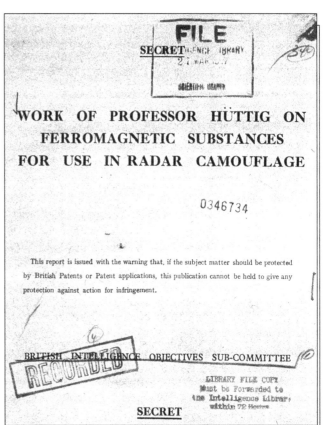

town of Lauf/Pegnitz in Bavaria (Dr. Franz Rother, inventor of the material used on the Type XXI U-boats).
5. Technische Hochschule Stuttgart (Dr. Fricke).
6. Technische Hochschule Praha (Prof. Hüttig).

There were even other signals, such as e.g. the report of a Polish soldier, performing his military service just after the war at the post-German airfield of Sorau (Żary) near Zielona Góra, who described how one of the aircraft was painted with some kind of grey, porous "kind of paint" found in one of the remaining barrels. It made a light Storch completely invisible to radar.

Regardless of this, in the Third Reich materials were developed that were to shield submarines against detection by sonic methods (asdic), i.e. absorbing acoustic waves radiating in the water. Here we also find a clear analogy to post-war continuations, among others the Polish submarine "Orzeł" ("Orzeł-III" —Soviet Kilo class), purchased in 1986 in the USSR, possessed an identical protective layer to some U-boats from the end of World War II.

The I.G. Farben consortium was also a pioneer in this field, and specifically its laboratory in Hoechst, in the suburbs of Frankfurt on the Main.[70] Research commenced in 1940 and in 1944 this bore fruit to the launching of 12 U-boats with an "anti-sonic" coating. This project bore the code-name "Alberich."

Although certain research work was carried out before 1940, it was only then that one could speak of the start of a "serious" programme. This date is associated with the exploitation of a completely new and fully executed conception, whose author was Professor Meyer from the Heinrich Hertz Institute. It relied on the application of many, suitably shaped layers of a specially well chosen rubber, at the same time an essential plus in relation to earlier ideas was the conservation of the external layer's smoothness, in connection with this hydrodynamic drag didn't increase. This was an effective and at the same time simple solution. One can describe the majority of tested variants in the following way: two thin (the order of 2-mm) plate panels of rubber were glued in turn onto a steel plate, constituting an imitation of the hull's plating. The external layer was a smooth plate panel. As one would expect, the layer adhering directly to the hull fulfilled therefore a fundamental role. It was perforated—a large number of smaller (up to 2 mm) and slightly less larger holes (the order of 5 mm in diameter) had been made in it. As a result of interference a large section of the acoustic wave was supposed to be damped in them, the diameter of the holes being inversely proportional to the frequency. The entirety formed therefore a soft, almost spongy shield, which was to some extent wash away the "acoustic signature" of the vessel.

In spite of its simplicity, this coating, in so far as it was well designed, could be very effective; the maximum degree of sound-damping, which was recorded (for several kHz) was 95%. This was however only one side of the coin. It was possible to achieve such a high effectiveness only then, when the diameter of the holes was ideally "well-matched" to the sound frequency used by the enemy. The Germans feared that after applying devices operating at different frequencies the importance of Meyer's invention would fall considerably (incidentally this is an exact copy of the problem observed in the case of the "stealth" coatings for aircraft, e.g. the F-117A "invisible" American bombers /target indicators are quite effectively detected by old types of radar, operating at medium wave frequency). In the case of "anti-acoustic" coatings yet another problem materialised. It was of course that they would be effective only to the depth to which "blisters" of air are not compressed to a density comparable to the density of water. Calculations and measurements carried out in a special steel tube, positioned vertically like a chimney revealed, that this limit was determined by a depth of approx. 70 metres. In spite of this it was however possible to regard the invention as valuable, as it constituted an essential and real defence in combat, when the vessel was situated at low depth and in proximity of enemy units (although the Type XXI could, as the first, carry out attacks without protruding its periscope above the surface of the water, it still had to be submerged shallowly). At greater depths this problem was not so severe first and foremost because the acoustic wave doesn't propagate, in defiance of appearances, just the same "downwards," as horizontally. The sea is by no means a homogeneous mass of water, but forms a certain structure, divided into many "layers," characterised by different salinity and oxygen content etc. This manifests itself in almost skipping differences in density, which results in partial reflection, refraction and in effect scattering of the acoustic wave. In the ocean (i.e. without the influence of the bottom) one may speak of such phenomena at depths an order of magnitude exceeding the height of the waves (mixing of water), i.e. below, let's say, 100-200 metres. The U-boats attained on the other hand a submersion up to approx. 300 metres. Of course at these depths they could be detected, but in general the submersion depth is one of the parameters most strongly influencing their detection. The second is of course the maximum speed developed underwater, the escape speed. In spite of everything, the described invention turned out therefore to be quite valuable and "decidedly worth possessing"—such a conclusion at least appeared in the American report.

The Germans fitted out twelve of their submarines with a layer of "Alberich." The majority were tested, only two taking part in combat operations. One of them was sunk, it didn't possess however the snorkel and was probably detected by radar. It wasn't stated what type of vessel it was.

While discussing these materials it is worth mentioning a certain important and widely unknown related field, the plastics of the Third Reich. Yet another intelligence report makes this possible.[71]

Plastics

Plastic substances are mainly associated with the 1960s, when they revolutionised industrial design and appeared in a range of objects of everyday use.

But as in the case of many "novelties" of this period, it concerned in reality the introduction of revolutionary achievements in science and technology from World War II to the mass market. Plastics were one of them…

The appearance of many of them is attributed to the Americans, the inventors of nylon (the Du Pont consortium), then called synthetic silk, which became fashionable with regard to new stockings, devoid of seams, although it showed its value above all as a cheap raw material for manufacturing parachutes. Now it was possible to produce it practically in any amounts.

The Germans also had nylon at their disposal, who mastered its production not much later than the Americans and at the same time created many other plastics—all in all a dozen or so types—most of what we currently know from everyday experience.

Of course the earliest to appear were chemically hardened on the basis of phenol, such as bakelite, which even before the war was widely used to manufacture the finishing components for small arms—the overlay for the grip, the foot of the butt as well as electric-isolating components. Production of it had started even before World War I, on a small scale, on the basis of a Belgian patent. If on the other hand work carried out in the Third Reich is concerned, one may differentiate the following turning points:

1. June 1938: the first composite (of highly durable woven fabric, laminated with plastic).
2. January 1939: synthesis of the first thermoplastic polymers. In this same period polymer bearings were developed, production of which was started in September 1939.
3. January 1942: the start of manufacturing of components from thermoplastics by the casting method. Four months later casting was replaced by pressurised injection moulding technology.
4. May 1942: specification arose concerning the production of plastic components by the pressing method.

In the Third Reich, the I.G. Farben consortium was a monopoly in the field of developing and producing polymers, and the greatest role was played by its plants and laboratories in Bitterfeld, Hoechst (near Frankfurt am Main) as well as in the Oppau district of Berlin, where the abode of "Stickstoff-Syndikat" was a division of I.G. Farben specialising in this field. This field was considered crucial, an expression of which was the appointment in Speer's ministry of a special plenipotentiary, keeping an eye on the realisation of the Government's war-time priorities ("Sonderbeauftragter für die Stickstoff Industrie"). In order to realise their ambitious assignments the Germans also employed a series of plants situated in the countries it had occupied in Europe, mainly in France, Belgium and Holland, among others the plants of Philips in Eindhoven and the town of Venlo, the company Cogebi (Compagnie Generale Belge d'Isolants) in the town of Loth near Brussels, and the "Institute of Plastics" in Delft.

Altogether the following types of plastics were developed:

The snorkel from a Type XXI submarine, covered with "stealth" material

A German advertisement from 1944, "The German quality product in German plastic"

- Polystyrene. This material, under the name "Trolitul" was produced on a small scale, and treated as not very competitive with regard to others.
- Carbazole of polyvinyl. A polymer supplied for injection moulding, i.e. technologically simple and at the same time, with regard to its fibrous structure, characterised by very

good durability parameters and small thermal and electrical conductivity. It was treated as perspective.
- Polyvinyl chloride. Many modifications were produced, among others in Venlo, for various applications (PCV).
- Plexiglas (polymethacrylate). Known universally as "organic glass." With regard to a limited tendency to cracking, glazing components for aircraft were made from plexi. An alternative was constituted by standard laminated glass (glued) with vinyl, manufactured by "Societe de Verreries de St. Gobain" in Lyon. The Germans had ambitions to make components for simple optical devices from plexi, ultimately ending in the production of lenses for torches.
- PCV acetate. A substance tested as a supplement to concrete of high durability. Perhaps this had a connection with considerations on the subject of possible concrete armoured linings.
- Polyamide (nylon). A material produced during the war in several different modifications.
- Polyethylene. Despite a considerable contribution to research work, it was only completed after the war. Polyethylene is a plastic known currently from disposable plastic bags, syringes and the like.
- Polyisobuthylene. A material used during the war as a substitute for caoutchouc/rubber.
- Aldols. A family of alcohols, being a derivative of polyvinyl. Cable insulation was manufactured among others from it.

In addition to the aforementioned, laboratory work was carried out on several new plastics, also developed from before the war. Various cellulose compounds were manufactured, among others photographic films from celluloid, packaging film from cellophane and the like.

The Germans used plastics mainly in the electrical engineering and electronics industries, with regard to their insulating properties. However, there also appeared the signs of these materials being universalised, plastic boxes, commonly used nylon camouflage nets and PCV "tiles" and buttons. High durability polyamide film was produced on a small scale, among others, for manufacturing audio tapes (mainly used by the Gestapo). The technology of producing synthetic textiles was mastered…[71]

Underwater Warfare

The next interesting and poorly known field, drawn attention to on previous pages is the technology associated with "underwater warfare" with the submarines. This problem relies simply on the fact that warfare at sea was waged by U-boats constructed before the war, in this connection the results of work from the period of the war have remained practically unnoticed.

There is therefore a certain kind of paradox that the Germans began to lose the war on the Atlantic, despite the enormous amounts expended on research.

Even such "innovations" like e.g. torpedoes homing on sound sources (the T-5 Zaunkönig torpedo, brought into armament in September 1943) didn't help, the Allies learned quite quickly how to counteract this danger. A device was conceived which towed behind the vessels produced a significantly greater noise than the submarines themselves. The German underwater fleet began to suffer ever greater losses. The cause of this was the use by the enemy of two already well known, revolutionary inventions—sonar and radar—capable of detecting not only submerged vessels, but even the periscopes and snorkels themselves. One should at the same time remember that the basic types of submarines were characterised by an underwater range in the horizontal of about 60 nautical miles (German designs brought into armament: Type VII, Type IX)—i.e. were rather "diving boats." There existed consequently the urgent need to accomplish truly radical quality modifications—the development of completely new submarines—able to wage war with measures against which the Allies wouldn't be able to defend themselves.

Such vessels arose and in fact became the peak of technology in this field, later many post-war constructions were modelled on them. These were types: XVII, XXI, XXIII.

The Type XVII vessel, despite being based on technical solutions that arose before the war, was one of the most interesting. A breakthrough in relation to older Type VII and IX units was the application of a new propulsion independent of atmospheric air, the Walter turbine. This was an invention already tested at the beginning of the 1930s. Its author was an engineer—a chemist from Cologne—Hellmut Walter. This was a turbine engine, in which classic fuel was used –diesel oil, combusted in a special chamber in an atmosphere of oxygen and steam. The oxygen originated from a catalytic decomposition of perhydrol—concentrated hydrogen peroxide H_2O_2 of around 80% concentration. This was therefore a "specific modification" of ordinary hydrogen peroxide solution, that which is bought in pharmacies, being characterised by a concentration the order of only 3%. The perhydrol for propelling the turbine was renamed by Walter to "Ignolin," in honour of his oldest son.

From the start he considered the use of concentrated nitric acid as an alternative oxidiser, but quickly relinquished from this due to its corrosive properties and toxic action of decomposition products, mainly nitrogen oxides, leakages of which e.g. into the vessel's interior could never be excluded. It is worth noticing that perhydrol of the required concentration was already produced in Germany on an industrial scale during construction of the type XVII vessels, being used to propel the pump turbines of propellant in the V-2 rockets. Since the production possibilities were limited and the cost of this substance quite high, the submarines as well as the torpedoes in which just the same propulsion was to be used had in the form of the V-2 rocket severe competition which was encompassed with the highest priority.

Work on the Walter's turbine engine as the propulsion of a perspective submarine commenced as early as 1933, leading to the construction of a 4,000 HP engine already in the middle of the 1930s. During operation a temperature of 450°C and pressure of 36 atmospheres arose in the combustion chamber. The combustion gases and steam were used to propel the turbine, after which they were cooled and carried away, the steam being condensed and left on the submarine.

Favourable results of work bore fruit in 1938 to the order of the first, research submarine with such propulsion, which was acquired by the Germania-Werft shipyard. The submarine, which was designated the V-80 was completed two years later. It had a displacement of merely 80 tons and three man crew, however it permitted the practical performances of the new propulsion to be determined.

It became evident that it gave in fact very good results; during one of the trial trips the V-80 achieved underwater a record speed of 28.1 knots (52 km/h), the remaining problem being chiefly the high cost of the oxidiser. The results of trials were however sufficient to take the decision of constructing a new, this time warfare type of submarine, the aforementioned Type XVII. Construction of the first, the U-791, was ordered in 1942, but for various reasons this order was soon cancelled. On the other hand a further four were constructed: the U-792, U-793, U-794 and U-795. They were relatively small units 52.1 m in length and with an underwater displacement of 330 tons. In comparison, the Type VII submarines had an underwater displacement up to 865 tons and Type IX submarines 1,232-1,804 tons (depending on the version). Apart from the Walter turbines (two, each with a maximum of 2,500 HP), conventional systems were also employed in them. Each was equipped with a diesel engine, operating at the surface, but of relatively poor power: 210 HP, and two "emergency" electric engines: merely 75 HP. Their main engines were characterised by a working pressure of 30 atmospheres and combustion chamber temperature at a level of 550°C.

These were the first submarines developing underwater a much greater speed than at the surface (approx. 26 knots) and at the same time with a record underwater range, approx. 150

The Type XVII U-boat, captured after the war by the Americans

nautical miles. But they could travel underwater so long and fast only once during a given trip and not after each charging of the accumulators as in the case of conventional submarines. They were not therefore ocean units and were to be on duty mainly in the North Sea.

They were introduced into the service of the Kriegsmarine at the beginning of 1944, but achieved operational readiness only in the final few months of the war, in this connection they never took part in combat operations. In the first half of 1944 a further three units of somewhat greater displacement were ordered, with an underwater speed reduced to 21 knots. They were completed still in the same year, but no longer found their way into service. At the end of the war they were blown up by the Germans themselves (the officer responsible for this operation was later condemned by a British court to seven years imprisonment!). Before it had come to this, the next three submarines were ordered: types XVII G and XVII K as well as 100 (!) units of a larger, oceanic Type XVIII, but they were never completed.

The Type XXI submarine in May 1945, in front of a Hamburg shipyard's bunker

One of the Type XXI submarines in Norway, April 1945

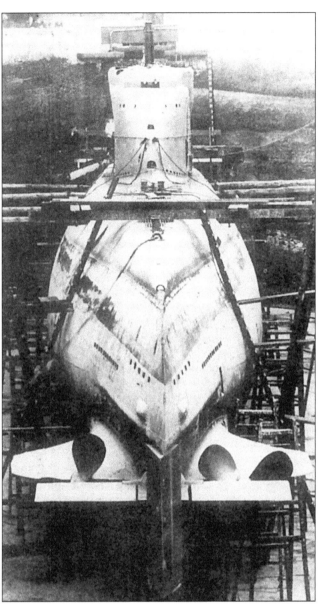

Their tested, revolutionary propulsion was developed after the war in the USA, Great Britain (where Walter himself was taken) and in the USSR, which as the only country ultimately brought submarines propelled by Walter turbines into service after the war.

The not large at any rate type XVII-A submarine took along 40 tons of hydrogen peroxide, which constituted ¾ of the propellants in general and was sufficient to cross only approx. 100 miles (admittedly with a speed of over 20 knots). The turbine efficiency is best testified by a comparison with the other propulsion unit, Diesel, which consumed the remaining one quarter of the total amount of propellant, ensuring a range on the surface approx. 20 times greater than that given above.

This propulsion was developed up until the very end of the war, developing several new types of submarine: the XVIII, XXIV, XXVI-A and XXVIII. The first two of the aforementioned were characterised by an oceanic range—e.g. the type XVIII could have travelled thanks to the Walter turbine (with a total power of 15,000 HP!) up to 350 miles, but had to take on board(?) over 200 tons of concentrated perhydrol. The type XXVI-A was to be a combination of classic solutions and Walter turbines; the latter would have been used only in a dangerous situation and obviously after an attack. None of these submarines were however completed, although intensive work gave fruit to an increase in turbine efficiency.[72,73] The above trials resulted from the fact that the propulsion simply implied the U-boats' basic limitations.

This was so for two reasons: a submarine (e.g. Type VII or IX) couldn't escape underwater from a pursuing destroyer, since it was markedly slower—even twice as much. Secondly, submarines were characterised by a short time of travel during submersion, the order of a few hours, in connection with this they were particularly susceptible to detection, e.g. by aircraft, continually patrolling the North Atlantic as well as the routes of passage from the ports.

above: The Type XXI U-boat, view from the rear, in dry dock
right: The Type XXIII U-boat

The parameters of the propulsion itself were improved in several ways. The very modern Type XXI (oceanic) and XXIII (coastal) submarines acquired classic propulsion systems (diesel plus electric engines), but of a completely new generation, which permitted a radical change in tactics. The type XXI developed underwater a speed of up to 17.2 knots, at the same time using purely the accumulators it could overcome up to 340 miles (640 kilometres) in one go! This was several times more than in the case of any of its predecessors, or competitors. It could not only simply escape from a standard destroyer, submerge to a record depth of 330 metres (so one of the trials indicated), attack targets without surfacing (using a passive range-finder, the so-called Balkon-Gerät) but was also particularly difficult to detect underwater. One of the specimens was tested in 1946 by the US Navy. Their ships couldn't even detect it at a distance of 200 metres. Covering the snorkels with "stealth" material also introduced of course a completely new quality on the surface, or rather at periscope depth.

In general this was a jump from the level of the 1940s into the 1960s…

This was a demonstration of how much can be achieved with the skilful use of radically modified, although as conventional as possible conceptions, already known earlier in general outlines.

The submarine possessed a completely revolutionary system of torpedo fire control, enabling it to carry out effective attacks even at complete submersion, the target positions being determined by creating three-dimensional co-ordinates of the noise's source through recalculating of delays received by various microphones placed on the submarine's hull. After an attack the Type XXI escaped at maximum speed, at which the enemy's sonar was totally ineffective (it maintained effectiveness up to approx. 12 knots). It gave therefore practically no possibility of detection of the attacking submarine, which was almost inevitable at the end of the war in the case of older types. They were detected by sonar and moreover before attacking they had to protrude their periscope, as rule detected by enemy's radars. In any case even at such a variant of attack, the type XXI had an enormous superiority over the enemy—detection e.g. of the protruding snorkel by radar was in practise impossible, since it was covered by a special substance, absorbing the radar beam.

A further variant of attack, which was to be used at greater distances was acquiring the position of targets with the aid of a protruding radar. The possibility was predicted of applying homing torpedoes—e.g. with magnetic warheads.

In addition the last of the serious weaknesses of the older submarines was removed, namely the relative susceptibility to attack from the air, in part caused by the lack of possibility of an adequately earlier detection of aircraft. This last problem was removed through the application of on-board radar, in addition, in contrast to its predecessors the type XXI submarine had a very high chance of victory in confrontation with a single patrol plane (the most typical situation). This was indebted to two anti-aircraft gun turrets with four quick-firing 20-mm guns. It was possible to couple the radar with fire control system of the guns. Even if they hadn't been sufficient, the submarine could also find itself completely underwater in a record time of 18 seconds.

The Type XXI constituted such a radical leap in quality that it made possible a further, it would appear already impossible, tipping of the scales in favour of Germany.

This was the first submarine characterised by a modular construction. The hull was divided into eight sections (plus the conning tower) and only their connecting took place in the dock. This simplified production but also and perhaps above all "pushed" most of the production stages away from the docks, most in danger of bombing. This was also the first submarine characterised by a single hull, the interior of which was furnished with all sorts of equipment, up till now submarines being built on the basis of an internal rigid hull (long and narrow), which was encased in ballast tanks and then only the external hull. This permitted the internal volume of the hull to be increased significantly and introduce among others numerous facilities for the crew, up to 17 m^3 of space fell to each man, air conditioning and lavatories also appeared, being only the object of dreams among the crews of older types. It was possible to take a normal shower, while sailors being on duty on the VII and IX types, filled with salty dampness, generally couldn't take a bath even for couple of months. This was the source of numerous diseases. Evidence of the value of the Type XXI is delivered by its post-war second life. The Soviets took control over some German shipyards, including the one in Gdynia/Gotenhafen, where the U-Boats were manufactured. Based on the design of the "Type XXI" they developed the "project 633" submarine, or the "Romeo class," according to a NATO specification. They were introduced into service in the 1950s and were operational until about the mid-80s, forty years after the war. The Chinese even went further with their copy of the "Type XXI," the Ming type. Most of them will probably put to sea until 2005-2010.

A complement to the fleet of the Type XXI submarines, were to be the new "coastal" Type XXIII submarines, matching the former with respect to their state of technology but approximately twice as small and of a smaller range.

German attempts at solving the problems associated with propulsion went in yet another direction…

It concerns the diesel engine operating without access to atmospheric air in a so-called closed cycle.

This was by no means a new concept. The first attempts at its realisation fell in the years 1915-1918, but a dozen or so years had to pass before the level of technology turned out to be sufficiently mature. In 1939 a vast research programme in this field was commenced in the Third Reich, in which the following companies were employed:

1. "Zeppelin G.m.b.H," where under the direction of Dr. Durr, diesel engines from Daimler-Benz were modified for this purpose.

The conning tower of a sunken Type XXI submarine. Visible are two anti-aircraft artillery turrets and between them: the radar antenna.

2. "Germania Werft" in Cologne, the shipyard responsible for the final stage of work.
3. "Forschungsinstitut für Kraftfahrzeuge," the institute realising the overall research work and calculations (Prof. Kamm, Dr. Huber).
4. The plants "Junkers-Bessar" and "Luftfahrt Akademie Sotow" (in Berlin), mainly specialising in new propulsions for torpedoes (Prof. Holfsleber) and in command of very valuable experience in this new field.
5. Engineers Dipling and Schlefler from Berlin, who earlier developed the special engines for fast attack boats and could contribute considerable knowledge on the possibility of increasing the efficiency of combustion engines.

What in reality however was this innovative type of propulsion?

Diesel engines operating in a closed cycle are, with respect to construction, approximate to classic diesel engines, except that the fuel is not mixed with atmospheric air, but with oxygen supplied from a pressure or cryogenic system. Gases arising as a result of combustion—carbon dioxide and steam are cooled—as a result of which the steam partially condenses,

Unfinished Seehund midget submarines

and the remaining CO_2 easily dissolves in sea water. Only in such form do the combustion products leave the submarine. The submarine has therefore only one propulsion unit (plus it may have a special electric propulsion unit for quiet "stealthy creeping"), which is used on the surface as well as underwater. In this way the time travelled in submersion could be extended even up to couple of days, and in the case of the latest contemporary designs even up to several weeks.

Such are the benefits resulting from the replacement of electric engines and bulky accumulators by considerably more efficient sources of energy.

At the beginning of the 1940s, the Daimler-Benz engines were modified for the aforementioned purpose, mainly the type DB-501. In 1942, two Type IXD U-boats were redesigned, in which the original MAN's engines (two per submarine) each 2,000-2,500 HP, were replaced by the new, though not yet modified Daimler-Benz engines. It was that they were considerably smaller and there were to be six of them in each submarine, in connection with this it was necessary to check how one could best redesign the propulsion compartment. Thanks to this, it was possible to check in practise the approximate configuration to the final estimate its performances and the like. It became evident that although the total power rose from 5,000 HP to as much as 9,000 HP, the maximum speed rose by merely 2-2.5 knots to approx. 23 (on the surface).

This was the first test of applying a new propulsion in mass-produced submarines, although before this work had been completed the decision was taken to modify one of the Type XVII submarines, which as the first would acquire the final engines working in a closed cycle (what however dragged on and was never completed; only certain components of the new system were installed, without the engines). It was designated the Type XVII-K. It would have taken on board over 23 tons of diesel oil and approx. 9 tons of compressed oxygen in 16 gas cylinders, which would increase the surface range up to 1,100-2,600 nautical miles, depending on the speed, while the underwater range would have been, at a maximum speed equal to 16 knots, the order of 110-120 miles. This submarine, which managed to acquire the tactical number U-798, was sunk by the Germans themselves in the first days of May 1945.

This was of course a purely experimental construction, devoid of armament. The first type in which the new engines were to be installed on a mass scale was the miniature "Seehund," not much bigger than a large torpedo. To this end its hull was to be extended by 1 metre, whereas the pressurised oxygen tanks were to be replaced by cryogenic ones, considerably easier to service, since the "Seehund" had a very small range and the loss of oxygen as a result of evaporation didn't pose a major problem. However this project also didn't live to see realisation, in connection with this no engine of the described type constructed in the Third Reich was ever used at sea…

Research work connected with the new idea however continued and in spite of everything gave some interesting results. Above all problems resulting from the use of a new fuel blend in the DB-501 engines were revealed, it was endeavoured to

The "Biber" mini-submarine in the Imperial War Museum

remove them, and all manner of accompanying devices were improved and developed.

This involved among other things the modification of the principle source of problems—the system regulating the injection of diesel oil and oxygen (it wasn't always under high pressure—at the beginning this was 400 atmospheres, but in measure of defueling the pressure dropped to 1 atmosphere). The proportions between the amount of injected fuel and oxygen had to be, despite this, maintained at a constant level.

Damage to the engine also occurred, caused by the presence of water in the compressed oxygen. Besides this, technologically very difficult to overcome was the pressurised oxygen system, in which, despite the painstaking precision of making welds, small leakages were detected. There were no problems on the other hand with the exhaust gas compressor, which was to remove the gases from the submarine, of course also at large depths.

A special oxygen compressor was also constructed, which would have allowed a section of the pressurised tanks to be eliminated, as its supply could now be supplemented at open sea.

This entire research programme was to enable the construction and bringing into mass production of several new types of submarines, designated as the Type XXIX-K (oceanic), XXXIII (coastal, but with an underwater range of up to 1,600 miles (3,000 km!)), and XXXIV (coastal). The underwater range of the first two of the aforementioned types would have been simply shocking during World War II, at least several times greater than the analogous parameter of any competitor. This emphatically attests to the value of the described conception.[72,74]

Recapitulating: how it did happen, that in spite of such specific technical breakthroughs, the Kriegsmarine didn't detect any improvement at sea? This is a very interesting issue, although few people know its explanation. After all as many as 118 (although in total 1,300 were ordered) Type XXI submarines were completed from June 1944 to April of the following year, and yet not one was sunk. The main problem rested on the fact that both modern types—the XXI and XXIII—were far more complicated than the previous. They required sailors, highly qualified experts. It was precisely them that was missing.

Training dragged on and was carried out by only one establishment—in Stolpmünde (now Ustka, in Poland). As a result the majority of these people, 1,100 individuals, were surprised by the Soviet January offensive. They were to be evacuated through a sea-lane, but the ship on which they were found was sunk. This was the famous "Wilhelm Gustloff"...

Concrete Ships

Another little known curiosity connected with the German navy is this, that in desperate aims to rebuild the fleet, even such weird (and probably absurd) conceptions were advocated like e.g. the production of concrete ships. This was not pure theory. The shipyard employed in this was in Rügenwalde (Darłowo) on the Baltic coastline. Still today part of the breakwater in Darłowo is composed of two hulls of such ships; on a military map one of them is marked with a dashed line as a block of dimensions 90 x 15 m. Only two concrete hulls were made, each having a displacement of 3650t. Goering once tried to urge a similar idea. At the time when the Allied air force concentrated among others on the destruction of locomotives—as one of the "bottlenecks"—the Marshal of the Reich announced his conception to mass produce concrete locomotives.[1] This actually didn't gain recognition, although this doesn't at all mean that the leaders of the Third Reich, as a general rule, were guided by rational criteria. This is testified at least by such dilemmas like the Me 262 bomber versus the Me 262 fighter, or the V-2 rockets versus anti-aircraft rockets. Super-heavy tanks constitute a no worse example.

I remind one, in connection with this, of the remarks included in the "Introduction."

Recoilless Weapons

Seeing that we are already at ideas of an odd character, let us still dwell on certain interesting concepts concerning barrelled weapons, ammunition and rockets, though already more sensible than those mentioned above. Many such examples are supplied among others by recoilless weapons. This doesn't rather attract attention, but the "Panzerfaust" was in the realities of the World War II something decidedly innovative, despite its simplicity. This was a weapon, on whose score one may record hundreds, if not thousands of destroyed enemy vehicles. It was also very cheap and simple, and mainly consisted of an overcalibre projectile (projectile's diameter larger than calibre of the launcher itself), throwing charge and barrel—steel tube. Despite this, the principle of operation alone—relying on the compensation of the projectile's forward motion by a motion of gases flying out of the other end of the tube, was something new and not exploited until World War II. It was

The Panzerschreck

likewise in the case of the projectile. It had never been possible before that a slowly flying projectile pierced 10-20 cm of tank armour. The change had come about due to a special type of explosive charge, the so-called hollowed charge. It has a conical hollow lined with the so-called copper or steel cumulative liner. The explosion causes it to be compressed, as a result of which a stream of liquid metal arises with a speed approaching 10 km/s. It is precisely this which pierces the armour.

All of this made possible the construction of a whole family of recoilless weapons.

The recoilless gun is a light, "specialist" artillery weapon. Such a conception appeared long before the outbreak of Second World II, however its period of turbulent development didn't fall until the turn of the 1930s and 1940s. They were developed in different countries (among others in Poland shortly before the outbreak of war by engineer Czekalski, who later continued his forgotten work in Argentina), however many types in particular arose in the Third Reich.

In the recoilless grenade launcher the barrel is simply a steel tube open at both ends, whereas the gun possesses a breech, which doesn't limit the propagation of gases to the rear. It is designed exclusively to block the cartridge in the barrel and initiate the primer. The cartridge case is made from combustible material or thin, perforated metal plate. The round chamber is constructed in such a way that the gunpowder gases pass round the rudimentary breech (the chamber has a diameter greater than the calibre) and end up in the rear section of the barrel, on the whole ending in a specially profiled nozzle.

The lack of force of recoil makes possible significant "slimming" of the construction, among other things, thanks to the elimination of the recoil mechanism. This is the chief virtue of the recoilless gun. Thanks to it one can offer a substitute to the classic artillery weapons of the light infantry—e.g. airborne, mountain or reconnaissance troops. They acquire in this way the ability to destroy fortified points of resistance (bunkers) and enemy combat vehicles (from the beginning of the 1940s that is from the moment of introducing hollow charge shells, since the recoilless gun is characterised by too small a projectile's muzzle velocity, to be able to effectively fire classical armour-piercing shells, penetrating the armour thanks to the kinetic energy).

We have thereby come to the main fault of recoilless guns, the small kinetic energy of the projectiles and in connection with this, small range of effective fire, not exceeding 1,000 m with direct fire. This results from the principle of operation alone—"leakage" of the round chamber and light barrel construction.

In spite of this it was then an avant-garde weapon and simply invaluable in many applications. This is at least testified by the "boom" which was possible to observe in this field directly after the end of the war. Not until a few decades later did the indications appear of a gradual fall in importance in connection with the introduction of mobile anti-tank launchers of guided weapons.

One may divide German recoilless guns from the period of the war into two fundamental groups – relatively conventional constructions, brought into armament and less conventional, left at the stage of firing range trials or plans alone.

A soldier with the Panzerfaust-60 grenade launcher

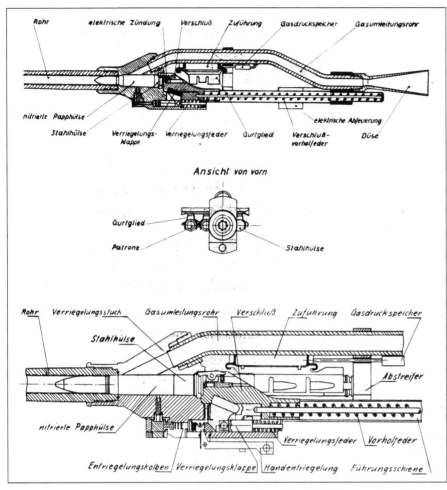

Original cross sections of the MK-115 automatic recoilless gun

The most interesting of course are the latter.[75,76,77]
They were the following designs:

1. A 150-mm gun, probably developed in 1942 (the designation indicates this) and supposed to be a complement to the 75 and 105-mm recoilless guns used by the German airborne troops. Two designations functioned in relation to it: the LG 42 and LG 292 Rh. They were developed by those same experts who "had on their account" the 75 and 105-mm guns, Engineer Wind and Dr. Biermann. The latter also designed a 45 kg round. Neither the overall mass nor the range is known. It is only known that the barrel had a length of 2,145 m and projectile's muzzle velocity that amounted in all probability to 290 m/s. Only several prototypes were made.

2. The 88-mm DKM-43 gun developed in the town of Sommerda for the navy, as a possible armament for gunboats and other small vessels. Its barrel was 2.8 metres in length and the total mass amounted to 350 kg. The range is unknown, but the high projectile's muzzle velocity (600 m/s) suggests that it could exceed 10 km. Test firings came off favourably, but mass production was never commenced. It was an unusual construction, completely differing from other recoilless guns.[78] It simply was about modifying the already existing, tested to perfection, anti-aircraft /anti-tank guns of the same calibre. Was it possible to transform them into recoilless guns without altering the basic structural features of the weapons?

It turns out that yes. A second chamber was simply placed behind the gun, in which a second charge exploded at the moment of firing, giving a recoil compensating the gun's recoil. An insoluble problem was constituted however by the situation in which it wasn't possible to achieve precise synchronisation of both explosions, or one of the charges didn't ignite.

3. The developmental version of the above construction was supposed to be aerial gun (!) of the same calibre and firing the same ammunition. Work was discontinued however, when it became evident that it was too heavy and the explosion of the shot too powerful in relation to the aircraft's strength (this mainly concerned of course the jet of gases behind the gun). Tests were carried out at a firing range near the town of Treuburg, and a modified Ju 87c aircraft served this purpose. It fired at various tanks interspersed on the firing range. In suffered serious damage during one of its flights, when the charge compensating the recoil didn't explode. Part of the streamlined gun housing was thrown backwards, damaging the tail of the aircraft.[78]

4. A heavy 280-mm recoilless gun, developed for the Kriegsmarine in 1944. It was designed to defend the coastline against the expected invasion in France, as an armament for heavy bunkers. It was supposed to have a mass of up to 28 tons! In connection with the fact that the landings took place before the completion of work, it was interrupted. This design bore the designation DKM-44 (Düsen-Kanone-Marine, 1944 model).

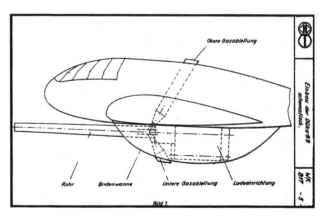

Original drawing illustrating the installation of the 88 mm recoilless cannon under the fuselage of an aircraft

In the case of the above, uncompleted designs, one may speak rather about searches for an optimal combination of the recoilless gun's merits with the merits of other types of artillery weapons, and not of real, crucial achievements. But there were those as well… It concerns the automatic (quick-firing) recoilless gun, developed in the final years of the war as an on-board aerial armament. It was supposed to be installed in the wings of aircraft. It concerns therefore the realisation of an altogether unusual conception and all the more crucial, that special rounds with the case partially burning (cardboard dripping with nitrocellulose) were developed for this weapon. Only the bottom of the cartridge case was metal, sealing up the rear section of the round chamber during firing. This gun acquired the designation MK-115.

The starting point for commencing work were the results of research of an earlier, experimental 50-mm, aerial recoilless gun ("single-loader") and the following drawn up demands of the Air Ministry:

- Calibre: 55-mm
- Muzzle velocity: 600 m/s
- Projectile's mass: 1.5 kg
- Minimum rate of fire: 300 shots/min
- The use of carbon steel
- Feeding: right- or left-sided
- Initiation of shot: electric
- Possibility of placing the bottom parts of the cartridge cases back in the ammunition belt

The company Rheinmetall-Borsig undertook the realisation of this task, which presented the design of a gun with the following characteristics:

1. Simplicity was fundamental, also with respect to technology. Hence the barrel was of a monoblock type, connected to a cast cartridge chamber by a thread.
2. The principle section of the gun was the cartridge chamber. Connected to it was the barrel, breech block, reloading and feeding mechanism, tube carrying away the gases to the rear and the like.
3. The breech was also of a monoblock type, cast in common carbon steel. Electrical firing of the cartridges was used.
4. One of the most original, innovative solutions was the feeding system of the weapon's automatics. It fired with a locked barrel, which was of course a necessity at this cartridge energy (locked in the classic understanding of the word, see: point 5). The so-called semi-free breech was employed. Its recoil was delayed by a special mechanism, operating under the influence of the gunpowder gases, carried away by a side opening in the cartridge chamber (since the cartridge didn't seal up the cartridge chamber for the whole of its length). In the gas conduit, connected to the cartridge chamber, was found a small metal piston, whose inertia gave the vital delay in unlocking the barrel. This system functioned therefore altogether differently (e.g. from carbines, among others the MP-43 carbine, brought into production at a similar time, in 1943) than the known to us system operating thanks to gases being carried away through a side opening in the barrel, where it is set in motion at a strictly defined moment, when the projectile passes the side opening in the barrel. In the MK-115 automatic gun the stimulus conditioning the unlocking appeared immediately and the rate of fire depended exclusively on the inertia of the piston and breech. This wasn't made just because the cartridge case (burning through) did not limit access to the cartridge chamber. It was about the cycle of reloading being executed relatively quickly, since as opposed to carbines the pressure of gunpowder gases didn't remain at a roughly stable level up to the moment the projectile left the barrel, since it was open from the very start. A tube, the gas conduit assured this, whose task was to compensate the recoil (this tube ended in a nozzle, placed at the extension of the barrel's axis). In this way recoil of the breech was caused directly by the bottom of the cartridge case, on which the gunpowder gases acted, and not by the piston rod powered by gases carried away through the side opening in the barrel (as in carbines). After firing, the breech expelled the metal cartridge case bottom and in a return motion, under the action of a spring, collected the next cartridge from the ammunition belt and inserted it into the cartridge chamber.
5. The problem of expelling misfires, or faulty cartridges from the barrel was also solved in an interesting way. In the tube connecting the cartridge chamber to the nozzle at the rear was found a small opening leading to a pressurised container, where a portion of the gunpowder gases found their way and were kept under pressure. If required this container fed a pneumatic mechanism, causing reloading. It was initiated electrically.

Only one prototype of the MK-115 automatic gun was made and partially examined before the end of the war. These were obviously only ground trials. It was merely determined that the weapon worked and in actual fact gave no measurable recoil. Only single rounds were fired, as with automatic loading it became evident that the cartridge case bottom was too weakly connected to its burning section and the round suffered damage. There was no longer sufficient time to remove this fault…

It was also not determined how the blast and shock wave, arising behind the gun after firing, acted on the construction and aircraft engines—the probably carrier of this weapon.

It would have constituted admittedly a complicated, but despite this attractive alternative to unguided rocket launchers. The expected merits were chiefly: higher accuracy (projectiles fired from a rifled barrel, with a greater muzzle velocity than in the case of rockets) and higher rate of fire, which would have rendered a greater probability of hitting the target.[75,76,77]

In general, the recoilless weapon turned out to be a very valuable invention. Only thanks to it was the German infantry, despite the constant rise in tank armour thickness, in com-

Technische SS- und Polizei-Akademie

Berlin, den 18. Januar 1945.

Geheime Reichssache!

Anlage. zu Tgb.z 966/45 (g.Kdos.)

A. **Abgeschlossene Arbeiten.**

1. **MG-Lafette mit Fahrgestell.**

Die Lafette mit Fahrgestell ist am 12. und 13. Dezember 1944 in Suhl dem Sonderausschuss Infanteriewaffen vorgestellt worden und hat sich beim Beschuss sowohl mit dem MG 34 als auch MG 42 bewährt. Z.Zt. wird vom Heereswaffenamt die Lafette auf Truppenbrauchbarkeit geprüft.

2. **Einstossflammenwerfer.**

Der von der Akademie zusammen mit dem Heereswaffenamt entwickelte Einstossflammenwerfer befindet sich in Grossfertigung. Die Akademie hat nunmehr die Entwicklung dahingehend weiter betrieben, dass der Einstossflammenwerfer statt aus Blech aus Pappe gefertigt wird. Die abgeschlossenen Versuche haben ergeben, dass die Ausführung aus Pappe allen Anforderungen entspricht.

Die Metalleinsparung beträgt pro Werfer 485 g bei einem Gesamtgewicht des Werfers von 1625 g.

3. **Panzerschreck.**

Angeregt durch die Entwicklungsarbeiten des Flammenwerfers aus Pappe hat die Akademie die Entwicklung betrieben, Rohr und Schutzschild des Panzerschrecks aus Pappe zu fertigen. Die Entwicklung ist abgeschlossen. Die aus Pappe gefertigten Geräte entsprechen allen gestellten Anforderungen durchaus befriedigend. Das aus Pappe gefertigte Gerät ist sogar gegen Deformieren durch Stoss oder Druck widerstandsfähiger als das Blechgerät. Durch die Umstellung auf Pappe wird eine Gewichtsverminderung von 2 kg und eine Einsparung von 5,5 kg an Metall pro Gerät erreicht.

4. **Mehrstossflammenwerfer.**

Die Entwicklung eines Einkessel-Mehrstossflammenwerfers für 8-10 Flammstösse wurde abgeschlossen. Es wurde an Stelle des mit Stickstoff gefüllten Druckbehälters eine Pulverpatrone verwendet, die sich im Ölkasten des Werfers befindet und durch einen Abreisszünder betätigt wird.

Dieses mittels Pulverdruck arbeitende Mehrstossflammenwerfergerät ist fertigungs-, bedienungs- und nachschubmässig (Fortfall der Stickstoffflasche) sehr viel einfacher als das Stosstruppgerät 41.

- 2 -

5. **Verdunklungszünder.**

Für den Sicherungsdienst wurde ein Lichtschalter für Sabotagezwecke entwickelt, der eine Sprengladung bei Eintritt der Dunkelheit zum Entzünden bringt (also zum Sprengen von Tunnels usw.).

6. **Entlastungsmine.**

Die Entlastungsmine ist eine Mine, die, wie der Name sagt, dann zur Auslösung kommt, wenn ein auf ihr liegender Gegenstand entfernt wird. Sie eignet sich besonders für Sabotagezwecke, Verminung von Häusern und dergleichen. Für die Verwendung dieser Mine ergeben sich, auf Grund ihrer Konstruktion, unzählige Möglichkeiten.

Die Entlastungsmine ist dringend von den SS-Jagdverbänden gefordert worden. Der Vorläufer dieser Mine war sprengstoffmässig gesehen leichter und hat sich bereits im Einsatz bewährt.

Von der Akademie ist die Fertigung von monatlich 100 - 200 Stück aufgenommen worden. Die ersten 100 Stück werden am 17.1.1945 ausgeliefert.

B. **In Entwicklung befindliches Gerät.**

1. **Barometrischer Zünder.**

Diese Zünderart ist seit langer Zeit bekannt, konnte aber wegen zu grosser Ungenauigkeit bisher nicht mit Erfolg eingesetzt werden.

Es gelang in wenigen Wochen in Zusammenarbeit mit O.K.L. einen voll brauchbaren Zünder zu entwickeln.

Verwendung: a) Bekämpfung feindlicher Pulks durch Abwurf von 1000 kg-Minenbomben.

Bedienung des Zünders ist derart vereinfacht worden, dass der Pilot lediglich die Höhe des feindlichen Pulks, aber in sicherem Abstand von diesem, einen elektrischen Kontakt zu betätigen hat. Er kann dann zu einer beliebigen Zeit und aus beliebiger Höhe seine Minenbombe abwerfen. Sie wird genau in Höhe der feindlichen Maschinen knallen.

b) Für alle Geschosse, die in einem bestimmten Abstand vom Erdboden zum Zerknallen gebracht werden sollen.

Z.Zt. finden Abwurfversuche statt mit einer Maschine des Baumusters Me 262.

2. **Stahlvergütung.**

Durch ein neu entwickeltes, sehr einfaches Wärmebehandlungsverfahren ist es gelungen, die Zähigkeit von chrom- usw. armen Stählen wesentlich zu verbessern.

2 cm-Geschützrohre, die nach bisherigen Härteverfahren bei 5 g Sprengstoff aufrissen, hielten nach dem neuen Verfahren bis 11,5 g.

Das Verfahren wird z.Zt. in Zusammenarbeit mit Ministerium Speer und Heereswaffenamt fabrikationsreif gemacht.

3. **Munitionswirkung im Ziel.**

Auf Veranlassung des Heereswaffenamtes und des O.K.L. wird das Eindringen von verschiedenen Geschossformen im Ziel mit neuartigen Mitteln untersucht. Wichtig für Verbesserung der Munition und für die Vereinfachung von Abwehrkonstruktionen.

Für Kaliber bis 10,5 cm und mittlere Auftreffgeschwindigkeiten werden Ergebnisse in Kürze vorliegen.

4. **Beton-Handgranate und Papp-Handgranate.**

Angeregt durch den Engpass Sprengstoff wurde die Entwicklung von Handgranaten aufgenommen, die bei gleicher Splitterwirkung mit weniger Sprengstoff auskommen.

1. **Ausführung Splitterbeton.**

Der Handgranatenkörper wird aus einer Mischung von Beton und Schrottabfällen gegossen und benötigt zur Zerlegung, wegen des geringen Energieaufwandes zum Zerreissen der Hülle und der bereits fertigen Splitter etwa 1/2 bis 1/3 (80-40 g) von der sonst in der Stielhandgranate benötigten Sprengstoffmenge (180 g). Beim Vergleich der Beton-Handgranate und der normalen Handgranate hat sich gezeigt, dass mit einem Splittergewicht von ca. 300 g bei 50 g Sprengstoff die Wirkung der Beton-Handgranate der normalen Handgranate überlegen war. Der Nachteil besteht im Gewicht der Granate. Bei Wurfversuchen lag die Reichweite etwa 10 m kürzer. Der Vorteil dieser Beton-Granate besteht darin, dass sie leicht in Heimarbeit herstellbar ist und so für Volkssturmverbände besonders gut geeignet ist. Sie lässt sich in Verbindung mit dem Zugzünder 42 auch als Schützenmine verwenden. Bei dieser Verwendung spielt das Gewicht keine Rolle.

2. **Papp-Handgranate.**

Zu der Ausführung der Handgranate in Pappe ist bis auf das Gewicht das Gleiche wie zu der Ausführung aus Beton zu sagen. Die Papp-Handgranate benötigt voraussichtlich bei gleichem Gewicht wie die eingeführte Stielhandgranate nur etwa 1/3 Sprengstoff bei gleicher Sprengund Splitterwirkung.

Durch diese Entwicklungen wird ausser der Einsparung an Walzblech- und Sprengstoffkapazität wegen der einfachen Fertigung die Produktion wesentlich gesteigert.

- 4 -

5. **Wesentliche Erhöhung der Anfangsgeschwindigkeit von Geschossen.**

Auftrag des Reichsluftfahrtministeriums. Es sind zwei Wege beschritten worden und zwar:

a) **Geschoss mit Kaskaden-Kartusche**, d.h. ein Artillerie-Geschoss wird mit mehreren Kartuschen versehen, die nacheinander im Geschützrohr zur Zündung gebracht werden. Die erste Kartusche wird wie üblich gezündet, die weiteren, die ja mit dem Geschoss mitfliegen, müssen durch die Rohrwand des Geschützes hindurch gezündet werden.

Die Versuche haben ergeben, dass durch einen Induktionsstrom eine einwandfreie Zündung durch die Rohrwand hindurch möglich ist.

b) **Antrieb durch Detonationsgase.**

Ersatz des Pulvers zum Antrieb durch Sprengkörper. Die Schwadengeschwindigkeit der Detonationsgase liegt bei 4-5000 m/sec.

Die Versuche sind sehr schwierig, es scheint aber mit Hilfe von Hohlraumsprengkörpern eine Lösung möglich zu sein.

6. **Kraftstoff-Kanone.**

Hierunter wird eine Waffe verstanden, die an Stelle von Pulver mit Diesel- oder Otto-Kraftstoff unter Zusatz von Sauerstoff genau wie ein Motor arbeitet. Nur wird an Stelle des Kolbens ein Geschoss vorwärtsgetrieben.

Gelingt es, eine ausreichende Beschleunigung des Geschosses zu erreichen, so sind die Vorteile sehr gross z.B.

einfache Konstruktion,
geringer Energiebedarf,
Wegfall der Kartusche,
Wegfall des Zündhütchens,
grössere Schussgeschwindigkeit.

7. **Automatische Pistole mit kartuschfreier Munition.**

Die Entwicklung einer solchen Waffe wird besonders vom Sicherheitsdienst gefordert.

Bei dieser Waffe handelt es sich im wesentlichen um eine Pistole, die voll-automatisch schallgedämpft schiesst. Bei den bisherigen Konstruktionen war ein automatisches Schiessen nicht möglich, da beim Öffnen des Verschlusses die Waffe unverändert knallt. Zur Erreichung des gewünschten Zieles musste also gesorgt werden, dass die Waffe in ihrer Funktion vollkommen geschlossen bleibt. Dieses kann nur dadurch erreicht werden, dass die Kartusche mit verschossen wird, die einen Rückstand (Kartusche) verlässt — kartuschfreie Munition entsprechend dem Prinzip der Raketenabschüsse.

Der Stand der Arbeiten ist z.Zt. so, dass bruchreif der Beschuss zwischenzeitlich erfolgen soll ein positives Ergebnis.

Police. It bears the date of January 18, 1945. It is an enumeration of many different types of modern weapons at the time, at the same time in section "A" were described those whose development was completed and in section "B" those that were still being worked on. The document has been reproduced in entirety, however my commentary concerns only the most interesting excerpts.

In point A-3 we find some rather odd information concerning the "Panzerschreck" anti-tank rocket-propelled grenade launcher. Here are the contents of this point:

> As a result of being inspired by research work on a flame thrower [point A-2] made of cardboard [impregnated?], the Academy carried out work with the aim of developing the tube launcher and safety plate of the "Panzerschreck" grenade launcher, made of cardboard. These works are finished. The devices made of glued cardboard fully meet all the presented requirements. They are more resistant to deformations (…) than the original, metal ones [the launcher was made of 2.5 mm steel sheets]. Thanks to switching onto the glued cardboard it has been managed to achieve reduction of weight by 2kg and saving of 5.5 kg of metal per launcher.

The "Panzerschreck" was a light, very effective infantry anti-tank weapon. A rocket-propelled projectile with hollowed warhead penetrated 220-mm thick steel armour, which ensured the destruction of any tank at that time (calibre: 88-mm). In addition it was simple and cheap, which enabled it to be used on a mass scale. This found confirmation among others in an analysis from 1951:[19]

> The whole projectile is made unusually lightly and cheaply from components not requiring any surface treatment (an exception is the fuse socket and screwed joints), the lack of any close fits permits the mass production of pressed parts.

In a further section of document R-2, in point A-5, we find a further curiosity. A fuse was described there, destined for sabotage groups, activating itself when darkness fell. Along with an explosive charge it was supposed to be fixed to a train and the explosion was follow on entering a tunnel. In this way it would be possible to destroy both the train and the tunnel. Whereas in point B-1 a different fuse was described, barometric (pressure). It was mainly to be used in one ton bombs designed for bombing … formations of enemy bombers. It was written that trials were being carried out using a Me 262 aircraft. I do not know if it was managed to put this concept into practise, however the one ton bomb could indeed destroy an entire formation of bombers. In this point information was found that it concerns a significantly more precise fuse than up till now.

A further section of the document contains not only a curiosity, but above all information about weapons fully deserving the term breakthrough. Point B-5 concerns a new type of

mand of a mass produced, portable but effective anti-tank weapon, right up until the end of the war. It took one's position in this respect as the best of the all the armies of that time.

A very important and equally effective complement to these weapons were the "Panzerschreck" anti-tank rocket launchers.

Unusual Ideas

A source of much interesting information about less conventional German concepts is the archive of the Reich's Scientific Research Council (Reichsforschungsrat).[79] In one of the documents was included among others, information about a version of the aforementioned weapon unknown until now.

It is exceptionally interesting, and found its way to the "Council's" archive from the Technical Academy of the SS and

R4A

BLATT 9

PATENTANSPRÜCHE

1. Waffe insbesondere zur Bekämpfung von Flugzeugen dadurch gekennzeichnet, daß das Geschoß aus einem eine Mehrzahl von Einzelgeschossen aufnehmenden stromlinienartig gestalteten Hohlkörper besteht, der aus einem Rohre mittels eines Treibsatzes, eines Raketenantriebes oder einer Treibladung ausgestoßen und nach dem Zielraum überführt wird.

2. Waffe nach Anspruch 1 dadurch gekennzeichnet, daß der die Einzelgeschosse aufnehmende Hohlkörper an seinem rückwärtigen Ende ein Leitwerk trägt, das in ein mit einem Treibsatz ausgestattetes Abschußrohr eingesetzt ist (Figur 1).

3. Waffe nach Anspruch 1 dadurch gekennzeichnet, daß der die Einzelgeschosse aufnehmende Hohlkörper an seinem rückwärtigen Ende einen Raketenantrieb mit Leitflächen trägt, wobei der Abschuß aus einem Führungsrohr heraus erfolgt. (Figur 2)

4. Waffe nach Anspruch 1 dadurch gekennzeichnet; daß der die Einzelgeschosse aufnehmende Hohlkörper an seinem rückwärtigen Ende ein Führungsstück und daran anschliessend einen Treibspiegel trägt, wobei der Abschuß aus einem Rohr mittels einer Treibladung erfolgt (Figur 3).

5. Waffe nach Anspruch 1 bis 4 dadurch gekennzeichnet, daß das Abschußrohr als Handwaffe ausgebildet ist.

6. Waffe nach Anspruch 1 bis 4 dadurch gekennzeichnet, daß das Abschußrohr auf einem Gestell mittels Richtmitteln einstellbar nach Art eines Werfers oder Düsengeschützes angeordnet ist.

7. Waffe nach Anspruch 1 bis 6 dadurch gekennzeichnet, daß der die Einzelgeschosse aufnehmende Hohlkörper einen Sprengsatz mit einem einstellbaren Zeitzünder enthält.

R4B

BLATT 10

8. Waffe nach Anspruch 1 bis 7 dadurch gekennzeichnet, daß das Gehäuse für die Einzelgeschosse mehrteilig ist und bei Erreichen des Zielraumes zwecks Freigabe der Einzelgeschosse zerteilt wird.

9. Waffe nach Anspruch 1 bis 8 dadurch gekennzeichnet, daß die Einzelgeschosse in an sich bekannter Weise mit eigenen Treibsätzen ausgerüstet sind, unter deren Einwirkung die Geschosse im Zielraum kreisende, spiralförmige oder hin- und hergehende Bewegungen ausführen.

10. Waffe nach Anspruch 1 bis 8 dadurch gekennzeichnet, daß die Einzelgeschosse bei evtl. Verfehlen des Zieles durch die Wechselwirkung ihrer eigenen Schwere und weiterer Kraftimpulse ihrer eigenen Treibsätze mehrfach erneut an das Ziel herangebracht werden.

11. Waffe nach Anspruch 9 bis 10 dadurch gekennzeichnet, daß bei Nichtzustandekommen eines Aufschlags durch Auftreffen des Geschosses auf das Ziel nach Beendigung des Abbrandes des letzten Teiles des Treibsatzes das Geschoß selbsttätig durch Entzündung seines Sprengsatzes zerlegt wird.

R4C

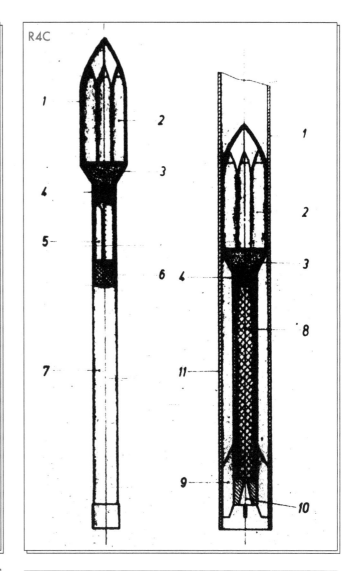

R4D

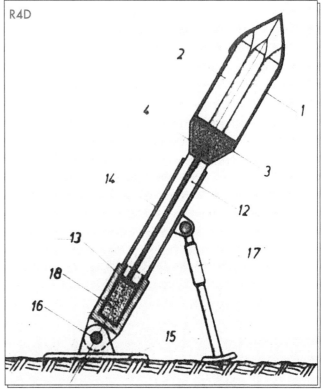

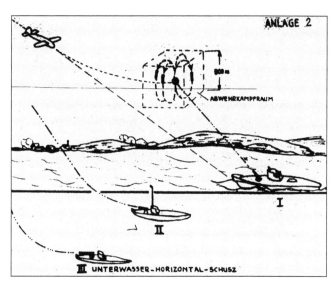
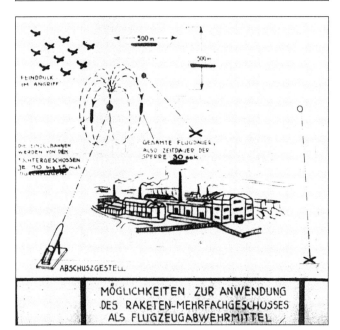
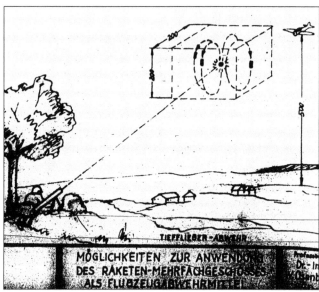

Drawings attached to the R-5 document

gun, which was supposed to be an alternative to the famous V-3, the multi-chamber gun from Misdroy. This complicated system didn't work, the construction was in addition badly calculated and didn't resist the "target" pressure.

Point B-5 describes something which gives the impression of being a considerably more rational solution. Outwardly this gun would have differed little from the classic long-range cannon, although there were also to be many powder charges which would have been initiated in succession. They would have moved along with the projectile in special containers and been initiated electrically, inductively (through the wall of the barrel). In order to increase the projectile's muzzle velocity it was even considered replacing the powder with a special shattering (detonating) high explosive, for in this way an increase in the speed of gas expansion to 4,000-5,000 m/s would have been attained. This is a practically unknown research project and it is a pity that more precise technical data was not given…

One of the enclosed documents (R-2F) describes moreover a cannon, in which the medium accelerating the projectile would have been hydrogen. This work was conducted by Prof. Reyner, in the same SS Technical Academy in Brünn (Czech Brno). It was expected in connection with this an increase in projectile muzzle velocity to 1,600 m/s. The first prototype was to be ready for the spring of 1945.

The continuation of point B-5 appears to be point B-6, in which a more long-range plan was outlined (although lying within the limits of technical possibility). It assumed the construction of liquid-propellant guns—oxygen and fuel being a derivative of petroleum. From a present-day perspective one may appreciate, that in reality this approach promised the greatest progress. In the document such virtues were emphasised like:
- simplicity of construction
- high power efficiency
- the elimination of cartridge cases
- the elimination of primer
- an increase in projectile's muzzle velocit

The "Fliegerfaust," anti-aircraft multiple rocket launcher

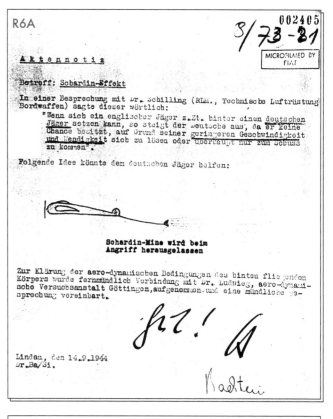

Point B-6 as well as point B-5 are in all probability the first known confirmation of the fact that this breakthrough to the core work, was carried out.

This is also the case in point B-7. It refers to construction work (by no means theoretical!) on a caseless cartridge automatic pistol. It was developed for an order by the Security Services (SD), as it enabled almost complete damping of sound. This is normally never possible in an automatic weapon, since in order to expel the cartridge cases, the cartridge chamber is opened when a relatively high pressure is still dominating in the barrel. This was to have such a high significance in the case of the weapon for the SD, in that was written: "In these operations the return of a second and third shot is often required…"

It was also written that:

The state of work is such, that test firing should be carried out in January or February [1945]. Initial research predicts a positive result.

Ultimately it was only in the middle of the 1980s that the successful weapon of this class was directed on a small scale for production. It was the G-11 carbine on caseless ammunition, from the company Heckler und Koch. Despite a fundamental superiority over the classic carbine (2-3 times greater effectiveness of fire with the consumption of a similar mass of ammunition, and smaller weapon's mass), it wasn't accepted for armament in consideration of a barrier in the form of the standardisation of ammunition in NATO. It had (and has) a calibre of 4.74-mm and eluded the name "machine pistol" only with respect to the high projectile's energy.

The next document describes the little known preparation for military use of 214-mm unguided air-to-air rockets. They had a mass of 100 kg and were equipped with 40 kg high explosive/incendiary warheads, each containing 400 pre-formed splinters, incendiary charges. These rockets were tested by a fighter regiment from Parchim (JG-10). The first batch of them found their way there at the end of 1944. I have enclosed

The Gustav super heavy (cal. 800-mm) cannon in Ruegenwalde. Certainly it wasn't the best solution from the point of view of the effect/cost coefficient, nevertheless it possessed some unique advantages. Probably the most significant one was the fact, that the shell exploded at an extremely great depth —usually 35-40 metres (!) below the surface of the ground. Normally it wasn't needed, but this feature was becoming priceless when attack on major fortifications was coming into account. Thanks to this, the Gustav has destroyed e.g. the ammunition depot of the Sevastopol fortress in 1942.

Another example of the super heavy artillery: the Thor mortar

a technical drawing of the rocket.

Document R-4 also refers to an unguided anti-aircraft weapon. It is a patent specification referring to different, proposed versions of surface-to-air equivalents to the "Panzerfaust" and "Panzerschreck" weapons. In Prof. Osenberg's original drawings can be seen three versions: rocket, recoilless as well as an analogy to a mortar. These drawings are interesting in so far as they represent completely different concepts to that which was regarded as final of the "Fliegerfaust" rocket launcher, introduced for limited production. I have also enclosed his drawing.

Professor Osenberg (Chief of the Planning Office of the "Council") was the author of even stranger conceptions, though not devoid of any sense, which were described and illustrated in document R-5. It mentions unguided, anti-aircraft circling projectiles! They would enable to create some kind of "anti-aircraft barrage," self-sustaining for some period of time.

The rocket would have been characterised therefore by a greater probability of hitting an aerial target than with a straight flight path all the more that simultaneously were being developed various warheads carrying submunition—"sub-projectiles" dis-

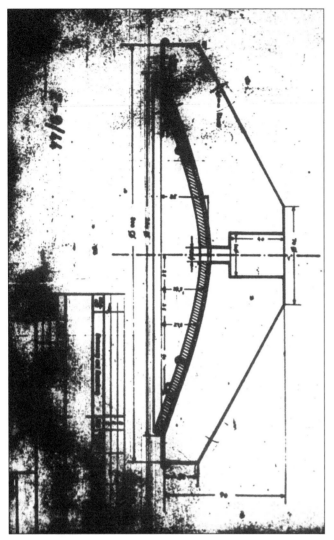

This drawing, of not too good quality, made by Professor Schardin, strikingly resembles projects from the 1980s.
(Explosively formed projectile)

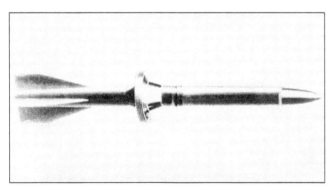

Another concept that proved to be futuristic, a fin stabilized sabot shell, developed for the 280-mm gun, which increased its range to 150 km!

loops. Nothing indicates that test firings of any version were managed to be carried out. On the other hand this conception wasn't developed after the war.

However it was different in the case of the idea described next, in a set of documents marked as R-6. They refer to the conception of "explosive forming (and acceleration) of the projectile," defined in German nomenclature from the time of the war as the "Schardin effect." This discovery had its origins in the United States, where in 1936 the physicist R.W. Wood observed the formation of a ball-shaped splinter—a minibullet from the concave front of a detonator, thrown in a furnace by accident (an investigation has been launched, since the splinter killed a certain person). However only the Germans exploited this effect during the war, developing a whole family of anti-tank charges. This research was already advanced in 1940, whereas in June 1944 Osenberg wrote of a very high initial velocity being attained—3,000 m/s and a reduction in scatter down to approx. 0.5 metres at a distance of 150 m.

Particularly little-known facts are among others the development of the "Schardin mine," which was to be towed by a fighter as a defensive weapon as well as the design of a two-stage projectile. A plate deformed as a result of an explosion was to accelerate the second projectile, stabilised with the aid of fins (see diagrams). An anti-armour warhead of this type was also developed to the order of Kriegsmarine.

However Professor Osenberg's invention didn't end at this! A set of documents designated R-7 presents another unusual (also developed after the war) idea, in the form of a patent specification along with a drawing. In this case it concerns an anti-tank weapon transported by an aircraft. This was a far-reaching, aerial derivative of the "Panzerfaust," differing from its predecessor among other things in that the projectile fired from a recoilless pipe consisted not only of a hollow charge (in this case developed mainly to penetrate concrete) but also of a second recoilless projectile, which was to fly inside the target through an opening made by the first charge. The second projectile possessed a delay fuse and the entirety was stabilised in

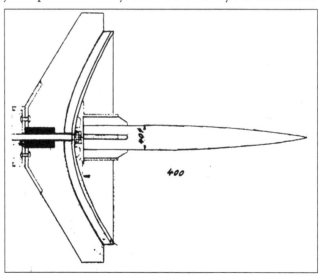

A modified project, relying on the explosive acceleration of a fin stabilized armour-piercing sabot projectile

persed in the target area and circling autonomously.

Osenberg wrote that the first version was to be ready for trials at the end of February 1945, whereas limited mass production of 500 pieces was supposed to commence at the beginning of May that year. This only concerned the air-to-air version, containing a "large number" of submunitions, performing in a time of the order of 30 seconds around 10-15

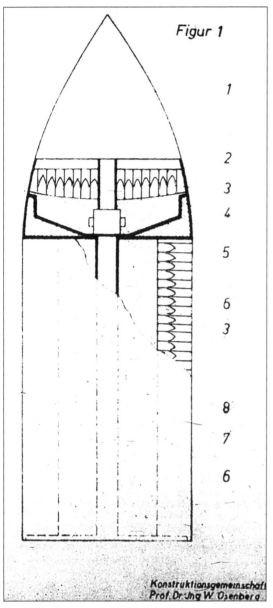

The drawing attached to a patent specification (R-7)

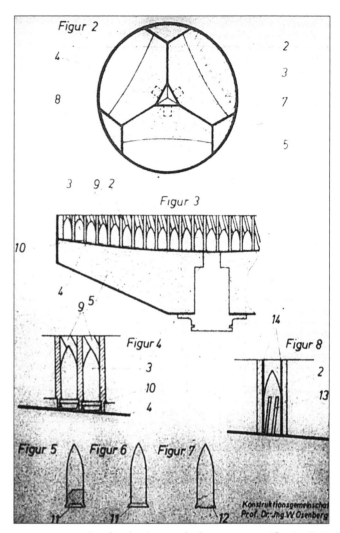

Another drawing attached to a patent specification (R-8)

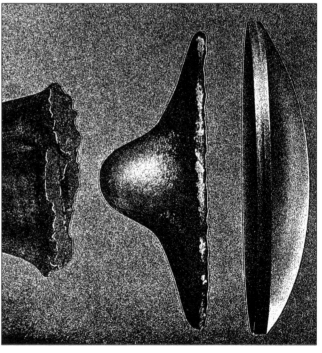

Successive phases of a projectile's explosive formation, a modern mosaic of photographs by Rheinmetall

flight with the aid of fins, unfolded after firing.

It is unknown if the Germans had pre-tested this weapon by 1945, but the results achieved after the war indicate that it was a very prospective idea. A whole group of very effective weapons arose designed to destroy bunkers, hardened aircraft shelters and runways, projectiles exploding under the concrete slabs of a runway as a rule inflict irreversible damage…

Another, and not at all the last invention of Osenberg was presented in a set of documents, bearing the symbol R-8. Once again we have to deal with a precise patent specification, this time to which a report was attached, describing the results of test firings. Described here was an artillery fragmentation projectile, but with a completely non-standard construction and differing with a significantly greater range and splinter armour-piercing ability than normally. Its shell is divided into four fundamental parts or plates: one frontal and three lateral. Each plate possessed from several dozen to hundreds of rifled

New Concepts for Conventional Weapons | 135

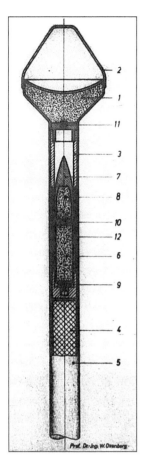

The drawing attached to a patent specification (R-8)

openings, in which pre-formed splinters were already located. In reality these were miniaturised projectiles with rotational stabilisation during flight. This was certainly a more expensive solution than a cast shell, but the significant increase in effectiveness in relation to certain types of targets also was significant. The practise of military technological development shows at the same time, that in spite of everything, the cheapest way to gain an increase in effectiveness of a given weapon system offers the upgrading of ammunition. Time shows that the expenses raised for this objective usually turn out to be justified.

On January 24 and 26, probably of 1945, the first trials of this ammunition were carried out near the town of Redlin. So as to be able to easier follow flight paths of the splinters, the surface of a frozen lake with 35-cm thick ice was chosen for this purpose. "Targets" were set on its surface: a vertically positioned 6 x 5 metre plate made of plywood and behind it, the wing from a Focke-Wulf 190, several 5-mm thick armoured shields each 0.5 m² in area and two empty rubber fuel tanks, with a wall thickness of 10-mm. At a distance of 150 m from this "set" was positioned a wooden block, 1 metre high and on it the aforementioned type of warhead designated 240/5. It had a mass of 39 kg. With the aid of an optical device it was aimed at the target and then electrically fired. The effectiveness turned out to be extraordinary.

Although in general the splinter flight paths formed a 26° cone, as much as 38 hit the vertically positioned "shield" alone. Two splinters penetrated the wing of the FW 190. Two pierced the wall of the fuel tank and remained inside, one of which was an incendiary. One of the armoured plates showed signs of being hit without penetration. At the site of the explosion itself on the other hand, a hole 1.5 m in diameter was made in the ice.

Later another four trials were carried out gradually decreasing the distance to the target, positioning a tank filled with fuel. A greater and greater number of direct hits were observed, but although part of the splinters were to have an incendiary action, the fuel wasn't ignited. The armoured plates also weren't pierced, although at the smallest distance, 45 metres, they were seriously deformed. The final test was the test of a rocket with the 240/5 warhead, fired from a FW190 fighter flying at low altitude. The rocket lost stability after several hundred metres of flight and spinning chaotically at a distance of approx. 1,300 metres from the launch site, has hit

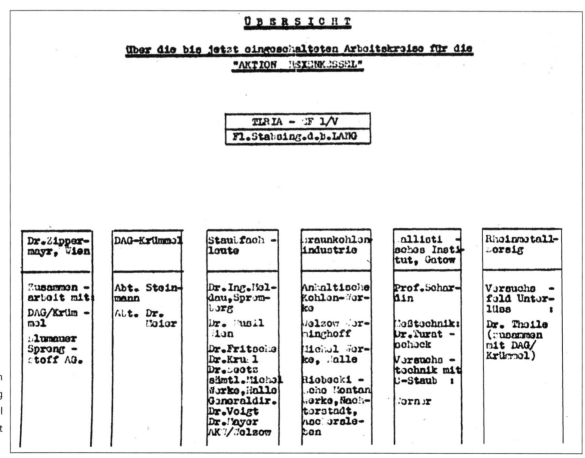

An organization chart regarding the Hexenkessel project

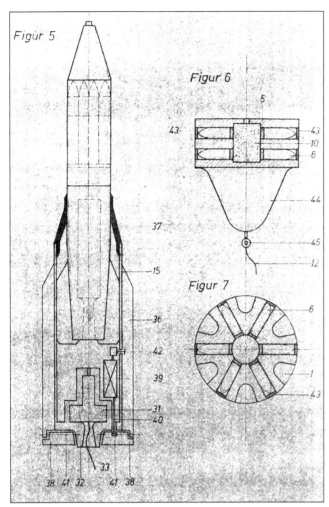

One of Prof. Osenberg's unusual concepts, the design of a two stage rocket, which in turn was supposed to fire smaller submissiles

In the German army the rocket artillery played a significant role, with the "WGr" being the main system

the ground, deflected and has wedged between the branches of a tree, the proximity fuse didn't work. So ended the story of one of the more interesting ideas of Prof. Osenberg...

The next document concerns the "Hexenkessel" project, work on coal fuel-air charges. It constitutes an excellent supplement to the description presented in one of the previous chapters. It is a report from a conversation with a certain engineer Lang on the subject of the work's progress. It mentions the necessity of encompassing it with a priority, for only then will it be possible to achievespecific results in a short time.

Since the document dates from February 17, 1945, it is obvious that the Germans had no chance of practically using this discovery before the end of the war. However, this document constitutes an important confirmation of the fact that the Reich's Scientific Research Council was deeply engaged in the realisation of the "Hexenkessel" project, currently so little known. To this document it has been attached the organisational scheme pertaining to this project.

Likewise one should deal with document R-10 from Feb-

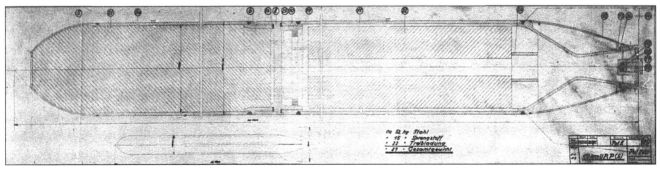

One of numerous German responses to the Soviet "Katyushas," a design

New Concepts for Conventional Weapons | 137

A salvo of the "WGr" rockets

ruary 9, 1945. It in turn proves the engagement of the "Council" in work concerning means of combating enemy bombers, disabling them by interfering with their engine' operation as well as with the navigational and radio equipment.

Much more interesting however is the document R-11; probably the most intriguing document reproduced in this chapter. It concerns an excerpt of an extensive set of reports, concerning information about new German weapons which appeared (e.g. were published or broadcasted) on the other side of the front. The Abwehr, military intelligence, regularly supplied the "Council" with this type of data.

On the reproduced page (see: Part 3) I found two interesting reports: J-9180 and J-9181. They deserve to be translated

The "Schräge Musik," the cannon's barrel fires upwards

in entirety. Here are the contents of the first, dated from December 7, 1944:

> *Allied pilots on the Italian front are exposed to the action of the new German weapon, with the objective of making bombing air raids impossible by causing icing of the bombers participating in them. The new weapon is described only as a fantastic "air icing vehicle" ["Eisluftwagen"], which forms clouds from "dehydrated, frozen air" before approaching Allied aircraft.*
> *This intention is fulfilled in such a way, that the "frozen air" mixes with the thinned atmosphere and leads to the fatal fall [of the aircraft] located in the midst of this, since the ice forming on the bombers disables their control systems and results in the aircraft falling into a spin.*

The above information tallies with the contents of Allied Intelligence reports apart from the fact that the causative factor was not "frozen air," but a special composition of gases, which generate a very low temperature during expansion. Similar confirmation in other sources, finds information on a "weapon" mentioned by report J-9181 (information from a German radio watch):

> *It has been notified that extraordinary silver spheres were seen on the western front today, which flew in the air. It is generally accepted that the Germans have employed a new weapon, though it is still not possible to determine how this secret weapon works. It will in all probability constitute a serious defensive measure.*

Very many innovative ideas of this kind were carried into effect in the field of anti-aircraft weapons, where the pressure of technological progress was exceptionally strong. One of the most crucial, though at the same time least known advances was the connection of anti-aircraft guns with radar, in other words the creation of a radar fire control system. Effectiveness of this solution was great and has been even more improved in connection with the so-called anti-aircraft towers (Flaktürme); giant concrete shelters, up to 20 storeys high, on top of which were installed 105 and 128-mm anti-aircraft guns, as well as the Würzburg radars. Because the towers formed a grid with known co-ordinates, it was possible to obtain very precise information about the positions of targets, in three dimensions, even when only the passive, thermal devices were used (when the enemy was jamming the radars). In Berlin alone, there were about 1200 anti-aircraft guns! The aforementioned towers were built in Berlin, Hamburg, Vienna, Bremen and Breslau (Wrocław).[129] This solution was realised perfectly. So recalls one of the Allied pilots:[97]

> *It was my second combat flight. I couldn't forget the grumbling of the older bomber pilots. "Usually after the second time they shoot off your ass." We had approached the target. The first projectiles exploded precisely between*

our bombers. This wasn't a barrage fire. They had fired the newest 105-mm guns guided by radar. (…)

A formidable fire ball exploded right in front of us. We plunged into it, feeling the crash of splinters lashing the fuselage. I saw how a projectile directly hit the leading bomber. The sides of the hit fortress bulged out like a balloon. The fuselage by a wonder returned to its previous shape, but great lumps started falling of it, and tongues of fire crept out of the ripped open junctures. The burning bomber began to fall away to the left.

Other ingenious solutions were used as an armament for fighters. At a certain moment the German pilots orientated themselves to the fact, that when they flew directly below a formation of enemy bombers, it was difficult for them to spot the attacking planes, particularly at night. They fell on the idea therefore that 20-mm guns be mounted on the Bf-110 fighters perpendicular to the fuselage's axis (behind the cockpit), so as to fire upwards. Thanks to this it was easier after all to hit a bomber; its silhouette then had a much greater area than e.g. with an attack from the rear. Such armament acquired the unofficial code-name "Schräge Musik," literally "sloping music." It was employed together with new on-board radar (SN-2 "Lichtenstein"), which were not very sensitive to jamming and completely didn't respond to "chaff," scattered by the Allied bombers.

The debut of such a combination coincided with the air raid on Nuremberg (March 30, 1944). As a result the Allies lost 95 of 795 bombers that night, and the Germans merely several night fighters.[96]

New Generation Small Arms

The conceptions described in this chapter to a large extent had a purely experimental character and didn't always show any predominance over existing constructions, what is in any case an inevitable "by-product" of all kinds of research. Obviously it wasn't always like this; after all only through research may one reach crucial solutions. I wanted to present such a weapon below.[80,81]

The Germans introduced many "historic" types of firearms during World War II, already well known. The MG-42 machine gun finds itself in a prominent position, regarded by many as the most faultless hand-held weapon of this class (although the key solution a roll-locking system was "scooped out" of the pre-war Polish patent that engineer Szteke developed for his semi-automatic rifle). The MG-42, under altered names—the MG-2, MG-3 and 7.62-mm calibre—is produced to this day in several countries and used in about ten armies around the whole world. Very successful were the Walther PP and PPK pistols, popular to this day. An original, successful design was the FG-42 automatic rifle designed for airborne troops. An avant-garde weapon was the aforementioned pistol for caseless ammunition…

I would like to devote however a little space to a different hand-held weapon, in my opinion equally crucial and less known, the MP-43 carbine, thanks to which completely new technological standards were introduced (under the pressure of difficulties with raw materials), being a true qualitative breakthrough. Being produced chiefly by a forging and pressing method it was many times less material-consuming and energy-consuming than previous types, manufactured chiefly by a machining method. At the same time it gave the soldier a new firepower as regards to quality. Let us begin however from the origins…

In the interwar period, the basic armament of the infantry in almost all countries of the world was comprised by magazine rifles (non-automatic). On the basis of experience gained in the stationary World War I, such solutions were considered optimal for several reasons. A merit of the magazine rifles was the high accuracy, range and penetrability of the bullets, a drawback, the small rate of fire. This was therefore a good armament in conditions of low-intensity conflict, where firing could be carried out for a longer time at distances the order of several hundred metres.

Sub-machine guns (machine pistols) were gradually introduced as a complement to magazine rifles, a light automatic weapon, enabling the infantry to increase its firepower during assault (when there was no possibility of accurate aiming) and more useful in other types of conflicts at small distances. It seemed that the magazine rifle/sub-machine gun combination was adequate. However after thorough analysis of the first encounters of World War II, it became evident that this was not the case. This became particularly evident after the Germans committed aggression on the USSR. On the eastern front the infantry fired from small arms mainly at distances the order of 100-200 metres. In these conditions a large magazine rifle range (about 1000 m) wasn't necessary, as was a high bullet penetrability. On the one hand the potential of this weapon was to a large extent unexploited, and on the other hand—de facto—it was small with regard to the small rate of fire. In turn a distance the order of 100-200 metres was too large for sub-machine guns. Their effectiveness at such distances was very small. Analyses carried out after the war revealed that for one killed or wounded German soldier fell approx. 40,000 pieces of used rounds for the PPS and PPSh sub-machine guns. It was likewise with the German MP-38, MP-40 and Bergmanns. For at such distances the wounding effect of sub-machine gun bullets (pistol bullets) was insufficient; a direct hit on a helmet didn't always result in it being penetrated, and a direct hit on a soldier didn't always result in him being eliminated from the battlefield.

The German commanders saw this problem particularly well, who on the whole came up against the numerical superiority of the enemy. The order drawn up after a thorough analysis of military needs led to the development of a new type of small arm—the so-called automatic carbine—combining certain features of the sub-machine gun (rate of fire) and of the magazine rifle (range of fire). Since it was considered that the key element determining the capabilities of the new

A comparison of the Kurtz Patrone round with a standard pistol round and with the Mauser riffle round

	Parabellum 9 mm	Kurz Patrone	Mauser's rifle round
round's mass [g]	10.5-12.5	16.5-16.8	24.1-26.2
projectile's mass [g]	7-8	7.9-8.2	11.53-12.83
mass of the powder charge [g]	0.32-0.36	1.57-1.59	3.15-3.25
length of the round [mm]	29.7	47.8	80.5
length of the case [mm]	19	33	56.8
length of the projectile [mm]	15.5	25.6	28
muzzle velocity [m/s]	390-400	690-700	765-911
initial energy of the projectil [J]	580-590	1880-2010	3374-5324

Comparison of three main German rounds for infantry weapons. From left to right: 1) 9 mm Parabellum, for pistols and machine pistols (e.g. MP-40), 2) "Kurz Patrone," the carbine round for the MP-43 (assault round or short round, according to English nomenclature) and 3) Mauser's rifle round

weapon and its construction was the round, its development was concentrated on in the first stage. A round was made, being a shortened version of the 7.92 x 57 mm Mauser rifle round.

Rounds of this type are currently defined by the designation carbine rounds or assault rounds. One should at this stage point out that the Germans were not the first to develop such a round. The Czech ballistics expert Karel Krnka accomplished this in 1892 for a weapon developed by the Swiss gunsmith F.W. Hebler. The rifle (carbine) was 1/3 shorter and lighter than the Swiss army magazine rifle of that time. Their work however was soon forgotten. In 1918, the German first lieutenant Piderit announced work including a project to bring into armament a round similar to the Czech one. In 1927, in the plants of Mauser a small number of 7-mm carbine rounds were developed and produced. In the 1930s small arms for this round were worked on in various production plants.

In 1938, the company Walther was the first to submit their prototype for testing. After the introduction of numerous modifications, the weapon acquired the designation MKb-42(W), Maschinenkarabiner-42. From this time on the name "carbine" began to be used (Karabiner). Although the MKb-42 (W) was withdrawn from testing (so as to appear later un-

der a different designation in a version adapted for a new round), research work on the new round continued. The basis for this was an official order addressed April 18, 1938, on behalf of the Army. Despite a reduction of 50% (in comparison with the rifle round) in mass of the powder charge, the effective range of the weapon for this round was considerably greater than the sub-machine guns for the 9-mm Parabellum round. In the years 1934-1938 the company GECO developed a version of the 7.92 x 33 mm Kurz Patrone round approved for production. This was the first carbine round in history to be brought into armament. Production of it commenced a year after the outbreak of war in versions with a leaden core or made from mild steel.

The second of the new carbines was developed for the new round from GECO: the MKb-42(H) from the company Haenel. The chief designer of the MKb-42(H) was the famous Hugo Schmeisser, which surely was instrumental in that this design turned out to be better than the MKb-42(W). Both types worked however on the same principle: the powering of the automatics followed through the use of the energy of a portion of the gases carried away through the side opening in the barrel. As in the case of the majority of modern carbines (such as the M-16, G-36, the British L-85 etc.), the gas conduit was situated above the barrel. Locking of the barrel ensued by sliding of the breech's front in a vertical plane. The weapon's large mass and high degree of complication in comparison with the MP-40 was unavoidable with regard to the bullet's high muzzle energy, forcing the barrel to be locked at the moment of firing and with regard to the large force of recoil as well as high gas temperature, forcing the use of a solid construction (more difficult for the barrel to overheat). Initially the MKb-42(W) and MKb-42(H) were produced in small numbers and sent for comparative testing on the eastern front. They were dropped by parachutes on an infantry unit encircled and cut off from supply. This unit managed to escape from the encirclement and dispatched its opinion about both types of carbines, from which it appeared that the Haenel's design was better. This company acquired in connection with this an initial order for 8,000 pieces renamed at this time the MK-43. The order was realised in a record time of 3 months.

Elite units were armed with the MK-43, chiefly the Waffen-SS. Later production was halted for a certain time, with regard to the high costs, which exceeded the production costs of the MP-40. The merits of this pioneering weapon as was the automatic carbine, were not so obvious then, than at present. However after analysing thoroughly reports from the front pouring in from users of the MK-43, the arguments of the sceptics were overthrown and production was again set in motion, this time on a large scale.

Because of Hitler's "reservations" or prejudices in regard

A soldier from a German ski-division armed with the MP-43, early 1944

The MG-42

to the automatic carbine as such, a new designation was introduced at the end of 1943: the MP-43, which was soon changed to the MP-44. Since the carbine wasn't, just like the abbreviation suggested, a machine pistol, the weapon finally acquired the "compromising" designation StG-44 (Sturmgewehr 44, assault rifle model 44). In practise all the designations were applied interchangeably, with the MP-43 becoming most generally accepted.

Infantry equipped with the MP-43 were in command of a significantly greater firepower and could apply different tactics. The Kurz Patrone round enabled the weapon to achieve a high accuracy, high "disabling power" of the projectile, and at the same time the weapon, despite being automatic, could be easily operated by a single soldier. The projectile had about three times higher kinetic energy, than the 9-mm Parabellum pistol bullet for the MP-40. Firing could also be carried out

The MP-40 machine pistol

on the run during an attack, without the necessity of resting the butt on the shoulder. The majority of produced carbines were equipped with an attachment for throwing antipersonnel grenades.

Trials were also carried out with curved barrels, fitted on adapted versions of the weapon. The StG-44V could use a barrel to make firing possible at an angle of 30-40° in relation to the weapon's main axis and in the case of the StG-44P version, even at an angle of 90°. Both devices bore the name Krummlauf and had additional sighting devices installed. However during trials it became evident that the studied conception was erroneous and didn't augur promising results. Both converted designs probably never found their way to front-line units and were not used in combat.

The MP-43 design was systematically improved in due measure of gained experience, among others, many produced specimens had the possibility of optical sights being installed, and from the beginning of 1945 also active night-vision sights(!), operating in the near infrared, called the Zielgerät-1229 Vampir. At the end of the war several modernised versions had been developed: the MP-45(M), Gerät 06-H and StG-45(M). While preserving the merits of the MP-43 its design was somewhat simplified with the aim of reducing the costs of production.

Work on the StG-45(M) was the most advanced. It was equipped with a two-stage breech with roll-locking. Just before the end of the war a group of designers and existing prototypes were evacuated to Spain, where the work was completed bearing fruit to the construction of the CETME-58 carbine, used in the Spanish Army and of the G-3 automatic rifle, produced by Heckler und Koch and in 15 other countries on license (it has been used in the armed forces of approx. 50 countries).

The family of weapons for the Kurz Patrone round presented a mature design conception. The MP-43 was highly prized by Allied soldiers as a captured weapon. Carbines captured by the Soviet Army were delivered to arm the East German (GDR) police; in connection with this the production of ammunition was maintained after the end of the war. They were in service for many years, right up until replacement by the Kalashnikov. In Finland the Suomi carbine was developed, modeled directly on the StG-44. The rounds were almost the same, despite the calibre being reduced to 7.62-mm. The pioneering MP-43 design was the prototype of modern carbines and automatic rifles such as: the AK-47, FN FAL and L1-A1.

During operation the MP-43 gained the ever greater liking of the soldiers. From their point of view not only accuracy was important. Thanks to the plastic treatment, replacing the "sculpturing," this weapon had almost the same mass as the basic German MP-40 machine pistol (4.9 versus 4.7 kg). Despite this, as opposed to poor pistol ammunition, a bullet fired from an MP-43 pierced any helmet of that time, even at a distance of 600 m. Meanwhile, thanks to the progress of technology, the cost of 1 specimen amounted to only 70 marks in 1944.

The exact number of produced specimens is unknown, but

it is common knowledge that this was at least 425,000 up until the end of the war.[80]

I have already mentioned in the "introduction" the so-called "Weissenborn report"—the testimonies of the former second in command of the weapons department in Speer's ministry. Weissenborn, according to whom the MP-43 was the best infantry' individual weapon in the world, has left us much information about continually little known facts, associated with its production:

One of the precursors of the MP-43; the MKb-42(H) from 1942

In the spring of 1943 the then Colonel Kittel, chief of the Heereswaffenamt's weapons research department delivered a lecture on the subject of the model 43 automatic carbine for the so-called assault round. This took place in the Technical Department of Speer's Ministry and the audience was recruited from the most inner circle clustered around the Hauptdienstleiter Saur, the chief of this department. Retaining the standard German 7.92-mm calibre, the automatic carbine constituted an important improvement with regard to the MP-40 sub-machine gun. The latter, used by the German army, fired 9-mm pistol ammunition and its effective range of fire didn't exceed 150 m. Fixing of the magazine was very unsatisfactory. If the magazine was moved during firing, it became unstable. Irrespective of this, the MP-40 was very sensitive to contamination(dirt). In contrast to this, the model 43 automatic carbine fired accurately up to 600 metres and its 30-round magazine was rigidly fixed. It was less sensitive to contamination. It could shoot with single or continuous fire, using the carbine round, a shortened version of the standard rifle round. On the basis of personal experiences from the eastern front and many conversations with regiment commanders, Col. Kittel became convinced that only the introduction of this weapon could halt a new Soviet offensive:

In the situation, where there is no possibility of strengthening the eastern front, one should at all cost increase fighting force of a single German soldier through the introduction of a hand gun with a greater rate of fire.

Experts in the field of carbine production, who took part in the discussion, had no doubt as to the progress which the automatic carbine represented and fully agreed with Col. Kittel. In spite of this rapid introduction of the new weapon, it was neither possible from a technical point of view, nor was there authoritative agreement, since the order from the Führer's Headquarters demanded the cessation of commended limited production and a further order forbid any discussion on the subject of the MP-43 during "conferences on the subject of the war-time situation" (Lagebesprechung) at the Führer's Headquarters (FHQ). What more, not long afterwards, the Chief Committee for Weapons, as the body directing the armament industry, received an order from Hitler demanding the immediate cessation of carbine production, whilst the MP-40, widely known for its faults, was to be manu-

The MP-43 automatic carbine

The MP-43

factured at just the same speed as up till now, in other words approx. 15,000 pieces monthly. The Chief Committee for Weapons however didn't carry out the Führer's order, and even increased production to 5,000 pieces monthly, at the same time classifying the carbine as a sub-machine gun (it was later named the "assault rifle" — "Sturmgewehr-44"/StG-44). The Chief of the Heereswaffenamt's Weapons and Armaments Department, as well as party comrade Saur from Speer's Ministry, demanded the execution of the Führer's order. However with regard to the persistent measures of Col. Kittel and officers from the front lines, the Chief Committee for Weapons also persisted in its opinion and maintained production at a level of 5,000 pieces monthly. This led to a debate with Hitler. Among others two soldiers took part in it, who had won the knight's crosses on the eastern front. As a result Hitler gave a new order, in which he demanded the immediate production of 30,000 carbines monthly. A few weeks later he had already demanded 50,000 and a few months later 90,000, and soon even 120,000 monthly. But before it came to this the potential of industry had been wasted for over a year. (…)

A consequence of the lack of technical management and supervision (in other words: planning) was the ever greater deficit of carbine ammunition for the MP-43, deepening from month to month. Supplies of standard small arms ammunition were also poor, which was an effect, among others, of the increased production of carbines. This soon led to the situation in which the cessation of mass production of the MP-43 became a necessity, with such an expenditure of energy increased to 50,000 monthly.

It was necessary therefore to scrap the many thousands of specimens of this weapon, manufactured untiringly day and night, even though the front lines begged on their knees for every one. Regardless of this, Allied bombers razed to the ground (at Easter 1944), the factory located in Poznan, producing the machines for the production of small arms ammunition.

Infrared

A classic example of searching for a way to solve certain problems, which ultimately led to the development of a whole new field of technology, was the question of infrared technology. Interesting results had already been recorded in this field before the war, but only during the war did the conditions exist for rapid development. One of the most important stimuli turned out to be attempts to create an alternative to radar, in the field of which the Allies had a slight advantage.

Therefore this isn't the first time we have met with a case when radically new technologies arose merely as a substitute to something that already existed. The new technologies used in the case of the MP-43 were also rather a result of necessity and not of the reason. Plastics were to a large degree regarded as a substitute…

Let us return however to infrared technology. I wouldn't like to describe all types, for there were very many of them. I will present therefore a certain summary. One may in general divide this equipment into three groups:
1. "Active" observation and targeting equipment requiring illumination of the target/area with infrared radiation, emitted by a special reflector.
2. Target locators and thermal cameras ("passive")—not requiring illumination of the target, simply receiving thermal radiation emitted by the target itself.
3. Heat seekers as homing warhead elements for bombs and missiles; these are described in the second part of the book.

The equipment mentioned in the first point comprised the principle group. Several dozen types existed, below I have described only selected ones.[64,82,83,84]

Most types of German active night-vision devices were bases on image converters (photomultipliers) developed and produced on a small scale in the Research Institute of the Post (Reichspostforschungsinstitut). Its laboratories were initially located in Berlin, but in connection with the danger of air raids they were evacuated to the town of Hassenbach.

The equipment produced there didn't yield to many post-war equivalents even up to the 1960s and 70s. Throughout this time of course the principle of operation had not undergone any change, the basic construction features also remaining as before. The description presented below of an image converter from 1984 to an equal degree refers to those from the beginning of the 1940s:[88]

The basic component of every active night-vision device is an electroscopic image converter. An infrared image of the observed subject is focused on its photocathode. A standard silver-cesium oxide photocathode, used in the images converters of the night-vision devices, is a receiver of short-wave infrared radiation in the range 800-1,200μm. The infrared radiation falling on the photocathode causes an emission of electrons, which—accelerated and focused by an electrostatic field—create an electron image of the subject on a screen covered with an appropriate luminophor. Under the action of the electron beam the screen emits a visible radiation, which may be observed through the night-vision device's eyepiece.

The equipment produced in the Reich's Postal Research Institute also possessed a silver-cesium-oxide layer sensitive to infrared. Externally the whole image converter resembled a large electron valve. It anode, i.e. the layer visible through the eyepiece, on which the picture was projected, was covered in a shining green luminophor made up of a composition of zinc sulphide and zinc selenide. Two basic types of converters were produced; 160-mm in diameter (600-700 a year) and 70-mm in diameter (over 200 a year). All of this work was carried out with relatively modest manpower—the personnel of the laboratory in Hassenbach numbered only 40 scientists and technicians.

The second producer of this equipment was the company A.E.G., where 400 image converters of similar dimensions to the aforementioned case were produced. However, only 186 passed quality testing with a positive result (the quality of the achieved picture was tested). It was affirmed that 78 pieces were fit for the production of sights, the remainder were to be used in various observation devices, chiefly for drivers. A.E.G. was a world pioneer in this field; the first image converter of this kind was tested there as early as 1934!

The name "Bildwandler" was used in relation to them, replaced in general with the abbreviation "Biwa."

Below I present descriptions of certain types of night-vision devices constructed in the Third Reich.

Z.G. 1221 Sight

This sight was designed for anti-tank guns, in which one of the converters from A.E.G. was used. The company Zeiss made the optical components (among other things 1,000 lenses were produced). From the markings figuring on the sight, it follows that it was factory aligned for shooting at a distance of 250 m.

The sharpness was set by the electrostatic focusing of a beam of electrons. An integral part of the system was an infrared searchlight 36-cm in diameter, designed to illuminate the target. Several hundred specimens from a test batch were delivered to ground forces with the aim of carrying out testing, after which the work was stopped. There is a lack of information concerning possible use on the battlefield.

The F.G.12/50 infrared system, mounted on a "Panther" tank's turret

"Vampire" Sight

The "Vampire" is one of the most interesting German devices from the group in question. This sight was designed for the MP-43/MP-44 automatic carbines. The night-vision device itself didn't exceed the dimensions of the larger optical sights; it was approx. 35 cm long and had a diameter not much exceeding 6 cm. Together with a small infrared searchlight (a 35 W lamp covered with a filter) it weighed only 2.3 kg. Heavier on the other hand was a battery/high voltage converter set carried in a backpack, which weighed almost 14 kg—this obstacle was not overcome until long after the war. Such a large backpack mass resulted not so much from the power consumption itself, but from the necessity of converting the battery current into high voltage current, in this case 11 kV, to power the image converter.

This equipment was introduced for armament in small numbers and behaved surprisingly well. It was indeed somewhat too heavy for soldiers of the first line, but was quite simply an ideal invention for guarding units. Under these conditions it was also no problem that the battery was sufficient for only 3-5 hours of continuous operation, as it was possible to quickly replace it. The company Leitz was the producer of the "Vampires," using small "postal" converters; all in all 300 complete sets were supplied. After the war the British examined one of the "Vampire" specimens, later concluding that the picture was "very clear and characterised by good contrast." In the conclusion of test results it was written that: "at a distance of 30 metres the silhouettes of men standing and lying were clearly visible, at a distance of 50 metres only people standing were clearly visible, people lying being difficult to differentiate from the background, at a distance of 80 metres people standing were difficult to see, especially when they moved."

Many of the solutions employed in the "Vampire" were became a starting point for analogous work carried out after the war in other countries, e.g. the Soviets copied this unit in entirety, later introducing it for the armament of its own army and other members of the Warsaw Pact.

The MP-43 with the "Vampir" infrared sight (weapon without magazine)

The F.G.12/52 device, mounted on a BMW car

The "Uhu" system, on a half-track scout vehicle

"Adler I" and "Adler II"

On the basis of the solution tested in the "Uhu" ground forces system, an analogous anti-aircraft device was created, meant to be some sort of equivalent to short range radar.

Small changes has been applied in the field of the night-vision device's optics alone, the focal length was shortened from 40 to 25 cm, which bore fruit to an increase in the field of view, instead of one eyepiece, two were employed (binocular type). Over 500 of these types of systems were produced, although the system was never introduced into armament. It lost the competition to radar due to the lack of possibility of determining the distance to the target, though the effective range was (according to the Germans) exceptionally large—the order of 25 km. It wasn't written however what kind of aircraft this concerned, although one should expect that it was large bombers.

"Seehund"

The "Seehund" was one of the most interesting and best German night-vision devices. It was developed for the Kriegsmarine as early as the beginning of the war, in Vendome, in France, where American officers found one of the devices, and also a service manual dated "October 1941."

This night-vision device was to be installed on ships (although in principle this was a mobile device) and in all probability its chief purpose was to detect enemy planes at night. Once again we have to deal with an attempt to create an alternative to radar—despite the fact that the Germans too had a significant merits in this field.

The night-vision device itself (receiver) weighed 11 kg and was to be powered from the ship's installation, through a 15 metre long cable and 20 kg stationary voltage converter. It was possible to apply various types of infrared searchlights—mobile ones ensuring a target detection range the order of 10 km, a searchlight 50 cm in diameter giving a two-fold increase in range, or a large searchlight 1.5 m in diameter, thanks to which the range increased practically to infinity ("bis zur Grenze der geographischen Reichweite"—as was written in the original instructions). Although it is common knowledge that mass produced equipment was concerned, there is however no certainty to where it was made. In an American report on the subject of the "seal" (for this is how the name "Seehund" should be translated) the statement appeared that a significant contribution to its development was probably undertaken by the company Goertz.

Before the war it was, next to Siemens, a pioneer in the development of television technology and the no less breakthrough electron microscope, whose first model arose in the plant of Siemens in 1938. By the way, it was just the leading position of the Third Reich in these fields, that enabled the

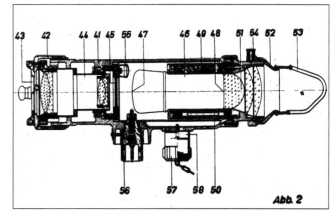

An original cross section of the "Seehund" active night vision device

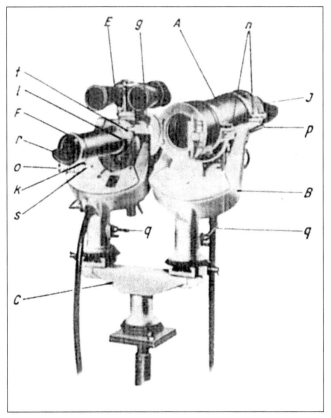

The "Seehund"

later rapid development of the night-vision and thermo-vision devices. In essence the fundamental principle of operation of the night-vision device' image converter—the creation of a picture as a result of the electrostatic or electromagnetic focusing of a beam of electrons, is just the same as in the case of the electron microscope.

Let us return however to the "Seehund"…

The Americans, in a report from the examinations of a "captured" specimen, described it almost in superlatives alone. Such features were emphasised, like the simplicity of operation, resistance to difficult conditions and in particular watertightness, good resolution, large field of view (22°), sensitivity and stopping of all visible light through a filter fixed onto the lens. The lack of possibility of quickly changing the image converter was regarded as the only disadvantage; such an operation took more or less an hour and required an expert. The "Seehund" received a modified image converter, with a beryllium cathode, permitting a significantly greater contrast to be achieved than up till now, which was crucial in the event of detecting aerial targets. This was in all probability the converter from A.E.G.— 75-mm in diameter. One of the derivative versions of this night-vision device was the "Seehund III," designed for submarines (radars for detecting aerial targets were only foreseen for Type XXI U-boats). In May 1943, 1,250 of these sets were ordered, although ultimately only approx. 400 were supplied.

The "Seehund III" differed considerably from its predecessor, above all it was smaller. The diameter of the lens amounted to only 5-cm, which probably negatively affected its range. It had however a greater field of view.

A large infrared reflector, 1.5 m in diameter, as used in naval systems

"Spanner"

As early as 1941, an attempt was made to construct a passive night-vision device (without searchlight), which was to react to hot combustion gases of the Allied bombers. It was used to equip night fighters, approx. 600 specimens being supplied in connection with this, as early as 1941. But this conception didn't fair well in combat conditions and the night-vision devices were withdrawn.

Thermovision (passive) devices constitutes an extensive group.

Many readers will be surprised by the fact that they were built on semiconductor components. In one of the British intelligence summaries I even found mention of attempts to use silicon![89] In the infrared detectors however semiconductors of a different kind were used.

They were sensitive to several micron wavelengths, which night-vision image converters didn't react to. Such a wavelength roughly corresponds to the heat emitted by objects with a temperature of several dozen degrees centigrade.

All of the German thermal detectors acted on the principle of the so-called internal photoelectric phenomenon. This relies on the fact that a semiconductor ceases to behave like an

The "Spanner I"

The "Spanner IIA"

insulator after absorbing photons and begins to behave like a conductor; the electrons are "knocked out" of the valence orbitals and pass to a so-called conduction band. "Holes" conducting electric current form in the semiconductor. The simplest detector therefore consists of a crystal of the appropriate chemical element and two electrodes. If it is part of an electric /electronic circuit, when the detector is illuminated by thermal radiation an electric impulse is formed, which makes itself known in the form of a picture on a monitor, or sound signal.

Currently single detectors are no longer used in observation-sight infrared receivers. A certain type of intermediary stage, dominating in the 1980s, was comprised by rows of detector systems. Such a row assured one picture dimension, the second dimension arose thanks to the application of a so-

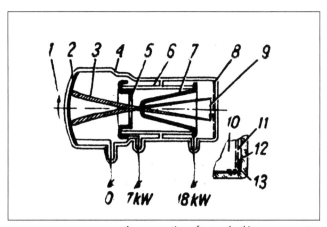

A cross section of a standard image converter. Description in the text.

called scanner, a system of rotating mirrors, thanks to which the row of detectors could deliver a two-dimensional picture, repeated after every revolution of the mirror..

A certain kind of analogy to such a device is comprised by the fax, but in it the object itself, and not the object's picture, was moved before a row of miniaturised detectors.

The newest generation of infrared receivers are constructed on the other hand like video cameras. There are no mechanical scanners in them, but instead two dimensional detector systems are employed (externally resembling integrated circuits). In light of this it would seem that construction of an infrared receiver giving even a somewhat "readable" picture, in command of a single detector, is unrealistic. But this is not so. The Germans attempted to compensate for this, by applying correspondingly "complex" scanning devices, thanks to which pictures were attained not differing significantly from those that lit the monitors of standard radars at that time.

Infrared receivers succumbed to the dominance of considerably more mature radar, however the Germans were very well aware that the time of their true rise was only to come and that it was a promising technology, in whose development it was undoubtedly worth investing.

At least several dozen types of detectors from the field of so-called medium infrared were developed. Their development and trial production was entrusted to two companies: Elektro-Akustic or Elac and Zeiss-Ikon from Jena, a branch of the Zeiss consortium, which incidentally to this day is a potentate in the production of infrared systems.

Elac brought into limited production among others (1,000 pieces were planned monthly) a whole series of detectors based on lead sulphide, PbS90 crystals. They achieved a maximum sensitivity to 2.5 micron wavelengths, i.e. to the so-called first atmospheric window—the atmosphere isn't transparent to the whole range of infrared and only to specific "bands"—wavelengths the order of 2-5 microns and around 8-12 microns. A whole range of sensors were developed of dimensions from 3 x 3-mm to round ones 30-mm in diameter. The larger were obviously more sensitive, but ensured a significantly smaller resolution of the completed device. In accordance with theoretical predictions the Germans quickly became orientated that the detector sensitivity could be increased by even fifty times, if it was cooled to a temperature of minus 40-50 degrees centigrade. In this way thermal radiation generated by the detector itself is eliminated. In connection with this a special cooler was constructed code-named "Eskimo," using carbon dioxide. Thanks to this enormous, to put simply, level of sensitivity was achieved for those times; the detector alone (that is without an optical system focusing the beam) even reacted to radiation of 25 millionths of a watt![82] By the end of the war around 500 semiconductor sub-assemblies of this kind were made in the laboratories of Elac alone, a no closer defined amount were produced moreover in the plants of the electronics company Kast und Ehringer in Stuttgart.

This last company was also engaged in the production of other kinds of detectors, developed in the laboratories of Elac.

Synthetic lead selenide (PbSe) crystals were used in them. They differed from the aforementioned mainly in that they reacted to the long wave range of infrared (the order of 4-5 microns), that is emitted by an object with a lower temperature. Their sensitivity was similar to sulphide crystals.

The key to success of the company Elac in this field, turned out to be in principle one person, the director of their infrared technology department, Dr. Kutscher. He had commenced work aiming to develop the optimal technology of producing sulphide detectors as early as 1930, de facto becoming a pioneer in the field of semiconductor technology.

The other leading producer of these devices, Zeiss-Ikon, also developed several types, all on the basis of lead sulphide. In general they didn't equal the detectors from Elac and were mainly used in less demanding systems—alarm devices—"keeping watch over" port entrances (the "Strahlungssperre" system) and in devices maintaining secret communication between ships, thanks to the use of a directed infrared beam (the "Puma" system). Work however was carried out on them quite intensively, improving the basic parameters, like the ratio of sensitivity to noise level, around 20 times. Despite this they were slightly inferior to the detectors from Cologne and Stuttgart until the end of the war. In the laboratories of Zeiss work was also carried out on a new type of detector, thallium bromoiodide, which was experimental in character. It is known that this detector was designated the KRS-5, but no technical data at all is known. This problem was also worked on independently at the Institute of Physics in Göttingen. Zeiss could on the other hand show off in another, equally important field—in the development of optical systems—lenses and corresponding optical materials (normal glass lets through electromagnetic waves from the visible range and e.g. ultraviolet, but not from the range of medium infrared, which we have the opportunity to be convinced of when in a greenhouse or sun-heated car devoid of air conditioning). Zeiss co-operated in this field with the I.G. Farben consortium.

In effect as many as eight types of "glass" were developed, devoid obviously of "classic" silica. Chemical compounds like bromide, iodide and thallium chloride as well as bromide and silver chloride were chiefly used. Altogether at least several hundred kilograms of such "glass" were produced, from which classic lenses were cut. They had however certain faults—above all they were very expensive and less mechanically and chemically resistant than ordinary glass.

Among other considerations, as well as in consideration of the generally greater simplicity of production, large diameter lenses were above all mirror lenses (concave mirrors), or hybrid systems. In the event of a lack of lenses, filters had however to be made from a special "glass" (chiefly interference filters).

Yet other work was carried out in parallel from the field of chemistry, likewise pioneering, with the aim of developing special paints masking in the field of infrared (they concerned protection against detection with the aid of night-vision devices). A significant potential was engaged in this objective, although the enemy never employed night-vision devices, at least on a significant scale. On the list of companies and institutions co-operating in this field were found among others: the Danish companies NVK and CPVA, a special cell designated by the letters F.E.P., probably operating within or in co-operation with the Reich's Scientific Research Council (Forschungen, Erfindungen, Pattente—research, inventions, patents) as well as the company Ludeck und Kohe from Berlin. This is in any case a very short list, if we compare it with the total number of companies and institutions engaged in the development of thermal and night-vision devices. Apart from the aforementioned, on it would have been found among others the Air Ministry (RLM), the Navy High Command (OKM), the companies Gema, Osram and Stohl, the Institute of Applied Physics from the University of Cologne as well as institutes of physics in Prague and Leipzig.

Let us return in the meantime to the description of masking coatings. Work in this field was crowned with success as early as 1943. Several types of special coatings were developed, among others code-named "Lattenzaun" and later also "Gartenzaun." They absorbed at least 96% of infrared radiation from the night-vision range (waves the order of 1 micron). It was intended to use them mainly for camouflage of submarines, although in time it became evident that a problem was their not so high resistance to the effect of sea water. However this had no greater significance, since the chief danger to U-boats was presented by radar and sonar anyway. The Germans didn't manage to make use of their achievements in this field during the war, only the next decades showed the accuracy of the accepted solutions.

After this introduction we may finally move on to descriptions of specific thermal (thermovision) systems...

The Germans considered that their most useful and best device from this group was the heat seeker named "Kiel." It was developed with night fighters in mind, obviously with the aim of making it possible for them to detect enemy bombers. In other words, it was designed to home in on the target and to make possible an attack without any illumination of the target (passivity of operation and resistance to the ever more frequently applied jamming, which radar was subjected to).

"Kiel" detected a single, four-engine "Lancaster" type bomber at a distance of 4-5 km. Trials carried out showed its usefulness in other applications; a small ship of 1,500 tons displacement was "noticed" at a distance of 7 km, factory chimneys on the other hand at a distance of 10 km.

"Kiel" used a relatively light and short mirror lens, a paraboloidal mirror 23-25 cm in diameter. In order to scan the field of view the mirror span at a speed of 100 rotations a second around the axis of the whole device and simultaneously at a speed of 2 rotations/s around the second, slightly inclined axis. In this way the focused infrared beam described a defined slanting rosette in the focal plane, covering the entire field of view. Several versions of this system were designed ("Kiel I" and "Kiel IV"), differing above all in precisely the field of view. The version considered the most successful, the

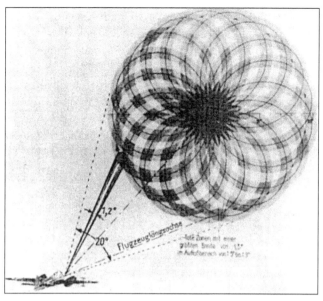

The "Kiel III," an original diagram, showing the field of view's scanning mode...

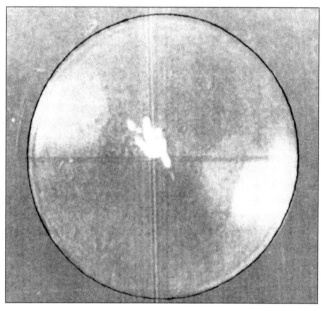

...and the target's image on the screen

"Kiel III," was characterised by a field of view 20° in diameter and resolution the order of one degree.

The "Kiel I" version also existed with a lens objective, being a development of an older heat seeker from Elac called the "Kormoran." Rapid work on it was however interrupted with regard to a fault possible to avoid, resulting from absorption of approx. one third of radiation by the objective's lens.

The "Kiel III" itself existed also in two modifications; in one a 3 x 3 mm detector was employed from Elac, whereas in the second (the "Kiel III Z") a similar detector from Zeiss, both being cooled by solidified carbon dioxide, transported in a special container. As a result of tests carried out among others by a Luftwaffe experimental unit (Versuchskommando) at Stade airfield and at a military training ground in Rechlin, only the "Kiel III Z" was classified for mass production. Production was carried out however at a modest speed. The only batch, approx. 20-30 pieces, was received at the turn of 1944 and 1945 by a squadron of night fighters, based at Goslar airfield. These devices were mounted onto the aircraft noses.[82,90]

Simultaneously with the "Kiel" were developed a whole family of considerably larger and on principle ground-based heat seekers, designed to detect aircraft as well as ships. They were named the "Wärmepeilgerät" (WPG) and "Nachtmessgerät" (NMG), at the same time the second fulfilled a complementary function in relation to the first, since it served as a stereoscope rangefinder. The whole system was the largest thermovision device that arose during World War II. The diameter of the WPG elipsoidal mirror lens amounted to as much as 1.5 metres (the elipsoid is a chunk of glass made by the rotation of an ellipse around one of its axis, on a similar principle arise the paraboloid and hyperboloid, according to which other mirror optical systems are made). With a large range, resulting from the mirror's large gathering area and high detector's sensitivity, as well as with the possibility of measuring the distance to the target, the WPG/NMG system was the fullest attempt at creating an alternative to radar. An alternative, which, one should remember, was distinguished by passivity of operation, and so was considerably harder to detect by the enemy and resistant to jamming. The most important component of the system, "Wärmepeilgerät," was, like the "Kiel III," developed simultaneously by Elac and Zeiss. Their performances were similar, although the prototypes from Elac were characterised by a somewhat larger range, in whose connection work on the Zeiss modification was stopped relatively quickly, probably as early as 1941, when it was carrying out testing of prototypes. The competition on the other hand received in the years 1943-1944 the first orders, implying the commencement of mass production. At that time 90 WPG sets were ordered—all of the ground-based version designed to detect ships. By the end of the war only 12 had been delivered, 3 of which were managed to be installed in bunkers on the west coast of France before the Allied landings. Such a device externally resembled to a certain degree

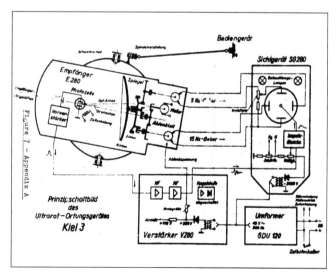

A diagram of the "Kiel III" system

radar, or rather a large anti-aircraft searchlight —above the roof of the bunker was located only the optical system—rotated around the vertical axis and inclined about the horizontal axis, the operator's post along with the essential part of electronic devices being placed several metres below.

At least five other types of heat seeking devices were developed in the Third Reich, based on semi-conductor components, at the same time in the case of two of them verified information is available, testifying that their development was completed during the war and were proved to be correct in practise. They were the: "Würzburg B" and "Armin."

The first constituted a kind of "thermovision appendix" to the large "Würzburg-Riese" radar detecting aerial targets. It was to be used in the event of strong jamming of the radar's operation. The heat seeker was mounted on the tip of its paraboloidal antenna three metres in diameter. Despite the small diameter of the lens (25 cm mirror) the range was relatively large—British "Lancasters" were detected at a distance of 15-20 km. In the focus of the mirror were located four detectors, which corresponded to four lights on the control panel. When all the lights flashed in uniform, this signified that the system was aimed directly at the target. This device was probably constructed in the Berlin company Gema, i.e. the same one that produced the radar. It was tested at the military training ground near Rechlin and near the town of Kuhlingsborn. The "Armin" was in turn a heat seeker developed by Elac for night bombers. It was to warn against approaching enemy fighters. It was distinguished by a very complicated, multi-lens optical system, whose task was to ensure a large field of view (120°) and simultaneously "accurate" enough scan-

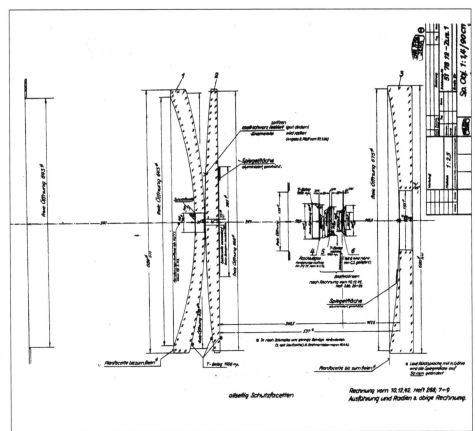

The "Kiel III," one of the head versions

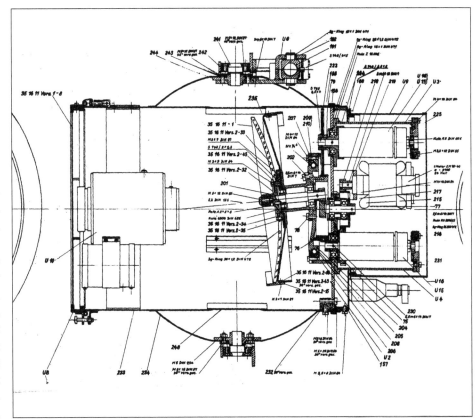

The "Kiel III," a cross section of the infrared telescope. Although it is only part of the system, which in turn constituted just a fraction of the work in this area, it gives some clue as to the vastness of potential engaged in the development of infrared technology.

The head of the "WPG"

A view of the optical system; after removal of the IR filter, the Kiel

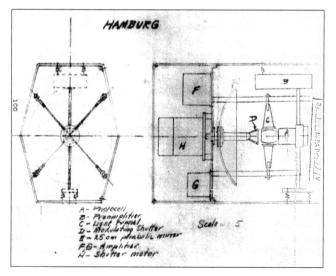

The "Hamburg" system

ning so that it would be possible to make out an approximate silhouette of the target on the monitor (despite the use of a single detector). However this was only feasible when the target was located at a distance no greater than around 2 km away. Two optical system modifications were tested in parallel and designated correspondingly the "Armin I" and "Armin II." Their tests were completed as early as 1944, but nothing was done to commence mass production. One of the reasons were the predicted very high costs of the optical components.[82,90]

Apart from the above several types of bolometric thermovision devices were developed, exploiting a different physical phenomenon, relying namely on the change in electrical resistance of certain materials after the absorption of infrared photons.

None of these devices ended up being mass produced, the majority not being competitive in relation to the heat seekers of the types described earlier. In this group however was found a truly revolutionary system, revolutionary, since it was possible to already define it as a (simple) thermal camera. It presented after all on the monitor not an "ordinary" spot, but a full picture of the objects located in the field of view. The "Potsdam-L," as so was it named, was a reconnaissance device, designed to be mounted in aircraft to detect ground-based targets at night.

A simple optical system and scanner with inclined (oscillating) mirror was employed in it. It projected an infrared beam onto a special, metallised foil fulfilling the function of a photoresistor. The current flowing through it powered in turn a lamp, whose focused light was being reflected in the reverse direction than the infrared beam. Reflecting off the reverse side of the aforementioned mirror, the beam of light was projected in turn onto a glass plate fulfilling the function of a screen, covered in a thin layer of phosphorus (glowing not only when illuminated by the beam, but still shortly afterwards). In this way a picture arose on the aforementioned screen, composed of lines. It moved at a speed proportional to the airspeed of the aircraft. The designer of this effective, though not very complicated device was not any large consortium, but one individual, a certain F.E. Leybold from the town of Clansthal in the Harz Mountains. From the contents of an U.S. Intelligence report, in which the "Potsdam L." was described, it follows that as in the case of other most breakthrough inventions, all prototypes were destroyed or effectively hidden before the arrival of the Allied armies. On the other hand the service manual fell into the hands of the Americans, who also had the opportunity of interrogating witnesses.[82]

Despite the spectacular character of the above construction, it wasn't at all the only thermovision camera that arose in the Third Reich. From the American report it follows that at least another two types existed.[82] They bore the names "Eva"

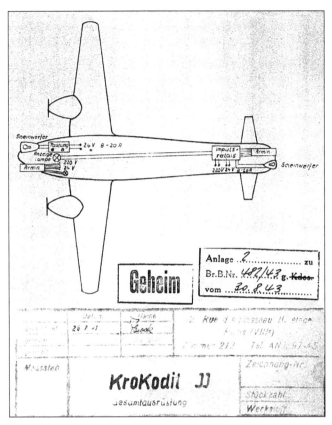

A diagram detailing the installation of the Armin/Krokodil thermal target detection system on a night fighter

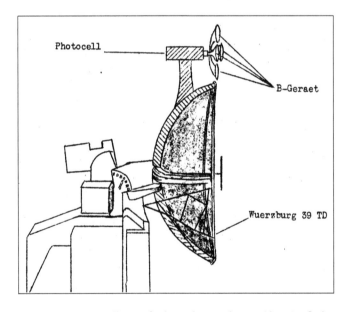

Installation of a thermal target detector/direction finder on a Würzburg radar's antenna

A German list of research projects (selected) regarding radars, homing devices etc.

and "Fernaktinometer" (they were described as "thermal picture-forming devices"). Unfortunately only with regard to the first do we have any kind of precise information at our disposal. Since the description is very modest, I quote it in entirety:[90]

Eva. Prof. Czerny from Frankfurt built this device. It gives a rough picture of a thermal object thanks to the emergence of interference colours on a thin layer of liquid. It was tested by the NVK and required 3-8 seconds for the observed picture to emerge. Its range amounted to 200 to 300 metres.

One may assume that the thin layer of liquid was situated between two glass plates, as in the majority of interference filters. However the field of possible guesses ends at this and the principle of operation of this invention remains de facto un-

Some German long-range radars: Würzburg…

…Mammut…

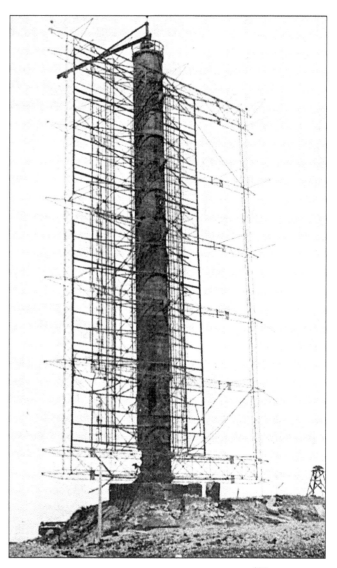

…and Wassermann…

known. This is quite an intriguing enigma, particularly if we take into consideration that the video camera probably possessed no movable components and simultaneously gave a colour picture, which in itself constituted at that time a shocking achievement. We also know nothing about how the Germans intended to use this device, though the obvious temptation surely existed to use it as a thermal sight for ground forces.

Despite the meagreness of available data one may however attempt one more, cautious assumption—in relation to the "screen" itself of "Eva." It is difficult to imagine that the picture would have emerged directly as a result of infrared absorption (the minimal power of the radiation). We know on the other hand that Prof. Czerny was a pioneer in the use of bolometry in this field; he tested materials changing their resistance under these conditions. It is therefore possible that electric current passed through the described thin layer of liquid—at a given moment the illuminated materials would change their resistance and therefore the illuminated part of the mysterious plate would have higher temperature. It would have operated therefore like some kind of "amplifier" of received heat. We would have had to deal therefore not only with the thermovision camera alone but with a prototype of a colour liquid crystal display (see: simple medical thermometers in the form of a plaster stuck onto the forehead, where a layer very sensitive to the heat of the liquid crystal changes its colour, "projecting" the appropriate number). The description of the device inserted above is only a very modest summary. In reality a widely conceived research programme was involved. It was a catalyst of progress in many areas, even in the physics of semiconductors, and in an area of work on new materials for producing optical components… Recently in one of the documents of the Reich's Scientific Research Council, seized by the American "Alsos" mission I found e.g. a reference to "light telephony," operating in the infrared band.[85] Could this have referred to some kind of precursor to present day optical fibre telephony?

Aircraft Carriers

The subject of this chapter is the review of new conceptions of armament development, advocated in the Third Reich, illustrating only by example the quests carried out there. I have resolved to supply one more question. This is equally little known, therefore I hope that it will turn out to be interesting.

It concerns plans referring to the construction of German aircraft carriers. On December 8, 1938, in the presence of Hitler, Goering, Admiral Raeder and other countless dignitaries, in the Kiel shipyard "Deutsche Werft" a sublime ceremony took place, which the German press gave wide publicity to. "The German Reich is striving to rule the seas," wrote that day the "Völkischer Beobachter." The first Kriegsmarine aircraft carrier floated into the waters of the Baltic, a ship of 21,214 BRT displacement. It was named the "Graf von Zeppelin" and was supposed to be the pinnacle of technology at that time. It was supposed to be, since in reality it was still from completion. However it indeed looked like it would be a very dangerous entity. It received e.g. the most powerful steam turbines to date, thanks to which it could develop a speed approaching 34 knots. Three years later it received modern for the conditions at that time radar, to detect aerial and nautical targets. The final artillery armament was impressive: 16 150-mm guns of range 27 km, 10 105-mm guns, 22 37-mm automatic anti-aircraft guns, and 38 large-calibre anti-aircraft machine-guns.[86]

After the victorious campaign in Poland, Hitler became personally interested in the fate of the "Graf von Zeppelin," for in his opinion it could play an important role in the planned invasion on Great Britain. At a staff conference convened regarding this, he recommended that all work be completed in record time and the ship be introduced for active service by mid-March 1940.

However, this deadline was not kept to, chiefly because of problems with the delivery of large-calibre guns and much other equipment. In addition the issue of aircraft assignment was continually undefined. As it became evident only after the war, this was chiefly the "merit" of Goering's making. Luftwaffe Marshal Erich Milch recalled at the time that one day in 1940 he was the witness to a conversation in which Goering blusteringly declared to the OKL Chief of Staff, General Jeschonek:[86]

Jeschonek, I declare to you Sir, that Raeder will sooner resign, than form under my nose his own Air Force. The Admiral should know that in the Reich only I exercise authority over the Air Force!

Hitler learned of the dispute and on March 12, 1940, decided to settle it, giving Goering specific orders. The Kriegsmarine was to receive 50 Bf-109F fighters, 4 Ju-87c "Stuka" dive bombers and 13 Fi-167 "Storch" reconnaissance aircraft. These machines were to undergo appropriate modifications.

In addition it was intended to begin construction of a sister ship, the "Peter Strasser." The Kriegsmarine commanding staff also advanced the offer to reconstruct the passenger ships "Europa" (54,904 BRT), "Potsdam" (19,293 BRT), former French "De Grasse" (20,396 BRT) and heavy cruiser "Seydlitz" to act as aircraft carriers. All of these plans however backfired—this time without the intervention of Goering. In the first days of January 1943, the entire expansion (reconstruction) plan of the fleet was re-evaluated, conferring highest priority to the construction of submarines. The uncompleted "Graf von Zeppelin" was hauled away to Stettin (Szczecin), where it lived to see the end of the war.

The above facts, however interesting, are relatively well-

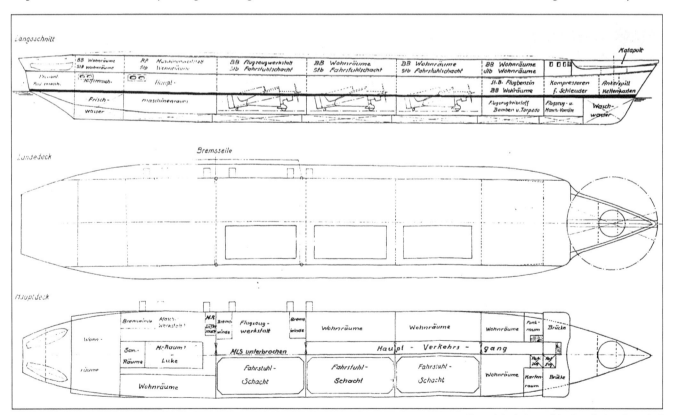

The original plans of the "small aircraft carrier"

known and I wouldn't have presented them if they didn't form the background to present a considerably less-known aspect of the whole story. Several years ago, while browsing through a file of the Reichsführer-SS' Personal Staff I chanced upon an extensive set of correspondence and projects concerning the plans of an alternative constructions in relation to the aforementioned aircraft carrier fleet![87]

In this case the initiator was Dr. Heinrich Dräger, the owner of the company and shipyard "Dräger," in Lübeck. His proposition bears the date of January 27, 1942. A certain kind of curiosity is constituted by the fact that Dräger found unexpected supporters in the individuals of Heinrich Himmler, and the Chief of his Personal Staff, SS-Obergruppenführer Karl Wolff. It concerned the construction of an entire fleet of "pocket" aircraft carriers each of displacement 3,500 tons, length 101.6 m, width 17 m and draught only 4 m. This proposition aroused the interest of the SS and expanded in wide circles for almost two years, although Dräger predicted each ship to be equipped with only 6-7 aircraft! The issue died a natural death only at the end of 1943, after the crushing opinion of the Kriegsmarine, indicating among others the too short flight deck (90m). Project KFT (Klein Flugzeugträger) had therefore so significance from a military point of view, unequivocally confirming the scale of Himmler's ambition.

Unusual Energy Sources

In the incessant quest for potentially ground-breaking technologies the verification of even much more extraordinary conceptions, than those mentioned so far was not avoided. The extensive issue of "new energy sources" supplies many such examples—alleged, or real. In a book of Professor Mark Walter, a historian engaged in the analysis of German work from the domain of nuclear physics, we find e.g. such an account (page 91 of the edition from 1999):[91]

> *Several members of the nuclear power project became involved in this irrational quest for wonder weapons. Werner Heisenberg and other prominent German scientists were called upon to judge proposals for inventions. While Heisenberg was plagued by inventors all his life, at this stage of the war such irritating contacts with self-styled scientists became dangerous. As the war worsened the National Socialist leadership took great interest in the inventive potential of Germans and especially of the front-line soldier.*

Here appear a few examples of inventions devoid of any substance, and further on:

> *However, in at least one case Heisenberg was not so easily rid of an inventor. In July of 1943 the Ministry of Armaments and Munitions asked Heisenberg for his opinion on the invention of a motor that ran without fuel. The Ministry admitted that as a rule such suggestions were considered to be perpetual motion machines, but nevertheless wanted Heisenberg to make a careful examination of the proposal, sent in by an engineer named Günther. Heisenberg replied two days later that the author's claim to be able to create energy from nothing was false and added that the proposal was so incoherent that Heisenberg found it very difficult to read. A few months later, Heisenberg heard from the Ministry once again. Günther had been left unsatisfied by Heisenberg's criticism and had complained directly to Adolf Hitler. The Ministry spokesman asked Heisenberg if he could perhaps reconsider the matter. Heisenberg was even asked to meet with Günther.*

There were many such reports of this kind and even if most of them were a waste of time, one may not overlook this phenomenon when analysing the technical progress of the Third Reich, for the simple reason that it throws light on the mechanisms governing this progress.

In the files of the SS Reichsführer's personal staff I found a classic example of such charlatanry. A whole folder was devoted to the affair of a certain Karl Schappeller, who maintained that he was able to obtain an unlimited amount of energy from ordinary water. Its source was supposed to be some kind of nuclear reaction not defined in detail. The documentation includes of course a description of the device, but is so obscure, that it is uncertain what it is at all about. I have reproduced an excerpt of this documentation in the book. The crucial fact in all of this is that despite the inconsistency with common sense, such ideas were not rejected beforehand. Schappeller's alleged discovery was passed judgement on by e.g. Professor Abraham Esau. He wrote that:

> *The material or ideological support of Schappeller would be something irresponsible—this is a pathological individual, whose ideas are completely muddled and purely fanciful.*[92]

Sometimes however it turned out that not rejecting fanciful ideas straight off was profitable, since the advantages resulting from allowing something authentically new to be realised completely covered the possible costs related to cases of sham.

An example of this type of case—in all probability an authentic and ground-breaking discovery—is constituted by the affair of the inventions of a certain Hans Coler. They were thoroughly examined not only in the Third Reich, during the war, but also after the war by British intelligence. In effect an extensive intelligence report arose, documenting this discovery and confirming its authentic character.[93]

It was entitled as follows: "The invention of Hans Coler, relating to an alleged new source of power." This report is currently available in the archives of Great Britain and the USA (although I have met with the suggestion that this is only part of it—the original was supposed to have over 150 more pages and also described its application—as a component in the propulsion of various weapon systems). Although the spell-

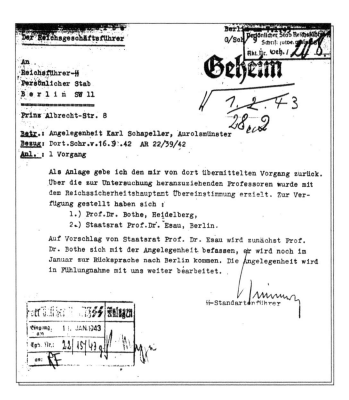

Selected documents from the files of the SS Reichsführer's Personal Staff, describing the work of Karl Schappeller

ing "Hans Coler" appears in the title and contents of the report, the German (and sounding more credible) spelling of the surname Kohler is also met. I will however use that which appears in the relevant documents.

In the first section of the report, being a kind of introduction, we read what follows:

Coler is the inventor of two devices by which it is alleged, that electrical energy may be derived without a chemical or mechanical source of power. Since an official interest was taken in his inventions by the German Admiralty [this clearly refers to the Oberkommando der Marine (OKM), the Navy command—note I.W.], it was felt that investigation was warranted, although normally it would be considered that such a claim could only be fraudulent.

*Accordingly Coler was visited and interrogated. He proved to be co-operative and willing to disclose all details of his devices, and consented to build up and put into operation a small model of the so-called "Magnetstromapparat" using material supplied to him by us, and working only in our presence. With this device, consisting **only** of permanent [bold by I.W.] magnets, copper coils, and condensers in a static arrangement he showed that he could obtain a tension of 450 millivolts [0.45 V] for a period of some hours: and in a repetition of the experiment the next day 60 millivolts was recorded for a short period. The apparatus has been brought back and is now being further investigated.*

Coler also discussed another device called the "Stromer-

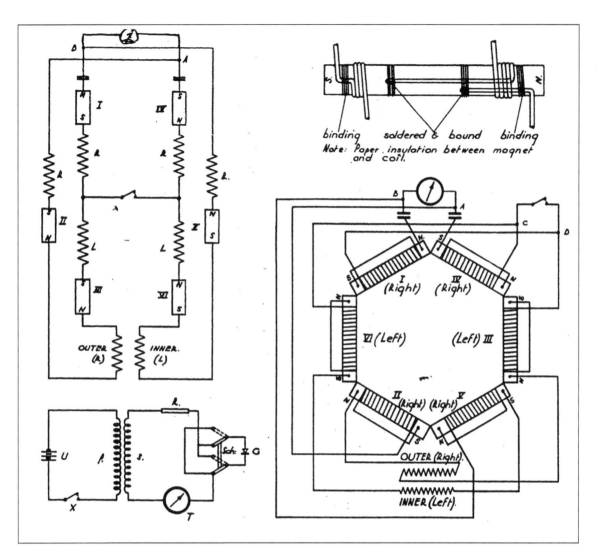

Diagrams originating from the B.I.O.S. report, regarding Coler's invention

zeuger" [literal translation: current generator], from which he claimed that with an input of a few watts from a dry battery an output of 6 kilowatts could be obtained indefinitely. No example of this apparatus exists today, but Coler expressed his willingness to construct it, given the materials, the time required being about three weeks. Opportunity was taken to interrogate Dr. F. Modersohn who had been associated with Coler for ten years and had provided financial backing. He corroborated Coler's story in every detail. Neither Coler nor Modersohn were able to give any theory to account for the working of these devices, using acceptable scientific methods.

In a further section of the report a summary of the basic technical characteristics of both devices appears (exact specifications along with technical drawings were placed in annexes). This information is the background to describing the history of both inventions. Here it is:

1. The "Magnetstromapparat" [this term is difficult to translate literally with regard to the absence, natural in the English language, of a synonymous relation between its elements. In any case what is involved is a device which is associated with current and magnets—note I.W.]. This device consists of six permanent magnets wound in a special way so that the circuit includes the magnet itself as well as the winding, (See Fig.1). These six magnet-coils are arranged in a hexagon and connected as shown in the diagram (Figs. 2 and 3), in a circuit which includes two small condensers, a switch, and a pair of solenoidal coils, one sliding inside the other. To bring the device into operation the switch is left open, the magnets are moved slightly apart, and the sliding coil set into various positions, with a wait of several minutes between adjustments. The magnets are then separated still further, and the coils moved again. This process is repeated until at a critical separation of the magnets an indication appears on the voltmeter. The switch is now closed, and the procedure continued more slowly. The tension then builds up gradually to a maximum, and should then remain indefinitely. The greatest tension obtained was stated to be 12 volts. The "Magnetstromapparat" was developed by Coler and von Unruh (now dead) early in 1933, and they were later assisted by Franz Haid of Siemens-Schukert, who built himself a model which worked in December 1933. This was seen by Dr. Kurt

Mie of Berlin Technische Hochschule [Berlin Polytechnic—note I.W.] and Herr Fehr (Haber's assistant at the K.W.I.), who reported that the device apparently worked, and that they could detect no fraud. One model is said to have worked for 3 months locked in a room in the Norwegian Legation in Berlin in 1933. No further work appears to have been done on this system since that date. [is this so???—note I.W.].
2. The "Stromerzeuger"
This device consists of an arrangement of magnets, flat coils, and copper plates, with a primary circuit energised by a small dry battery. The output from the secondary was used to light a bank of lamps and was claimed to be many times the original input, and to continue indefinitely. Details of the circuit, and a theory as to its mode of operation were given (summarised in Appendix I).
In 1925 Coler showed a small (10-watt) version to Prof. Kloss (Berlin), who asked the Government to give it a thorough investigation, but this was refused, as was also a patent, on the grounds that it was a "perpetual motion machine." This version was also seen by Profs. Schumann (Munich), Bragstad (Trondheim) and Knudsen (Copenhagen). Reports by Kloss and Schumann are translated in Appendices II and III.
In 1933 Coler and von Unruh made up a slightly larger model with an output of 70 watts. This was demonstrated to Dr. F. Modersohn, who obtained from Schumann and Kloss confirmation of their tests in 1926. Modersohn then consented to back the invention, and formed a company (Coler G.m.b.H) to continue the development. At the same time a Norwegian group had been giving financial support to Coler, and these two groups clashed. Modersohn's connection with Rheinmetall Borsig, and hence with the official Hermann Goering combine [properly: with the Hermann Goering Werke consortium (after the war renamed to Salzgitter)—note I.W.] gave him an advantage in this. Coler then in 1937 built for the Company a larger version with an output of six kilowatts.
In 1943 Modersohn brought the device to the attention of the Research Department of the Navy Command [O.K.M.]. The investigation was placed under the direction of Oberbaurat Seysen, who sent Dr. H. Fröhlich to work with Coler from 1.4.43 to 25.9.43. Fröhlich was convinced of the reality of the [examined] phenomena, and set about investigating the fundamentals of the device. He apparently concentrated on a study of the energy changes which occur on the opening and closing of inductive circuits. At the end of the period he was transferred to BMW to work on aerodynamic problems and is now working in Moscow.
In 1944 a contract was arranged by the O.K.M. with Continental Metall A.G. for further development, but this was never carried out owing to the state of the country. In 1945 the apparatus was destroyed by a bomb, in Kolberg, whither Coler had evacuated. Since that time

Coler had been employed, sometimes as an engineer and sometimes as a labourer. Modersohn had severed his connection with Rheinmetall Borsig, of which he had been a director, and was working for the Russian authorities as a consultant in chemical engineering.

In a further section of the British report, in the conclusion of the official record of Coler's interrogation, it was also mentioned that according to him the intensity of the magnetic field emitted by the magnets didn't fall as a result of the device's operation. In other words, they had been subject to no "consumption" whatsoever. Coler asserted that what was involved was a new, previously unknown type of energy, which he aptly called "space energy" (*Raumenergie*). Without doubt the most interesting section of the British special services report is however a description of an examination carried out by them of a copy of the generator, by now constructed in Great Britain. This was a device devoid of any kind of housing so that would be beyond question that any power supply was concealed in it. The British also placed great pressure on eliminating the possibility based on the fact that the generator in reality would obtain energy from external, artificial electromagnetic fields as a result of induction (e.g. by cables set up in the surroundings). In connection with this it was placed far from live cables so that remaining fields couldn't give sufficient induction—to this end it was suffice to carry out simple calculations. In spite of this the generator operated faultlessly. The British acknowledged this result as "unexplained."

In a section of the report headed "conclusions" we read:

1. It was judged that Coler was an honest experimenter and not a fraud, and due respect must be paid to the judgement of Fröhlich in the matter as deduced from his report to Seysen.
2. The result obtained was genuine in so far as could be tested with the facilities available, but no attempt has yet been made to find an explanation of the phenomenon.
3. It is felt that further investigation by an expert in electromagnetic theory is warranted, and that Coler's offer to construct a model of the "Stromerzeuger" should be taken up.

Shocking?

The question of Coler's inventions constitutes a curious and important example of carrying out a strange conception, in spite of its inconsistency with the scientific knowledge of that time; something which today would be almost impossible. It is worth pointing out here that inconsistency with the knowledge of that time does not signify inconsistency with current knowledge. A number of discoveries in recent years allows us to assume that in Coler's case what was involved was the harnessing to work of the so-called quantum fluctuations of the space-time continuum. They are a source of the so-called vacuum energy. Here is an excerpt of a contemporary popular scientific article on this subject:[94]

One may interpret the existence of a cosmological constant different from zero [the "cosmological constant" is a parameter in physics defining the magnitude of the vacuum energy—note I.W.] as the presence of exactly such a homogenous medium, which, despite being invisible, stores a certain quantity of material or its energy equivalent. This energy if often called the "energy of the vacuum." Quantum field theories stipulate that the vacuum energy may be equal to zero, or very large. Since in the latter case the cosmological constant would have a value several times exceeding all observational limits, the assumption that it is equal to zero appears to be most natural. (...) The latest observational data however says a different thing.

It says a **completely** different thing...

As our press announced at the end of 1999 (soon after the British monthly "Nature")[95], the vacuum energy makes up as much as 70% of the Universe's entire energy! Therefore it becomes evident that an unimaginably great sea of energy surrounds us, of whose existence only a few so far had appreciated and that in defiance of the school of thought from decades before, it is a dominating form of energy in nature. In this situation it is difficult not to ask oneself the frankly rhetorical question: Can one call the discovery of a way to derive energy from this infinite sea the breakthrough of the millennium?

Work in this field is currently being continued in different countries.

PART TWO

WEAPONS WHICH COULD HAVE CHANGED THE COURSE OF WAR

THE TURBULENT DEVELOPMENT OF GUIDED WEAPONS

One of the fundamental arguments of the size of scientific and technical breakthrough that was triggered off by World War II, is the enormous number of guided weapons arising in this period. In the Third Reich at least 20 types of homing warheads alone were designed (the Allies also produced individual designs). This signified the beginning of a completely new age. In short, the V-1 and V-2 were only the proverbial tip of the iceberg. A very small portion of these weapons was used on the battlefield, although technical problems were not at all the fundamental barrier. First and foremost it was the projection of the ignorance of Hitler alone, who pushed for the mass production of the V-1 and V-2, at the cost of guided weapons truly important from the point of view of the war's progress. "Surface-to-air" rockets constituted the leading position in this group, which could halt the waves of Allied bombers that were destroying the German armament industry. Anti-aircraft rockets were one of the most dangerous trumps of the Third Reich. It is a paradox that Hitler delayed work on them. It was a similar case with the Me 262 fighter and the nuclear research, whose revolutionary nature the Führer simply wasn't able to grasp. In any case the remaining types of guided weapons could also have had a noticeable influence on the course of the war whereas they were used on a small scale. I present their specifications below, beginning however with anti-aircraft rockets

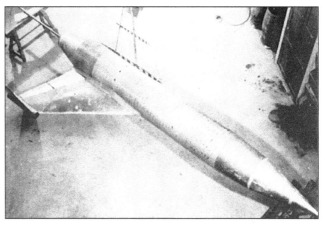

The Feuerlilie F-55

Work on the "Feuerlilie" began in 1942—almost simultaneously with work on the "Hecht 2700" rocket—but which flew along a trajectory already programmed before launching and as such didn't entirely deserve the term guided rocket. In 1943 further work was given up, developing instead the rival design.

The "Feuerlilie"

The "Feuerlilie" missile (Fire Lily) was the first guided anti-aircraft rocket, which was designed in the Third Reich (paradoxically without the participation of the specialists from Peenemünde). A whole series of institutions were engaged in this work, which was led by the Hermann Göring Institute of Aerial Research (Luftfahrt forschungs-anstalt Hermann Göring) in the town of Völkenrode.

The F-55 version of the "Feuerlilie," complete and disassembled (in the foreground). Visible is the powerful "Pirat" engine and very small in comparison Walter engine, shaped like a jar.

In actual fact the "Pirat" consisted of a battery of four smaller rocket engines

Preparing the F-55 prototype for launching

The aforementioned institution managing the work, of a somewhat lengthy name (the abbreviation LFA was also used), was generally one of the leading institutions in the development of German rocket weapons. Among other things the majority of work from the field of aerodynamics was carried out here; computations were carried out and rocket models tested in a wind tunnel. Two institutes assigned from the LFA were engaged in this: gas dynamics (under Prof. Busemann, who also worked on the ramjet engines described further on) as well as aerodynamics, chief of which was Dr. Blenk. Also located here

Work on the homing system for the F-55 ended at the research stage in a wind tunnel. The nose control surfaces can be seen in the photograph.

was the wind tunnel "operated" by a team under the management of Dr. Zobel. In the Institute of Aerodynamics there also existed a department engaged in the design of guided rocket homing systems, managed by Dr. Braun.

Next, a team subordinated to Prof. Busemann from the Institute of Gas Dynamics was engaged in the dynamics of supersonic flow (Dr. Guderley), where among other things rocket engine nozzles were designed (Dr. Winkler, Dr. Grumpt). However trials of the rocket prototypes were not carried out here. In accordance with an established regulation, launch trials of liquid propellant anti-aircraft rockets were carried out on the grounds of the centre near Peenemünde, whereas those with solid propellant engines were launched from the Luftwaffe "experimental station" rocket range near Leba. At the moment, when work commenced on the "Feuerlilie" the LFA entered a pioneering phase, although it had modest experience at its disposal, gained during initial trials of the experimental "Hecht" rocket. As I have mentioned, it didn't possess any homing system, but was equipped with movable rudders and its own control system (keeping to an earlier assigned course). This rocket was tested in flight under conditions which would allow a detailed comparison of its predicted and actual behaviour in flight. It was dropped (without engine) from an aircraft flying at a ceiling of 2000 m. At a designated altitude a special fuse caused a parachute to be released, thanks to which the "Hecht" prototype could be repeatedly modified and tested. These experiments were so crucial, that as it turned out, the aforementioned prototype often surprised the constructors while in flight to a considerable degree, especially during the initial phase of flight. Rich theoretical knowledge had thus become enriched, in short, with a pinch of practice.

The design of the "Feuerlilie" took place in three stages—first a small model with a fuselage amounting to only 5 cm in diameter was tested, next an "intermediate" missile 25 cm in diameter and 2.08 m long was constructed, and finally prototypes of the final version with a fuselage 55 cm in diameter. The "intermediate" version was characterized by a launch mass of 120 kg, carried 17 kg of explosive in the warhead and was propelled by a Rheinmetall-Borsig 109-505 solid propellant engine. This missile, which was given the designation F-25, was characterized by an aerodynamic system typical for a fast aircraft, among other things it possessed trapezial wings with a large slant of the leading edge. At the back were located tail fins ending in a tail unit, although only the elevons (horizontal) were connected to a simple gyroscopic control system. The purpose of its testing was solely to determine the link between angle of attack and aerodynamic drag—at a large range of speeds. Around 30 test launches were carried out, but the value of the obtained data was so limited, because the missile never exceeded a speed of 220 m/s (792 km/h), that is to say it didn't approach supersonic range, in which the final F-55 was to operate.

Contrary to intentions, this last missile was therefore also doomed to be purely experimental.

Two versions of the F-55 were designed. The first was a

single-stage surface-to-air missile with a launch mass of 473 kg, propelled by a solid propellant type 109-515 Rheinmetall-Borsig engine. The second and final version was a two-stage rocket. The first stage was a powerful (230 kg) bank of solid propellant engines, which was given the code name "Pirate," whereas the second stage acquired an engine from the company Walter from Kiel, powered by liquid propellants. The total mass of this unit is unknown, however one should expect that it considerably exceeded half a ton.

At the beginning of 1944, one prototype of the first version was made, which was launched without any control system (reportedly it travelled 77 km), after which further work on this version was given up in May of that year.

Somewhat later two prototypes of the two-stage version were made, but they were also devoid of homing systems. They acquired only simple flight calculators, based on gyroscopes. The sole purpose of their existence was to calculate optimal algorithms for the control system, since this had been the main source of problems up to now. Trials of the two-stage version of the F-55 were the first during which it was expected to obtain useful information from the point of view of the problem of homing, concerning the behaviour of the missile in supersonic flight. The Germans were however to be unlucky…

The first prototype hit the ground right after launching and the second fell victim to a British bomb on the grounds of the centre in Peenemünde, while it was being prepared for launching. In spite of this severe blow, it was still endeavored in the LFA to finalize the design of the F-55. Final modifications of the wings were introduced, in the wind tunnel a nose section equipped with small rudders was tested (the "canard" system with frontal control surfaces), which were to be operated by the homing system located in the warhead. It appeared that finally, after three years of work, the "Feuerlilie" design was close to completion. But at the beginning of February 1945, the decision reached LFA to completely discontinue all activity connected with the aforementioned code name.

The reason for this was as follows: the LFA as a scientific-research institution didn't build prototypes and all the more wasn't taken into consideration as a future producer. This task was entrusted to the company Ardelt-Werke from Breslau (Wroclaw), having at their disposal a huge underground factory (incidentally continually clouded in mystery) on the grounds of the present Mæelice district. The aforementioned decision was given because after the beginning of the January offensive by the Soviets the real threat appeared, that this super secret production plant, together with its no less well guarded contents would fall into the hands of the enemy. In connection with this the order was taken to destroy parts of the rockets and documentation.[102,107,109]

The "Wasserfall" (C-2)

The second project in terms of importance that was worked on in Peenemünde and in nearby Karlshagen (EMW) was the development of an anti-aircraft guided missile significantly smaller than the V-2, which was given the code name "Wasserfall" (waterfall).[101,102,105,106,107] It also presented the highest level of technology at that time, although as a defensive weapon (as opposed to the V-2) it wasn't favoured by Hitler, which as we know, very seriously weighed on the position of the Third Reich.

The "Wasserfall" was surely the heaviest and most complicated among all German "surface-to-air" missiles, in addition many innovative solutions were employed in it. Its launch mass reached up to 3500 kg. This was however still 3.5 times less than in the case of the V-2. The second fundamental difference resulted from the application of completely different propellants—noncryogenic. The liquid oxygen familiar from the V-2, was kept in this state (at a low temperature) thanks to vaporisation, and was from a chemical point of view the ideal oxidiser, however for the aforementioned reasons was unstable, which disqualified it from a military point of view when it involved defensive weapons, expected to be constantly ready for use. In other words it was necessary to find an oxidiser which would ensure the right propulsion efficiency and simultaneously could be constantly present in the rocket's fuel tank, at the same time not requiring any additional expenditures associated with this. A mixture of strong oxidising acids was chosen of composition: 90% concentrated nitric acid and 10% concentrated sulphuric acid. This fulfilled the aforementioned requirements, creating however other problems "in exchange." This was of course a strongly caustic material, which made completely new demands for the fuel installation designers as well as maintenance personnel. Moreover during launching and testing it caused strong toxic nitrogen oxides to emerge, which

The W-10 version of the "Wasserfall"

The "Wasserfall W-5" ready for launch

alcohols, such as benzol and xylol.

As much as one can successfully regard the fuel mixtures as good from a chemical point of view, the oxidiser was clearly the fruit of compromise. This is even manifested by the applied proportions—for one part of fuel fell over three parts of the oxidiser (76-77%)—although de facto the fuel defined energy value of the mixture. The "military" goal was however achieved. It was predicted that the "Wasserfall" filled with propellants could be stored without maintenance for half a year, and in the long term perhaps even for a year.

The first technical problem, which in connection with this had to be overcome, resulted from the design of the oxidiser fuel tank itself. Practically until the beginning of 1945 various fuel tank variants were tested, made from the following materials: common steel covered on the inside with a layer of aluminium (aluminium doesn't react with nitric acid), manganese and chromium steel, as well as common enamelled steel. The propellants were forced into the rocket engine thanks to the use of compressed gas (nitrogen). Right behind the warhead, in the frontal section of the fuselage was located a spherical fuel tank—containing 235 litres—70 kg of nitrogen, held at a pressure of 260 atmospheres. During the engine's operation this pressure fell to around 90 atmospheres, this difference was partially balanced by a pressure reducing valve. During storage these fuel tanks were sealed by aluminium membranes, which at the moment of launching were torn apart by the electrical detonation of pyrotechnic charges.

One of the most important elements of the rocket was of course the engine itself. In the final and smallest, from among the tested versions of the missile (the W-10) it ensured a speed in vertical flight reaching almost 2900 km/h. The G-force during flight varied from approx. 2.1 G just after launching to 4.5 G in the higher reaches of the atmosphere. The engine operated for 41 seconds.

The engine itself was made of common, mild steel and was cooled as in the V-2 engine, but by an oxidiser, and not by fuel.

in any case clearly betrayed position of the launch pad (the exhaust gases had the form of a dense, yellow-brown smoke). With respect to a different oxidiser it was also necessary to use a different fuel, which would easily react with the acids; this was to be a self-igniting mixture.

As a result of long-lasting analyses and testing, two different fuel mixtures were designed. The first of these was "Visol," based on saturated and non-saturated ethane derivatives (C2H5-OC2H3), and the second was "Optolen." It consisted of approx. 50% Visol, and also aniline (10-20%), refined coal tar dissolved in the remaining constituents as well as heavy

It was initially estimated that the internal temperature of the combustion chamber would reach up to 2800°C, but it turned out that a large portion of heat was carried away by combustion products, in connection with this the temperature didn't actually exceed 1800°C. The upper section of the engine was comprised of one large injector, connected to the combustion chamber by a circular plate, in which dozens of openings had been made for the injection of fuel and oxidiser. The mixing of both constituents ensued mainly in the combustion chamber. The engine didn't possess any form of ignition system; the fuel ignited automatically in the presence of concentrated nitric acid. In spite of large thrust, reaching up to 1800 kg, the pressure in the combustion chamber remained at a level of "only" 20 atmospheres. This combustion chamber had a capacity of 75 litres with an internal nozzle throat diameter amounting to 192 mm.

The problem of control was solved similarly as in the case of the V-2—aerodynamic rudders were located behind the tail fins, whereas closer to the axis—thrust vectoring plates deflected gas streams expelled from the engine. As employees of EMW (Elektro-Mechanische Werke/Karlshagen) testified after the war, after the first series of trials the latter were given up, since "they negatively influenced the rocket's performance."

During the war three basic versions of the "Wasserfall" were in turn designed (the W-1, W-5 and W-10), differing in dimensions, mass and homing systems. The first of them, the W-1, was already ready for testing at the turn of 1943-1944. Work was formally begun on it in 1940, i.e. soon after the outbreak of war. This rocket differed externally from later developmental versions in having relatively large "winglets," with a small slant, but with a wingspan significantly larger than the fins. Its launch mass amounted to 3500 kg, whereas the warhead mass was 235 kg (such a large explosive charge would have been sufficient enough to destroy group targets). Control was assured by a radio system code-named "Kehl/Strassburg," based on commands generated with the aid of a "joystick" by an operator on the ground, who visually observed the target. Such a system was also employed in the prototypes of later versions, although a whole series of considerably more modern devices were simultaneously developed. The closest to realisation was

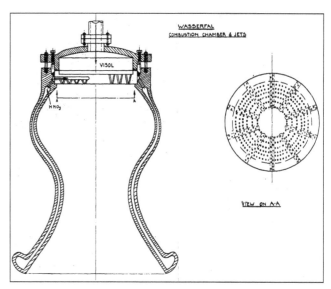

Longitudinal and lateral (at the height of the injectors) cross sections of the "Wasserfall's" engine. The nozzle cooling circuit is clearly visible. The oxidizer was pumped in from below.

a homing command radio system, in which two radars were to substitute the ground observer/controller: one for detecting and tracking the target and the other for tracking the missile. From among all the guided anti-aircraft rockets that arose in the Third Reich, the "Wasserfall" was however distinguished in that ultimately it was to be equipped with a state-of-the-art guidance system: a heat-seeking warhead and proximity fuse. A new generation warhead with a significantly increased explosive force was also designed—based on a fuel-air charge. The first test launching of the W-1 variant took place on January 8, 1944. This trial however ended in failure. Only the second flight, on February 29, proceeded according to plan and the missile gained a maximum speed of 2772 km/h.

A few months later the first W-5 modified version was tested. This missile was somewhat longer (7.765 m in comparison to 7.450 m), and had significantly reduced "winglets" at the cost of enlarged fins. The launch mass rose from 3500 to 3810 kg. Earlier faults in the radio command system were removed. The W-5 was characterised by a range (in the horizontal) amounting to 26.4 km and a maximum altitude of 18,300 metres.

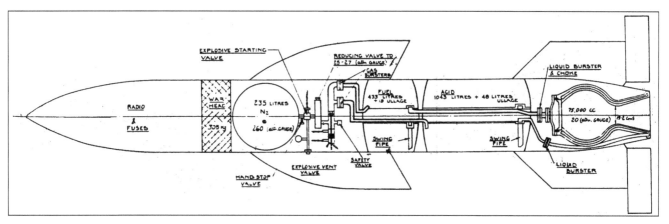

A diagrammatic cross section of the "Wasserfall" from a US report. If it reflects the proportions at least approximately, then it conveys the startling conclusion that the warhead occupied much less space than the homing system ("radio and fuses").

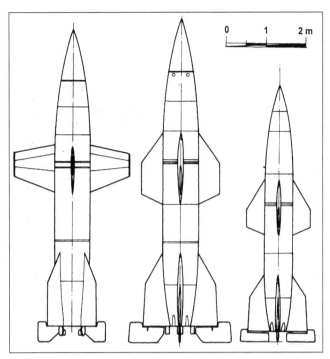

The Wasserfall missiles. From left: the W-1, W-5 and W-10.

The most state-of-the-art version, the W-10, designed in the latter half of 1944, was characterised by just the same mass as the W-1, but as a result of the construction's "rationalisation" had smaller dimensions. In relation to the W-5 the length was reduced by over 1.5 m to 6.128 m, and the diameter reduced from 86.4 cm to 72 cm. The "winglets" and fins were also smaller, the former characterised by an even larger slant of the leading edge. This had a significant influence on the aerodynamic drag and enabled a record speed of ascent of 2855 km/h to be achieved.

Up until the end of the war tests of only 40 specimens of various versions of the "Wasserfall" were carried out, which however didn't interfere with the development of plans of mass production and the design of a rings of anti-aircraft rocket defences, which were to surround the main cities and areas of Third Reich armament factory concentrations. It was intended on the basis of the underground space of a former mine near the town of Bleicherode to create a factory protected against attacks from the air, where initially 900 missiles were to be produced each month. In the long run this number was to be increased. It was assumed optimistically that the cost of mass production for one specimen would not exceed 10,000 RM.

Albert Speer, the Armament Industry Minister, in his "Memoirs" showed the sabotage of the "Wasserfall" project to be one of the greatest leadership mistakes of the Third Reich. This happened in spite of the numerous "voices of reason" already reaching Hitler. These are the words of Speer himself:[1]

> *Disregarding Hitler's arguments, what stood in the way of this reasonable stance was the fact that Peenemünde produced for the ground forces, and protection against air attacks was a matter for the Luftwaffe. The result of dividing the activities of the ground forces and the air force as well as the result of the ambition existing in the armies of the Wehrmacht, meant that in no way would ground forces be willing to give their equipment built in Peenemünde to competitors. As a result of the division between the Wehrmacht's armed forces, even testing and construction work was impossible. The "Wasserfall" could sooner have entered production had the full construction possibilities of Peenemünde been made use of at an earlier time. As late as January 1, 1945, characteristic of the division of priorities, 2210 scientists and engineers in Peenemünde were busy with A-4 and A-9 long range missiles, whereas only 220 were engaged with the "Wasserfall" project and 135 with other anti-aircraft rockets ("Taifun").*
>
> *Professor C. Krauch, plenipotentiary for Chemical affairs, advised me in a detailed memorial of June 29, 1943, less than two months before we made our wrong decision: "Supporters of the fast development of aerial attack measures, i.e. counter-terror, assume that the best weapon is attack and that our counteraction with the aid of rockets directed at England must lead to a decrease in aerial attacks on the Reich. Even with the assumption, unfulfilled so far, that one could use long range rockets in unlimited amounts and then enable destruction on the largest scale, this proposal appears wrong in the face of experience so far. On the contrary, after our rocket attacks on England, even up till now opponents of aerial terror towards the German people ... will demand from their government the maximum intensification of aerial terror against clusters of our population, against which we are still almost defenceless ... These considerations support the further drive for measures of anti-aircraft defence and C-2 "Wasserfall" equipment. We should use them immediately in the largest amounts possible ... In other words: every specialist, every worker and every hour of work devoted to the maximum development of this program will have an influence many times greater in deciding the outcome of the war than any effort in favour of other programs. Any delay in executing this program may have consequences which will have an impact on the settlement of the war.*

The "Taifun"

The "Taifun" was the second anti-aircraft missile made at EMW. It was linked to the "Wasserfall" by liquid propellant propulsion, all similarities however ended at this. It had no guidance system and was in general considerably simpler and smaller, which reflected the situation in which the German armament industry found itself in the final stages of the war.

The "Taifun" exemplified an attempt to substitute quality (a complicated guided rocket) with quantity. It was to be launched from multiple launchers, in fast bursts of 60 rockets each, leaving the launcher every 0.025 s. The launch of a whole batch/salvo would last therefore approx. 1.5 s. At the same time

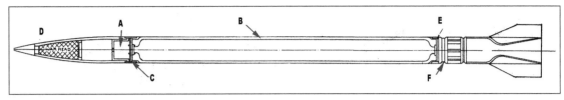

The "Taifun-F"
A) pyrotechnic charge B) external fuel tank
C) membrane, disrupted at the moment of launch D) warhead
E) membrane, disrupted at the moment of launch F) injector

it was endeavoured to ensure the smallest scatter possible; the missiles were stabilised not only by fins but also rotationally. Designed for this were spiral rails positioned inside the launch runners and on the flight path rotational movement was maintained thanks to appropriate profiling of the fins. The conception of deploying the "Taifun" relied on the ability to create a defined zone in which the probability of hitting the targets/bombers would be very high. Trials carried out showed that the rocket scatter was indeed small and e.g. at a ceiling of 10,000 m they created a field of fire 250 metres in diameter in which the probability of hitting a standard bomber would be the order of 10-20%. The author of this conception was General Dornberger, directing the ground force research establishment in Peenemünde and in his honour the multiple launcher was given the code name "Dobgerät."

The premises therefore existed to regard a weapon of this type as an alternative even for the complicated "Wasserfall" rockets. Despite its simplicity and low cost the "Taifun" had however a very significant defect under the conditions of that time; it required the consumption of a disproportionate amount of raw materials in relation to its effect.

It was a classic rocket in terms of appearance with the fuselage in the form of a steel pipe 10 cm in diameter. The total length amounted to 1.90 m and the launch mass 19 kg, of which 10 kg was propellant and 0.5 kg warhead. A single salvo of 60 missiles "weighed" therefore 1140 kg.

As in the case of the "Wasserfall" missile, the design of the fuel composition, oxidiser and determination of the optimal proportion between fuel and oxidiser was chiefly the work of the BMW laboratories in Munich. According to one of the scientists employed there (Dr. Hemmersath) up to 6000 various types of compositions were examined.

A new feature in the case of the "Taifun," resulting from its small dimensions, was the substitution of the fuel pump with a simple gas generator, exploiting a burning pyrotechnic charge. This forced the entire propellant into the combustion chamber in a time not exceeding three seconds. As a result at the moment of launching the missile gained an acceleration of 35 G, which in the course of a few seconds rose further to around 60 G. The gas generator was placed at the front, right behind the warhead. Two tanks, of fuel and oxidiser, were situated behind. These were two steel, co-axially positioned (one inside the other) pipes. The external one, being simultaneously the frame of the fuselage held a fuel composition called "Tonka." Its internal wall was the second pipe, in which was situated a mixture of acids called "Salbei."

It was predicted that production of the rocket would mainly take place in the underground factory of "Mittelwerk" near Nordhausen and in February 1945 the first orders were placed for 20,000 pieces. But since the testing of prototypes was continually prolonged and the rocket de facto wasn't ready for production, a certain substitute was opted for: a hurried design of a simplified version with solid propellant engine (compressed powder). This version, of which 50,000 pieces were ordered, received the designation "Taifun P," the original version was marked with the letter "F" for distinction. The "compromise" rocket generally had a similar performance to its more complicated ancestor, apart from one thing, namely the scatter was significantly greater, which obviously left its stamp on the conception of this weapon's use (although in the existing situation it was still only a purely theoretical concept). However almost 20,000 pieces of the "Taifun P" were produced, 2,500 till the beginning of March and 15,000 till mid-April 1945. On the other hand the "Taifun F" never left the production line.

It wasn't the only weapon of this class developed in the final years of the war. The main "competitor" of the "Taifun" was a light solid propellant rocket missile, designed by Henschel, the "Föhn-73" (or Hs 217). It had a mass of only 3 kg and achieved a ceiling of 11 km. The warhead, weighing 400 grams constituted the equivalent of a medium calibre anti-aircraft cannon artillery projectile, the accuracy of this weapon was however considerably worse. The "Föhn-73" was to be launched from simple multiple launchers, up to 48 rails. In spite of these bold plans only 59 such launchers were made till the end of the war. The "Föhn-73" was also to be the basic armament of a simple rocket fighter, the Bachem Ba-349 "Natter."

There were other designs of this type, such as the "June-Bug" solid propellant rocket, or the liquid propellant TE-5. There also existed an attempt to create an interesting anti-aircraft equivalent of the famous "Panzerfaust." This was a hand-held launcher of six light unguided rockets. None of these however played a role on the battlefield.

The postwar development of military technology has shown that the conception of unguided "surface-to-air" rockets was completely erroneous.[105,106,108]

A test drop of the Hs 117 from a bomber's wing

The Henschel Hs 117

The Hs 117, also known as the "Schmetterling" (at a certain time it was even informally designated the V-3 weapon, until it turned out that this symbol had been reserved for the multi-chamber cannon) constitutes the next example/reflection of this part of the German rocket programme. As in the case of the "Wasserfall" or R-1 this involves a project which was for a long time completely sabotaged. Hitler, in defiance of all rational premises, had no comprehension whatsoever of defensive weapons. On September 11, 1941, he gave the order to halt all

The Hs 117C on the launcher

work on guided anti-aircraft rockets. For the Third Reich this was as equally catastrophic as the later halting of work on the Me 262 jet fighter for a year. That these weapons were at all constructed results simply from the fact that Hitler's decisions were not very conscientiously carried into effect. Admittedly in 1944 the situation gradually underwent changes, "better" priorities had begun to be awarded to guided surface-to-air rockets, but it was already too late. The "Schmetterling," just like its relatives was practically completed at the moment of capitulation. Even mass production had been started, but at a point when the war was already de facto lost. The majority of these rockets never left the workshops of the enormous underground factory code-named "Hydra" near the town of Woffleben in the Harz mountains. American soldiers found them there. In any case the factory itself was also far from completion.

Design of the Hs 117 was formally begun in the spring of 1942, after forming "an anti-aircraft rocket department" in the Luftwaffe headquarters (March) and after the official start of a parallel research programme one month later. So things were taking place relatively early, what is more plans of the uncompleted Hs 297 anti-aircraft rocket were used in the design of the future Hs 117 (such a designation only appeared in the spring of 1943). It was designed by Professor Herbert Wagner from the works of Henschel as far back as 1941; until Hitler gave the memorable decision to block all work in this field.

In spite of this the "Schmetterling" exemplifies frankly a classic example of a project, which despite the availability of technical knowledge and the urgent need of introducing such a weapon to production (forcibly emphasised by the bombing of the Reich), wasn't carried through due to organisational reasons.

According to preliminary arrangements from the spring of 1943, the new rocket was to be characterised by a maximum ceiling of target interception of the order of 8 km at a horizontal range of up to 20 km. It was to have a launch mass of 330 kg, although the warhead mass was to amount to only 5 kg. Propulsion was predicted in the form of a liquid propellant engine from BMW (nitric acid plus a mixture of "Tonka"), aided by two compressed powder boosters. Ultimately these parameters were to undergo further changes—the launch mass rose to 440 kg and warhead mass to 25 kg, at the end of 1944 "Schmetterling" could, in the course of 60 seconds achieve a ceiling of up to 11,000 metres. It was 4.3 m long with a fuselage diameter equal to 33.5 cm and "wingspan" of 1.98 m. The aforementioned information refers to the SI version.

Despite the fact that, as mentioned, research work dated back to the beginning of 1941, the change of priority which enabled the project to be moved onto a more clear-cut path didn't come until September 1943. But this didn't solve all problems, since it turned out that due to the earlier blocking of work it was necessary practically from scratch to assemble a team of specialists: scientists, technicians and qualified workers. Up to the end of September only 100 people had been assembled, although 546 were judged necessary as the bare minimum. Supplies of raw materials also in reality looked different, than on paper. Thus it was out of the question that launching of the first prototype would be carried out according to plan, that is on January 1, 1944.

Only on February 15 could one formally state that the first trial launch had taken place. "Formally," for in reality it involved something which only externally resembled the "Schmetterling," since that what ascended over the rocket range in Peenemünde had neither a target cruise engine, nor control system, nor even a warhead. The only propulsion consisted of two, discarded solid propellant engines. All other missing components were still to be found at the design stage.

In spite of this situation as early as May 1944 ambitious production plans were drawn up in the Air Force Ministry. Initially it was intended to produce 265 pieces solely for research purposes, ultimately mass production was to encompass 24,500 (!) rockets. Soon these figures were modified—200 instead of 265 research rockets and 23,650 combat. Monthly production was to approach 3000 pieces, although it was realised that this wouldn't be possible earlier than mid-1945. Apart from Henschel, Askania, Bosch and Siemens were to receive orders for production.

It was only in the summer of 1944 that the first specimens with homing system devices (command type, operating on a similar principle to the "Feuerlilie" missile) were launched and dropped from the He 111 bomber. This system was dependent on the ground operator's good eyesight, due to which even in theory it could operate efficiently only in daylight, in conditions of good visibility. Initially the device mounted on board the missile bore the code-name "Colmar," later this was renamed to "Strassburg" and its ground equivalent to "Kehl." The first versions however didn't fare very well; the electron tubes used were unusually susceptible to damage, and were often defective when installed. It even reached the situation that equipment installed in the rockets was missing certain cables or they had been torn off. As a result Dr. Sichling, operating one of the launch pads in Peenemünde, affirmed that "it was a fortunate case, if a rocket in working order was placed on the launch pad, after which it turned out that the guidance system actually worked."

Such unusual cases did however occur and the rocket then behaved in accordance with expectations—its aerodynamic properties were not a further source of unpleasant surprise.

Yet the Hs 117 still didn't have cruise engines. The first three specimens with target propulsion were only ready at the beginning of September (in the first rocket the engine didn't work at all, the other two had various faults). Once again it turned out that this "unfortunate" project, having being carried out all the time for several years, still found itself at an astonishingly early stage. In this situation carrying out production preparations "by force" was at the very least premature.

It is a little known fact that the Hs 117 was tested not only in Peenemünde, but also on occupied Polish land, specifically at a rocket range code-named "Nord." It was situated near the Mława-Ciechanów road (then: Mielau-Zichenau) and occupied almost 300 km². It wasn't only a military training ground, but a perfectly prepared research complex, with a developed road network and fully developed premises, with its own cinema and

Launch of the Hs 117C

sports stadium, and enough buildings to house 15-20 thousand people. In the spring of 1944, agents from Soviet Intelligence had been dropped by parachute on a reconnaissance mission and photographed among other things "successive versions of the Schmetterling rocket" (!) As a result somewhat later the Soviet air force bombed "Sector 14A," where research was being carried out.[114]

During this period, in November 1944, the finishing touches had already been put to the first mass-production "assembly shop." It was situated in one of the tunnels of the Berlin underground (code-name "Sperling") and for the time being was to satisfy the needs of the centre in Peenemünde—this was before production was transferred to "Mittelwerk," In March 1945, production was to rise to 150 missiles monthly and in November to 3000.

Up until evacuation of the research centre in Peenemünde only 38 launches from ground launchers and 21 rockets drops from aircraft had been carried out. Out of this number (59) 28 trials had gone "satisfactorily"—less than half. At this stage the program was scuttled, although with the "force of inertia" the administrative decisions of its further development were still taken. For example on February 6, 1945, Himmler ordered SS-Obergruppenführer Hans Kammler (the SS had also taken control of this field of armaments) to design the Hs 117H version with a television homing system—in the final stage of flight the operator was to observe the target with the aid of a camera located in the rocket's nose. Up until the end of 1945 it was intended to create a special, independent system of target detection (radar?). In January of that year Professor Wagner

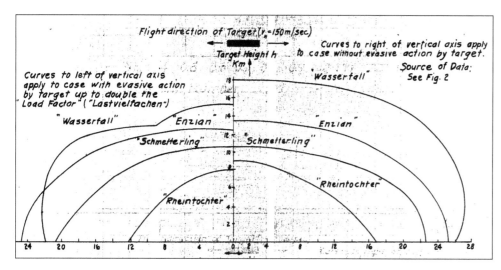

A comparison of predicted anti-aircraft rocket parameters according to the state from March 1944. The range has been shown as a function of ceiling (in kilometres). The right section of the graph refers to the situation when the target does not carry out manoeuvres, on the left the opposite situation is depicted.

went as far as to propose a completely modified version of the rocket, which was given the code-name "Project SII." This was in fact two modifications, completely redesigned, with much better aerodynamics (looking much more "refined"). Each had not two like so far, but four additional solid propellant boosters. As a result both modifications, the larger SIIa and smaller, more tightly-packed SIIb (an analogy is visible to two modifications of the "Wasserfall:" the W-5 and W-10) were to develop a greater speed, in addition their range would also increase.

These were however only designs on paper, Kammler's "Rüstungsstab" still wasn't in a position to carry through the development of even the basic SI version.

At the turn of January and February preparations were completed to continue the testing of prototypes in an area of the town of Karlshagen; several Hs 117 missiles were tested up to February 19, dropping them from the He 111 bomber. An improved "Kehl/Strassburg" guidance system was tested. In the second half of March production of the SI was finally to begin in the underground factory "Hydra." It is estimated that altogether 150 of them were made, of which approx. one third were practically abandoned in the underground workshops after assembly, as there was no longer any chance of getting a supply of cruise engines for them, production of propellants also came to a standstill.

So bearing this in mind, the fact that the "Schmetterling" was placed in a leading position in the German anti-aircraft rocket defence plan is surprising. This plan, which was provided with the code name "Vesuv." was developed in 1944. According to this plan weapons of the discussed class were divided into two groups. The first one formed the operational rockets (there are three levels in the army: tactical, operational and strategical), which were to create a continuous barrage of defensive zones, deployed in general within a given Theatre of War. The second group, tactical rockets, were designed to defend specific "points," e.g. factories, airfields.

The operational echelon was to consist of 1,200 "Wasserfall" W-10 rocket batteries (altogether 96,000 launchers!) as well as 1,300 "Schmetterling" batteries. At the beginning of 1945, the "Wasserfall" was replaced by the "Taifun" in these plans. The tactical echelon was to be based in turn on the "Enzian" and "Rheintochter" missiles described next.[107,110,114]

The "Enzian"

In the case of the "Enzian" we have a fairly curious case to deal with, as the construction of this surface-to-air rocket exemplified the far-reaching development… of a fighter aircraft, obviously a rocket propelled fighter, namely the Messerschmitt Me 163. In addition the huge bulk of this construction was wooden. The

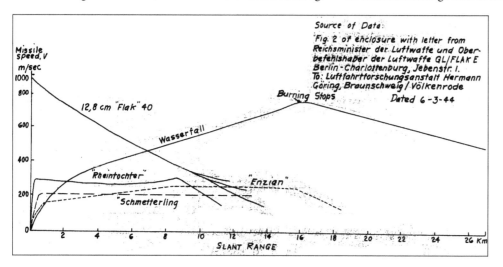

The airspeed of German surface-to-air missiles (on the vertical axis) as a function of range. This diagram is based to a large extent on estimates, since it bears the date of March 6, 1944.

"Enzian" in its basic version (E-1) was characterised by a type of homing standard for this class of rockets at that time, using commands generated by a ground operator and sent to the missile by radio. The first design of this version arose in June of 1943. Its fundamental, characteristic features, resulting from the aforementioned origins, was a large launch mass—1900 kg and a supersonic flight speed—the order of 850 km/h. These parameters constituted the basic "burden" of the design, and were the source of its low competitiveness, in connection with this the chief effort of the constructors aimed to overcome these "hereditary faults." The basic version was from the start only treated as a transitory stage. The design of more perfect developmental versions lasted incessantly therefore up until the beginning of March 1945, when under Kammler's order the project's execution was discontinued.

The E-4 "Enzian" on the launcher

Almost throughout the whole of 1944 work headed simultaneously in different directions; various aerodynamic systems were analysed, and at the same time many electronic systems designed, which were to render the missile independent of the ground operator. These were among others: the "Kehl/Strassburg" system already known to us (the first was the transmitter, the second the receiver), treated as a temporary, parallel system to the "Kogge/Brigg." Three alternatively dealt with homing warheads were designed especially for the Enzian, of which two operated passively. One of these bore the code-name "Madrid" and homed in on sources of heat (a heat-seeker)—it was designed in the company Kepke from Vienna. It contained a mirror lens and single (?) ELAC detector, chilled by liquid oxygen to a temperature of -80°C; therefore theoretically it should have been fairly sensitive. The second type was an acoustic warhead—in a sense the equivalent of warships' sonar—code-named "Archimedes," which arose in the works of AEG-Telefunken. The third warhead, bearing the code name "Moritz" was a semi-active, radiolocative homing system, comprised of a pulsed radar, to "illuminate" the target as well as a receiver on board the rocket. The receiver in turn consisted of

The E-4 "Enzian" during launch

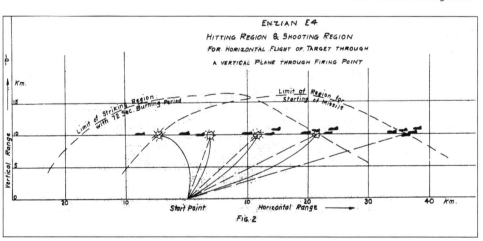

The "Enzian," a diagram illustrating interception of the target

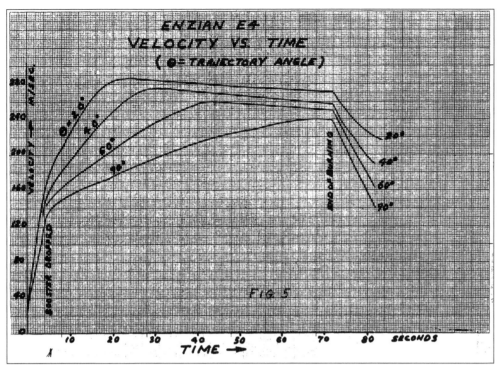

A diagram referring to the E-4 versions, showing the velocity (vertical axis) as a function of time. Specific curves correspond to various angles of ascent. After about 5 seconds the take-off boosters were discarded, after 70 seconds the main engine ceased operation.

a directional receiving antenna and a simple electronic system, working out commands for the control system. It calculated the position of the target knowing the direction of the antenna's inclination as well as the time difference between receiving the original signal and the echo, thanks to which the distance to the target was determined.

At the same time that the electronics were being worked on, the construction of the rocket itself was perfected as well as the engines (1 cruise and four boosters). This work lasted practically throughout the whole of 1944 and it was only at the turn of 1944/45 that Messerschmitt could present the designs of three versions which it regarded as final. These were the E-4, E-5 and E-6 versions. The first of these was selected for production, although it constituted the smallest departure from the fighter's construction (in all probability just because of this). As opposed to the E-4, the E-5 missile was to be characterised by carefully finished aerodynamics—finished from the point of view of supersonic flight, as it was to develop a speed of Mach 2, whereas the E-4, somewhat below Mach 1. With its slender profile and slanting, symmetrical fins it resembled some kind of futuristic missile, as opposed to the E-4, which was rather more suggestive of an unmanned aircraft. Better performance, including an increased range from 25 to 30 km, was to be the chief merit of the aerodynamics as well as of the improved engine, since the launch mass and propellant mass were practically the same.

The last of the aforementioned three, the E-6, was a "surface-to-surface" guided rocket design lighter than the V-2. As in the case of the E-5, it remained solely on paper.

Provisionally qualified for mass production (provisionally, since none of the tested prototypes was fully equipped, although "us much as" 24 were launched), the E-4 was a missile which essentially differed little from what had been asserted as the objective as far back as the summer of 1943. Obviously the general situation of the Third Reich in 1944 and earlier wrong decisions of Hitler blocking access to the latest technology weighed

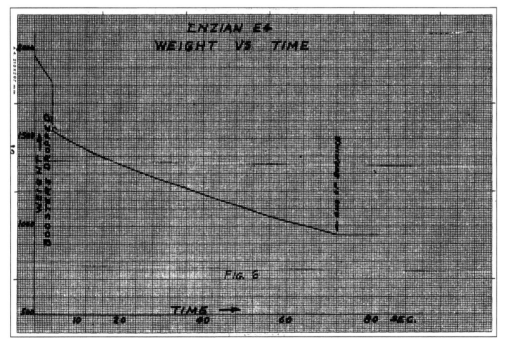

The E-4, the missile's mass as a function of time, from the moment of launch

heavily on this, but equally... probably the too large a dispersion of work on guided anti-aircraft rockets. Even in the existing political-organisational situation the effects would have been without a doubt much more clear-cut, if one promising type of rocket had been concentrated on. These considerations equally determined that homing warheads remained in the case of the E-4 a fairly remote prospect.

Not much better than the E-1, it could develop a speed of up to approx. 1000 km/h and to a similar degree wood "not in short supply" was used in its construction—overall 150 kg plus 45 kg of steel (mainly in the warhead). The liquid propellant type FKF-613 engine was regarded as the simplest engine of its kind, that had ever been employed in an anti-aircraft guided rocket. It developed right after launching a maximum thrust of 2000 kg, which after 72 seconds fell to around 1000 kg. The launch was aided by four solid propellant boosters acting for 4 seconds, giving a total thrust of 6000 kg. The missile had a fuselage 4.08 m long and 0.876 m in diameter, as well as two modified (mainly shortened) wings "inherited" from the Me 163, with a wingspan of 4.05 m. The maximum ceiling of target interception amounted to 13,500 m. The 225 kg warhead could have theoretically struck even group targets, however with the engine's limited thrust such a large mass would probably have been excessively extravagant. However this has remained solely in the domain of theory, since no "Enzian" was ever used in combat. Work on this construction was discontinued in February 1945.

This was quite an odd rocket, full of compromise and "inherited burdens," which in the version accepted for further realisation didn't hold much hope of success in battle.[102,106,107,109,111,112]

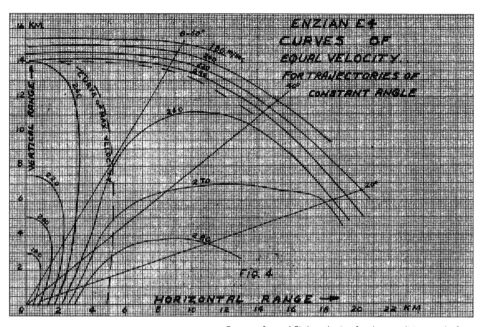

Curves of equal flight velocity, for the conditions as before

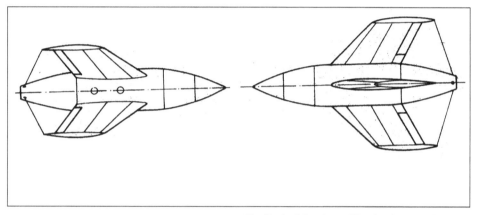

The "Enzian," the planned E-5 developmental version

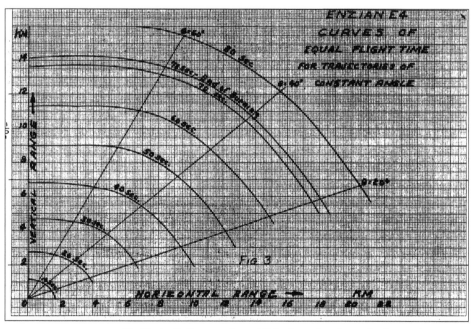

The "Enzian E-4," curves of equal flight time, when the angle of ascent is constant. The horizontal axis corresponds to the horizontal surface, the vertical axis: altitude.

The Rheintochter

The rocket described below was probably the most successful construction of the discussed group, which appeared during World War II (i.e. in Germany).

The origins of work on the "Rheintochter" (translated: "Rhine Maiden") date back to September 1942, when the Chief Anti-Aircraft Defence Inspector, General Walter von Axthelm, turned to industry with a memorandum in which he offered to cover the cost of developing several guided anti-aircraft rocket designs. One of the recipients of this was the company Rheinmetall-Borsig, which in a record time of about two months presented a preliminary design of their "Rheintochter" missile. Rheinmetall had little experience in the field of electronics, however problem solving connected with this could be sub-contracted out and in principle concerned the rocket's equipment, on which dedicated companies and scientific institutions worked anyway. Rheinmetall had another and as it turned out crucial trump. Namely the company had very much experience in the field of the design and production of solid propellant rocket engines, including large ones (among others the "Rheinbote" multi-stage surface-to-surface missile was drawn up simultaneously with the "Rheintochter," which in spite of a mass of 1656 kg was propelled entirely by engines of this type). In those times this constituted a real sensation. Guaranteeing the stability (i.e. linearity) of powder combustion was quite a serious problem, since substances of this type are by their nature unpredictable, particularly in production practise and obviously at a high engine/combustion chamber volume. So if the aforementioned company was able to build the required rocket much simpler and cheaper than the competition, as well as faster, then to begin with its path to success had been made significantly easier. With this one should bear in mind that a typical liquid propellant engine consisted of hundreds of parts and the "military success" of the rocket depended to a smaller or greater degree on every one of them.

Both stages of the "Rheintochter" possessed six steel cylindrical rocket engines. In the first, the launch stage, they were comparatively small, were to operate for only 0.6 second and in practise their task consisted only of accelerating the rocket on the launch rail. During this time they gave a total

The R-1 at the factory

The "Rheintochter R-1" on the launcher

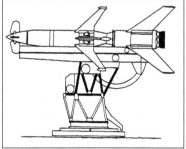

The R-1 on the launcher

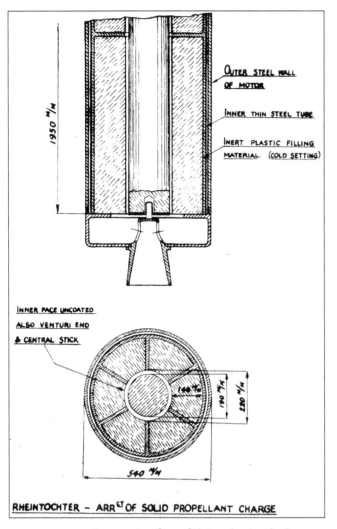

Cross-section of one of Rheintochter's rocket boosters

Remnants of the Rheintochter launcher near Leba

thrust of 73.5 kN, that is 7.5 tonnes. Afterwards this stage was discarded and ignition of the cruise engines followed. These operated for 2.5 seconds, giving a significantly larger thrust of 24 tons. In the course of only 3 seconds from the moment of launching the rocket gained a maximum supersonic speed of 1,300 km/h. As one can see, for a fundamental part of the flight trajectory the missile ascended solely thanks to kinetic energy, without propulsion (the maximum ceiling amounted to 7000 m, at a horizontal range of up to 40 km). The version in question was characterised by a launch mass of 1750 kg, length of 6.29 m and possessed a 150 kg warhead. This version received the designation R-1. Its design took over a year and a half and the first prototypes were ready for in-flight trials at the beginning of the summer of 1944. Overall several dozen specimens were supplied for testing. During August-December 1944, they were launched from the Luftwaffe rocket range near Leba. These trials didn't reveal any fundamental design faults. As a result from a technical point of view the "Rheintochter R-1" was already ready for production at the end of 1944.

It was characterised by a relatively simple type of control, radio transmitted commands from the ground. The radio signal receiver was located in the nose section, just behind the servo-mechanisms steering the aerodynamic rudders situated in the

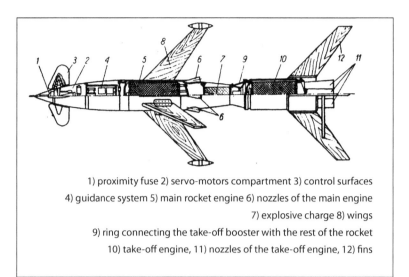

The Rheintochter, a cross section

1) proximity fuse 2) servo-motors compartment 3) control surfaces
4) guidance system 5) main rocket engine 6) nozzles of the main engine
7) explosive charge 8) wings
9) ring connecting the take-off booster with the rest of the rocket
10) take-off engine, 11) nozzles of the take-off engine, 12) fins

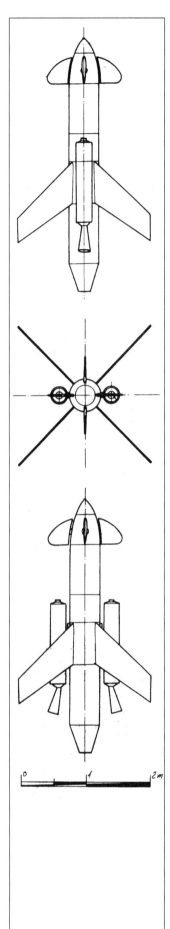

The design of the R-3 version

nose itself. The warhead was located however in the rear section of the second stage, that is behind the engines, whose nozzles were drawn aside on the outside, located on the fuselage's circumference.

In 1944, yet another, upgraded version of the "Rheintochter" was drawn up, designated the R-2. Instead of the first stage this was given four unshielded boosters, fixed between the wings of the second stage used so far. These were discarded directly after launching. This was in all probability only an "experimental" design, which was designed to verify a new configuration before designing yet another version: the R-3. This was outwardly similar to the R-2, however ultimately was to receive a liquid propellant cruise engine (in practise this never followed and the existing prototypes were launched with powder propellant engines).

The R-3 was generally somewhat smaller than the R-1, the removal of the first stage and replacing it with additional boosters mounted on the sides decreased the length of the missile from 6.29 m to 5 m. The launch mass was reduced to 1500 kg, but with an unaltered warhead mass. In the target production version, propelled by the liquid propellant Konrad engine, up to 336 kg of oxidiser (a mixture called "Salbei") as well as 90 kg of fuel ("Tonka") was to be located in the fuselage. The additional launch boosters had a total mass of 173 kg.

Thanks to this a maximum ceiling of 14,700 m was to be attained, significantly exceeding the flight ceiling of American and British bombers, the flight speed was however to be subsonic, up to 1080 km/h. Most publications from past years concerning German rocket testing have devoted some passages to descriptions of the initial combat launches of the "Rheintochters." This would appear not to be controversial, however in an endnote to a very good article on this subject, which appeared in 1999,[107] it was announced, that:

According to older sources at the end of 1944 production started of the Rheintochters R-1 and R-3. Part of them were to even find their way to anti-aircraft artillery units within the confines of tests under combat conditions. In January and February 1945 these rockets had the opportunity to demonstrate their effectiveness against American bombers. At that time 82 R-1 and 88 R-3 missiles were supposedly launched. From among the R-1 as much as 51 were to have hit the target, while in the case of the R-3 only 5 didn't hit the target! As to that effectiveness a section of sources give radically different information. According to these only 8 R-3 hit the target. One may however not believe a word of this information, as the latest research shows that the Rheintochters were launched only within the confines of rocket range testing and never managed to enter either production or arm military units.

It is a pity only that the author of the endnote didn't give the source on which it was based, which surely renders the previous information doubtful, but does not in the least completely settle this issue.[101,106,107]

The Natter found by the Americans

The Natter during pre-launch preparations

The "Natter"

One may regard the "Natter" (translated: "Viper") missile as a peculiar type of anti-aircraft rocket, although it was guided by a pilot. The Germans classified it however as an anti-aircraft rocket, and not as an aircraft which in any case more accurately reflected its nature. This weapon resembled very much the "Ohka" Japanese suicidal rocket missile, used by the kamikaze, and was in all probability its prototype.

The "Natter" was classified for production after comparing it with several other competitive types. These were: the aircraft-missile "Julia" designed in the Heinkel production plants, the Junkers "Walli" as well as the Messerschmitt P-1104. All of them had a similar propulsion, however the "Natter" surpassed them all in terms of performance. A similar manned missile, the "Eber," was also rejected, designed in D.F.S. This was however to be launched from under the wing of a flying bomber and was devoid of… any armament. Its task was simply to ram the rudders of the enemy bomber (the construction was strengthened respectively), or dive into it like a spike. The Bv-40 glider from Blohm und Voss was modified to fulfil a similar role. Its equivalents were: the air-to-air aircraft-missile "Taran" from Zeppelin as well as the Me 328 from Messerschmitt. A version of this last one was even designed with folded wings, which was to be an anti-aircraft weapon for submarines. These weapons never entered production and this was not only due to the difficult situation of industry. Their military usefulness was doubtful and apart from this the fact that this armament was semi-suicidal

The Natter, getting into the cockpit, as can be seen, was somewhat troublesome

The Natter, the rocket launchers are clearly visible

in nature aroused both controversy and opposition.

The "Natter" was thoroughly examined after the war and described among others in an aforementioned Polish technical analysis of German weapons (incidentally excellent).[102] Here are its excerpts:

> *Ba-349 A. The first production modification of the "Natter" propelled by the Walter HWK 109-509 A engine achieved a maximum speed of 880 km per hour, climbing with a vertical speed of 10.8 km per minute and gaining a ceiling of 14.7 km. The fuselage construction was made of wood; this is a construction possessing profile stringers, forming a skeleton covered with plywood; it was composed of two main sections. The pilot was placed at the rear directly behind the main armament composed of explosive rockets, built into a quadrangular or hexagonal frame, depending on the rocket calibre. The whole lot, together with the frame was shielded by a "Plexiglas" windshield, fixed to the body's airframe with the aid of explosive bolts. In order to increase the airframe's firepower, 2 cannons were placed on some manufactured types of airframe, the MK 108 30mm calibre, apart from the typical armament composed of 33 R4 explosive rocket missiles or 24 Föhn type 73mm calibre rockets. For the pilot's protection a durable shield and 2 armoured walls were located in front of and behind the pilot. The sides of the cockpit as well as part of the fuselage designed for the fuel tank were protected with a 3/16" armoured mantle. This armour along with the narrow frontal area as well as the great speed of the "Natter" made it beyond the reach of bombers' defensive weapons. The steering device was operated by pedals, transmitting devices acting on the lower fins, throttle valve switching and the auto-pilot. On the instrument panel were located 2 buttons, one launched the rocket missiles, the other ejected the pilot.*

The "Natter" airframe components are unknown. Technical details of the "Natter" rocket

Span	5.486 m
Length	6.50 m
Wing area	5 m²
Take-off weight	2200 kg
Wing load	440 kg/m²
Maximum speed	1000 km/h
Speed of ascent	11275 m/min
Flight time	2 min

In order to easily disengage part of the airframe, a hand lever was built in. Behind the cockpit were located two fuel tanks. One for 190 litres of "C-stoff" fuel (upper), the second below the first for "T-Stoff" fuel, containing 415 litres. Looking from the front of the airframe, we can see a parachute case fixed to the left side of the fuselage's airframe, while on the right, turbine pumps, powering devices and a double fuel type Walter HWK 509 A-1 auxiliary engine. The engine itself, built in centrally, filled the rear section of the aircraft-bomb's fuselage. The wings of a short span, fixed permanently to the fuselage, were made from individual wooden longerons passing between the fuel tanks as well as wooden ribs. The wing tips were protected by protective metal plates. There were no ailerons in the wings. The construction of the airframe tail consisted of a fin and rudder above and below the fuselage. The horizontal fin was positioned above the fuselage's centre line; in it were located ailerons operated by a control gear; they served two purposes: as an elevator and as the ailerons. In order to improve control the aileron control rods were connected to winglets situated in the rocket's nozzle, washed with the stream of exhaust gases and directing the outflow. In this way not only was a more sensitive response to any longitudinal and lateral movements of the airframe achieved, but equally the violent response of the nozzle winglets adjusted by the auto-pilot was balanced, thanks to the influence of the ailerons. This method was also introduced into the V-2 rocket during testing in Peenemünde. The introduction of this method to the "Natter" undoubtedly resulted in their improvement and increase in stability.

Launching took place on an almost vertical launcher under the thrust action of A.T.O. solid propellant rockets situated in the tail end of the fuselage's airframe. The launcher was 24 m high. 4 A.T.O. rocket units operated for 6 s releasing a thrust of 500 kg each. Next a Schmidding 553 rocket was applied giving a thrust of 1000 kg for 12 s. In the final stage only these units were used for the BP-20 prototype. The initial acceleration was greater than 2 g, at the same time the boosters were detached at an altitude of 1.5 km at full operation of the main double fuel engine. The launcher could revolve on a vertical axis in such a way, that it could be positioned horizontally which would allow the "Natter" airframe to be loaded from a conveyor onto the launcher. The airframe's wing tips and tail fin were guided exactly into the launcher's rails. At the signal of approaching enemy aircraft the pilot got into his cockpit, and the missile together with its launcher were turned in the direction of the arriving bomber squadrons. The bomber course was transmitted and verified by a radar instrument, and position transmitted to the autopilot's receiver by an electrical connection broken at the moment of launch. Directly after launching the machine was controlled automatically on a fixed course, until the pilot himself made a correction to the flight path on account of a possible change in bomber course, as soon as they had noticed the airframe being launched. The bomber course was transmitted by radio from ground control to the pilot, which made it easier for him to quickly set the correct flight course. On approaching the bombers the pilot fired the whole load of explosive rockets, his machine then experienced a recoil as a result of the gas streams. The whole missile batch reached the centre of the formation and destroyed the bombers by flame, blast and splinters.

When the pilot moved the lever disconnecting the frontal segment of the airframe from the fuselage, the machine received a powerful impact of gases and air. For a while the pilot was subject to enemy fire, but a successive movement of the lever caused him to be ejected from the airframe. Within 2 seconds his parachute opened automatically and the pilot descended slowly to the ground. At the same time the rear section of the fuselage's airframe, encompassing the engine and equipment, also detached and descended to the ground with the aid of its own parachute.

(…) The "Natter" rocket could be launched from any not very well secluded place such as city parks, factory grounds, field artillery positions, through the use of a light launcher. Great promise was held for the anti-aircraft defence of Germany through the application of these rocket missiles in fighting foreign squadrons. However the bombing of production plants, workshops and assembly plants prevented the Germans from applying the winged "Natter" missiles in a mass operation.

The X-4 missile under the wing of an Fw 190 fighter

Air-to-Air Rockets

The company Henschel Flugzeugwerke A.G. from Berlin-Schönefeld should be regarded as a pioneer of German work in this field, which began design of the first missile in 1941.

It concerns the Hs 117 H, a derivative of the surface-to-air missile, which in its mature form only appeared in the last

A "joystick" in the fighter's cockpit to control the missile

The X-4 – a specimen captured by the Americans

months of the war. As with the remaining types, the Hs 117 H was never used in combat. In this case it was restricted to only 20 trial launches carried out from the Dornier Do 217 bomber; 6 rockets hit their targets. This was a single-stage missile, but with the greatest range among this group of weapons, which amounted to 11 km. Like the others it was mainly designed to destroy enemy bombers and therefore large and slow-flying targets. It was one of many types of weapon which were to hold stop the devastating, massed air raids of the American and British air forces.

In order to accomplish this task as quickly as possible, in the first instance the already verified surface-to-air Schmetterling (Hs 117 A) missile was adapted. The additional boosters were abandoned. It was also endeavoured to replace the engine running on troublesome and dangerous liquid propellants (58.6 kg of an oxidiser designated SV as well as 12.4 kg of propellant designated RZ) with an engine using compressed powder which would eliminate ground-based propellant installations for the rockets and make them always ready for use, however this had to be abandoned as a result of a visible imperfection of the powder-propellant engine.

The missile was guided by radio commands sent from a carrier-plane, the development of a wire-guided system resistant to jamming was also predicted.

In the basic version the warhead contained 36 kg of 643-BS explosive composition probably initiated by a proximity fuse, the application of warheads containing 60 to 100 kg of explosive was also predicted, which were designed to destroy group targets such as the "dense" formations of enemy bombers.

The first specimens of the Hs 117 H originating from the first, short production batch reached the Luftwaffe in January 1945.

Almost simultaneously with work on the rocket described above another air-to-air missile, the Hs 298, was constructed in the Henschel works under the direction of Professor Wagner. It was smaller and propelled by a significantly simpler solid-propellant rocket engine, on a base of diglycol, however this yielded markedly to other constructions. The initial launches carried out on December 22, 1944, with the Junkers 88 revealed its small effectiveness. Out of three launched rockets only one destroyed the target. In the short time which remained for the construction's improvement it wasn't possible to improve the effectiveness of the Hs 298, as a result of which it never managed to seriously interest the Luftwaffe. The unfinished, 7 kg Kakadu proximity fuse was a serious burden on the poor performance of the missile, which revealed significant operational faults under conditions of vibration caused by the engine's operation.

Considerably better on the other hand was the X-4 missile developed between 1942-1944 by Dr. Max Kramer from the works of Ruhrstahl A.G., in any case it aroused the greatest hope of the Germans. It was also the first guided air-to-air rocket missile referred for mass production, although as has already been mentioned, it was never used in combat. The X-4's good reputation was first and foremost a result of applying a successful though innovative, combined guidance and homing system.

The missile was wire-guided for a significant section of the flight path. After the signal to launch two powder charges caused membranes sealing the compressed air tank to be pierced (forcing propellant) as well as inlet into the engine. The propellant, Tonka-250, being a mixture of organic substances underwent mixing with an oxidiser (concentrated nitric acid with the addition of 4-5% iron chloride as a catalyst), which triggered off self-ignition in the combustion chamber. This ensured an initial thrust amounting to 150 kg dropping to 33 kg after 35 seconds of flight, shortly before the engine stopped working. At the same time that the missile descended from the launch rail, flares located at the tips of two of the four wooden fins were ignited (enabling the operator to opti-

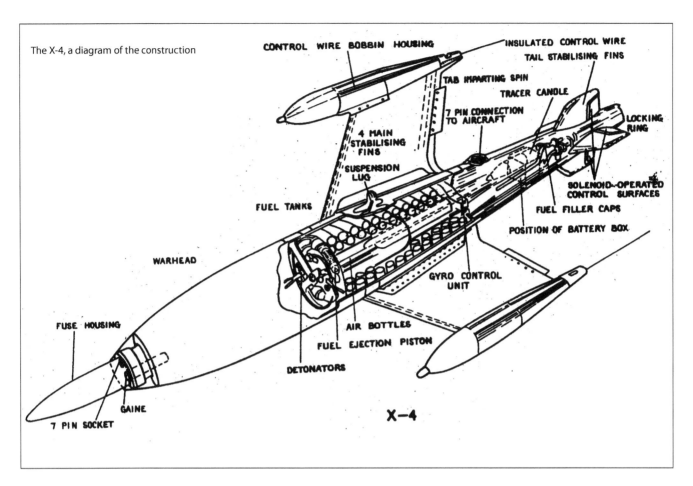

The X-4, a diagram of the construction

cally track the rocket), and cables for wire-guiding the missile were unwounded from two larger containers at the tips of the remaining fins. They were each 5.5 km long and 0.2 mm in diameter. On approaching the target the homing system took control, guiding the missile to sources of engine noise from the bomber. In this way it was possible to achieve satisfying precision at a relatively large distance.

A new feature was the use of propellant tanks in the form of two light alloy pipes winded like two thick springs in the central section of the fuselage. At one end they were connected to the compressed gas tanks and at the other to the engine's combustion chamber.

The whole propulsion system was the work of designers from BMW and although it was the source of their biggest problems with the X-4, the management of BMW decided on the basis of this to construct their entirely own missile. Kramer intended for the meantime to replace the BMW engine with a new solid propellant engine, this however never took place. All in all 100 specimens of the X-4 were produced; this batch was reserved for putting into effect a program of in-flight trials, which commenced at the end of 1944.

These revealed a high hit probability, the warhead exploded at an average distance of 7 metres from the bomber. 20 kg of plastic explosive based on pentryt, enclosed in a steel shell 10 mm thick guaranteed in these conditions complete destruction of the target. An explosion at a distance of up to 15 m away caused heavy damage. Chiefly Fw 190 fighters were to be armed with the X-4 missile.

The fourth type developed in Germany, the aforementioned BMW missile, which was designated the "Gerät 3378" was externally very similar to Kramer's missile. It came in two versions: 1540 mm and 1690 mm in length. Although up to 1540 of these rockets were produced in total, a series of unsolved problems didn't foresee a very favourable future for them. Apart from the engine's imperfection, which was also intended to be replaced by a solid-propellant engine (but this never happened) up until the end of the war a final homing system was never chosen. Acoustic, optical and other systems were tested, but the war's end interrupted this work, in any case a fundamental problem was the lack of a suitable number of combat aircraft and the fuel for them.

The BMW rockets lost the competition to the X-4. On February 6, 1945, the decision was made to begin mass production of them in the Ruhrstahlpresswerken works in Brackwede near Bielefeld.

The Hs 293A missile under the wing of a Do 217 bomber

Tactical and technical details of the basic guided air-to-air missiles

	Hs 117 H	Hs 298	X-4
Length [m]	3.7	2.0	2.0
Wingspan [m]	2.0	1.29	0.725
Total mass [kg]	260	95	60.5
Fuselage diameter [m]	0.35	0.205	0.222
Engine mass [kg]	143	26	22.7
Explosive mass [kg]	40	25	20
Max. flight speed [m/s]	250	234	248
Range [km]	11	1.6	5.5

Air-to-Surface and Surface-Surface Rockets

Apart from the "V" series missiles described in the first section of this book, many smaller guided weapons were developed in the Third Reich.

Let us begin with aerial armaments...

Henschel from Berlin was also to take the lead in this field. The Hs 293, of which many different versions were built, was by far the most important weapon, developed under the direction of Professor Herbert Wagner. Initially it was to be a guided bomb, however after the addition of a rocket engine the Hs 293 became a "normal" guided rocket. The mass produced basic A version was also undoubtedly the most important German guided missile from a military point of view. Already brought

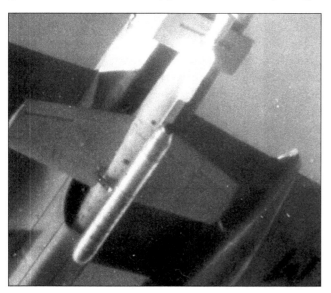

The Hs 293A

A derivative of the Hs 293: the Hs 294 air-to-sea missile

into mass production in November 1941, it inflicted the greatest Allied losses. The total tonnage of sunken vessels and ships alone amounted to 440,000 tons. It was characterised by a relatively high hit probability under combat conditions, of the order of 45% (under rocket range conditions approximately 90%). Overall 1,900 specimens of the Hs 293 A were produced, deploying them for the first time in battle on August 25, 1943, in the Bay of Biscay. They were mainly deployed against ships, as targets relatively easy to destroy with a precise hit even with a single large calibre rocket.

The basis of this construction was to be a modernised, conventional SC-500 bomb weighing 603 kg, containing 295 kg of high explosive. This became the missile's warhead. The A-1 production version was remote-controlled by an operator situated on board an aircraft by radio commands. Soon after launching the missile developed a maximum speed of almost 900 km/h, which during further horizontal flight gradually dropped to around 560 km/h. It went into a dive over the target, which made it considerably more difficult to shoot down. Approximately half a second after hitting the target (after penetrating the deck in the event of a ship) the fuse detonated a trialen charge, the most powerful explosive used in World War II, composed of RDX, TNT and aluminium dust, which increased the heat of the explosion. The rocket's thrust was ensured by a Walter HWK 109-507 engine slung under the basic fuselage. It had a mass of 134 kg (of which the liquid-propellant weighed 66 kg) and for 10 seconds gave a thrust equal to 590 kg. Besides the engine, at the moment of launching five pyrotechnic tracers were ignited, burning for 100-110 seconds. These made it significantly easier for the operator to track the missile, especially at night. The particular method of deploying the Hs 293 A resulted in a certain fundamental problem, which was the necessity of guiding it on a flight path not covering the line of sight between the aircraft and the target. This enforced very high requirements regarding the operator's training. His experience was highly significant.

As the missile was deployed on a relatively large scale, one can give several examples of its use in combat:

The first "serious" air raid on a large group of enemy vessels and ships (all in all 67 units) took place on November 21, 1943. 22 He 177 Greif and Fw 200 Condor long-range heavy bombers took part. 18 Hs 293 A specimens were used in total, which sunk one 4,500 tonne freighter and heavily damaged another of 6,080 tons, at the same time four He 177 aircraft were lost. In spite of the rockets' small dimensions the ships' anti-aircraft defence shot down one and damaged another leading to a change in its trajectory.

A somewhat smaller air raid was carried out almost two months earlier, on September 30, on the Corsican port of Ajaccio. It was also not very successful. Seven of eleven Do 217 bombers didn't return to base, "in exchange" for only one destroyed ship. To make matters worse two missiles went into a glide and alighted on water in the port. Almost undamaged they found themselves later in the hands of the enemy. The Luftwaffe had somewhat better luck over the Aegean Sea, where still before the end of 1943 seven destroyers were sunk or heavily damaged using the Hs 293 A.

On the basis of this experience the missile's construction was improved, developing in 1944 several new versions. With respect to the possibility of the guiding signals being jammed, in the first instance alternative homing systems were concentrated on.

Soon the first (and only) batch of 200 Hs 293 version B specimens were referred for production. The only new feature was the introduction of wire-guiding. Two spools of cable each 16-20 km long were fixed onto the missile, in addition cables each 12 km in length were unwound from on board the aircraft. This was a very accurate version—its average direct hit error amounted to, in the case of a well trained operator, the order of 5 metres. During one of the trials 12 missiles were launched, aimed at a circle 25 metres in diameter. All of them hit the target.

Next was the C version, no longer introduced for mass production and armament. Only a test batch numbering around 60 specimens was delivered, used in testing. The only difference in relation to the Hs 293 B was the installation of a warhead which was supposed to hit the ship below the waterline from a horizontal flight, originating from the larger Hs 294 missile, weighting over 2100 kg.

Potentially the most crucial, although imperfect development of the Hs 293 was the D version, characterised by a television homing system, operating on just the same principle as the modern American Maverick missiles. A television camera was mounted in the nose section whose lens had a field of view of 25°. The camera's installation required the fuselage to be extended by as much as 748 mm. In addition an aerial transmitting the picture from the warhead to the aircraft was installed at the rear on the clampings designed so far for the tracers. In the camera's first variant (Tonne-1) detection of the target was only possible in the final stretch of the flight path, from about 3,800 metres, obviously only in daylight and in good atmospheric conditions.

This still imperfect version was referred for rocket range testing in the autumn of 1943, carrying out the first launching at one of the lakes near Stargard. The tests showed the missile's small effectiveness, they revealed moreover problems with keeping the missile on its flight path in its initial phase. Despite small improvements the Hs 293 D was for a long time a weapon not yet fit to deploy in combat. During the next series of tests, carried out in April 1944 in Jesau near Königsberg out of 12 missiles only one hit the target directly, the rest hit at a distance of around 80-100 metres away.

Soon after a new, significantly better camera, the Tonne-2, became available with a resolution almost doubled from 224 to 441 lines. Radical progress was however achieved only in August 1944, along with the introduction of the Tonne-4a camera.

Overall around 250 "mass production" Hs 293 D were produced, which however never went into armament. Further work on this missile encountered difficulties resulting from Hitler's orders, concerning the "concentration of armament production," although in spite of this several newer versions were still made.

An increase of the missile's dimensions forced the use of a "more powerful" propulsion. In this way among others the G version was created with a propulsion in the form of a powder-propelled WASAG 109-512 engine of thrust 1200 kg (instead of the 590 kg so far), but guided directly by radio commands as well as the H version with the television system from version D and liquid-propellant Schmidding 109-513 engine with concentrated nitric acid acting as an oxidiser. It gave a maximum thrust of the order of 1000 kg.

Prototypes were also made of two derivative versions renamed the Hs 295 and Hs 296; the first with a huge 1260 kg warhead for destroying large and particularly important targets and the second being a modified Hs 293 H missile with a camera based on the latest available technology. This camera, developed by Dr. Rombusch from the Institute of Physics in Dressenfeld was characterised by a mass of only 2.5 kg (!).

The Hs 293 should be regarded as the most significant (in terms of use in combat) guided aerial weapon, which arose during World War II. A feature in a sense symptomatic is however this, that considered one of the trumps of the dying Third Reich, at the end of the war it was evaluated for evacuation (i.e. concerning all of its documentation) beyond the reach of the Allies within the confines of a secret and to this day little known operation with the aim of securing Germany's most crucial assets.

Information concerning its derivative versions found its way namely to the "Instituto de Investigaciones Cientificas y Technicas de las Fuerzas Armadas" in Argentina, where work was continued, bearing fruit with the introduction in 1958 of a missile based on the Hs 293's construction to the armament of the local armed forces.

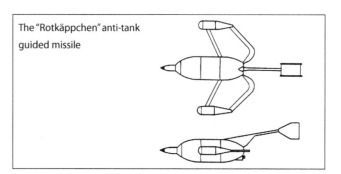

The "Rotkäppchen" anti-tank guided missile

The Hs 293 fell also into the hands of the Soviets, where it became the basis to developing the "Shtshuka" family of missiles.

It is a little known fact, that the Germans also developed an "air-to-air" version, designed to destroy bomber formations. It was to be transported by the Ar 234 as well as Do 217 K-2/U-1 aircraft. One of them was equipped with a "Hamburg" heat-seeking warhead and "Pinscher" proximity fuse. Work on these versions of the Hs 293 was never completed.[105,115] It is estimated that in the Third Reich around 1,900 specimens of all versions of the Hs 293 were produced (not much less than the number of guided weapons used by Americans in the "ultra modern" war for Iraq in 2003).

A similarly large technical leap like, which took place in the Hs 293 was made by constructing in the Third Reich the first anti-tank guided missiles in the world. The design of the first of them, called the Rotkäppchen (Little Red Riding Hood) arose in DVC under the direction of the aforementioned Doctor Kramer in 1943. This was better known as the X-7 missile. It was a light missile weighing 9.08 kg with a range of around 2.4 km propelled by a two-stage (launch and cruise) solid-propellant engine using diglycol. The first version was radio guided, however in the version referred for production wire-guiding was employed, with the aid of two cables unwound from containers on the wingtips. The warhead was comprised of an altered hollow charge grenade weighing 2.5 kg and 140 mm in diameter enabling the penetration of monolithic steel armour 200 mm thick. Only a test batch was produced numbering around 300 specimens, plans to start mass production in Brackwede and Neubrandenburg were written off by the end of the war. Ground launchers, operated by two soldiers, as well as aerial, for the Fw-190 aircraft, were to be produced for the X-7.[116,101]

At the same time Dr. Kluge's team from the research works of the AEG-Telefunken consortium developed a parallel missile, of very similar dimensions to the X-7, but distinguished by the innovative sending of commands with the aid of light signals. This missile was named Rumpelstilzchen (Rumpel-Stilt-Skin). Although a test batch of around 100 rockets were produced, very little information is available on this subject.

Tactical and technical details:

	Hs 293 A-1	Hs 293 D	X-7
Length [m]	3.58	4.35	0.765
Fuselage width [m]	0.47	0.47	0.14
Launch mass [kg]	975	1040	9.08
Warhead mass [kg]	603	508	2.5
Explosive mass [kg]	295	–	–
Max. flight speed [m/s]	265	ap. 170	ap. 100
Range [km]	ap. 6	ap. 10	ap. 2.4

Guided Bombs

During the war several types of guided bombs also arose in Germany, of which two deserve special attention: the PC-1400X ("Fritz-X") and the BV-246. The first even had on its record many military operations ending in success. Its production was started in 1943. It weighed 1570 kg and was designed first and foremost for attacking ships, as targets relatively small, manoeuvrable and often armoured, and so difficult to destroy in a different way. In the basic version the Fritz-X was guided by radio commands however in the event of jamming a cable could be unwound from the bomb to serve this purpose through wire-guiding.

The Turbulent Development of Guided Weapons | 185

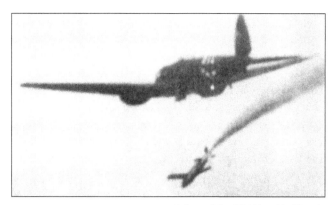

A drop of the BV-143 guided bomb

The BV-143 guided bomb

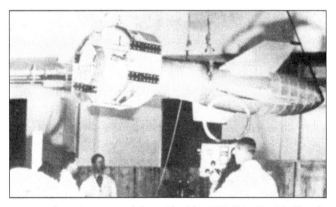

A US intelligence report, analysing the possibility of the Germans using guided bombs and air-to-surface missiles to counteract the invasion in Normandy from April 1944

Opposite page and above: The PC-1400X / Fritz-X guided bomb

First blood fell in the summer of 1943, among other things after the capitulation of Italy, on September 9, Do 217 aircraft from flight squadron KG-100 sank the Italian battleship "Roma" and heavily damaged her sister ship the "Italia" with these bombs. During the allied landings on Sicily the battleship "Warspite" was seriously damaged as well as three cruisers: the "Uganda," "Philadelphia" and "Savannah."

At the same time two more modern homing versions were being developed: on sources of radio emissions/radars (with the ZSG "Radieschen" warhead) and sources of heat (the ZSG "Offen" warhead). It is common knowledge that this first version was tested in August 1944 at the Luftwaffe rocket range in Leba. The target was a 500 W transmitter; two dropped bombs fell at a distance of 30 m away. The "antiradar" version was to enable the destruction of a network of British radio beacons, which ensured navigation to Allied bomber groups, whereas the thermal warhead version was to be used to destroy British steelworks. These plans came very close to being implemented.[115]

The BV-246 guided bomb was also a very interesting construction, which was given the code-name "Hagelkorn" (Hail

Stone). Thanks to a very streamlined shape and unusually wide wingspan spanning over 6 metres, its range reached as far as 200 km! So in this respect it resembled the V-1 missile. Of course in practise such a large range excluded any form of command guidance. In its initial version the BV-246 possessed solely a gyroscopic autopilot and with a scatter of the order of 10-20 km was suitable solely for attacking area targets (and not point), such as cities. The warhead mass was similar as in the V-1: 500 kg. Admittedly it was endeavoured to install the "Tonne" cameras from the Hs 293D, but ultimately homing (passive) warheads were employed—the same as in the case of the "Fritz-X"—plus the thermal "Netzhaut" warhead. They would switch on after reaching the target area. British and Soviet steelworks were also considered the main target in this case. About 400 copies have been manufactured, wchich were used mostly against the atlantic convoys.[115]

Although the guided bombs described in the previous pages make up only a small part of German work on guided weapons, they clearly make us aware of the scale of technical and military breakthrough which was made as a result of creating guidance and homing systems. The number of types of these systems clearly shows, that it came really close to a type of warfare appearing in Hitler's arsenal, that would have forced the other side to at least re-evaluate its methods of waging war. In other words: if the development of the guided weapons mentioned in this chapter had taken place, without the obstacles imposed from above and without redundant competition in the form of the "V" weapon, the course of the war would surely have been different.

This is such a significant field, that I have allowed myself to devote several pages to the description of guidance systems, all the more that the source materials (documents), which I have made use of are not commonly known.[82,104,112,113,8]

I present therefore, made on this basis a short summary of the most interesting designs. Let us begin with the groups already partially described.

Homing Warheads

Warheads of this type were obviously the derivative of work on heat-seekers, carried out since the beginning of the thirties. In the majority of cases their construction was based on lead-sulphide semiconductor detectors.

None of these warheads ever "caught a glimpse" of an allied aircraft, although the development of certain types was practically completed. Not all were in any case designed for surface-to-air rockets. Some of the first arose e.g. with "explosive boats" in mind. This referred to a certain type of "above the waterline" equivalent of the torpedo, fast, armoured motorboats, designed to attack enemy landing craft or ships in port. The Germans had already used such a weapon during the Allied landings in Normandy, although then it was still guided by … a diver, who after guiding the motorboat to the target jumped into the water and returned on his own. However, during this time "automatic" versions had already been designed—homing on the target without any man taking part. Two thermal homing systems were made in connection with this code-named "Tasso" and "Linse;" the second was a derivative of the first. These devices originated in the Berlin works of Gema, with the co-operation of CPVA (Danische Nienhof), AEG (Berlin) and Elac (Kiel).

The older of the systems, "Tasso," on which work had been interrupted in 1943, was relatively simple, even for the simple reason that as opposed to a rocket the field of view could only be scanned along one plane, along the line of the horizon. The "Linse" system developed in 1944 was more complex to such a degree that the "Schwarzschild phase comparison method" was employed to separate the target's signal, i.e. a certain kind of processing.

The optical system had a diameter of only 50 mm (a mirror lens). This miniaturisation obviously had certain advantages, but the lens's small diameter meant a small amount of gathered radiation which certainly impinged on the range of target detection, although their considerable dimensions partially compensated for this. In the autumn of 1944, the completed system was tested on a captured floating unit (a minelayer)—the "Studebaker," displacement 600 tons—i.e. small. This ship was detected at a distance of 2 km, in connection with this it was established that in combat conditions the homing system would be activated at a distance of approx. 1,000 m from the target. Until this moment the skimming boat (developing a speed of 126 km/h!) was to be guided with the aid of a two-channel radio system. The principle charge was to be placed under the hull, underwater and detached from the hull shortly before hitting the target.

For the "Linse" system (the code name "Teichlinse" also appeared) the Tietjens shipyard from Potsdam designed a special type of boat, characterised by an exceptionally stable "ride," however there was still no doubt that the new weapon would only be effective on a calm sea. A fundamental limitation was the relatively small field of view of the lens, 3°. The risk simply existed that the warhead would lose its target, though this wasn't such a serious problem as would appear at first glance. After all it only assured guidance correction in the final phase of the "ride," up to 20-30 seconds before impact.

In the "Linse" system a single, 8 mm detector was used from the company Elac.

The next of the described systems is the by now classical thermal homing warhead (or rather: family of warheads) designed for rockets. This concerns the diverse, successively developed versions of a device known by the code-name "Hamburg," which originated in the laboratories of Elac.

The first of these versions, the "Hamburg I," was distinguished by an exceptionally large 30 mm detector, which was located in the focus of a mirror 25 cm in diameter. The entire mirror was the scanning element, which spirally sweeped the whole field of view, and had a diameter of 60° at a beam width (resolution) of the order of 10°. The trials that were carried out revealed that these values were insufficient in the case of anti-aircraft rockets, among others of the "Wasserfall" type, on

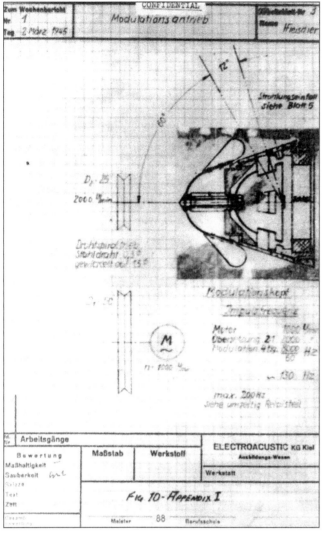

A diagram of a thermal homing warhead, designed for the "Wasserfall" missile

account of this the authorities specified requirements regarding possible modifications; an increase in the range of target detection to 2-5 km as well as resolution was demanded, i.e. a level of homing accuracy to half a degree. In the opinion of Dr. Kutscher directing the work, fulfilling these requirements would be possible if the optical system was completely redesigned and a more complex method of scanning was applied, as in the "Kiel III" system. However work in this line wasn't carried through.

Instead a smaller version of the "Hamburg I," the "Hamburg II" system, was developed, in which the sulphide detector was replaced by an analogous detector, utilizing cesium oxide. The "Hamburg II" lens had a diameter of 14 cm. This was practically the only plus in relation to the older version. The device continued to be unfit for anti-aircraft rockets, however could successfully fulfil its role in relation to "less demanding" targets, such as ships. It was designed for gliding, anti-ship guided bombs. In order to carry out trials 10 prototypes were ordered, out of which only two were supplied and tested before the war's end. One of these was installed in an aircraft's nose, which carried out flights in the area of the Bay of Gdańsk.

The first conclusion which these trials bore fruit to concerned the necessity of ensuring some form of stabilization of the warhead's optical system. At this stage military operations forced work to be discontinued.

It was a similar case with other types of warheads of this kind, although in their case we have a much more modest resource of information at our disposal. However it is known, that the following were worked on:[103,112,113]

1. The "Emden," a warhead for rockets and bombs, developed jointly by Elac and AEG. It was to constitute an alternative to the "Hamburg II," however no prototype was ever completed.

2. The small company Kepke from Vienna created a similar system. It is common knowledge that it received the codename "Madrid" and was considered as equipment for the "Enzian" rocket and BV-296 gliding bomb. The detector produced by Elac was used in it.

3. Not much is also known about what was surely the most state-of-the-art warhead which originated during World War II, the "Armin-2." Its chief designer was the aforementioned Dr. Kutscher from the Elac laboratories in Namslau. Why the most state-of-the-art? It was a warhead fully deserving the term "thermovision," that is creating a thermal picture of the target. As was written in one of the intelligence analyses, "its development was almost completed," the warhead worked, but didn't have a sufficient range, of the order 1.2-1.5 km, in relation to the He 111 bomber.

4. In the company Kepke a large (as it turned out: too large) homing system with a lens 28 cm in diameter was made. As much as 50 prototypes were supplied for testing, however mass production never began. A virtue of the large diameter was the pretty good range of target detection, of the order of 3 km in the case of a standard bomber.

5. Several derivatives of this type of warhead were produced in secret in the works of AEG. It is known only that they bore the code name "Widder," "Netzhaut" and "Krebs."

6. In the Wrocław (Breslau) branch of Rheinmetall arose the interesting and successful "Gluhwürmchen" warhead. Two versions existed differing in the type of detector (in both cases from Elac) and field of view, which had a width from 3 to 8°. Spiral scanning was employed. As early as 1944, 50 prototypes were supplied with the aim of carrying out trials. This warhead had a mass of only 3 kg. There is a lack of information on its range.

Although work on none of the aforementioned warheads was brought to the stage of mass production, much hope was however placed in them and among the whole group of homing systems they took first place. This probably doesn't surprise anyone, especially when we take into consideration the present role of these devices. On the other hand it may surprise one that... acoustic homing warheads were to be found in second place. This simply resulted in all probability from their accessibility and degree of technological mastery and not from their considered military superiority over other types.

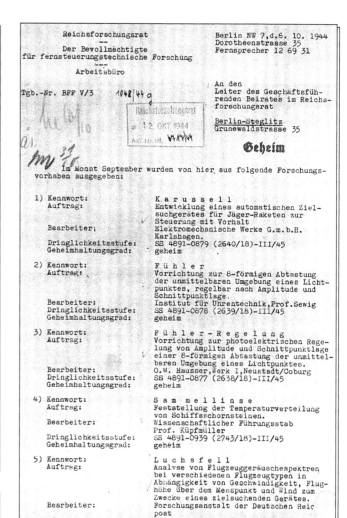
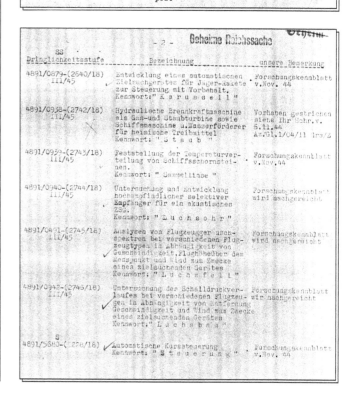

Documents from the Reich's Scientific Research Council, containing code-names and short descriptions of certain research projects, related to new homing systems

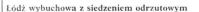

A page from the magazine "Signal" (1944), regarding "Linse" boats

As many as eight types of these devices were developed...

Available sources supply only scant descriptions[112], this was in any case a typical dead end in the development of military technology (this field was soon forgotten), however it is worth acquainting oneself with even rudimentary information on this subject. I have given therefore a short list:

1. "Baldrian," developed with the co-operation of Telefunken and Messerschmitt (the plant in the town of Hallein). This was a relatively complicated warhead, with a valve amplifier containing 12-16 valves and 4 sensitive microphones, in which mica crystals had been applied. In spite of this its range was at best moderate and didn't exceed 350 m. An exception was the Me 262 jet, in whose case a greater range was recorded. A plus, like in the case of other warheads operating on this principle, was on the other hand the large "field of view," the order of 180°.

2. An parallel system was developed (approx. 60%) in the works of Elac. It never managed to receive any name, and it was estimated that it would have a range reaching up to 2 km.

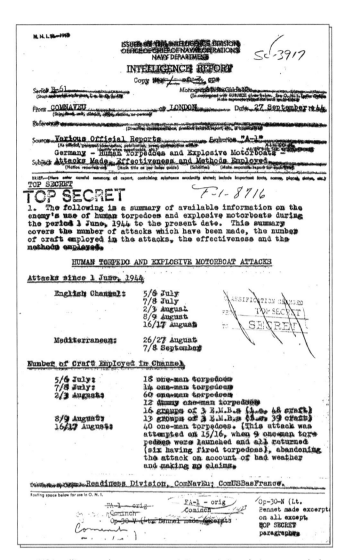

US intelligence document, containing statistics relating to attacks by Linse boats and the Neger—"living torpedoes"

3. In Ruhrstahl A.G. an acoustic warhead was developed under the leadership of Dr. Kramer, which was to be employed in a modernized version of the X-4 air-to-air rocket. It was given the code-name "Pudel." It was relatively simple, based on four resonant membranes and 2-3 vacuum valves. No specimen was ever completed, range was estimated at 500-1,000 m.

4. A similar system, although with four microphones (the direction was determined by the measurement of the phase shift between them) developed in the laboratories of the post, under the direction of Dr. Trage. This warhead was designed for the "Rheintochter" rocket. No parameters are known.

5. Still less is known about work on four successive devices from the described group, both their names and code-names are unknown. It is only known that they were worked on in the "postal" laboratories (Dr. Schops, Würzburg), in an establishment designated by the abbreviation AVP (Göttingen, Prof. Kussner), and in laboratories in the area of Darmstadt, engaged in measurement technology, and in particular in oscillation measurement. The last type was developed by Prof. Lubke from Braunschweig.

The next group, although modestly interesting, constituted the designs of radar warheads: two semi-active and one active (both transmitter and receiver were to be located in the missile). The first two of the aforementioned bore the codenames "Licht" and "Blaulicht," whereas the active was codenamed "Dackel." In this field the "postal" research establishments had the monopoly, the "Licht" system was developed by Dr. Pressler, "Blaulicht" by Dr. Heymann and "Dackel" by Dr. V. Octingen. 9 to 12 vacuum valves were included in all of them, at the same time in the account of the "Dackel" type we find some intriguing information, that two diodes were "also" employed; could it have referred to successive semiconducting elements?

All three warheads of the aforementioned type were to have a target detection range of 1-2 km, however no trials were carried out under rocket range conditions, even though work was seriously advanced at the moment of the war's end.

A supplement to the aforementioned designs was an active warhead based on the Doppler Effect. Although it was given the code-name "Max," this work didn't go far beyond a preliminary design. "Max" was designed by Dr. Gullner from the company Blaupunkt.

An American intelligence report[112] mentions another two "similar" types of warhead, namely passive radar warheads, i.e. according to modern terminology: anti-radar (homing on the

The "Neger"

enemy's radar). These were the "Windhund" and "Radieschen" types originating in the postal research establishments. Information about the first construction is very scant, more is known about the "Radieschen" warhead, whose development was completed and several completed specimens were installed in the "Fritz-X" guided bombs. This warhead weighed 15-17 kg and was based on a "Strassburg" system receiver. It was distinguished by a very large range, which reached as far as 100 km; in the case of a large target, emitting approx. 1 kW of power

The completion of the above mosaic of the most diverse missiles is constituted by the optical warhead, with a single detector and spiral scanning, developed in the private research company of a certain Dr. Rembauske. It bore the code name "Pinsel." Several prototypes were referred for testing shortly before the end of the war. This permitted the practical range to be determined, up to 2,000 metres. The width of the field of view amounted to 40°.

Television warheads constituted a very important group, characterised by a relatively large effectiveness.

The introduction of a system of this type onto the battlefield (an amazingly mature system) would of course not have been possible without earlier, pre-war successes. Thanks to this fundamental technological problems were overcome long before the outbreak of war.

Work on using television in guided weapons, described among others in one of the very extensive American intelligence reports,[113] began as far back as 1939. A whole series of companies and institutions took part, among others: Blaupunkt, Siemens as well as Fernseh GmbH, a company appointed exclusively with the aim of carrying out tests in this field. The laboratories designed to carry the discussed design into effect were located, at least towards the end of the war, on the grounds of the complex near Peenemünde. This resulted first and foremost from the fact that the television warhead developed for the Hs 293 was also adapted for a second, unspecified type of rocket; another design was therefore involved, production of which didn't commence until September or October 1944. The development of this warhead was never completed and its short description has been placed in a further section of this chapter.

If on the other hand the version designed for the Hs 293 is concerned, then its basic parameters are known. These apply first and foremost to the television system itself, because the employees of the aforementioned companies knew very little about the method of transmitting the image and commands between the missile carrier and the missile itself. This constituted a strictly guarded secret. During trials carried out by a Luftwaffe unit only a representative of Fernseh GmbH was present, as only one person, Prof. Wagner from the works of Henschel, knew the technical details connected with the aforementioned issues and control system. Counterintelligence protection in this area was exceptionally strong.

In practise however controlling the missile was relatively simple. The operator situated on board the bomber had among other things a "joystick" (control lever) at his command as well as a miniature, modular television monitor, whose screen

The "Tonne-2" camera, one of the greatest achievements in the field of electronics from the period of World War II.

A modular TV monitor to be placed on board a bomber

Comparing the overall dimensions of the camera for the Hs 293 with those of a TV camera from the 1930s gives some idea of the rapid technical leap that occurred during the war; about 100-fold reduction of weight was recorded

dimensions were 8.8 cm by 8.8 cm. The aforementioned report describes one of the warhead versions ("Tonne"), characterised by unimaginable, for those times, resolution of 441 lines. This is practically the same resolution as of the Hi8 or SVHS system, almost two times greater than the resolution of standard, domestic VHS (around 250 lines). This information relates to the "Tonne-2" version.

At this standard the whole system, including the equipment installed in the missile, but not including the missile's transmitter, aerial and battery weighed around 50 kg, this mass however dropped in the case of later modifications, among others the "Tonne-4a" from the turn of 1944-45, or a variant developed by Dr. Rombusch from the Institute of Physics in Dressenfeld, in which the camera (being a part of the warhead) weighed scarcely 2.5 kg!

In the "Tonne-2" version, and probably also in later versions, the picture repetition frequency was high enough: 25 Hz

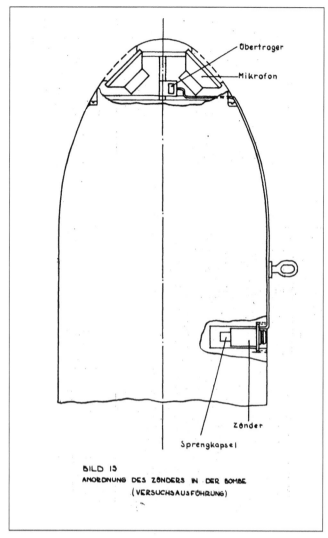

A diagram showing an acoustic proximity fuse in an aerial bomb

risk was created that in the event of losing contact it wouldn't be regained, however practical trials didn't confirm these fears. This solution was as equally good as the previous ones and was accepted as the norm.

It follows from available information that all in all around 60 television version specimens of the Hs 293 were launched.

A further step was the development (uncompleted) of a developmental version of the television homing system, designed for surface-to-air rockets. Of course the warhead itself underwent the main changes. It was to be "not much bigger than the proximity fuse" (length 60 cm, diameter 14 cm, not counting the energy source and aerial). The development of this warhead, which was given the code-name "Sprotte," mainly relied on a reduction in dimensions of the components used so far. Miniaturization of the camera was followed by a reduction in resolution to 220 lines, the battery was replaced by a small generator, driven by a propeller protruding from the rocket's fuselage. As a result it was possible to reduce the mass to 10 kg, in comparison to 50 kg in the case of the basic version of the Hs 293. It is not clear for what type of rocket the "Sprotte" warhead was designed for, this probably wasn't specified up until the moment that work was discontinued, although "attempts" were made to employ it in the V-1 missile.

* * *

Playing a secondary role in the described field were the so-called proximity fuses, causing the warhead to explode not as a result of a direct hit, but when it was situated close to the target (at a certain distance). They had a particular significance in the case of anti-aircraft rockets, since their warheads were as a rule "powerful" enough that they would destroy the target/s even with an explosion at a distance of a few dozen metres away, and on the other hand it was incomparably easier to "hit" an area several dozen metres in diameter than the aircraft itself. Therefore proximity fuses were a precious, if not essential supplement to homing systems.

So as not to bore readers with the less interesting technical details, I have presented only one selected type of such fuse, which in my opinion was the most successful construction. This was an electrostatic proximity fuse from Rheinmetall. It exemplified the rare combination of construction simplicity (and alternatively production) with high effectiveness.[117]

Above all, it was resistant to high G-forces and could be used in the shells of anti-aircraft artillery cannons. Irrespective of this it was more reliable—at first its effectiveness was at a level of 70-80%, only to finally rise to as much as 95%, and this was in the case of shells fired from anti-aircraft cannons (88-mm calibre). For trials aerial bombs weighing 50 kg or more were also used, as well as projectiles fired from experimental Rheinmetall 75-mm calibre cannons (probably designed for the E-50 tank).

This fuse operated in such a way that it was electrically insulated from the rest of the missile, in connection with this the entirety was comprised of two parts electrically insulated from

to ensure a smooth picture even in the case of a moving target. The width of field of view amounted to 40°.

The remaining "component" of the system, radio links, enabling transmission of picture and control commands, was as equally as modern as its television part. A fundamental problem that had to be overcome during its design was the resistance of the links to jamming. A whole series of solutions were developed in connection with this, in any case very similar to modern ones and relying on fast automatic changes in frequency. It was intended from the start to ensure synchronization between the transmitter and receiver by applying an additional radio signal, coding the transmitter as well as the receiver. It was however equally susceptible to jamming.

Therefore a system coded "from the inside" was made. Accurate quartz resonators—in other words quartz clocks—tuned before launching fulfilled the synchronizing function. This solution passed its examination perfectly in practise, but it was expensive. So a different method was tried—this time with a source of code, i.e. a source of information of occuring changes in frequency was the television signal itself. Theoretically the

each other, between which existed a precisely defined electric potential. In air this wasn't significantly disturbed, as long as a large conducting object didn't appear at a small distance—by this we mean an aircraft. Then the difference in potential diminished, which thanks to a simple electronic circuit led to ignition of the detonator.

Tests (shootings) were carried out mainly on steel nets, stretched out on special insulators. The projectiles as a rule exploded at a distance of 1-2 m away. Later this distance was increased to 3-4 metres; a circuit realignment turned out to be sufficient. It was determined that by carrying out certain modifications it would be possible to achieve an increase in detection radius even up to approx. 10-15 m. This proximity fuse, which possessed only the name "Influenzzünder," was probably the best construction of this type during World War II (the Americans also developed proximity fuses). In total around 1,000 pieces were produced, which were used exclusively for test shootings. Work on them began as far back as 1935, was interrupted in 1940 and re-started at the beginning of 1944 only to be interrupted once again a year later. Firing range testing occurred during this final period. At this time it was in principle ready for production (and could have been much earlier), but as in the case of night sights it never aroused any interest from the bureaucratic side of the Wehrmacht's elite.

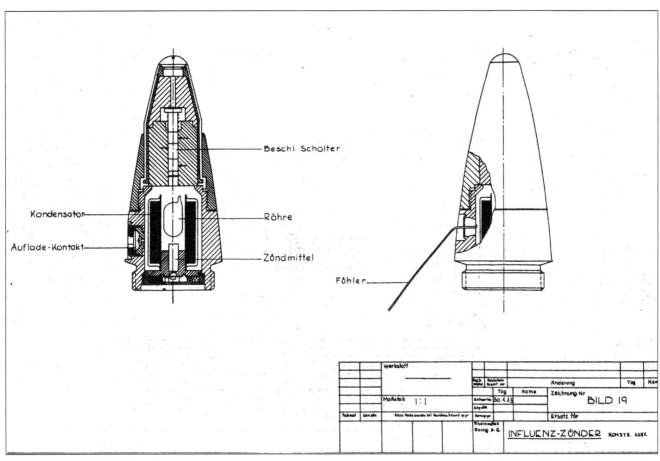

A cross section and semi-cross section of the German electrostatic proximity fuse

The launch of the Wasserfall missile. It was the only German anti-aircraft missile that could be guided automatically, by a ground fire control system.

In 1944 the Germans have developed and subsequently tested a state-of-the-art, semi automatic target detection and fire control system for anti-aircraft missiles. It was code-named Egerland and consisted of several sub systems, including the visible target detection and fire control radars: the FuMG-74 Kulmbach and FuMG-76 Marbach (on the left), the latter was relatively resistant to jamming, mostly because it worked on centimeter waves. The visible specimen of the Egerland was actually used in the outskirts of Berlin.

Above: One of the best German missiles—the Bv-246 gliding bomb. The photo shows the version with the Radieschen homing warhead, which locked onto a radar signal, mostly from vessels. Despite lack of propulsion, the bomb had an effective range of up to 200 km (comparable with V-1's!). Around 190 of them were used against Allied convoys.

Left: The plans for the creation of missile anti-aircraft defenses around the main cities involved semi-automatic coupling of input signals from an array of radars (launching batteries) dispersed over a large area. For this purpose some interesting systems were developed, such as a system generating coded signals sent to the missiles (Erstling), and an interface (visible on the photo) code-named Landbriefträger-1 which was to transmit coded data from a radar to a distant command and control post. This one was actually coupled with the Jagdschloss radar. The Munich air defense district was the first one which actualy was "wired" for such an integrated system.

RAMJET PROPELLED FIGHTERS

One of the most important and at the same time least known (!) technical breakthroughs which was accomplished in the Third Reich is constituted by ramjet propulsion (Staustrahlantrieb) and the attempts to exploit it. In relation to turbojet engines it offered a similar qualitative jump as turbojets did to piston engines. And this breakthrough was not accomplished only on paper!

This issue can only be described to a small degree, based on generally accessible publications. I have chiefly made use of four technical analyses, specified in the bibliography.[109,102,118,119] Let us begin however from what ramjet propulsion is in general.

Such an engine is in a way an evolutionary development of the pulse engine, which found application in the V-1 missile. Their principle common feature is simplicity, both being devoid of a turbine and compressor. The pulse engine pretty much consists of a long pipe, combining the functions of a combustion chamber and nozzle, at the same time it doesn't run in a continuous way, but cyclically. The truss is situated at the front, resembling a grid, attached to which are tabs, performing the function of valves. At the moment of ignition of the fuel-air composition in the frontal section of the pipe, the tabs under the influence of increased pressure close the air intake, which causes intense expansion of the exhaust gases to the rear. This gives a recoil, but due to the pipe being quite long (the order of 10 diameters) the inertial force of the exhaust gases causes them to shift to the rear longer than the high pressure is maintained in the section acting as a combustion chamber. They cause therefore the tabs (valves) to open, the next portion of air to be sucked in, and thus the execution of one full cycle. The pulse engine may even operate at the zero airspeed of its carrier, but on the other hand tests carried out during the war showed that it quickly lost effectiveness in due measure of approaching the speed of sound, in this connection it wasn't fit to propel the future generation of military aircraft.

The Germans therefore started to seek ways of overcoming the deadlock and recalled an already pre-war idea, such as the ramjet engine (called at that time the "Lorin duct," and in English nomenclature, the "athodyd," currently "ramjet"). At that time it was still only a curiosity, not arousing much interest and considered as having no future, since (paradoxically) it wasn't fit for the propulsion of aircraft of that time; they were too slow for it, and the ramjet could only operate effectively at speeds close to the speed of sound or higher.

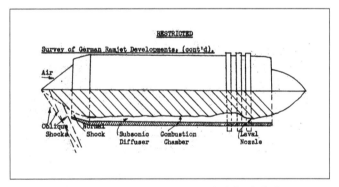

One of the ramjet engines representing a high level of advancement, probably designed by Oswatitsch.
A semi cross section from a US intelligence summary.

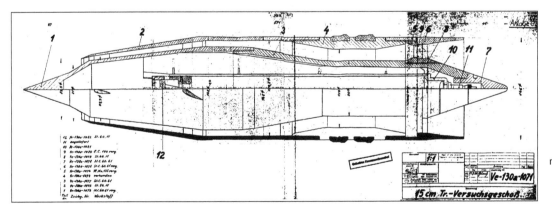

Original plan of the 150 mm calibre "Tromsdorff-Geschoss." This projectile developed max. velocity of 5255 km/h. Atypical fuel, carbon disulphide, was used.

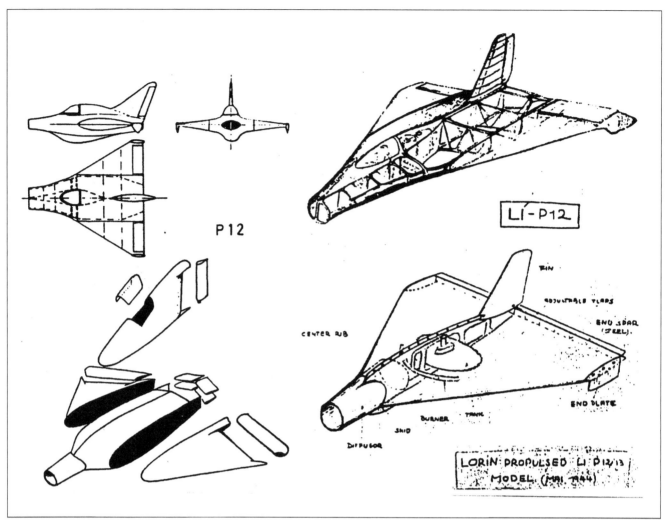

One of the preliminary designs of Lippisch's flying wing with ramjet propulsion. It was designated the P-12. Work on it was interrupted in May 1944, simultaneously commencing development of the P-13b of the final design, completed in turn on January 7, 1945.

This results from the method of air compression. In this type of engine there is no "standard" compressor. It is a housing in the approximate shape of a pipe, inside which a core-conical diffuser of relatively large diameter is placed. The space between the "pipe" and the "core" is the compression chamber, combustion chamber and at the rear: the exhaust nozzle. In most types the compression is ensured by the frontal, conical section of the "core"—but not on the basis of the "conventional" laws of aerodynamics. When the air flowing round the cone reaches the speed of sound, it generates shock waves, which should overlap each other in the vicinity of the ring-shaped air intake (the nature of the shock waves has been described in the middle of the chapter pertaining to electromagnetic weapons).

The "Tromsdorff-Geschoss," cal. 105 mm

The shock waves cause the gas to be compressed to a density comparable to that of liquid. This is precisely the key to the ramjet's principle of operation, since it enables not only the compressor to be eliminated, but also the turbine (behind the combustion chamber), which propels it.

This made possible the design of a jet engine almost devoid of moving components... The Germans had the courage to gamble on this card and thanks to the employment of many outstanding scientists, in the course of only a few years had overcome the fundamental technological barriers. Many of these achievements remain practically unknown, although in the most extensive Polish scientific analysis (from 1951) they were assessed very highly, calling to attention among other things that:[102]

> *Sänger carried out experiments with the use of special pressed blocks made from fine coal dust and plastic, which lined the internal walls of the combustion chamber to protect it from damage. Pabst, a Focke-Wulf employee, redesigned the duct, obtaining a short length [only approx. two diameters—note I.W.] and Tromsdorff invented*

high-explosive [H.E.] shells propelled by athodyd, which possessed a high airspeed and could be used for long-range bombing. The liquid fuel used in them was self-igniting. Theoretical estimated range—450 km.

In connection with the development of ramjet propulsion the most outstanding minds of the aviation and rocket industry were employed, among others:

- Dr. Ing. Eugen Sänger, the father of German rocketresearch.
- H. Walter and his laboratories in Kiel.
- Prof. Alexander Lippisch, one of the most outstanding aerial designers of the Third Reich. He designed aircraft built in the "flying wing" configuration.
- Dr. Klaus Oswatitsch, an employee of the so-called Prandtl Institute in Göttingen.
- Ing. Pabst from Focke-Wulf.
- Dipl. Ing. Peter Kappus and Dipl. Ing. Huber from the BMW company—they designed ramjet—propelled missiles, but probably no prototype has been completed.
- Dr. Rheinlander from "Versuchsanstalt Heerte" near Brunswick (Braunschweig).

The contribution of Eugen Sänger was undoubtedly the most important, who was the first to commence serious research in this field in the Third Reich, thus creating the foundations for future achievements. As early as 1938 his extensive report on the future development of ramjet propulsion had already been published, which was an in-depth theoretical analysis. It was used almost as a handbook by all future researchers, who after all eventually overtook him.

Sänger considered ramjet propulsion above all as an alternative, or supplement to rocket propulsion (and not to turbo-

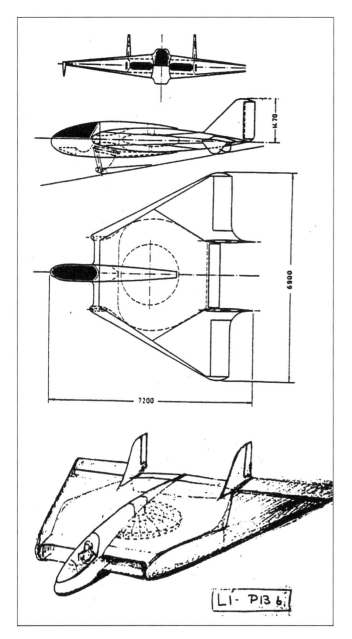

The P-13b was a special fighter in many respects. Attention is drawn to the large air inlets on both sides of the cockpit, exhaust gas outlet on the wing's trailing edge and disc-shaped, horizontal combustion chamber.

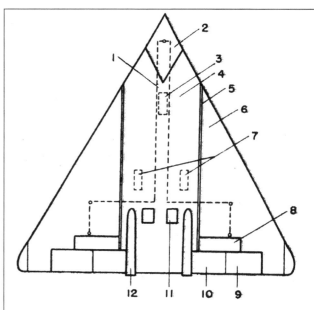

Another version of the P-13b, described in a US intelligence summary from April 8, 1945. This variant had much sharper nose and leading edges, than the trapesium version and was much better suited for supersonic flight. According to other researchers (Ryś et al.) it was exactly the triangle, that "passed comprehensive trials" during the final months of the war.

jet engine, as in the case of other designers). He was mainly interested in the rate of climb, which thanks to this could be achieved by a fighter taking off. The economics of fuel consumption therefore stood in the background, since this engine was only to operate in a strictly defined and short-lasting stage of flight. The key parameter which Sänger paid attention to, was the combustion temperature. He was aiming to achieve as high combustion temperature as possible, the order of 2,000°C, but in practise, after carrying out trials, it became evident that the temperature usually did not exceed 600°C. His designs were characterised by an unusual length to diameter ratio, generally the order of 10 : 1, although one of the prototypes had a length of up to 4 m. This was because Sänger had assumed from the very beginning, that the full increase in gas pressure must take place in the combustion chamber (as was to become evident at the beginning of the 1940s, thanks to the work of Pabst among others, it was possible to overcome this limitation by applying a different approach to the problem).

Sänger built many prototypes. They were fixed to cars and tested during driving at high speed. In at least one series of trials a bomber was employed for this purpose. But no working ramjet engine of his design was ever tested in a wind tunnel. In spite of this, "initial" data was obtained, describing the basic dependencies regarding this new form of propulsion.

The basic feature of his designs was simplicity. The engine consisted of a long pipe with a conical diffuser at the front (improving the sucking in of air). Inside the frontal section was situated a grid made of steel pipes, in which a dozen or so holes or nozzles had been made, being the fuel injectors. They were directed to the front, to obtain the best possible mixing of fuel with air as a result of entry whirls.

During the years 1941-42 the aforementioned airborne trials were carried out on a Do 217 bomber. They involved one of the large engines, with a length of approx. 3 m. The engine was not equipped with any measurement instruments, but the pilot stated that after it was turned on the airspeed only increased by about 10-20 kilometres per hour (at an average speed of 400 km/h). Taking into account fuel consumption and the size of the ramjet engine itself, the results were decidedly not very promising. In later years Sänger tried again to introduce certain modifications, not recording however any kind of breakthrough. Probably he himself lost confidence in his own conceptions, since he started to treat this work as of minor importance—among other things he started the development of a rocket engine with a 100-tonne thrust. In connection with this the ultimate trials on the Me 262 jet fighter were not realised, while the Skoda and Heinkel aviation companies, which before completion of tests on the Do 217 had shown "initial interest" in Sänger's conceptions, soon lost it.[109,118]

In spite of his "initial supremacy," the aforementioned designer was outdistanced by his competitors during the war. One of the most respected was Professor Lippisch. Although little is known about work on his extraordinary engine, in American documents (reports) there is explicit information about airborne trials being carried out, so research must have been well advanced.[118] Polish analysts (scientists and engineers) described them in the following way:[102]

Prof. Lippisch had designed a remarkable shape of aircraft strongly resembling a bird's wing, with ramjet propulsion. But due to difficulties with its construction the prototype was continuously modified. The flying wing developed a basic airspeed of 2,500 km/h, while the pilot was in an inclined position, to exploit its diving capabilities. The wings had been swept back by the designer; the air entered through an opening in the front part of the aircraft. In order for the aircraft to reach adequate speed, necessary for the ramjet engine's operation, solid propellant rockets were used. The air was compressed by a revolving compressor [around the vertical axis!—note I.W.], possessing sufficient revolutions, so as to acquire sufficient compression at high altitudes. The compressor was not a turbine, because in Lippisch's machine blocks of white-hot coal were used just before take-off, being loaded into the combustion chamber to increase the temperature and generate the initial thrust. Other systems had an injection of liquid hydrocarbons applied to increase the air temperature. Through the use of ATO rocket boosters to realise take-off an initial speed was acquired, while inflowing air heated the coal even more, increasing the magnitude of thrust.

An organizational diagram of the DVL's research sector and the location of Lippisch's post

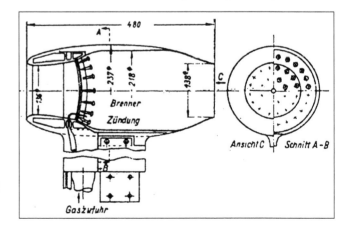

Original cross-section of one of the Triebfluegel's engines

One of the "arms" of the Triebfluegel, in a wind tunnel

Hot gases flowed out of a narrow opening on the trailing edge. The applied method permitted a useful force to be maintained for a stretch of 45 minutes. In the final months of the war the material's heating power was increased two-fold by an injection of liquid fuel above the coal.

Prof. Lippisch's ramjet fighter (the P-13b) was also accurately described in one of the American analyses (Intelligence Summary).[119] The technical drawings published there show a small difference in comparison with the "trapezoid" version of the P-13b, also presented in this book. The air intake edges are not perpendicular to the aircraft axis, but are a continuation of the wings' leading edge, so that the aircraft is shaped like an almost ideal isosceles triangle. Since in the report there was mention of at least two similar versions being designed, it can be assumed that it concerned this kind of difference which wasn't a result of inaccuracy.

In the report it was mentioned that work was carried out by various establishments of the Vienna Research Institute LFW (Luftfahrtforschung Wien), and namely:

1. A wind tunnel in Tülln near Vienna.
2. The LFW plant at Wiesenfeld near St. Veit, where most of the components were produced.
3. A hangar belonging to the aforementioned institute, in Ramsau on the Danube, where final assembly was carried out.

Lippisch's office was located in Tülln, where he worked from the beginning of 1943 till the end of the war. The P-13b had a mainly wooden construction. Most of the elements were made from light plywood 18 mm thick, laminated with a special plastic called "Dynal." In all probability the skin was also composed of this laminate; in any case it had to be thick, if it was to withstand flight at supersonic speed. The participation of metal parts in the supporting structure was reduced to a minimum, mainly to two longerons running parallel to each other, from the external edge of the air intakes to the fins. Obviously the engine was also metal, as was the cockpit housing, control rods and tri-cycled

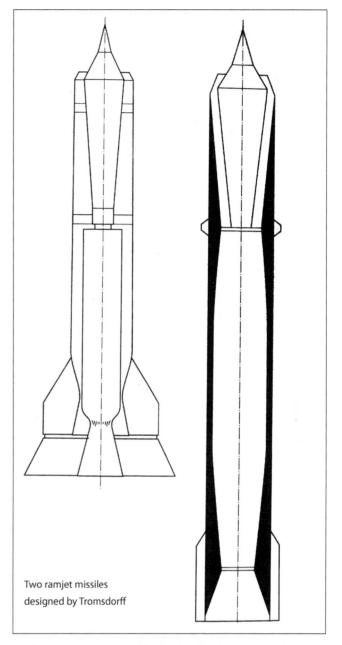

Two ramjet missiles designed by Tromsdorff

undercarriage, retracted hydraulically (interesting, how was the problem of hydraulic drive solved?).

Of course the main engine itself was the most interesting and odd. The Americans gave some information about its fuel. It consisted of specially standardised coal briquettes, produced by Siemens in Berlin. Their weight varied from 250 to 280 g. This was a coal with the calorific value of 8,000 cal/g, desulfurised and leaving an unusually small amount of ash. Before the flight this "fuel" was heated in a special furnace. After reaching a target temperature of 950°C "refuelling" followed, in other words the briquettes were thrown into the combustion chamber (up to 800 kg was predicted).

The American source of information was "cut off" in the first days of January 1945, when the P-13b prototype was being prepared for its first flight.[119] Due to the fact that it was expected to achieve a "magical" speed of 2,440 km/h, the Americans mentioned that there was no shortage of test pilots, eager to undergo this extraordinary experience.

Daimler-Benz "F"
© Marek Ryś 2002

The Daimler-Benz "F" was an ideal candidate to acquire ramjet propulsion

Fortunately the gap in the American report is filled perfectly by an "infallible" Polish critical analysis from 1951 (incidentally, the best work devoted to German rocket weaponry and research on ramjet propulsion, that I have ever come into contact with, and at the same time an official publication of an institute researching German technology, almost 1,000 pages).

I found there the following description (partially repeating the information quoted previously).[102]

Take-off of the "flying wing" was to take place with the aid of A.T.O. rocket boosters on an inclined launcher. At a relatively high speed air hit the propelling duct's intake (air-passage-ramjet engine), was compressed and then reached the chamber in which the heated coal blocks had been placed. The burning coal raised the gas temperature and pressure. (...) The device was very simple, not demanding any valves that constantly deteriorated, or other mechanical devices. The thrust magnitude depended only on the temperature inside the chamber and the wing's airspeed. The jet-streams from the fuel ducts were leaving through the rear trailing edge of the wing. This model of a rocket [ramrocket?—I.W.] fighter was made at the end of the war and during the final months before capitulation passed comprehensive trials. In the original type only coal was used for the creation of thrust, which lasted for 45 minutes. But later an injection of liquid paraffin was used on the coal and an almost two-fold increase in thrust duration was achieved. The drawback of this propulsion system lay in thrust's invariability. It was attempted to remedy this through the application of two solutions:

a) regulating the size of the exhaust nozzle's cross-sectional area;
b) providing the fuel ducts with a controls enabling the cross-sectional area of the nozzle throat to be adjusted,

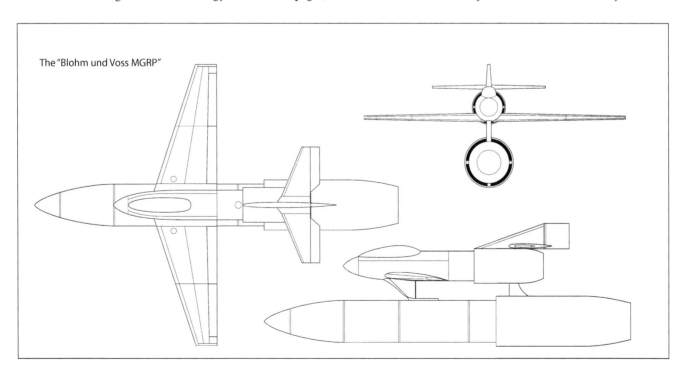

The "Blohm und Voss MGRP"

through which it would be possible to achieve a variable thrust magnitude.

It doesn't take any kind of special imagination to realise that the introduction of the supersonic P-13b to production would have completely revolutionised the aerial war. Roughly speaking one can ascertain that a jump had been made from the era of the 1920s to the 1970s within a few years—and therefore approx. half a century!

The subsequent fate of the P-13b has been presented in the chapter referring to operations "Paperclip" and "Lusty."

This was not at all the only fighter of this kind…

Two others were designed in the factories of Focke-Wulf. They too, from a technical point of view, were the only ones of their kind. Let us start from the vertical take-off and landing (VTOL) "Triebflügel" fighter. This name could be translated as a "propulsive wing," reflecting its unusual principle of operation. During take-off and landing the lifting surfaces performed the function of a helicopter's rotor, whereas during flight at high speed they "transformed" into wings.

The aircraft possessed a metal, cigar-shaped fuselage, in the axis of which a huge, specially constructed bearing, connecting the fuselage with three rotating wings, had been mounted, close to its centre of mass. Three ramjet engines were mounted on the wing tips, each with a maximum thrust of 840 kg. During take-off they were boosted by three Walter rocket engines, accelerating the wings to a speed enabling the ramjet engines to be started. Take-off and landing were to take place with the fuselage in a vertical position—during this time the pilot, located in a cockpit on the nose, was turned 90° backwards.

The aircraft possessed four large fins, on the tips of which had been mounted small undercarriage legs. The main leg with a large single wheel and powerful telescopic shock absorber of large lead, was situated on the other hand in the fuselage's axis. This object took off in the following way:

First of all the wings along with the engines situated on their tips were positioned in the plane of rotation (with a minimal angle of attack). As they were "warmed up" and accelerated by the rocket engines, this angle was gradually increased until at a suitably high speed the primary engines were started. The take-off and climb demanded an angle of attack of the wings of the order of tens of degrees. During climb the fuselage gradually levelled out into horizontal flight. This occurred simultaneously with a further increase in the angle of attack and reduction in the wings' rotational speed. The aircraft's airspeed on the other hand increased. Further flight did not differ from that of a conventional aircraft—both the wings as well as engines performed exactly the same function as in any other aircraft.

Contrary to many other unconventional solutions, the Triebflügel proved itself in practice—airborne trials carried out during the final months of the war turned out to be very successful.

The concept represented by "Triebflügel" was ideally suited to the advantages and disadvantages of ramjet propulsion. As is common knowledge, the main problem is ensuring its efficiency (of thrust) at low airspeeds. This ingenious concept fully solved the problem, since even at zero airspeed the blade tips could move at sufficient speed. At one stroke not only had the limitations of ramjet engines been overcome, but even the range of tolerable flight parameters (so called flight envelope) had been significantly widened in relation to a "standard" jet fighter. After all the "Triebflügel" could even come to a dead stop!

It was designed by a number of specialists under the direction of Dr. Pabst, von Halem and Multhopp (probably the best aerodynamics expert from this period), while the concept itself was advanced by Professor von Holst, reportedly inspired by insect flight analysis. There is no doubt that the inspiration was to at least some extent supplied by a certain unconventional helicopter, designed in Wiener-Neustadt in 1942. It was designated the WNF 342 and its author was a certain Baron von Doblhoff. The rotor of this helicopter was not powered by a standard engine, but by small rocket engines, placed on the rotor blades' tips. The WNF 342 possessed a 60 HP piston engine, but its main function was to power the propellant pumps for the rocket engines.

Von Doblhoff's inspiration could even have consist in a ramjet engine being combined with a rocket engine in the same housing.

A source of information about the practical performances of the "Triebflügel" fighter is constituted by an infallible analysis from 1951:[102]

Just before the end of the war an "athodyd" propelled helicopter was developed in the design bureau of Focke Wulf. The fuselage was streamlined with the cockpit shifted forward. The tail possessed four fins. Three rotors [three blades/wings—I.W.] protruded from the fuselage. Each one was an independent unit. Before take-off the arms were set at an angle that did not give any resultant lift to the helicopter. The time to get the engines up and running to maximum thrust did not exceed 5 seconds.
After the engines had been set in motion the pilot slowly rotated the arms and the aircraft moved along the direction of resultant force.

In a further section of the quoted specification the following little known details were given:

The maximum vertical speed did not exceed 124 km/h. After climbing to a sufficient altitude, the aircraft commenced horizontal flight with the adjustment of control surfaces and ailerons. (…) In horizontal flight the aircraft reached a speed of 1,000 km/h. The rotor operated at 520 rpm, which after conversion gave a rotational speed of the tips of 1,500 km/hr.
The initial rate of climb amounted to 7.5 km/min. Rotor working time—42 min., range 640 km. At an altitude of 11 km horizontal speed amounted to 800 km/h (…)
The system of descent relied on a gentle rotation of the engines, which enabled sideslipping. The head resistance only amounted to 20% of that which would have been obtained in the case of a standard fighter with the same fuselage size.

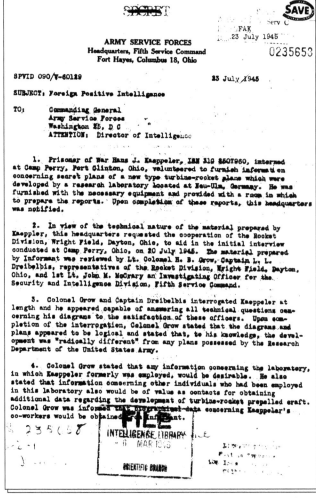

The US intelligence report pertaining to Hans Kaeppeler…

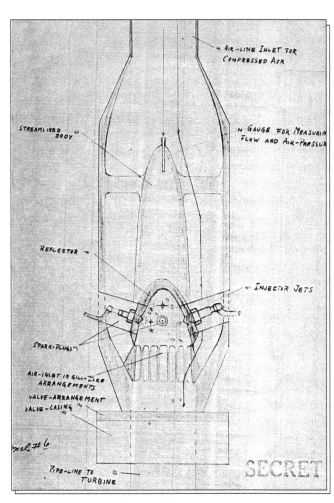

…the diagram of the hybrid engine supplied by him…

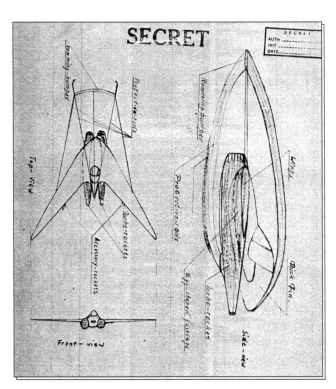

…and sketches of other projects, on which he worked

Technical data of the "Triebflügel" fighter

Length (height)	9.15 m
Diameter with rotating wings	10.8 m
Take-off mass (max.)	5,175 kg
Fuel mass (total)	1,500 kg
Max. speed	approx. Mach 0.85
Max. rate of climb	7,500 m/min. (125 m/s)
Rate of climb at altitude 14,000 m	120 m/min.
Endurance (depending on ceiling)	0.7-3.4 hr
Armament	two MG-151 20-mm guns
	two MK-108 30-mmguns

In a further section of the text we find mention of yet another ramjet propelled aircraft:[102]

> *The tailless Heinkel P 1080 aircraft with swept wings possessed "athodyd" type propulsion; the engines were built into the wing roots.*

The general conclusion is also worth making a note of:

> *It became evident during testing that the possibilities of employing athodyd propulsion at the speed of sound and*

supersonic speeds were great. During tests on this method excellent results were obtained for a range of speeds from Mach 1-Mach 3.[102]

The unprecedented successes described above, like approaching the speed of sound (if not exceeding it), or the fact that a vertical take-off and landing (VTOL) high-speed jet fighter had been tested, were mainly due to the already forgotten Dr. Pabst from Focke Wulf. In any case his discoveries also remain practically unknown, despite being breakthroughs in aviation development! These achievements become even more astonishing, when we take into consideration that Pabst had commenced work on ramjet and ramrocket (hybrid) engines no earlier than 1940 (inspired by Sänger's experiments) —as was stated in one of the American reports.[118]

Dr. Karl Oswatitsch, working in Göttingen, could also boast of interesting results, although his engines were never tested in flight. During the years 1943-1945 he was engaged in the development of engines, which could be used to propel supersonic aircraft and guided missiles. He was interested in speeds of the order of Mach 3. He carried out many very complex analyses, which convinced him that the construction of an intercontinental aircraft, travelling at a ceiling of 20-30 km and speed of 2,000-3,000 km/h, was within reach of the technological possibilities at that time. The high standard of these analyses and Oswatitsch's expert qualifications in general allow him to be regarded as one of the most avant-garde in the described group. Undoubtedly he would have had a great future, if his activities had not been interrupted by the end of the war.[118]

The next ramjet propelled fighter was the Ta 283, designed by Kurt Tank. This forgotten today design was one of the most promising in the Third Reich. It did not remain only on paper! In the already quoted extensive analysis, the flight trials programme of the Ta 283 prototype was described:[102]

"It was a fighter similar to a standard one, possessing two characteristic features: the first referred to the shape of the fuselage, which was unusually strengthened in its rear section and crossed into a fin with rudder; the second relied on two ramjet engines of the "athodyd" type, mounted on both sides of the fuselage on relatively long arms. This kind of positioning demanded strengthening of the fuselage.

The cockpit hidden in fuselage had a canopy made of "plexiglas," with good visibility. The cockpit was located more or less half-way along the fuselage. The fighter's wings were swept back at an angle of 45°. The "athodyd" engines were 1,320 mm in diameter. To accelerate the aircraft to a speed necessary for the "athodyd" engines to commence operation, a Walter dual-propellant rocket engine with 3,000 kg thrust was built into the rear section of the aircraft.

Fuel in the form of sprayed Kerosene was applied to the "athodyd" engines, injected into the combustion chamber. A stream of air was compressed at the intake, released oxygen with the burning of kerosene in the combustion chamber, increased its pressure and temperature and flowed out of the nozzle at high velocity. Thanks to both "athodyd" engines the aircraft acquired an average speed of 1,100 km/h at sea level and a ceiling of 9.3 km with a rate of climb of 90 m/s. If however the aircraft reached an altitude of 11 km, its maximum horizontal airspeed was reduced to 960 km/h. It was determined by way of tests that rocket propulsion should be used in this machine to reach operational ceiling, but was not necessary on landing.

At sea level operation of the "athodyd" engines lasted 13 minutes, but by taking off and climbing to an altitude of 11 km through the use of rocket propulsion it was possible to obtain an endurance of 43 minutes.

Technical data of the Ta 283 fighter

Empty weight	2,700 kg
Total weight	5,450 kg
Wing lifting surface:	18.5 m²

The trumps of ramjet propulsion—and above all its simplicity—allowed a whole range of aircraft-missile systems to be designed representing some kind of next step with respect to the "Mistel" system. An expression of this trend was e.g. the "Blohm und Voss MGRP," where both segments possessed such propulsion. This was then one of the developmental routes of strategic weapons, since it was difficult to reconcile a large range with accuracy—with the "complete" lack of a pilot.

* * *

Reviewing an archive from American Intelligence I found documentation affirming that German research for new aerial propulsions had gone down yet another path ...[120]

This concerns a group of documents referring to a number of avant-garde designs that arose in an unmentioned by name, aerial design office in Neu-Ulm. Part of the plans were captured and one of the designers was interrogated.

From these materials it follows among other things that some kind of combination of a ramjet engine with a resojet and conventional turbine engine was developed, at the same time the turbine was located behind the combustion chamber of the ramjet unit. The specification is incomplete, yet the design deserves special attention if only for this reason—that as was stated, its documentation was regarded as "crucial" and evacuated at the end of the war to Japan, and most of the German originals were destroyed. Among other things it was stated that this hybrid engine was powered by quite an unusual fuel. It was a milky-white liquid with a very acrid smell, evaporating very quickly. This was accompanied by a sudden drop in its temperature. Its composition remains unknown.

Above: Focke Wulf's Ta 283, subsonic ramjet fighter
Below: The Triebflugel VTOL fighter

AN ARSENAL MORE POWERFUL THAN AMERICAN NUCLEAR BOMBS

I was not the first to draw attention to the very existence of the Third Reich's "ultramodern" chemical weapons. Most authors however overlook the fact that operational plans and delivery systems must have been developed in parallel with production itself. This was precisely the case and is testified by a series of until now "scattered" facts (new ones occupy a further section of this chapter). Professor Mieczysław Mołdawa, who by virtue of his work in the so-called Technical Office at camp Gross-Rosen—engaged in many Lower Silesian projects with access to much unique information—stated for example that the German foray into Brzeg Dolny and the Tabun factory had a somewhat different objective to what is generally considered. I briefly call to mind: after the factory had been captured by the Russians a group of Wehrmacht commandos received an order to penetrate the rear of the Soviet Army and seize the aforementioned facility for at least several hours. In an interview he gave me recorded on film Professor Mołdawa spoke not only of the necessity of destroying the chemical reserves, but above all clearing out the factory safe which held top secret plans for a new phase in the war!

A rather credible and significant reflection of such plans is that at the end phase of the war practically every one of Hitler's headquarters or industrial-command mountain bastions ("Riese," Thüringen) were built taking into consideration on the one hand the possibility of surviving a conventional (or nuclear) enemy attack. On the other however, one can see connections with German weapons of mass destruction. This theme comes and goes in the case of the following locations:

1. Thüringen—in Jonastal Valley a monstrous conglomerate of central command posts were built, deep bored into the mountains (S-4, Amt 10 and "Jasmin"). Being underground they were much better protected than for example older bunkers in East Prussia, but moreover the theme of German weapons of mass destruction is present—perhaps as an arsenal assigned to the "redoubt" itself. Key nuclear laboratories were also transferred to nearby Stadtilm. Furthermore lethal chemical weapons were found throughout the valley region after the war.[29]

2. The "Riese" area in Lower Silesia can be treated as a second redoubt where a connection existed between a command centre (at Ksiaz Castle) and research and development centre (the Sowie Mountains, Ludwikowice, a facility at a second castle in Ksiaz, and the research post Strughold connected with strategic weapons in Szczawno-Zdrój). One is struck on the one hand by the choice of unusually hard rocks and localization of the central sector of "Reise" in a natural Gneiss dome-like structure approximately 200m thick, but on the other by the intertwined theme of German weapons of mass destruction. Two reliable witnesses exist who state this was the case—including the aforementioned Professor Mołdawa. In an interview given several years ago he spoke of a chemical weapons arsenal, as part of a strategic weapon, which the complex was designed for.[245] The other witness who supplied information practically at first hand was Dr. Jacek Wilczur.[130] At the beginning of the 1960s on behalf of the Chief Investigative Committee of Nazi War Crimes he penetrated an underground facility linked to "Reise" situated a little to the south of (and beneath) the recognized facility "Osówka," where sappers securing the penetration found the remains of radioactive ore from a railway. Statements suggesting similar connections were also made by the former chief power engineer of the complex—Anton Dalmus.

3. A new trend (1944) in the construction of central command facilities was represented by a complex of underground bunkers in Pullach near Munich. It allows one to observe elements not present for example in "Riese" or S-4, in the sense that it was actually completed. Directly after the war it was examined by Allied officers, who were shocked by the technical standard it represented. But they were most surprised by unusually advanced chemical defence shelters, which they had never before come into contact with.[131] In a report they wrote about the internal ventilation installation being equipped with air conditioning (at that time this represented a certain sensation). The entire system generated excess pressure, which combined with a modern assembly chemically neutralizing (!) possible toxic vapours, enabled operation during deployment of state-of-the-art chemical weapons "for a practi-

cally unlimited period"—as stated in the report. Areas for the chemical decontamination of people also existed along with the respective protective suits.

Hence one can see undoubtedly that Hitler's sceptical attitude, or lack of decision—still observable in the summer of 1943, had given way to fundamental revaluations. One should also realize that this could not only have involved the fear itself of an Allied chemical strike. For the Allies were aware of German achievements in this field and surely could have linked this to the V-2 rocket, against which there was no defence. Third Reich preparations for mass chemical warfare could only follow from its own offensive plans, and from the fact that this was the only area where it held a crushing advantage over the Allies. It was simply a monstrous arsenal…

So what was Hitler's attitude to this problem?

He was initially intrigued and full of hope, all the more as the eastern front shifted further westwards. On May 15, 1943, Hitler summoned a conference at the "Wolf's Lair" in East Prussia, where he summoned among others the Chief of the so-called "Committee C" in Speer's ministry (responsible for chemical warfare preparation) as well as board member of I.G. Farben, Dr. Otto Ambros.[132] Hitler demanded full explanations on the possibility of an analogous arsenal being used by the Allies. He was supposedly to have asked: "What is the other side doing about poison gas?" After the war Ambros testified at the Nuremberg Trials that he had consciously misled Hitler, aiming to prevent such a destructive phase of the war being unleashed. He allegedly said: "With regard to greater access to ethylene the enemy probably has a greater possibility of manufacturing mustard gas." The Führer reportedly interrupted him, pointing out that he never had in mind traditional poison agents: "I understand that states which have oil are able to produce greater quantities, but Germany has a special gas, Tabun, over which we have a monopoly." Ambros then engaged in quite a risky game in declaring that the secrets associated with its production had got through to the West before the war. He added: "if Germany deploys Tabun, it must count on the possibility that the Allies manufacture it in greater quantities." Hitler supposedly then left the meeting…

I repeat—its contents only result from the post-war statements of a man who wanted to avoid bringing a death sentence upon himself, in not admitting that he had directed the aforementioned preparations—and this in effect was the case! These testimonies do not sound entirely credible, especially in that according to Ambros the source of the information leak was to have been pre-war official German publications, containing production specifications. Such statements, if they had in reality taken place, would have been precarious to the extent that such publications … had never existed.

Whatever the case Hitler had no reason to yield to similar "discouraging" reassurances. He had the whole intelligence service at his disposal, which confirmed that the Allies (and the Russians) had no idea whatsoever of phosphor-organic weapons.[132] In reality they only learned the truth until after the war! The Russians presented a particularly poor level in this field. I actually remember some Russian documentary, in which someone recalled that during the war a factory that filled chemical ammunition was located in Moscow. The workers included … handicapped children, who used kettles for decanting poisonous liquids! Every fortnight the workforce was replaced.

On March 1, 1944, Dr. Ambros reportedly warned Hitler again, this time without his previous confidence (!) that the

Dr. Otto Ambros at the Nuremberg trial. He was responsible for industrial preparations for a chemical war in the Speer's ministry.

Allies *could* be working on nerve agents. This again doesn't appear entirely credible as preparations were intensified at roughly this time in the Reich and Ambros kept his position. The scale of preparations is clearly indicated among others by the data presented further on.

However before I move on to presenting new information from documentation, allow me to return again to the aforementioned, secret episode of World War II—the commando foray into Brzeg Dolny and a description of the (key) factory itself…

The aforementioned factory, codenamed "Hochwerk," was built using 120 French POWs in 1940. They were soon assisted by approximately 800 Italian workers and 80 Germans—specialists, who were given the status "u.k."—"unabkömmlich." This meant that these people became in effect "secret bearers" and could no longer for example be enlisted in the army… Later the Gross-Rosen sub-camps "Dyhernfurth I" and "Dyhernfurth I" were formed. Initially only several hundred prisoners worked there, in time they were expanded and at their peak of operation held at least 2,500 people. As one of the articles revealed:[132]

The prisoners worked in a separated room, with restricted access to outsiders. The level of production secrecy was so high that SS guards patrolling the camp, despite procedural declarations on maintaining secrecy, could not observe the prisoners at work. During the unloading of shells ready to be filled with gas, high screens were erected near the railway wagons. Thick hoses transferring the gas from tanks were hauled up to the isolated factory room, where the shells were filled [Tabun was a liquid, note I.W.]. The chambers in which chemical reactions took place were coated with quartz and silver. The area where production was finished was sealed with double glass-lined walls, between which a cushion of pressurized air was maintained. The room's doors and windows were hermetically sealed, and for ventilation channels laid into the floor blew air inside drawn from very high chimneys. The whole area was periodically decontaminated with steam and ammonia. For the safety of the gas transporters, cases damaged in transit were repaired in the factory joinery and leak stoppers were prepared while immobilising prepared gas carriers in the wagons. Further on was situated a bomb filling station, an area for filling artillery shells and glass containers as well as marking and checking posts. Manufactured gas was stored in a bunker warehouse, adjoining the production building. Machines decanted liquid Tabun into shells and part of the production was removed from the factory in cisterns. (…)
A further two [factories] were built by the end of the war, managed by the SS, producing monthly around 500 tons of so-called "N-Stoff"—blistering trichlorofluoride. Col. Ochner portrayed the new gas as being so powerful and effective, that only employment of the so-called "Grünkreuzmaske," protecting against the most powerful gases, could avoid death. In stating this he was unaware that they were so lethal that even employing the best gas masks would help little in the event of being enveloped in a cloud of Tabun or Sarin. The huge gas factory codenamed "Hochwerk" also had an underground outpost in Krapkowice (Krappitz), in the form of a production room for filling artillery shells with gas.

In the aforementioned citation a curious and little-known theme appeared—chemical weapons production managed by the SS. This contributes to one of the following chapters about Himmler seizing control of one of the most promising fields of weaponry and therefore worth expanding on. In short, the above example is not the only one which illustrates this phenomenon. In the case of this armaments field ties with the SS are exceptionally clear—despite the lack of full information—and I will again take the liberty to make a brief calculation:

1. Facilities such as "Riese" or other underground factories were generally supervised by the SS (Kammler), if only on the grounds that they were located precisely underground. Moreover the workforce consisting of prisoners were in the hands of the SS.
2. The same refers to locations associated with strategic weapon delivery systems. The "Zement" complex in Austria, designed for researching and manufacturing intercontinental weapons, was at the disposal of Kammler. Likewise the research team working on a new propulsion system for strategic weapons in Pilsen (described later) was subordinate to the SS-FHA technical office, or the SS. Work connected to this and carried out in Lower Silesia, presented in Part Three, was also the domain of the SS, despite collaboration with the Luftwaffe.
3. In his book "Slave State" Minister Speer mentioned Himmler's ambitions in this field. Characteristic is the following excerpt:[133]
"*Himmler ultimately wanted to bring the production of not only N-Stoff [mentioned earlier, I.W.], but also the nerve gas Sarin into the hands of the SS. Sarin was our most modern weapon, many times more effective than any previously manufactured war gas [approximately twice as powerful as Tabun, but not as long-lasting and more difficult to manufacture, I.W.]. Moreover, there was no defence against it because the gas masks and filters known at that time offered no protection. Sarin could thus someday be an inestimable factor, at least for extortion, in an internal German struggle between the Wehrmacht and the SS. And after July 20 [following the attempted assassination of Hitler—I.W.] such considerations could no longer be viewed as absurd. In early 1944 Hitler had stated that he had made up his mind to assign both the testing and the manufacturing of N-Stoff to the SS. I pointed out to Hitler "that the operation of a chemical factory [should] if possible remain within*

the overall chemical industry." Hitler changed his mind. But he wanted "to charge the SS Reichsführer with testing and evaluating N-Stoff and only then deciding with me whether the production of N-Stoff will remain in our hands. On July 7 [1944?] Hitler again ordered General Buhle, the head of the army staff, to have "the SS Reichsführer speedily perform further experiments with N-Stoff. Three weeks later I spoke to Hitler about the SS's intention of simply going ahead and manufacturing N-Stoff without first testing it: "I convinced the Führer that for the time being the Waffen-SS should not take over production. (...) Another reason why I cannot agree to the Waffen-SS taking over the Falkenhagen [the factory described several pages back—I.W.] production is that a facility for manufacturing a crucial chemical warfare agent is located next to and in connection with N-Stoff. A Tandem factory management does not seem reasonable. Sarin, the combat agent manufactured in Falkenhagen, is the most valuable and most modern of any chemical combat agents and is six times as effective as any previous one. (...) Hitler no longer spoke about having the SS take over the manufacture of N-Stoff. But despite the negative findings of the highest experts in the SS, the SS had unhesitatingly appropriated the factory, which was valuable because of Sarin production. In early November 1944, Schieber, who was about to be dismissed because of Himmler's intrigues, performed a not undangerous duty by making me aware of SS machinations: "Problems have arisen in Falkenhagen because the Waffen-SS is already manufacturing N-Stoff there, the workshops and general manufacturing facilities are overburdened to excess, and these machines are now to be moved into bunkers to boot. All this adversely affects the expansion of Sarin production... We most sharply protest against the SS measure." For as soon as the SS had established itself in the factory, it would appropriate the Sarin manufacture "for operational reasons."

4. Himmler undoubtedly wanted to play a key role in preparations for chemical warfare, on which all hopes were pinned for the Third Reich's salvation. One can observe this tendency not only at top-level, but first and foremost after scrutinising numerous, smaller undertakings carried out far from Berlin—of which Speer, jealously guarding his influence, may not have even been aware of. Such a "cross-country" example is illustrated by the research centre at Lubiaz (Leubus). According to numerous accounts work on weapons of mass destruction was carried out there, including biological and chemical. Some interesting accounts are found in the publications of Ms Sukmanowska and Anna Lamparska. The latter's book for example includes the following quotation:[134] "one could detect a strange smell in the air irritating the lungs and eyes, some kind of gas..." One witness reported underground rooms, where the Germans worked in white lab coats; also present were the glass walls, already familiar to

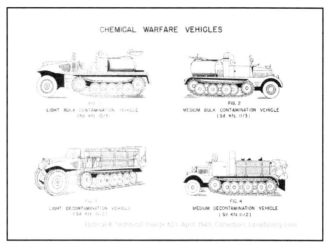

German chemical decontamination vehicles

us from the factory in Brzeg. Directly after the war small vials or rather ampoules were found among other objects in one of the buildings in neighbouring Krzydlina Wielka, resembling those used for injections, but smaller. As some of Lubiaz's new inhabitants included rats, appearing "out of nowhere" in unexpected places, particularly when the level of the nearby River Odra rose; somebody once hit on the idea of throwing a few of those ampoules, useless anyway as they were unlabelled (marked only with coloured bands), into the "holes" from which the rats emerged. It must have been a very effective agent, as in an unforeseen way all the rats were poisoned, also by chance cats and dogs ... (so-called phospho-organic warfare poisons manufactured in nearby Brzeg on the River Odra, then Dyhernfurth, caused symptoms of serious poisoning in doses of the order of one thousandth of a milligram, so even a small ampoule of Tabun or Sarin solution could kill all living things in a large, sealed building). Everything indicates that this outpost also came under the jurisdiction of the SS.

5. For comparison and to give a more complete picture: the SS had no ambition to seize control of the chemical arsenal alone, not in the least! It is worth taking note that the only significant research outpost in the Reich working on biological weapons—in Pokrzywno—was managed by the SS! Allegedly because the Wehrmacht would not have anything to do with human experimentation... A further element is the nuclear research described later—in Czechoslovakia. This was carried out in such secrecy that even Speer hadn't the faintest idea of its nature, and—as he admitted after the war—was even unaware of the existence of the special SS department that these research groups were subordinate to[133] (SS-Führungshauptamt, Amtsgruppe "A"—T.Amt VIII FEP)—which of course we will return to later...

These little-known facts enable one to put forward the argument that if the Third Reich had actually deployed its (existing!) weapons of mass destruction arsenal—not a very far off perspective—this would have been to a large extent an "SS War." Likewise it is not difficult to come to the conclusion that if the SS had seized control of and used on a strategic scale the chemical arsenal (and other possible weapons of mass destruction)—it would have undoubtedly become not only the dominating force in the Reich itself, but perhaps the entire world!

The significance of chemical warfare preparations is also emphasized in some way by the aforementioned German special forces raid on the Brzeg factory in February 1945. Once again I refer to the already-quoted article, where a superb account of this curious episode is found:[132]

The Russians had occupied villages and towns without securing the widespread area. They only remained in key positions, observing the terrain in the event of a counterattack being launched. German front line units in the area were represented by a jumbled group of war veterans, soldiers who had survived the defence of the Vistula line, the Volkssturm and the Hitler Youth. They all had the required weapons, were "uniformed" and well integrated to act as operational teams. A reconnaissance mission had reported that enemy positions could be overrun with ease. These reports were quickly "confirmed," when a general along with two front line officers approached the river. A Russian [Soviet] light machine gun immediately opened fire from the other side of the river, wounding the two officers. Continuing his reconnaissance, the general slowly retreated back to his position. He observed a dynamited bridge, which on the German side hung in shreds over ice floes piling up on the water. At "their" end of the bridge, on either side, the Russians had positioned two 20mm machine guns [rather cannons, I.W.]. The next visible defensive position, 200 m away, was located behind the machine guns. The area between them was presumably mined. Just beyond a railway line veered sharply to the left, in the direction of Dyhernfurth [Brzeg Dolny]. From here ran a supply line, leading directly to the factory. A perfect way for people who didn't even know the area to find it in the dark. Much depended on the Russian crew of both machine guns being quietly and quickly eliminated…

(…) The youngest general in the German Army [Sachsenheimer], was respected by his soldiers with whom he had on more than one occasion endured the January offensive from Puławy to Wrocław, including suicidal infantry assaults against endless attacks of Russian tanks, before reaching the area of fortress Breslau and the Odra—to later take command of fortress Glogau [Głogów], a few dozen kilometres away from Dyhernfurth. On 3 February the Chief of Staff of the 4th Panzer Army, Oberst Knüppel summoned the general to a meeting at his headquarters in Lauban. The raid on the factory in Brzeg had to end unconditionally in total success. Oberst presented him with all known details and ordered immediate work on planning the operation, handing him the written order:

Form an attack group to carry out an unexpected raid on the chemicals factory in Dyhernfurth. The objective of this operation is to secure chemical weapons with the aid of civilian volunteers, two scientists and eighteen factory workers, and destroy the top secret poisonous gases stored there. These materials are located behind enemy lines in underground tanks, in liquid form. Engineers have suggested pumping the material into the Odra, which should be simple using factory pumps and equipment. Further, Army Group has ordered all remaining chemicals to be made unidentifiable. Dynamiting the tanks is not a practical solution and discouraged. Blowing up the tanks may have consequences for personnel involved in the raid and leave a suitable quantity of material behind for the enemy to analyze. Next the attack group should destroy all remaining material and nine bunkers—warehouses, using explosives. The amount of remaining chemicals may be too large for their evacuation.

In the mean time Corps General Staff had already formed the appropriate units necessary for the operation. They immediately made their way to their bases near the railway bridge. General Staff had also promised additional units and specialist weapons. Two paratrooper companies, two 88mm artillery batteries and one pioneer company [so-called sappers] along with 81 landing boats were to take part in the raid. The general knew that the mission must succeed, although he feared one thing—he wasn't in any way familiar with the units he was in command of. He was also worried by the presence of civilians, unsuitable for such a mission. He could only pray that the Russians occupying the castle at the other side of the river remained drunk as long as possible. He looked over his decisions and with a new idea made his way to the General Staff of the 4th Panzer Army, not waiting for any reports. The General Staff accepted his plans and it was agreed not to drop paratroopers from the air, but move them closer to the area of departure as a reserve. This message was telegramed from AOK-4 to Headquarters. In complete silence and darkness Major Joos and his landing force passed under the bridge and silently despatched the Russian machine gun posts. Further on, avoiding mines on both sides they ran bunched into a group across the top of the railway embankment to pounce on the next Russian position. These, taken completely by surprise, surrendered without a word. 65 minutes from the raid's start, factory technicians and chemical weapons experts were working on starting the generators and factory pumps.

Work proceeded faster than expected. The sounds of dispersed battles reverberated around the factory. It was not until 13.00 that the Russians became aware of the why the Germans so fiercely harassed them there. Moments later they commenced a concentrated counterat-

tack. From the north, from the direction of the village Seifersdorf, eighteen tanks—several T-34s, the majority T-52s, formed a wedge in the shape of the letter "V." The general decided that there was insufficient time to move anti-tank PaKs from the river, so he positioned two Hetzers on the riverbank, which immediately began to shell the Russian tanks. Returning German sentinels brought back information on the attacking tanks to camouflaged covering troops [anti-tank squads, I.W.], assigning them new targets. The Panzerjagdkommando moved into action. All tank killers were volunteers, always a shortage in the army. For each destroyed tank they were entitled to a week's leave. The killers' effectiveness at the end of the war was significant. Using tried and tested methods many could destroy a large number of tanks in one attack. After a short time the terrain was reduced to a flaming and smoking area, occupied by both sides. All tanks had been immobilized. Just before evening seven more tanks appeared from the direction of the village Kranz, firing continuously at the railway embankment leading to the factory. A Russian victory would signal the operation's failure and no way out for all units taking part. The Russians eventually became aware of the operation's objective and knew just what to do to stop the factory's contents being destroyed. The 88mm cannons had been installed near the bridge precisely for this possibility… The barrels were raised above the river embankments and firing continuously from settings preset at a range of 500-700 m, quickly destroyed almost all tanks still firing at the embankment. Only one managed to turn round and escape in the direction of the redeeming forest wall.

The brief, extremely effective shelling completely paralysed the Russians, who carried out no further attacks. Battles around the factory gradually died out, leaving dead infantry everywhere and tanks burning in the cold twilight. The raid's participants prepared to retreat—retrieving equipment, wounded and several dead colleagues. The operation would have ended in total success, but for an accident in the factory. While the sappers were preparing explosive charges, one of the containers holding residual gases burst under pressure spraying several soldiers and civilians, blinding them immediately. Headquarters maintained constant, direct contact with the team working in the factory! Marshal Schörner personally demanded that General Sachsenheimer come to the microphone, wanting undoubtedly to congratulate him and remind him of the mission's importance… The general meanwhile ignored his order and along with two scientists attended to the operation's progress. Later he asked his aide-de-camp to sit at his typewriter and write a report informing that all poison gases, important materials and documents had been properly "secured." This document was signed by two professors participating in the raid.(…) The paratroopers promised by General Staff never arrived, even as a reserve. The furious Russians torched Dyhernfurth castle as well as the adjacent cloister, containing a priceless library.

This was February 1945—when the Third Reich was already on the decline. Let us go back in time however to the preparatory phase of chemical warfare—to its "apogee." The number of research posts and plants harnessed in its top secret cogs testifies to its hitherto unknown scale. The information presented in several of following pages comes from the already cited extensive American intelligence report—the vast majority of which has never before been published.[135] In any case it is worth re-emphasizing the conclusions presented in the report's preface:

Findings, point 6: Manufacture of poison gas is intensive and decentralised."
As well as:
Conclusions, point 3: Preparations for chemical warfare, both offensive and defensive, are well advanced.

Let us begin with a list of locations, or rather a complement to all that we have known to date:

Garmisch-Partenkirchen
Chemical preparation of the flying-bomb was done by the Chemical Works in Garmisch-Partenkirchen.
(source: Allied Govts. London, No. 2036, 29 Jul 1944, S.)

Gelsenkirchen
Launching site at Gelsenkirchen of the self-propelled plane determined in 1943/1944.
(source: Allied Govts. London' No. 2036, 29 Jul 1944, S.)
Previously reported that a factory between Gelsenkirchen and Essen is making some sort of gas.
(source: Allied Govts. London, No. 1501, 4/18/1944.)

The following report may have no direct connection with chemical weapons, but with delivery systems:

Monachium, Bayerische Motoren Werke, Plant No. 1.
This plant produced 400 airplane motors per month of a total German production of 12,000 a year. Motors of E & I groups of BMW Type 801. A & D groups produced elsewhere. They are experimenting with a new 35,000 hp motor. This experimental motor, BMW Type 806, is for use in a super plane to fly nearby to USA coast to release a robot bomb. (…) BMW Plant No. 1 completely destroyed. Production transferred to BMW Plant No. 2 in Muenchen-Allach.
(source: Cable: Stockholm to State, 2 Sept 1944, Minister Johnson, S)

Tilleur (Belgium), Angleur-Athus works
This works, on the left bank of the Meuse at Tilleur, has been requisitioned by the Germans and is an important

center for the manufacture of liquid air. Installations above and underground, are 1km long and are situated between the bridges at Seraing and Ougree, starting 750 meters east of the Seraing bridge.
(source: OSS, SO-1257, 2 Aug 1944, Eval: B-3. S.)

The Angleur-Athus factory is engaged in loading bombs either with toxic gases or with liquid air.
(source: OSS, SR-796, Feb. 1944, Reliable Belgian source; C.)

Tanwald (Czechoslovakia)
a report particularly important with regard to the following chapters!
Parts of the German secret weapon are being manufactured in Tanwald. The exact location of the plant has not as yet been ascertained. The weapon, referred to as the 'V-3' [?], is said to be an aerial torpedo spraying incendiary material. The spraying instrument is being manufactured by Skoda at Pilsen.
(source: Allied Govts. Rpt No. 2232, 18 Aug 1944, Eval: C-3. Czechoslovakian Intelligence Service. S.)

Comment: it seems quite probable that a rocket was indeed involved, perhaps the V-2 (often the term "aerial torpedo" was used in reference), however an "incendiary device"—devoid of any meaning in the case of a large rocket—could have masked something that in the military is termed "chemical weapons dispersal system." This is important in that it is known from other sources about the connection of the outpost in Pilsen with strategic weapons! Once again an SS outpost was involved...

If one was intrigued by my mention of hitherto unknown research in Czechoslovakia, then I have good news: the present book is devoted to a significant degree to precisely this forgotten aspect of the war. This country actually had a specific link with the development of German strategic weapons, chemical weapons probably being its most important component!

Anvers (Belgium)
17/6/44 the garrison at Anvers received a new mask. The inner composition of the cartridge will be the following: anti-arsenic filter paper strongly folded, wadded 2 mm; metal gauze; 3 layers of active charcoal 20.20 & 18 mm padding [?].
(source: Allied Govts., Rept No. 1931, 20 Jul 44. Encl: Belgian Mil Intell Bulletin No. 203.)

Berlin/Dahlem, Kaiser Wilhelm Institute
This institute carries out continuous research in all fields of physics and chemistry and allied sciences. It is largely independent and is controlled by Professor Planck. It is organised in numerous branches, some of which are secret and closely guarded. One of its branches will undoubtedly be responsible for C/W research, according to P/W. The institute is widely scattered and has many branch offices and laboratories, but P/W cannot give locations.
(source: Report No. 70675, 21 Jul 44. Interr. P/W. S.)

Brno (Czechoslovakia) Waffen Union Brünn
This factory has received an inquiry from Germany as to the possibility of manufacturing rocket missiles filled with gas.
(source: M/A No. 3592, Nr. Allied Govts. 1995, 27 Jul 44.)

Cakovica (Czechoslovakia)
Produces fuselages for ARADO 96.B aircraft and fits engines. Since the beginning of 1944 and until March, 8,000 special funnel-shaped steel castings were produced. These had a square base, 50 × 50 cm, tapering to a point at a distance of 50 cm, from the base, with an orifice at the point. They were constructed to withstand a pressure of 40 atmospheres and it was believed they would be used for releasing gas or liquids. 5,000—6,000 workers employed.
(source: Encl. w/OSS 2297, 19 Jul 44. "War Industries in the Prague Area." Info. dtd end of War 44. Unstated reliability.)

Czechoslovakia
Indications are that war gases are being transported on the Vienna-Moravska Ostrava Railroad; instructions have been issued as to protective measures, stating that the danger zone in the event of an explosion of gas in transit is 2 to 3 km against the wind and 20 km downwind.
(source: Allied Govts., No. 2089, 5 Aug 44. C. Eval: C-3)

Decin (Czechoslovakia)
Large poison gas factories located here.
(source: S & I Div. 4th SC PFI-2, 1 Mar 44)

Fürstenberg [near Gubin]
Fürstenberg (52° 07′ N.—14° 40′ E.) a busy gas plant is located here.
(source: rpt frm CP & M Br. BX-94, 24 Aug 44. S. Info frm very reliable source in enemy occupied territory covering the period 13 Apr-30 Jul 44.)

Hamburg (Germany)
Gas Warfare Preparation. Professor Keeser of the Pharmacological University Institute at Hamburg has written a new circular on gas warfare defence which has been distributed to doctors. It makes mention of Lewisite, Gelbring, Phosgen, Gelbkreuz and Stickstofflost.
(source: OSS, undated, Org. Rpt. RB-18121.)

Pyrennes Mountains
The manufacture of gases is reported in this area. Attempts are being made to obtain additional information.

(source: Digest of Chemical Warfare Intell. Rpts. No. 7, 20 June 1944.)

Kolin (Czechoslovakia)
Located at Kolin are sodium syenite (base of a poison gas) works. The director of the surrounding potassium works of Kolin was a German.
(source: FS/S&ID 6SC, rpt # 1675, 29 Jul 44. S.)

Landsberg
There is a laboratory in this works in which all kinds of infectious microbes are cultivated to fill the bombs of the V-2.
(source: OSS-SO-1524, 12 Aug 44, British source.)

This document was found perhaps by accident in a report on chemical weapons, but is nevertheless extremely interesting. It reminds me of a story once described by Mr Kordaczuk from the Regional Museum in Siedlce—a tireless researcher of V-1 and V-2 test sites near Siedlce (launched from the region of Blizna). He once challenged me on the following: numerous craters in a field, the result of rocket strikes, tens of metres in diameter that usually filled with water after some time forming artificial ponds. Farmers gained in this unusual way scattered watering places for their animals. In the years directly after the war the area of one of the towns was plagued by a strange epidemic, which wiped out many herds of cattle. A veterinary inspection team appeared on the scene which began to check the source of the animals' drinking water. A sample was taken from one of the rounded "ponds." It emerged that it was full of a suspension of … anthrax germs! It is intriguing if a chemical warhead version of the V-2 also existed—for there is no doubt that such warheads were manufactured for the V-1…

Marpingen
Chemical plants at Marpingen are manufacturing poison gas.
(source: OSS, RB-17949, 21 Aug 44. C-3. C.)

Munich-Allach
BMW plant No.1 (at Munich) completely destroyed. Production transferred to BMW plant No.2 at Munich-Allach, experimenting with unknown type of poison gas
(source: Cable: Stockholm to State, 2 Sept 44, Johnson Minis. S)

Comment: interesting connection! So this is the same plant mentioned in the context of an undetermined in more detail strategic guided missile?—with which the Germans planned to attack the USA! And as to the "unknown type of poison gas"—Tabun and Sarin were not known to the Allies at that time… The delivery system was described as a "super plane to fly nearby to USA coast to release a robot bomb." It is also worth remembering that in this same intelligence report pertaining to chemical weapons, there is also mention of certain research activities at the University of Munich—but in physics! It was described as the most important project carried out in the Third Reich. As we shall see further on, the project was carried out at Professor Gerlach's university and faculty, officially code-named "decisive for the war." This is the description from the report presented in the chemical weapons report:

University of Munich
P/W finds it difficult to access which are regarded as the most important scientific research jobs being carried on in Germany at the present time. P/W says that undoubtedly the fields of physics and chemistry are regarded as the most urgent and important spheres of research.
(source: P/W interr rpt at Cherbourg 14 Jul 44. by M/A London No. 70675, 22 Jul 44.)

Neuberg (Germany)
At Neuberg, there is a special training field where reserve infantry troops receive a two-week's training in gas warfare, during which mustard gas is used, as well as Weisskreuz, the antidote for which is Losantinoasta. Weisskreuz eats up vegetation, turning it white.
(source: OSS, undated, Org. Rpt RB-18121.)

Pardubice (Czechoslovakia)
Poison gas factory.
(source: S'I Div., 4th SC. PFI-2, 1 Mar 44.)

Pieve Vergonte (Italy)
Much poison gas is being manufactured at the "Rumianca" factory at Pieve Vergonte (D-5729). It is liquid of the FT type which evaporates when in contact with air.
(source: OSS, 21 Aug 44, Org rpt No. J-2193, B-3.)

Radebuel, Heyden
Representative (Mr. Prosch Jr.) said in Nov. 1938 that his firm was then working night and day on the manufacture of poison gas. Such a strong and penetrating gas that it rendered the gas masks being used by all nations at that time useless. Heyden formerly made chemicals and pharmaceutical products for general commercial use.
(source: 9 SC No. 280, 31 Aug 44.)

Vilbel (near Frankfurt am Main)
In the town of "Jilbel" (probably Vilbel) there is a large chemical factory supposedly making medicinal products. Actually, they make nothing but poison gas. There is only one factory in the town so it is unmistakable.
(source: OSS—SO-1524, 12 Aug 44, British source.)

Zamky (Czechoslovakia), Pyrotechnische und Munitionsfabrik Ing. F. Janecek GmbH.
The company is capitalized at 200,000 kc, with a muni-

tions plant in Prague proper and this pyrotechnical plant at Zamky in the outskirts of the city. Before this plant was engaged in producing rockets, fireworks and signals for military purposes. It is thought that explosives are stored on the premises. The factory is located in a valley off the Moldau River and is hidden in the hills. From the air the plant would show up only as a series of 1-storey wooden sheds. Subject stresses that this is one of the largest gas mask producers in Czechoslovakia, second only to Bata in Zlin.
(source: Mr. Albert Klauber, Speedry Products Corp. 19 Rector St., N.Y.C., Chemical Engineer and Czech citizen. NY MID 12488, 1/14/44.)

Zlin (Czechoslovakia), Bata
In building No. 34 muzzle attachments for gas bomb throwers are being produced. [Note I.W.—the text is unclear and certainly contains some mistake—suggestive of an aerial dispersal system, although from further reading it follows that they were mobile throwers] The gas container is carried on the back; the compressed gas is thrown a distance of 50m [!!!—I.W.] and has a dispersal arc of 28m. The gas cannot be blown back by the wind because it streams out in rotations [see chapter: "Rotational Weapons"]. The kind of gas has not been ascertained.
(source: OSS—A-31218, 1 Jun 44, Czech, London.)

The mention of vortices not being blown back by the wind suggests that they used so-called solitons—donut-shaped vortices, probably like those in the "Windkanone," described earlier in this book (see also: www.zerotoys.com).

Vorarlberg (Austria)
Orders for 500,000 masks for the end of September have been placed with 5 plants at Vorarlberg. This is five times their monthly production. Only 10% of the German population is believed to have masks.
(source: M/A Bern, No. 1736, 9 Sept 44.)

Karlsruhe
Use considered probable of gas called Lewisite which attacks filters and tissues of mask. 800 party members gathered for special instruction course at Karlsruhe on 3 Sept 1944.
(source: M/A Bern, No. 1736, 9 Sept 44.)

Further reports do not give accounts of precise locations, but general preparations for chemical warfare—considerably more interesting![135]

A British broadcast monitored by NBC said Polish Patriots in Warsaw have reported the capture of "gas grenades" from Germans fighting inside the Polish capital. [Comment: SS units actually threw this type of grenade into the sewers insurgents used to escape from surrounded areas of the city. The type of gas remains unknown.].
(source: AP, London. 30 Aug 1944).

The production of gas masks in Germany is currently afforded first priority rating although it had been given third rating in the past. This information comes direct from an official of a plant which is turning out gas masks. The official adds that his plant has increased production 100 percent over the past year. [Comment: clear evidence proving that indeed, as Speer stated, feverish preparations for a chemical warfare offensive were being carried out. One of the following reports contains even more shocking figures!].
(source: OSS, RB 18014, (pt). 22 Aug 1944, F-3, C.)

Hans Lazar, Press Attache at Germany's Madrid Embassy, says Nazis will go down to defeat at as painful and high cost as possible both in internal and external enemies, including chemical warfare.
(source: Cable: Madrid to State—30 Aug 1944. Ambassador Hayes)

Gas mask shortage confirmed. N. J. Temschutzmasken made by N. S. Frauenschaften are now being passed out.
(source: Bern to State Dept. No. 5778, 2 Sept 1944).

Pamphlet called "Kampfstoffisches Werkblatt fuer Aertzte" ["Instruction manual for doctors on chemical warfare"] has been issued to all German physicians in the past few days describing First Aid for war gas victims—and such materials as phosgene acetate, Lewisite, "Lost" and "Gelbring." All communities over specified size are to practise "passive defence" against attack and mass production of gas masks has been ordered.
(source: Cable, Minister Johnson, Stockholm to State, 2 Sept 1944. S.)

According to Italian economic circles, delegates of the German Armament Ministry visited the few chemical factories still operating in Northern Italy and placed orders which are absolutely secret. Informants regard this information related to alleged German preparations for gas warfare.
(source: Telegram from Bern to State Department, 2 Sept 1944. No. 5771.)

On August 25 "Libera Stampa" stated Germans placed with Italian chemical industry large orders of chlorine and activated carbon used in large-scale production of war chemicals and gas filters.
(source: Telegram from Bern to State Department, No. 5592, 26 Aug 1944).

The Turkish Home Service quotes the editor of "Vakit" as writing that Germany will make an all-out effort to

defend her frontiers. It was predicted that when the German war system enters this new phase, different kinds of gases and means of destruction will be used.
(source: FCC Daily Report, 10 July 1944—13 July 1944)
Authorities have ordered all German citizens to attend compulsory lectures on behaviour in case of gas attacks. The lectures are to be given by district chemists in all cities of southern Germany, reports from the German frontier disclosed 11 August.
(source: N.Y. Times, 12 August 1944, Berne, Switzerland.)

The Germans are supplying all factory and other workers with gas masks. Mysterious cases whose contents evidently are associated with gas warfare are being dispatched to the front. Earlier reports said German industry was working on the completion of a priority order for 60,000,000 gas masks. The entire population of Germany should have been equipped by 25 July, but various shortages had delayed the program.
(source: Zurich, teletype, 17 Aug 1944).

Informant, previously reliable, was told by the Director of Chemical Industry that negative reports on future gas war have been made by I.G. Farben and other chemical firms. Although cyanide gas achieved results, it has not been possible for Germany to manufacture superior gas. It is impossible to mass produce complicated valves and containers. Failure also attended experiments with microbes. [Comment: I.W.: this is obvious, deliberate disinformation, aiming to dull the enemy's senses—and direct his attention away from facilities connected with preparations—possible bombing targets! The Germans not only managed to mass produce Tabun and Sarin without difficulty, but by the autumn of 1944 already had sufficient supplies to strike decisively at the Allies—for this involved no "bottleneck," but the availability of strategic delivery systems! Manufacturing containers and valves also presented no problem! Cyanide gas was however never considered a possible chemical weapon—there was no need...].
(source: OWI Official Dispatch, American Legation, Berne, 29 VI 1944).

According to information from the Balkans, the Germans are said to be transporting large quantities of poison gas from the Reich into Hungary with a view to using it against the Soviet Army from which they dread a new and powerful offensive. It is pointed out that the Reich official paper in its edition of June 2, published detailed instructions concerning the transport and handling of poison gas.
(source: FCCL, Radio France, Algiers, 6/20-506A)

It is widely felt in Britain that gas will be used in this war only if extreme rule-or-ruin Nazis prevail in plans for a bitter end resistance.
(source: AP, London, 31 August)

The Daily Mail's Geneva correspondent today quoted reports "circulating freely in the Reich and neutral countries" that the Germans were planning a poison gas offensive as a last ditch measure.
(source: AP, London, 30 August 1944)

In March 1944, POW's unit was instructed in the new Schwarzkreuz (Black Cross) gas [probably Tabun—note I.W.], which is colorless and odourless. The gas penetrates the old type gas mask and new masks were issued.
(source: P/W Branch, No. 1318, 5 Sept 1944)

According to the underground, in about two weeks the Luftwaffe will appear over the German capital (garbled?). This is about the same date that gas warfare on the west front is scheduled [and so the Allies already knew the specific date!—note I.W.]. Information from a well known Geringer manufacturer and a designer of dresses who visited Stockholm recently is author of information similar to above.
(source: OWI, Stockholm USINFO, 2 Sept 1944)

A robot bomb [cruise missile?—I.W.] filled with poison gas may be the German's next secret weapon, according to Norwegian Intelligence Service. Gas mask production has risen greatly in Germany and the people are being made ready for gas warfare.
(source: Moscow, No. 3614, 12 Sept 1944, Minster Johnson).
[Brief author's comment: this brings to mind among others the previously described work by BMW in Munich].

The above embodies all that I have taken the liberty to present—raw, unprocessed information from so-called "primary sources"—striking one however with their certain authenticity. It represents in any case a huge quantity of various reports, from different sources. They together form an entirely different picture of undercover operations from the second half of 1944—completely different from the "textbook" standard. One can clearly see that the world was a step away from a monstrous, chemical apocalypse, which to all extent would have annihilated entire cities. For example: "only" one hundred tonnes of Tabun (hypothetical attack on a single, large city, less than 1% of all resources) represents approximately 1 billion lethal doses and 100 billion incapacitating doses—resulting in blindness.

Very few historians are even aware of the scale of German preparations in this high priority area. Tens of millions of gas masks were manufactured at a time when the raw material rubber was so in deficit that India rubber was even brought by U-boats from the Far East! Without doubt the victims of this more decisive phase of the war would have numbered in the millions! In all probability the change in decision was not due to problems with the chemical weapons themselves, but the lack of their certain and in some way dependable supply to the enemy's rear areas—the lack of a strategic arsenal enabling a mass strike to be carried out on a range of different targets. Work in this field was delayed… In the aforementioned plans distant targets must have been involved, since existing delivery systems would have sufficed for the initiation of this phase—e.g. V-2 rockets. This reminds me of a certain remark made by Professor Mołdawa, who once overheard a conversation in the technical office of Gross-Rosen, where (in the present context) he heard the words: "London, Moscow, New York…". Then, in the second half of 1944, the leaders of the Third Reich came to the conclusion that only by such measures would they succeed in bringing the enemy to his knees. One must concede that even from today's perspective this judgement appears justified…

This also recalls the redeployment of German operatives to the USA in 1944, equipped with … beacon transmitters. In the context of preparations for a strategic offensive (also against the USA) I encountered the argument that the Third Reich did not possess weapon guidance systems of such a range. However this argument is not quite justified. The inertial navigation system found for example in the V-2 warhead could deliver an "object" with inaccuracy the order of tens of metres. Precise guidance in the final section of the flight path could however have been successfully intercepted by a receiver using the signal from a ground transmitter, placed in the city centre by an operative. Warheads existed that could be guided by the signal from a ship's radar over 100 km away. It is worth bearing this in mind concerning the account of Pilsen written later—where work was coordinated on avant-garde propulsion for the German strategic weapon!

But this is just the beginning of new information on the Third Reich's preparations for unleashing chemical warfare!

As I have already mentioned, strategic delivery systems were the deciding factor here, but not the only one. A weapon with which one could carry out strikes on a tactical level would need to be taken into consideration. After the English first edition of "The Truth about the Wunderwaffe" was published, I counted on some response from readers at the time, for information that could help in unravelling certain issues connected with secret weapons of the Third Reich. The response was not great, however a few interesting signals reached me thanks to the publication. The most important actually concerned preparations for chemical warfare—namely the manufacture of rocket-propelled missiles to be launched from submarines.

I entered into correspondence with a Briton whose father had uncovered this trail in 1945 as a serviceman! His name is Keith Sanders and during the long aforementioned correspondence he presented the altogether unusual story of his father. In 1945, he supposedly reached a secret underground factory, under the authority of the SS—at Espelkamp in north-western Germany. It is defined in his recollections as MUNA—simply an abbreviation of ammunition plant. Work there was engaged in filling rockets …namely the "Feuerlilie"—with chemical weap-

British children. In the German strategic concept the chemical war would be aimed against the civilian population like no other.

ons, specifically Tabun! These rockets did not resemble known prototype versions in any way. They had solid propellant engines, as they were intended to be launched from submarine decks—against ground targets. Today there is no way of proving the authenticity of this story, among other things because of the shocking degree of difficulty in gaining access to documents on this subject. My subjective feeling however is that this story is true, in any case Mr. Sanders gives the impression of being a serious and responsible person. I submit this case simply in consideration of my Readership. Ultimately the whole theme of preparations for this "super-total" phase of the war was until recently still cloaked in complete secrecy, thus one shouldn't be surprised by the existence of such unexplained issues. On the other hand however—as I mentioned—the manufacture of poisons themselves must have accompanied work on delivery systems as well as operational plans—of which we still know surprisingly little (although in a further section of this publication I endeavour to at least partially explain these issues).

Keith Sanders also gave account of his post-war "adventures" with the aforementioned rockets, or rather attempts to unravel the secret surrounding the previously unknown "chemical" version, class ship-to-ground:

(3 III 2006)
In 1957 I attended the College of Aeronautics at Cranfield in Bedfordshire for interview and examination intending to study for a post-graduate degree in aircraft design. Following the exams we were let loose in the splendid museum supporting courses. Being then an establishment funded and run by the Ministry of Supply it also housed an incredible array of captured weaponry, rockets and aircraft. To my everlasting shame I carried a 35mm camera and used it on all sorts of aircraft, never a rocket!
"Feuerlilie" was displayed on a trolley at chest height to allow me to touch it and examine the two filler plugs set in stainless steel landings looking like a Porsche filler cap housing today. I now realise the stainless was sil-

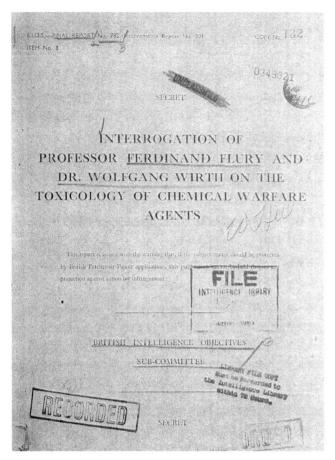

The cover of an allied report on German chemical weapons, quoted in the text.

ver to cope with the aggressive nature of binary Tabun. [Note I.W.: a "binary" chemical warhead is where two substrates or more often—semi-finished products—mix together only upon leaving the launcher, reducing danger during transport and launch. Also eliminated is the essential problem of how less durable agents are manufactured (though in the case of the German arsenal this concerned rather Sarin and Soman—Tabun to a small degree). The missile can be transported with a "full" warhead and the crew made up of draftsmen has the hampered possibility of "ruining" it all or being killed, which could occur for example while filling in the combat zone. Besides, and perhaps first and foremost: even the smallest problems with the air tightness of a normal warhead would threaten catastrophe on a submarine deck, which cannot be excluded due to the continuous change in air pressure! Silver was also used in Dyhernfurth.]
When I researched the story for the military tribunal in 1995 I contacted the head of the department Aeronautics at Cranfield now Cranfield University. That Feuerlilie was sent years ago to The Military College of Science at Shrivenham he said, they in turn assured me it was now with the Aerospace Museum Cosford. Maybe, as they have made a crude mock-up of Feuerlilie on display, confirming to me something has been hidden. One must bear in mind the actions of the Wilson Government in 1966 when all papers relating to this story were destroyed. I have this confirmed by several sources.
By March 1945 Feuerlilie was in its version F165, i.e. a range now of 165 km and a two stage solid fuel rocket. The first outer boosters got the rocket once ejected from the U-Boat launch tube up to altitude 40,000 feet and being winged it was now capable of gliding. The second set of four rockets ignited, the difference these rockets had been developed to pulse burn so providing sustained power to cover 160km at least before the warhead parachute deployed and at altitude the barometric switch closed to detonate the Azide explosive. [Note I.W.: according to Mr Sanders the Germans were convinced that detonation of the conventional explosive destroyed part of the Tabun charge, so they manufactured a special material that emitted less heat]. (...)
Chairman (a barrister) of the Tribunal giving the final verdict in favour of my mother July 1998: "It is very rare in history for an ordinary soldier to have had such a profound effect on the outcome of a war."

There is yet another intriguing theme in the materials sent by Keith Sanders. An enclosure to one of his letters includes a photocopy of a rather surprising article published in the renowned British newspaper "The Times." It concluded that in 1983 the wreck of a German submarine was recovered differing significantly from all known designs! The article's author suggested that a special "evacuation" version was involved (what were they supposed to do off the coast of the USA?), but the account in effect brings to mind associations with post-war submarines—namely ballistic missile carriers![136] Here are some excerpts:

The discovery of an unrecorded German U-boat from the Second World War. Lying sealed and intact in Caribbean waters, gives a new twist to the theory that Goering commissioned nine U-boats as a means of escape for high-ranking officials of the Third Reich. [Note: I.W.: in the case of work on what is conventionally defined as strategic weapons associated with a chemical arsenal, there exists the common, frequently repeating motive—close SS collaboration with the Luftwaffe. But the Luftwaffe never carried out strategic evacuations!!!]. The submarine, which was found by an American salvage operator, has only a brass plaque saying in German: "Hamburg, Germany. Commissioned 1944" There is no visible registration number, and the craft bears no resemblance to any known design during the last war. Nor is there any record in the Imperial War Museum of a German submarine being sunk, scuttled or mined near its location and library staff agree that from sketches the design appears to be unusual." It has been found by Mr Roger Miklos, aged 41, who runs his own business, Nomad Salvage, operating off Florida, and who has spent the past few months scouring German museums for any record of its type.

MISCELLANEOUS REPORTS CONCERNING CHEMICAL WARFARE

Meanwhile a British broadcast monitored by NBC said Polish Patriots in Warwaw have reported the capture of "gas grenades" from Germans fighting inside the Polish Capital.
(AP, London, 30 August 1944.)

The production of gas masks in Germany is currently afforded first priority rating although it had been given third rating in the past. This information comes direct from an official of a plant which is turning out gas masks. This official adds that his plant has increased production 100 percent over the past year.
(OSS, RB 18014, (pt) 22 Aug 44, F-3, C.)

Hans Lazar, Press Attache at Germany's Madrid Embassy, says Nazis will go down to defeat at as painful and high cost as possible both in internal and external enemies, including C/W.
(Cable: Madrid to State - 30 Aug 44. Hayes, Ambassador.)

Gas mask shortage confirmed. N. J. Temschutzmasken made by N. S. Frauenschaften are now being passed out.
(Bern to State Dept. No. 5778, 2 Sept 44.)

Pamphlet called "Kampfstoffisches Werkblatt fuer Aertzte" has been issued to all German physician in the past few days describing First Aid for war gas victims -- and such materials as phosgene acetate, Lewisite. "Lost" and "Gelbring". All communities over specified size are to practive "passive defense" against attack and mass production of gas masks has been ordered.
(Cable, Johnson, Minister, Stockholm to State, 2 Sept 44. S.)

According to Italian Economic circles, delegates of the German Armament Ministry visited the few chemical factories still operating in Northern Italy and placed orders which are absolutely secret. Informants regard this information related alleged German preparations gas warfare.
(Telegram from Bern to State Department 2 Sept 44. No. 5771.)

On 25 August LIBERA STAMPA stated Germans placed with Italian chemical industry large orders of chlorine and activated carbon used in large-scale production of war chemicals and gas filters.
(Telegram from Bern to State Department, No. 5592, 26 Aug 44.)

MISCELLANEOUS REPORTS CONCERNING CHEMICAL WARFARE

The Turkish Home Service quotes the editor of VAKIT as writing that Germany will make an all-out effort to defend her frontiers. It was predicted that when the German war system enters this new phase, different kinds of gases and means of destruction will be used.
(FCC Daily Report, 10 July 44 - 13 July 44)

Authorities have ordered all German citizens to attend compulsory lectures on behavior in case of gas attacks. The lectures are to be given by district chemists in all cities of southern Germany, reports from the German frontier disclosed 11 August.
(N.Y. Times, 12 August 44, Berne, Switzerland.)

The Germans are supplying all factory and other workers with gas masks. Mysterious cases whose contents evidently are associated with gas warfare are being dispatched to the front. Earlier reports said German industry was working on the completion of priority order for 60,000,000 gas masks. The entire population of Germany should have been equipped by 25 July, but various shortages had delayed the program.
(Zurich, teletype, 17 Aug 44.)

Informant, previously reliable, was told by the Director of Chemical Industry that negative reports on future gas war have been made by I. G. Farben and other chemical firms. Although cyanide gas achieved results, it has not been possible for Germany to manufacture superior gas. It is impossible to mass produce complicated valves and containers. Failure also attended experiments with microbes.
(OWI Official Dispatch, American Legation, Berne, 29/6/44.)

According to information from the Balkans, the Germans are said to be transporting large quantities of poison gas from the Reich into Hungary with a view to using it against the Soviet Army from which they dread a new and powerful offensive. It is pointed out that the Reich official paper in its edition of June 2, published detailed instructions concerning the transport and handling of poison gas.
(FCCL, Radio France, Algiers, 6/20-506A.)

It is widely felt in Britian that gas will be used in this war only if extreme rule-or-ruin Nazis prevail in plans for a bitter end resistance.
(AP, London, 31 August.)

The DAILY MAIL'S Genevia correspondent today quoted reports "circulating freely in the Reich and neutral countries" that the Germans were planning a poison gas offensive as a last ditch measure.
(AP, London, 30 August 1944.)

CONFIDENTIAL

Some of the documents quoted in this chapter

He said: "At first my only interest was in the mercury that was aboard for ballast" [Note I.W.: how did he know it was there and then consider it ballast? There existed must cheaper and safer materials. U-boats never embarked upon such risky voyages just to transport ballast!] But after several dives and some research, he realised the submarine differed from the usual fighting class. He says the U-boat is of the type VII-C class which was extensively modified during the war. But none, he maintains, has the features of this one: extremely large tail fins, a conning tower positioned well forward of mid-ship and linked to the bow with a large, reinforced jagged ripping bar. The 250 ft craft, which weighs some 200 tonnes, is 80 ft down but hidden from the surface by a reef which forms a shell over it, and by the breaking foam. It is perfectly preserved. Mr Miklos says, and with no damage, because of the peculiar non-corrosive properties of the waters it is in. He is convinced the U-Boat is sealed, with the crew and 18 passengers inside [why so many?—I.W.]. The torpedo, deck and conning tower hatches are all closed and in the locked position, he says, and sonar tests show that the escape chamber has air in it.(...)

Admittedly this text in reality explains little, but at least shows that hitherto unknown secrets of World War II are still being discovered!

The last of the sources, which contributes some new information to Third Reich preparations for this new phase of the war, is a British Intelligence report on "toxicology," in words: chemical arsenal toxicity[137]. It is based on the interrogations of several leading scientists in this field. However it does not appear so interesting at first glance. The interrogated experts, not confronted with any evidence (German operational plans and their research results were not preserved) shirked carefully from any association with something that could be treated as yet another crime against humanity. They would have in effect only incriminated themselves… They stated that they had had nothing to do with phosphor-organic agents and when questioned about human experimentation it goes without saying kept their mouths firmly shut and threw up their hands in despair—such attitude is broadly outlined in comments from the report's authors. Despite this it is worth familiarizing oneself with at least a summary of the report. It does actually contain some useful information:[137]

This BIOS trip was primarily laid on to interrogate Professor Ferdinand Flury, formerly of the Pharmacological Institute at the University of Würzburg, who was known to have worked both on the physiological effects of the war gases and on the general principles of industrial toxicology and hygiene [brief note: I.W.: the notion of "war gases" should obviously be treated as strongly conventional, in reality almost none of this type of agent was a gas under normal conditions!]. The latter subject is being dealt with in a separate report, graded "Unclassified," so as to enable its circulation to industry [I.W.: so it is obvious that the Allies were simply concerned with repeating the "success" of the Germans!].

In addition to Flury, the team was afforded an opportunity of interrogating Dr. Wolfgang Wirth, head of Group VII of Wa. Prüf. 9 [the equivalent Wehrmacht research post—I.W.] and of Sanitäts Inspektion, as well as being head of the Toxicological and Therapeutical Section of the Military Academy of Medicine in Berlin.

Flury was still in very poor health and, in the absence of his files, was unable to recall more than the barest outline of his work [!!!—I.W.].

He stated, and it was subsequently confirmed, that he had never worked with the "nerve gases," and the main bulk of his work of interest to the present investigation had consisted of a large number of screening tests ("Vo-

---

```
          MISCELLANEOUS REPORTS CONCERNING CHEMICAL WARFARE

          In March 1944, P/W's unit was instructed in the new Schwarzkreuz (Black Cross) gas, which is
colorless and odorless.  The gas penetrates the old type gas mask and new masks were issued.
          (frm P/W Branch, No. 1318, 5/9/44.)

          Libera Stampa, Lugano, August 24, from Chiasso.  "German authorities have ordered Italian
industries to deliver large quantities of chlorine which will supposedly be used for the production of
poisongas."

          According to the underground, in about two weeks the Luftwaffe will appear over the German
capital.  (garbled?)  This is about the same date that gas warfare on the west front is scheduled.
Information from a well known Geringer manufacturer and a designer of dresses who visited Stockholm re-
cently is author of information similar to above.
          (OWI, Stockholm USINFO, 2 Sept 44.)

          A robot bomb filled with poison gas may be the German's next secret weapon, according to
Norwegian Intelligence Service.  Gas mask production has risen greatly in Germany and the people are being
made ready for gas warfare.
          (Moscow, No. 3614, 12 Sept 44.  Johnson, Minister.)
```

One of the documents shedding a new light on the last phase of the war

runtersuchungen") on candidate chemical warfare agents carried out under a series of "extra-mural" contracts with the Heereswaffenamt (HWA). Several weeks after the visit, his files were made available for study, and a list of the compounds screened in his laboratory is attached as Appendix III. The compounds, with the exceptions of Excelsior referred to in more detail in other reports, and some of the already standardized agents, were of no value. In most cases no attempt was made to determine even the approximate MLD [Mean Lethal Dose—causing death in 50% of test organisms—I.W.] (…)

Enquiries regarding tests on humans

As was perhaps to be expected, none of the personnel interviewed admitted to any direct or indirect knowledge of tests of C.W. agents on humans other than the so-called "subjective tests," i.e. determination of limits of detection by smell etc. and the numerous skin tests of candidate vesicants or agents for the treatment of vesicant contamination of the skin. It was stated that no chamber experiments designed to measure skin-burning dosages directly had been carried out, and that tests in this field had been limited to arm exposures or the like. Under repeated questioning, estimates for the human L(Ct) 50's for AC and Tabun of 2,000 and 2,300 mg.min/cu.m respectively were obtained; it may be doubted if these figures are in fact anything other than what they purport to be, that is, guesses.

So far as the groups directly controlled by Wa Prüf 9 are concerned, it seems quite likely that no actual determinations of human median lethal dosages were made. Wa Prüf 9, indeed, appears to have been strangely unwilling to carry our tests on humans even where the risk of death or serious injury was negligible; for example, there appear to have been no exposures of unprotected humans to varying Ct's of the various vesicants. (vide C.O.I.S. Rept. XXI—86, dealing with Raubkammer; independent evidence to the same effect was obtained on this trip). Furthermore, there is clear evidence that the Government departments concerned were on the whole opposed to such tests (vide C.I.O.S. Rept. XXVI—37, page 25, which quotes a latter from Himmler protesting against this attitude), and considered that any individuals taking part in them would expose themselves to the risk of being charged with murder under the existing German law. (vide C.I.O.S. Rept. XXVIII—50, page 34). For these reasons, it was deemed essential that work of this sort should be confined to organisations under 100% SS control. [Comment I.W.: an analogical mechanism operated as in the case of biological weapons—see relevant chapter. In effect control over a large section of the weapons of mass destruction arsenal, potentially decisive for the war, was "thrust" into the hands of the SS—not the outcome of some internal struggle or scheme of Himmler, but at the request of Wehrmacht Generals themselves!!! Confronted with this it is not surprising that the SS became the chief advocate of using such weapons, all the more that they already controlled certain key resources—concentration camps and underground factories. And vice versa—the reluctance of certain circles to invoking chemical warfare could have had its origins in fears that the SS would have been promoted to preeminent master! Besides this was surely how it was in Speer's case.].

At the same time, the tendency to effusiveness and evasiveness on the part of Professor Wirth on this subject gave rise (in the absence of knowledge of his character) to the suspicion that he knew more about this than he was (for reasons not unconnected with the Nuremburg trial) going to admit. But, as was verified during this trip, rumours even about top secret matters seem to have circulated amongst the senior personnel (for example Flury learnt of the existence of Tabun etc., at a time when he was definitely not entitled to know anything whatsoever about it) and it therefore seems highly unlikely that someone in the rather central position of Wirth could have failed to catch some rumours at least of the executions by gas, which appear to have been quite numerous. For this reason, it is a matter for serious consideration whether further pressure should not be placed on Professor Wolfgang Wirth, to see if he can be "encouraged" to recall more information, perhaps even the location of relevant documents—as has occurred in several similar but unrelated cases; alternatively Generalarzt Asal, Chief of the Militärarztliche Akademie might be questioned.

The impression gained from the reports of previous investigators was that the individual most likely to have some information on this subject, and who is currently available, is Professor Richard Kuhn of Heidelberg. He held a responsible consultative position as regards the party research organisations, apparently as an advisor to Dr. Osenberg, and was therefore probably in an unusually good position to speak on SS-sponsored research throughout Germany. Furthermore he was personally interested in C.W. problems; he took up the study of the action of Tabun, Sarin etc on enzymes in 1944, and was led by this investigation to the discovery of Soman. But it must be recorded that interrogation of Kuhn, subsequent to the investigations reported here, brought no such data to light.

Development of the Nerve Gas series

An attempt was made to secure a fairly detailed history of the development of the nerve gas series, partly in the hope that any theoretical considerations which might have guided the German work would be brought to light, and partly to secure additional background information to facilitate enquiries as to whether this series of agents had ever been tested on human subjects.

Very little arose during the various interrogations which had not already been elicited by previous investigators and put on record in the numerous reports which have appeared on this subject during the past few months. It

Part of the British preparations for chemical war. They were based on experiences from World War I. In fact, in a confrontation involving tabun or sarin, the countermeasures visible here would be worthless.

was disappointing (though by no means unexpected) to find that no one questioned had any ideas of a theoretical nature to offer concerning the constitutional factors determining the pharmacological activity of these agents; in particular, Schräder [considered the discoverer of Tabun—footnote I.W.] contended that at the present stage there was no sound basis for any theory designed to predict the degree, or even the nature of the activity of any new compound in this series, and he insisted, not without pride, that his method was simply to prepare every possible variant of the original compounds and to discard only those lines which involved uneconomic yields, or needed intermediates which were never likely to become available in the required quantities. This procedure had been laborious, entailing some 600 syntheses, mostly new, but it must be conceded that it had not proved unprofitable. (...)

It is clear that the original decision to adopt Tabun as a standard agent was taken in the summer of 1939 when the I.G. were asked to draw up estimates for the construction of a 1,000 ton/month plant [i.e. Dyhernfurth—I.W.] These estimates were provided verbally at a meeting in September 1939, further discussed at a meeting between I.G., Wa Prüf 9, and Wa J. Rü (Mun 3) on November 7, 1939, and formally accepted in December, during which month the advance "order to proceed" was issued by Wa J. Rü 9/IX, under reference 9/IX-240-9018-39 dated December 18, 1939.

Most SS operations involving the killing of inmates of concentration camps, institutions for incurables or the insane, etc, under various euphonious titles, seem to have started well after the outbreak of war, and it therefore seems more likely that the original Wa Prüf 9 recommendation in favour of Tabun was not supported by quantitative experiments on humans. A good deal of non-quantitative data, arising from accidental exposure of the workers concerned, must have been available even in 1939, (though no deaths appear to have occurred before the commencement of large scale production in 1942) and it seems probable that these experiences, combined with the absence of any indication of marked variation of sensitivity with species...sufficed for the 1939 decision (...). It is known from German documents that a total of 324 accident cases (mostly of a minor order with no fatalities) were encountered up to the end of 1941 and the question therefore is: was this, together with the animal data, sufficient to confirm the German decision to embark on large-scale production in May 1942? A total of 10 fatal cases occurred after production got under way, so that if no experiments were carried out on humans, and the preceding summary is correct these would have been the first deaths caused by any agent in this series. But taking into account the Himmler letter referred to above, it does seem to be a matter for serious doubt whether the higher Party organisations would have agreed to the diversion of considerable effort, in difficult circumstances, to the production of a chemical warfare agent which had not been shown unequivocally to be capable of killing men; it will be realised that this point carries the more weight in that Tabun was not being offered to the Services as a "harassing" agent, but as a quick-acting lethal agent. (...)

Concluding this chapter, I again encourage readers to familiarize themselves with "Part Three," since the whole story of the German super weapon presented there closely matches plans to unleash chemical warfare on a strategic scale. Let us take for example a secondary, seemingly enigmatic excerpt, the report of a witness who overheard some SS men talking: "only those in the forests or high in the mountains will survive, for no filter in any shelter will stop Tabun."

BIOLOGICAL WEAPONS

Little is currently known about German preparations for biological warfare, although there is no doubt that this type of weapon of mass destruction was incomparably less developed than e.g. chemical weapons.

It was never prepared for use on the battlefield, although towards the end of the war various plans were made—e.g. it was probably intended to attack New York in this way, by launching missiles from submarines (the so-called operation "Elster").

German research and development posts worked mainly upon bacteria, viruses and funghi transmitted by animals, at the same time work on the use of anthrax, botulin (bacteria growing in rotten meat) and parrot fever viruses was at an advanced stage. It is known that biological weapons were tested among other things on female prisoners from the Ravensbrück concentration camp.

Anthrax was probably the most dangerous, first and foremost due to a very large and often long-term pathogen vitality, mortality rate close to 100% and rapid action—death follows several days after infection.

Botulin is somewhat less dangerous than anthrax. In the event of infection it results in death in 1-10 days, but the mortality rate is lower—the order of 80%. In the event of survival however up to 3-4 months must elapse before full recovery is reached.

The Germans came relatively close to using them—a spilling system was prepared for mounting on aircraft, which were to disperse the bacteria in the form of an aerosol. This would have been one of the most destructive weapons of World War II —theoretically 1 gram of botulin is enough to result in the death of 5-8 million people.

It was predicted that the so-called parrot fever viruses would have been much more effective, though considerably more difficult to "mass produce." They are characterised by a lower mortality rate but are quite vital and with one gram it is possible to infect up to 20 million people. So as one can see biological weapons don't have to yield precedence at all to nuclear weapons.

In the Third Reich work was carried out on other germs as well (among others pathogens and funghi causing diseases in plants and animals), but information on this subject is still quite scant. But the Germans for sure were not pioneers in this field, the first to use biological weapons being the Japanese—in 1939 in Mongolia, somewhat later the British acted similarly in battles for the Malakka peninsula in Indochina.

Laboratories working on biological weapons—either small or large existed in all countries that were involved in the war: in the USA a centre was opened in 1942 in Camp Detrick (in the state of Maryland), in Great Britain a special "research station" came into existence in 1940—in the town of Porton, and in the Soviet Union the organisation of work in this field was commenced as early as 1938.

The Germans carried out work on biological weapons in strict secrecy, in connection with this it is difficult to find any specific information on this subject. At the beginning of the 1970s the problems associated with this issue were engaged with in Poland by the Chief Commission for Investigating Nazi War Crimes, and specifically by Dr. Rafał Fuks, co-operating with it. A by product of his work was a cycle of articles, unravelling a presently little known aspect of the German biological programme.[121]

From these materials it follows that as early as the end of 1943 preparations for the production and mass employment of biological weapons had been put into effect on a wide scale. Various institutions participated in the realisation of this plan, first and foremost the Reich's Scientific Research Council and the Institute for Military-Scientific Research—subordinated to the pseudo-scientific organisation "Deutsches Ahnenerbe," of the SS.

Initially Dr. Leonard Conti was to direct the programme, Secretary of State in the Interior Ministry and Director of the Office for Social Health of the NSDAP. Eventually however the aforementioned function was entrusted to his deputy, Professor Kurt Blome, at that time Chief of the National Socialist Association of Doctors. Officially he was Plenipotentiary of the Reich's Scientific Research Council for research in fighting cancer, however this was only a convenient cover-up. In May 1943, one way or another the true character of Professor Blome's mission became clear—when Goering nominated him "Plenipotentiary for Biological Warfare" ("Bevollmächtiger für biologische Kriegsführung").

This position wasn't a result of the need to commence preparations—since these were already in progress, but the necessity

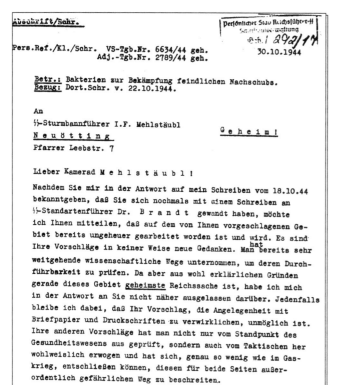

A selected document from the Personal Staff archive of the Reichsführer SS, confirming the standing of work on biological weapons

of co-ordinating the activity of different institutions independent of each other. Employed in this area were among others:
- The Scientific Division of the Armed Forces' Supreme Command (Wehrwissenschaft in OKW/We-Wi).
- The Ground Forces' Sanitary Inspectorate (Heeres Sanitätsinspektion) as well as:
- The Ground Forces' Veterinary Inspectorate (Heeres Veterinärinspektion).

The circle focused on Kurt Blome included needless to say supporters of using biological weapons. Despite the existence of a specific lobby, this zeal was pushed aside for quite a long time to the margins of the Third Reich's armament effort. A change in the situation was evoked by a secret intelligence report, which arrived in September 1943. It described a meeting of American and Soviet biologists, which was to take place that very month in Cairo. It followed from it that the Allies were co-ordinating preparations for a "biological offensive" against the Germans.

As a result at one of the councils, on September 25, 1943, a staff doctor from the Ground Forces' Sanitary Inspectorate along with a consultant of the Heereswaffenamt, Professor Kliewe, mentioned:

We shouldn't look on in a heedless way, but also make considerable efforts in the field of employing biological agents on a mass scale. First of all America should be attacked with various pathogens, both people as well as animals. It is necessary to win Adolf Hitler's support for this plan.

According to settlements established at this council, the chief means of transport was to be the bomber air force and in the case of the USA—a fleet of submarines. These formations were to accommodate their equipment for the new role accordingly.

In September 1943, the Wehrmacht took over supervision of the whole programme. At the Armament Office of the Wehrmacht (Waffenamt-Wa-Prüf-9) a special committee was called into being, with the task of co-ordinating research regarding the most promising armament programmes, including those connected with biological weapons. It acquired the code-name "Blitzableiter-Ausschuss." At the forefront of the committee stood the chief of a department in the Armament Office, Col. Hirsch, and it included Prof. Blome, Prof. Kliewe, Dr. Standien, Dr. Nagel from the Ground Forces' Veterinary Inspectorate, Prof. Gerhard Rose, a Luftwaffe consultant in the field of tropical diseases, as well as a counter-intelligence officer keeping watch over the security of the work's secrecy. The insufficient amount of data concerning the propagation of pathogenic factors and their infectiousness etc was considered a fundamental problem. In a report from the first sitting of the "Blitzableiter-Ausschuss" among others was recorded:

"Since it hadn't been determined so far if atomised particles cause illnesses in people and if so under what conditions, Prof. Blome proposed that experiments be carried out on humans."

It was recognised that an obstacle in gaining experience in this field was constituted by the lack of a specialised research institution, adequately equipped and applied—among others in conducting experiments on people. But this aroused the opposition of the Wehrmacht, Field Marshal Keitel affirming that the Army shouldn't participate in this type of research. In the meantime months had passed since the last and still the only "Blitzableiter-Ausschuss" council. Given this situation the SS took the initiative, which Blome himself had been striving for, for a certain time. Reichsführer-SS Heinrich Himmler was not only an enthusiast of biological weapons, but he had been conducting research for a long time and already had definite results (among others research on typhus fever at Buchenwald concentration camp). Himmler demanded that the speed of work be accelerated. So without any problems he approved the project to build the appropriate institute and as opposed to Keitel, intended to commence experiments as fast as possible. Work associated with the institute's foundation was commenced by the Wehrmacht as early as 1943, but was speeded up only after the SS had taken over control. It arose in occupied Poland, in the town of Pokrzywno (Nesselstadt) –presently suburb of Poznań, on the grounds of a former Educational Institution of the Ursuline Sisters. Existing buildings were exploited and the construction of new ones was commenced. Everything was surrounded by a two metre high wall, and bunkers were situated in its corners with gun openings. The Gauleiter of the "Wartheland," Greiser, was vividly interested in the construction and at Himmler's recommendation also the Higher SS and Police Führer (HSSPF) in this region, SS-Obergruppenführer Wilhelm Koppe. In addition Himmler sent his representative, Kurt Gross, permanently to Pokrzywno, who was later to supervise experiments on people with plague pathogens.

Obviously efforts were made to keep everything in strict secrecy, but a lack of consistency relying on the employment of over one hundred Polish workers on the construction site only resulted in the SS guard and high walls attracting the enemy's attention. The Poles quickly conjectured the true purpose of the facility. For example one of the buildings was officially named an experimental station for animals (Tierversuchsstation). It consisted of an undressing room, from which a passage led to some bathrooms with bathtubs and showers. A successive passage led to five toilets, designated on the plan as a stall for animals. Access to this section was sealed by armoured doors and a grid. One of the walls in the "stall" was made of thick, bullet-proof glass. Behind it was situated a room for personnel. In addition tables in the autopsy room clearly revealed that it wasn't animals that were involved. Other features in turn threw light on the purpose of the complex, e.g. the complete airtightness (sealed hermetically) of one of the buildings, the existence of disinfection chambers adapted for temperatures of 80-120°C, poison gas chambers (Vergiftungskammer) and crematoriums ...

The entire complex was only finished in October 1944, when there already existed the real threat of it being captured by the Russians. Directly after the completion of work, 115 Polish workers were transferred therefore deep into Germany—to Thuringia, where they were to raise an identical post. Despite the efforts of Himmler and the support of the

Pokrzywno, fragment of the wall surrounding the former laboratory

Pokrzywno, armoured door in one of the buildings

biological programme by the Waffen-SS' Institute of Hygiene and the "Deutsches Ahnenerbe" organisation (concentration camps), research was barely begun. They concerned mainly the pathogens of malaria and plague. After the Soviet January offensive there was no longer any possibility of carrying the work through to completion. This is how Prof. Walter Schreiber from the Military Academy of Medicine described this episode:[121]

In March 1945 I was visited in the Military Academy by Prof. Blome from Poznań. Exasperated he asked me to accommodate his team of co-workers as well as make it possible for them to continue experiments in laboratories in Sachsenburg. The rapid movement of the Red Army had forced them to leave their Institute near Poznań. Blome's worry was that they hadn't managed to blow up the Institute on time. It was possible to easily recognise that the equipment located there had been used for conducting experiments on people. They had even considered bombing the Institute, but had come to the conclusion that this was no longer possible. Prof. Blome asked me to enable him to carry out experiments in Sachsenburg using plague cultures, which he had managed to save.

NUCLEAR WEAPONS

Contrary to my own expectations, this chapter is completely different from the rest of the book. Perhaps this will surprise many readers, but I will not describe the history of German work on the breaking of the atom, since in available publications there are so many missing elements and even contradictions, that a consistent representation is difficult.

Initially I intended to begin this chapter with the statement that what one usually understands by calling to mind the "German nuclear programme" slogan, in reality had very little in common with work on an atom bomb. After all, the resources designated for this objective (which we know) were very modest in comparison e.g. to the USA; in contrast to this country the Germans never passed to the industrial phase. In short, any mention of a German atom bomb seems to be a misunderstanding and suggests an unawareness of the basic facts, mainly of the fact, that there was no equivalent of the American "Manhattan Project."

It was through such optics that I saw this aspect of history, before I began to collect and analyse materials on my own. When it had already come to this I found that I would never take full responsibility for the above statement—my optics had been subject to change... Instead of writing on what the "nuclear programme" was, I decided to write rather on what it was not and to clarify some common misunderstandings.

Above all, one should take into account that German research concerning nuclear technology was carried out by many **independent groups** of scientists, acting within the confines of various institutions (from particular institutes right up to the Post Ministry). A reflection of this was the large number of laboratories and research establishments, scattered all over the Reich. The problem lies among others in the fact that for certain we don't know about many of them and because of this, we do not know the full picture of German work, and certainly will never do so. There are simply too many of these "blank spots."

From talks which I once conducted with people who had been analysing this problem for many years on the basis of intelligence materials, I recall that the town of Torgau fulfilled a very important role in the German programme, where in 1944 in all probability a plant for enriching fissionable materials was constructed. It was "legended" as a water purification station. This issue finds in the meantime no reflection whatsoever in contemporary publications.

Similarly nobody mentions the role of the underground facility in Książ (Fürstenstein bei Waldenburg), although on the German plan the designation for that time for fissionable materials appeared—three circles overlying each other.

Completely omitted is the role of the nuclear research laboratory in nearby Kowary (Schmiedeberg), where an electrolytic installation for the production of heavy water was built at the end of the war and in a nearby underground facility lead plates 20-cm in thickness were found (!) as well as a number of tiny pipes, probably made of cesium.

A similar, unexplained element is contained in a report from Polish Home Army Intelligence reproduced in part three. It suggests associations with some form of nuclear weapon—the term "lead chambers" appears, for assembly of some unspecified devices (mention of production carried out in the "Mittelwerk" facility in Thuringia even before assembly of the V-2 was commenced).

Another case: in 1995 the periodical "Przegląd Techniczny" ("Technical Review") reproduced a list of post-German facilities, located on Polish territory, mainly underground.[124] This was probably a document from the Ministry of Industry and Trade. It dates from 1953. Apart from such curiosities like e.g. the description of an underground facility, in which "lighting was arranged with the aid of phosphorizing walls," mention was made of: "An underground ammunition factory, in which the Germans conducted experiments on **atomic weapons**."

This is position no. 42.

In a column referring to the location was written: "administrative district Nowogard, town Marty (Sobótka)." Nowogard suggests the region of Szczecin, but I was unable to find either the town of Marty or Sobótka in this region—in 1953 many unofficial names still functioned, sometimes being changed several times. This case may be a good example of how difficult the unravelling of such mysteries may be. Later on I managed to clarify that the town of "Marty" never existed, but the name "Sobótka" was in use for a short period after the war, with regard to the present day village of "Mosty" (previously: Speck). The underground facility is however completely inaccessible—being flooded.

A similarly secret nuclear research laboratory was mentioned in the files of the so-called operation "Lusty," described further on.[125] It was destroyed shortly before the end of the war and located in the town of Linnessrabe or Linnesgrabe (one letter is almost illegible). Short description of this "target" suggests work on thermonuclear fusion—there is a mention of "20 MV deuterons cyclotron." In the report it was emphasised only that a "fierce" officer would be needed to interrogate possible personnel. This laboratory is also not described in any generally available materials and it is unknown what activity was taking place in it. Who can assure that there were not significantly more unknown elements in the German nuclear "programme"?

Of course all of this doesn't yet have to necessarily mean that the results of this "programme" were different from what is universally considered, i.e. that work on the bomb was completed. It doesn't have to—although in one of the reports a description is found … of a German nuclear explosion. It concerns an intelligence summary prepared by the command of the American Strategic Air Forces in Europe.[126] At that time this report was regarded as top secret. The description of the explosion comes from a German pilot, a specialist on anti-aircraft rockets, by the name of Zinsser. He was admittedly the only "available" witness, although his description is quite detailed and fully conforms to the characteristics of an authentic explosion. The report bears the date of August 19, 1945, in other words it was made almost two weeks after the explosion over Hiroshima, although the report itself was for sure written down before August 19. Anyway Zinsser couldn't have known all the details of the "Japanese" explosion—even after August 6—after all he couldn't have seen it on satellite television, and these matters were clouded in secrecy. Here is his report:

At the beginning of October 1944 I flew from Ludwigslust (South of Luebeck), about 12 to 15 km from an atomic bomb test station [?! I.W.], when I noticed a strong, bright illumination of the whole atmosphere, lasting about 2 seconds. [the given time of existence of the fire-ball corresponds to a 5-10 kiloton explosion, note I.W.].
The clearly visible pressure wave escaped the approaching and following cloud formed by the explosion. This wave had a diameter of about 1 km when it became visible and the colour of the cloud changed frequently. It became dotted after a short period of darkness with all sorts of light spots, which were, in contrast to normal explosions, of a pale blue colour. After about 10 seconds the sharp outlines of the explosion cloud disappeared, then the cloud began to take on a lighter colour against the sky covered with a grey overcast. The diameter of the still visible pressure wave was at least 9,000 metres while remaining visible for at least 15 seconds.
Personal observations of the colours of the explosion cloud found an almost blue-violet shade. During this manifestation reddish-coloured rims were to be seen, changing to a dirty-like shade in very rapid succession. The combustion was slightly felt from my observation plane in the form of pulling and pushing. The appearance of atmospheric

Fragment of documentation regarding the "Operation Lusty" with a short description of one of the nuclear laboratories.

disturbance lasted about 10 seconds without noticeable climax.
About one hour later I started with an He-111 from the A/D at Ludwigslust and flew in an easterly direction. Shortly after the start I passed through the almost complete overcast (between 3,000 and 4,000 metres altitude). A cloud shaped like a mushroom with turbulent, billowing sections (at about 7,000 metres altitude) stood, without any seeming connections [to this place? I.W.], over the spot where the explosion took place. Strong electrical disturbances and the impossibility to continue radio communication as by lightning, turned up.
Because of the P-38s operating in the area Wittenberg-Merseburg I had to turn to the north but observed a better visibility at the bottom of the cloud where the explosion occurred. Note: It does not seem very clear to me why these experiments took place in such crowded areas.

I am inclined to consider that the above description is a misunderstanding, however let us not delude ourselves—we do not know everything about German nuclear physics from the time of the war.

By the way, in 1995 the newspaper "Krasnaya Zviezda," the official press organ of the Russian Ministry of Defence, revealed the results of an investigation carried out shortly after the war by their intelligence. According to it, the Americans had found in an unspecified Bavarian mine a prototype of the German atomic bomb. However let us treat this just as a curiosity, I still do not think that the Germans "made it."

However, what I think has to be emphasised, is that our "vision" of their "programme" is very incomplete, mainly because there never was a unified, centrally controlled "programme." There are a lot of elements that hardly match the rest. A frequent rule in such cases is that there are several "sub-worlds" that don't necessarily know about each other. This is the first association that crosses my mind after realising that there were several laboratories, previously completely omitted.

Apart from this one has to realise that the term "nuclear physics" is very capacious and includes domains not having anything in common with bombs—e.g. the project described in the third part of this book also fully deserves the term "nuclear." We know, that nuclear research laboratories in command of particle accelerators (betatrons, cyclotrons…) were used to a large extent in research on "death rays." Some consumed possessed stocks of radioactive isotopes for the production of fluorescent paints, which were used to paint the needles of clock-indicators in aircraft, and cover the aforementioned "phosphorizing walls."

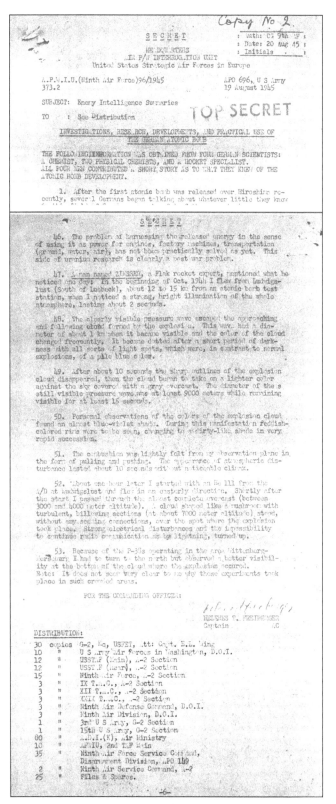

Fragment of the U.S. report cited in the text

Research was also carried out in the field of controlled thermonuclear fusion (doomed to fail)—e.g. in Prague and in the Berlin' Kaiser Wilhelm Institute, continued after the war on the Argentine island of Huemul.

Therefore one cannot exclude that what used to be labelled the "German nuclear programme" was in reality a "conglomerate" of completely different research.

The fact that there were many more nuclear research laboratories, than appear in books concerning the "mainstream" of German nuclear research, and that these groups do not appear to be linked to each other, isn't at all a manifestation of organisational chaos—as some might suggest, but rather the display of a sophisticated management system over the vital branches of science. It proved to be so effective, that on the one hand even today the situation from 1944 appears hard to embrace, and on the other hand this very system became a blueprint for the Americans (to a lesser degree for others too…), when they organised the management of their own black projects. This is called compartmentalisation and means that there are very strict divisions between particular segments within a certain branch—i.e. existing for the same purpose or related purposes. The Germans exercised this problem rather sharply, among others because so many forced labourers, prisoners or companies from the subjugated countries had been incorporated into the war economy. That is why they worked out such a system, which in some respects fulfils its role very well even today.

That is why so relatively many places or facilities are either known only superficially, unknown, or their role is misunderstood. Apart from the several places listed above, I will give yet one more example of how compartmentalisation worked.

Well known is the story, repeated in countless publications, of the tireless German efforts to build a nuclear reactor operating on heavy water. Well known, but ultimately insignificant. A story regarding the almost simultaneous attempts to construct a reactor operating on graphite, somehow evades attention, although the decision to build a full-size reactor of this type was taken as early as 1941. The main challenge however was the production of suitable graphite blocks. This task had been commissioned to a "carbon electrodes' factory" in the occupied Polish town of Racibórz (then the "Plania Werke" of Siemens in Ratibor). In the last weeks of 1941, Professors Bothe and Harteck ordered one hundred unusual, thick rods, each 3 m long and 0.6 m wide, with a very strong emphasis on the purity of this material. It happened however, that the chief technologist of this plant was… an anti-Nazi, a fact extremely unlikely, if not almost impossible in the Third Reich, at least at that time. For him the strange specification seemed very suspicious and he knowingly used a raw material contaminated by pyrite as well as by the compounds of calcium and sulphur. This very professional sabotage, committed by Erwin Schmidt, because such was his name, went unnoticed and the Professors withdrew from the graphite track for two and a half years. This was the very moment, when they lost superiority over the Europeans working in America—a turning point in German nuclear research and one of the most crucial moments in the entire war. Yet the rules of "compartmentalisation" provided that this "episode," along with all other elements of that part of the puzzle (the project itself), just sank in the deep sea of information noise, as if it had never existed. That is why some of the blank areas, in what has been left of one of the most secret domains of Nazi war science, will never be solved.

GERMAN PROJECTS IN LIGHT OF THE AMERICAN DRAINAGE OF TECHNOLOGY

(OPERATIONS "PAPERCLIP" AND "LUSTY")

At the end of 2001, I had the rare opportunity to familiarise myself with the archives of the U.S. National Air Intelligence Center at Wright-Patterson Air Force Base, where after the war was located one of the centres where captured German documentation and copies of the most interesting constructions were transported to (then it was Wright Field base). One should after all know, that American Intelligence didn't plunder just anything. The tasty morsels alone were brought over, with greed and marvel, which constituted the best certificate of German technical achievements during the war. Obviously others did the same…

At the aforementioned base were found among other things files, accompanying the technical know-how drainage of operation "Paperclip," which I made use of.[127] This operation relied on bringing the most outstanding experts of the Third Reich to the USA, those connected with aerial technology found their way to Wright-Patterson, near the town of Dayton in the state of Ohio. They were the strict elite of German and Austrian science and technology—people at the highest level.

American General Joseph T. McNarney, Supreme Commander of Air Materiel Command, wrote on February 3, 1948, that the German experts had a significant contribution in research and development programmes carried into effect over the ocean and have permitted "the very critical shortage of highly qualified scientific and technical personnel" to be alleviated. On February 27 that year, the Secretary of the Air Force defined operation "Paperclip" as a "complete revelation." It was estimated that American scientific and technical progress thanks to this operation alone was accelerated by 2-10 years, and savings in the rocket programme alone were estimated at a "minimum of 750,000,000 dollars."[127] The Americans certainly didn't realise, or didn't want to realise, that thereby they were deriving profit from the blood split by forced workers, concentration camp prisoners as well as patents robbed by the Germans in the subjugated Europe…

By December 4, 1946, 270 experts had been brought in, and under the influence of the first assessments of their work, it was decided to recruit another 700 people. At Wright-Patterson air base at the turn of 1946-1947, worked around 100 people. They received wages at a level of 2.70 to 11 dollars daily—by definition lower than civil servants in the American administration. Within the confines of this operation 1,200 tons of technical documentation were seized in the Reich, from which after selection 150 tons were sent to the USA. The Americans conveniently turned a blind eye to the fact that on very many of the documents the words "Heil Hitler" were visible above the authors' signatures! The documentation as well as the people were later concentrated mainly at three American bases—

A German He 177 heavy bomber,
after being transported to the USA in 1945

The Focke-Wulf Fw 190, captured by the Americans

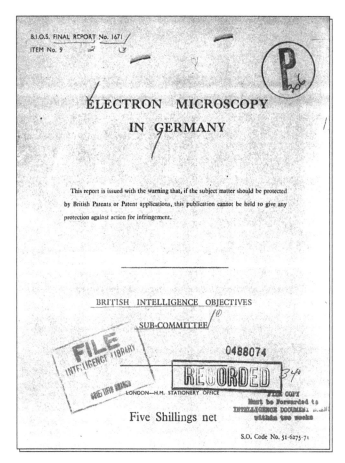
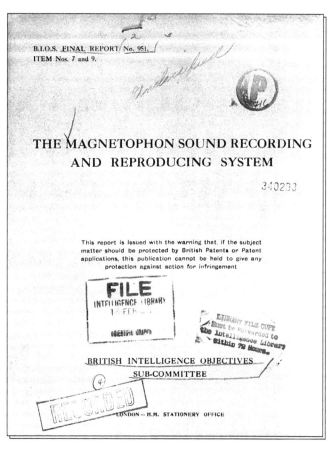

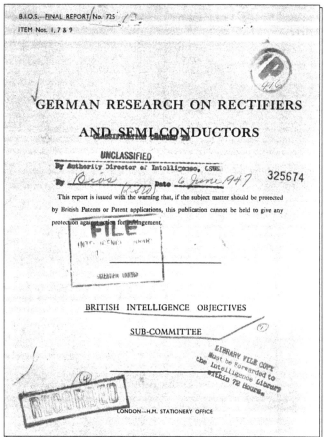

The Allies were interested mostly in avant-garde German technology. Above are reproduced covers of selected British intelligence summaries referring to electron microscopy, a magnetophon sound recording and reproducing system, quartz clocks and semi-conductors.

The German "Tonschreiber-B" tape recorder from the late 1930s—a forerunner of the birth of an entirely new field electronics. The lower part of the set is the amplifier

Wright Field/Wright Patterson, White Sands in New Mexico (a missile test ground) and a section of German aerial designs—this the least known, and their authors, ended up at Freeman Field base in the state of Indiana. Part of the scientists were "scattered" over various military design and research establishments and universities.

Apart from the prototypes of various aircraft, e.g. the Me 262, Ju 290, at the base in the state of Ohio were to be found among others:

- Dr. Hans Mayer, former director of the Berlin Siemens und Halske plant, the only unquestionable anti-nazi, who had been in five concentration camps.
- Dr. Hans Eckert, thermodynamics expert, previously working at "Luftfahrt Forschungsanstalt, Braunschweig." He had a crucial contribution in work on turbojet and ramjet engines, developing the high temperature resistant alloys used in them.
- Dr. Heinz Schmidt, co-designer of German turbojet engines.
- Professor (the Americans mentioned "Dr"—?) Alexander Lippisch. As it became evident in the USA, he had designed not only the Me 163, DM-1, P-12 and P-13, but also had his share in the design of … a German orbiting space station. The Americans were very interested in this field. In the materials referring to operation "Paperclip" it was mentioned that Lippisch was brought over along with a prototype of his advanced fighter—it wasn't mentioned which, only from a file from operation "Lusty" described further on, does it follow that it was the aforementioned P-13. At Wright Field Lippisch was a centre of interest, recognised as the greatest authority in the field of supersonic aircraft, giving lectures and conducting seminars on this subject. According to American Air Force officers Lippisch's achievements, by far exceeding their own, caused a complete revolution in the American understanding of this field. Senator Harry F. Byrd wrote of "revolutionising the very character of aerial warfare" and in an interview for the press, enclosed in the files, General Donald L. Putt stated that:[127]

*Their research progress in jet and rocket propulsion, aerodynamics, thermodynamics, supersonics, and other fields was clearly far ahead of anything of the kind we had done. I don't mean that the Germans were any abler than our best American scientists and technical men. After all, ours produced the atomic bomb. The difference, aeronautically speaking, was that **while we had made great progress along conventional lines, the Germans had explored and exploited entirely new roads into the aviation future.** [bold by I.W.]*

By the way: of about a dozen key scientists working on the American atomic bomb, only Feynman was American. In addition part of these bombs was comprised of German fissionable materials, originating from seized evacuation transports.

The next scientists "imported" to Wright Field were among others:

- Fritz Doblhoff, the young, at that time 30-year-old designer of the WNF-342 rocket propulsion helicopter. He arrived along with his prototype.
- Dr. Helmut Heinrich, previously an employee of "Graf Zeppelin Forschungsinstitut." On the basis of the vertical wind tunnel located there he created the so-called "ribbon parachute"—resembling a standard one, but sewn from separated gaps of ribbon. It behaved perfectly at high airspeeds and loads. It enabled a heavy aircraft to be retarded shortly before landing, pilot ejection equally at high speed and low altitude (incidentally, the Germans were the first to apply catapulting with the aid of an ejection seat). Thanks to Heinrich's parachute the experimental specimens of guided rockets could be recovered—among others the V-2 with undamaged electronic elements, which the Americans couldn't entirely cope with.
- Dr. Theodor Zobel. During the war he directed the wind tunnel in Braunschweig, where among other things he perfected the so-called "Schlieren interferometer." This device made possible a precise measurement of the pressure distribution around the aerodynamic profile by measuring the interference of light. Thanks to this device it was also possible to remotely determine the temperature distribution in a rocket's nozzle. According to Zobel himself, in 1/1000 of a second it was possible to obtain data equivalent to 250 classic point measurements of pressure and temperature. In the opinion of one of the American scientists from Wright

Field, Dr. Wattendorf, this invention accelerated American work in the field of aerodynamics by approx. 5 years.
- Dr. Rudolf Hermann, one of the employees of the centre in Peenemünde, who had a large contribution in designing the rockets made there. He developed a number of supersonic wind tunnels and at the end of the war constructed even in Bavaria a tunnel of flow velocity exceeding 11 km/s. Obviously for those times this was an absolute breakthrough. Hermann participated significantly in the development of the A-9/A-10 intercontinental rocket, among others he tested its model in his tunnel. He arrived in the USA along with a group of seven co-workers and large number of developments still unknown to the Americans.
- Dr. Ernest Steinhoff and Dr. Martin Schilling, rocket experts, co-designers of the A-9 rocket.
- Dr. Bernhard Goethert, aerodynamics expert, who developed the swept wings for jet aircraft, and whose knowledge was regarded as the key to work on supersonic fighters.
- Dr. Richard Vogt, chief designer from Blohm und Voss, author of many unconventional jet aircraft designs.
- Dr. Rudolf Edse, in the final years of the war he directed research on new rocket propellants (Braunschweig). This was a field in which the Americans were only taking their first steps.
- Dr. Otto Gauer, he was the first to put into effect a research programme conceived on a wide scale, concerning the biological effects accompanying flights at high altitudes (hypothermia, oxygen deficiency) and at high speeds (G-loads, tunnel vision effect). Soon another authority in this field joined him, Prof. Hubertus Strughold.

The names given above are merely examples of a domain that was only part of operation "Paperclip," and this wasn't the only operation of its kind. It is worth at this point putting the question to ourselves of how it looked against the wider background—compared to the operations of other countries.

Well, the Americans had seized neither the greatest amount of technology, nor the greatest number of experts, but with respect to quality had surely taken first place. This doesn't only follow from the effectiveness of their technical intelligence but simply from the fact that the leading scientists, having the freedom to move, preferred to find themselves at the moment of capitulation in the American zone rather than the Soviet. So it was with Von Braun, who after being evacuated to Oberammergau, gave himself up to the disposal of the Americans along with stocks of documentation and the large group of designers accompanying him. So it was with Dr. Anselm Franz, Chief Designer from the Junkers factories in Dessau. On the other hand neither the Americans nor the British employed any German nuclear physicist, fearing that this would endanger the work's secrecy. The British in any case fared quite modestly in this respect. The relative weakness of their technical intelligence services had an influence on this, but above all it was the serious economic crisis, which enforced drastic limits on scientific research. On the other hand they employed probably the most outstanding German aerodynamics expert, Hans Multhopp.

The Soviet drainage of technology had a completely different character than the American. They didn't succeed in capturing many high-class experts. One may count as exceptions Dr. Ferdinand Brandner, jet engine designer, (the Jumo 004 and BMW-003 engines were copied as the RD-10 and RD-20), Brunolf Baade, jet aircraft designer from Junkers, or engineer Hans Ressing from Siebel. The Soviets seized on the other hand entire laboratories and factories and after physically disassembling them transferred them to Russia, along with the greatest number of personnel as was possible. In consideration of the underdevelopment of many areas in the USSR, the Soviets were even interested in medium and low qualified personnel. In this way, with the confines of operation "Ossawakin" (this codename appeared in the files of operation "Paperclip"), a greater number of people were brought over than all the remaining countries combined had. The Berlin "Deutsche Forschungsgemeinschaft" in the latter half of the 1980s assessed, by way of arduous research, that there had been around 5,000 of them.

Contrary to the Americans, for the Soviets their most valuable "catch" was formed by documentation, hardware and scientists related to the nuclear programme (among others: Bewilogue, Döpel, Geib, Hertz, Vollmer, Wirth, Herrmann, von Ardenne, Thiessen, Timofeeff, Zimmer and Riehl). Many belonged to the "second league," but some technicians were virtually priceless—instead of theories they mastered technological processes—often extremely difficult. For example, Riehl was considered the best expert in the production and purification of uranium.

Many high-class experts ended up on the other hand—by being evacuated or just leaving—in Argentina. Their leaders were made up of 60 academic lecturers, around 100 aviation experts (Reimar Horten, Kurt Tank, Ludwig Mittelhuber, Ulrich Stampa, Otto Pabst, Julius Henrici, Klage…) and a dozen or so nuclear experts—above all Ronald Richter, working on thermonuclear fusion, Heisenberg's assistant Dr. Guido Beck, co-author of the German atom bomb project, Dr. Decker, Dr. Walter Seelmann-Eggebert, working during the war on controlled thermonuclear synthesis, engineers Gans and Hellmann from AEG and many others. It was also intended to bring over Heisenberg, but this upset the Allies.

Some German experts remained after the war in Spain, where as in the case of Argentina, they continued their work from the time of the war. They were among others Prof. Willi

Doblhoff's helicopter in American hands

Messerschmitt (Malaga), Dr. Claudius Dornier (Madrid), and a group of MP-43 carbine designers. I encourage those further interested in these issues to read my book devoted to the German evacuations of 1945.[128]

* * *

As I have mentioned, "Paperclip" was not the only American intelligence operation connected with the drainage of technology from the Third Reich. At the beginning of 2002, my co-worker from Great Britain, Nick Cook from "Jane's Defence Weekly," sent me a copy of extensive documentation concerning a parallel operation in relation to "Paperclip," which bore the code-name "Lusty."[125] I have to admit that this is such an absolute revelation that it gives the impression of being a story from another planet.

I have been engaged with these issues for a long time and know people who once had access to top secret materials on this subject, but in spite of this the contents of the documentation from operation "Lusty" are something completely new to me, as well as to everyone with whom I have come into contact. These materials consist of a descriptive section as well as a list of intelligence facilities/"targets" in the occupied Reich. In the descriptive section, at the very beginning, mention is made for example (unfortunately in shortened form) of seized German evacuation transports—U-boats. This concerns facts that not only shed a completely new light on the end of World War II and the issue of the Third Reich's scientific and technical achievements, but above all are shocking with the awareness that they are still clouded in a curtain of secrecy!

I present below excerpts of the aforementioned documentation, with my own commentary limited to a minimum:[125]

[Frame 590]:
At a medieval inn near Thumersbach near Berchtesgaden, early in May 1945, the German General Air Staff patiently awaited the outcome of surrender negotiations taking place in the North. They had arrived by car and plane during the past weeks, when the fall of Berlin was imminent, and had kept in contact by radio with Admiral Doenitz at Flensburg. Through the interception of one of these messages, their location, which had previously been unknown, was discovered. Within twenty-four hours Lt Col. O'Brien and his small party, representing the Exploitation Division of the Directorate of Intelligence, USAFE, had arrived, located the party and conducted the first of a series of discussions with General Koller, who was then in command. All documents and records that had been brought by the High Command were immediately turned over, and the first unearthing of buried records and documents, in and around Berchtesgaden, as well as the initial interrogation of the staff officers present, took place.
A casual remark made by a technical engineer, who stated that he had recently been offered a position in Japan, led to his being thoroughly interrogated for significant

technical information. As an aside, and what he probably considered a relatively unimportant incident, he stated that less than a month ago, about the middle of April, **ten submarines heavily loaded with the latest German equipment** *relative to aerial warfare, were dispatched from Kiel to Japan. When Lt Col. O'Brien was thus informed he immediately advised the Directorate of Intelligence, USAFE, who in turn notified the Japanese Intelligence Section of SHAEF.* **A cable was then dispatched to all commands in every theatre of war. All vessels in ports and at sea were notified, and one of the biggest searches ever undertaken during the war for submarines was initiated.** *[bold by I.W.] What route they had taken, whether they had gone alone or together, no one knew. But so extensive was the search and so carefully was it executed by warships of all Allied nations, that by the end of June, six of these ten submarines had been captured intact, some a relatively short distance away from their bases, others perilously close to Japan.*
In a mountain side near the camp of the German Staff Officers, an air raid shelter had been blocked up [the entrance to it? note I.W.] and then carefully covered and concealed with dirt. Its presence was eventually revealed by the officer who had directed this concealment, but only after he had noticed that a hole, large enough for a man to crawl through, had appeared on one of the sides. Thinking that the cache had been discovered, he explained to the USAFE party the location of the shelter. (…)
[Frame 591]
When the contents were eventually removed, one important document after another was laid out and carefully examined. One file contained correspondence from 1 Gruppe/6 Abteilung, German Air Ministry Intelligence, dated January 1943 to March 1945, concerning the supply to Japan of all types of equipment for aerial warfare, including models of the Me-262 and Me-163, quantities of V1 equipment, high explosives, incendiary bombs,

```
study and development.  The Messerschmitt aircraft series 1101,
1106, 1110, 1111 and 1112, a series particularly interesting in
that it illustrates a phase of coordinated aircraft design into
which American aircraft is only now entering; seven rocket-propelled
piloted aircraft specifically designed for anti-bomber inter-
ception work; a jet-propelled helicopter; Flettner 282 helicopter;
Horton 9, a flying winged glider; Ju 88, a radar equipped twin-
engine night fighter; Ju 290, four-engine long range transport;
seven Me 163s, rocket-propelled interceptor fighters; ten Me 262s,
twin jet-propelled fighter-interceptors; HE-162, single place
fighter powered by jet engines; flying bombs, type V1 single and
dual piloted; Lippisch P-13 Jager, a tailless twin rocket-propelled
wing for supersonic speeds; designs and models of small rocket-
propelled piloted aircraft created for bomber interception work;
three sets of FX-1400, a radio controlled bomb, and seven com-
plete A-4 rockets (V2s).   Numerous types of aerial equipment
and instruments of all models of latest designs were obtained
and likewise quickly dispatched for evaluation and study.  A
specimen of the German secret weapon, the X-4 rocket-propelled,
winged, flight-controlled anti-aircraft missile, intended for
launching from fighter aircraft against United States heavy
bombardment daylight formations, and the new anti-aircraft
missile HS-117, which was launched from the ground, were found
and sent to the British Air Ministry for examination.

However great in quantity and extensive in scope the captured
equipment, documents, and records of industrial concerns and
technical research laboratories were, greater still was the
extensive scope of information and salient facts gathered from
eminent German scientists, technicians and factory managers
through personal interrogations.   During the early phase of
```

A page from a report referring to Operation "Lusty," on which captured types of German aircraft and other aerial equipment has been listed bomb sights, radar apparatus of all description, including models of the Würzburg and Freya radio and signals installations, telephone, teleprinters, and so forth, and all types of aircraft parts.(…)

A report reached Lt Col. O'Brien's party that a "strange aircraft" had been seen in a mountainous retreat near Salzburg. Investigation quickly determined that this "strange aircraft" was a jet-propelled helicopter, the only one of its kind in the world. The inventor and his entire staff, who had laboriously worked ten years to perfect it, were present, guarding his invention as one would a precious jewel. The helicopter was examined, and a preliminary superficial interrogation of the staff was sufficient to reveal its tremendous importance. It was carefully loaded in a large truck and taken to Munich. From there it was sent across Europe to France, placed on a boat and shipped to Wright Field, together with the confiscated notes, drawings, and meticulous records of experiments conducted by the scientist and his assistants. After lengthy and detailed interrogations of these persons by technical experts assigned to the USAFE, the men were sent to prisoner of war cages. [literally! note I.W.] After ten years of labour they were left with only their memories of a remarkable technical achievement. The only jet-propelled helicopter in the world had been found and disposed of in such a way that it would prove advantageous to American scientists and the government.(…)

Further on, in frame no. 593 a curiosity of a completely different kind appears. A high-ranking intelligence officer, reporting the conduct of operation "Lusty" describes the character of field groups, combing occupied territory. He writes, that each group consisted of:

"Officers (…) fluent in German, Austrian, Russian and the Slavonic languages." The question arises of how can it be that the officer conducting intelligence reconnaissance of the given region had no particular idea which languages the inhabitants of this region were in command of? After all he must have been engaged in this for a long time. Something like this is possible probably only in America. Probably nobody is in doubt why it wasn't the Americans who designed the American atom bomb. We return however to the essence of the matter…
[Frame 593]

In six weeks of operation they had exploited more than five hundred important targets and interrogated hundreds of eminent German scientists, research professors, technicians, and workers.(…)
[Frame 596]

Team members often found, after initiating the exploitation of targets, that there were clues at each target which lead to others unknown: these usually were investigated at once. Quite frequently it was discovered that the Germans had almost always abstracted the documents they considered most valuable, generally hiding rather than burning them. Under pressure, however, their presence was revealed, quite frequently in lakes, swimming pools, mines, barns, buried in closed over shelters, tucked away in attic corners or in cellars of houses scattered over the country, in jails, insane asylums, or even grocery stores. (…) The Hermann Goering Aeronautical Research Establishment, located near Brunswick, yielded the greatest return in the field of research. Members of the Exploitation Division arrived on the 22 April to organise and conduct the scientific exploitation of this establishment. Dr. Von Karman, General Arnold's personal aeronautical advisor, and his Group, remained at this place on several occasions for periods ranging from several days to a week. According to Dr. Von Karman, seventy-five to ninety per cent of the technical aeronautical information in Germany was available at this establishment, and that information on research and development which had not previously been investigated in the United States would require approximately two years to accomplish with the facilities available there. Information obtained on jet engine developments available at the Goering establishment, it was stated, would expedite "the United States development by approximately six to nine months.(…)
[Frame 597]

(3) A document covering a complete source of instruction in the handling of plastic welding ["plastic," although reason suggests that this concerns aluminium alloys, remark by I.W.], a process which had been employed by the German aircraft industry, was located at Halle. This novel method of fabrication provided for the joining of plastics by flame gas welding, and enabled the sections joined together to possess the same strength at juncture as the original material. This information was reported to the AAF for joint study with the Office of Scientific and Research Development.

(4) An acoustic-controlled guided missile research development programme, together with operating personnel, was located at Bad Kissingen. The experimental control system developed there contained four electrical circuits that are activated by sound with the intended purpose of launching a rocket-propelled missile into the space occupied by a heavy bombardment formation, and constantly correcting the missile's course by means of incoming sound waves from the aircraft engines. The group of scientists who were engaged upon this development were detained in American custody at the laboratory to develop the programme for Allied use.

(5) Athodyd (Lorin Engine) units that developed thrust in excess of 1,500 kilograms were uncovered, and sufficient data was in our possession in May 1945 to permit immediate application in the field of high-speed aircraft production.

(6) High-altitude engine test beds, the most elaborate in the world, which were capable of supplying refrigerated low-pressure air both for engine cooling and combustion, thus

[Frame 598]
simulating atmospheric conditions at approximately 40,000 feet, were found at the BMW plant – the largest German plant engaged in manufacturing aircraft engines. (...)
(7) Complete information on the Freya and Riese "G" Würzburg and of the Jagdschloss radar equipment was uncovered at the research laboratory located at Koethen.
(8) After the Aerodynamic-Ballistics Research Station at Kochelsee was discovered about May 15, 1945, the Directorate of Intelligence, USAFE, assigned personnel to exploit fully this important target. Over one hundred and ninety German civilian research specialists under their original director, Dr. Hermann, continued their work, the results of which, however, were turned over to us. More than one hundred detailed reports concerning the station were prepared. The "Kochel Wind Tunnel" located there, had the largest testing sections and the greatest air flow of any known supersonic wind tunnel. It was considered by Army Air Forces of such exceptional importance for research in connection with jet fighter and fighter bomber priority projects that the War Department directed that it be dismantled at once and shipped to the United States.(...)
(9) Documents of all descriptions and nature were eventually discovered. The records of the German Patent Office, for instance, were found buried 1,500 feet underground in a potash mine near Bacha. **There were approximately 225,000 volumes, which included secret files.** *(...) Eventually, the files were evacuated and studied. (...)*
(11) The records of a department of the Speer' Ministry, the German Ministry of War Production, which dealt with the secret weapons programme, particularly V-weapons, rockets and jets, were seized and evacuated through air channels (...)
[Frame 599]
(13) Practically the latest type of every German aircraft, some of which never saw combat, eventually were located intact, or a sufficient quantity of available parts discovered for German mechanics to assemble a certain type. Usually these were sent across Europe to France, where they were shipped to Wright Field. Occasionally some were flown back. At least one, in some cases as many as ten, of the following, which represent only a fraction of the types, were located, some only after extensive searching throughout Germany, and forwarded to the United States for extended study and development. The Messerschmitt aircraft series 1101, 1106, 1110, 1111 and 1112, a series particularly interesting in that it illustrates a phase of co-ordinated aircraft design into which American aircraft are only now entering; seven rocket-propelled piloted aircraft specifically designed for anti-bomber interception work; a jet-propelled helicopter; Flettner-282 helicopter; Horton-9, a flying winged glider; Ju-188, a radar equipped twin-engine night fighter; Ju-290, four-engine long range transport; seven Me-163s, rocket-propelled interceptor fighters; ten Me-262s, twin jet-propelled fighter-interceptors; He-162, single place fighter powered by jet engines; flying bombs, type V-1 single and dual piloted; Lippisch P-13 Jager, a tailless twin rocket-propelled [for take off only, note I.W.] wing for supersonic speeds; designs and models of small rocket-propelled piloted aircraft created for bomber interception work; three sets of FX-1400, a radio controlled bomb, and seven complete A-4 rockets (V-2s). Numerous types of aerial equipment and instruments of all models of latest designs were obtained and likewise quickly dispatched for evaluation and study. A specimen of the German secret weapon, the X-4 rocket-propelled, winged, flight-controlled anti-aircraft missile, intended for launching from fighter aircraft against United States heavy bombardment daylight formations, and the new anti-aircraft missile Hs-117, which was launched from the ground, were found and sent to the British Air Ministry for examination.(...)
By the way: Frame 635 contains information about the max. flight speed of the P-13: Mach 1,85, which is about 2100 km/h.
[Frame 601]
In three months time, alone, over **111,000 tons** *of such documents were flown from Germany to the Centre for processing before being sent elsewhere to one or more agencies interested in the subject.(...)*
[Frame 602]
By the time the first Air Force officers arrived from the United States in June to prepare staff studies on aspects of the German Air Force, practically all of the necessary material had been collected and categorised so that work could immediately commence.
Eventually, over two hundred officers, chosen, for the most part, by HQ Army Air Forces, from various air force commands
[Frame 603]
in the United States, were engaged in scholarly research to fulfil their assigned requirements. Working from offices located in London and under the direction and guidance of the Assistant Chief of Staff A-2, USAFE, the officers exploited every available pertinent document located in repositories in London. (...) It was soon discovered, however, that in spite of the tremendous quantity and excellent quality of captured material available, too many questions remained unanswered and too many enigmas concerning various aspects of the enemy's air force unsolved. (...)

In the next section the documents referring to operation "Lusty" contain a range of dispersed information, referring to various research posts and issues. I have already written about the nuclear research laboratories mentioned here in the town of Linnessrabe (Linnesgrabe), but obviously it does not end here.

In frame 958 of the microfilm, information is included about the development of a guided missile in Völkenrode, carrying a gas stopping the engines of enemy aircraft and even tanks ("engine—stopping gas missile to be used against tanks or aircraft"). I have to admit that theoretically this should surprise and shock me, although in the report itself from the realisation of operation "Lusty" there are descriptions of so many unusual achievements and facilities, that the above "invention" appears to be something altogether ordinary…

Somewhat further (the frame number is missing) an even stranger description is found. It is written that in a building at 87 Weimarstrasse in Vienna, was situated a laboratory where "anti-aircraft rays" were worked on. The work was carried out in such strict secrecy, that the personnel were completely forbidden to leave the building, and it was sealed. In another town reference appears to a research post of Daimler-Benz in the town of Unter-Turkheim near Stuttgart, where a device was constructed "paralysing" petrol engines at a distance approaching 2-3 km. This facility was completely destroyed as a result of bombing before the end of the war.

In frames 1419-1420 the importance of the Munich-Innsbruck region is emphasised, as a concentrated area of electronic industry of particular significance. Here was supposed to be made among others centimetre-range radar.

In frames 887-932 attention is in turn drawn to the scale of transferring key plants and research institutions underground. It is written that it had been managed to start production in as many as 143 underground factories before the end of the war. A further 107 were built or planned—although from the wider context it follows that this quantity is not complete and probably does not refer to adapted mines among others. As I have already written, this issue is directly linked to the most secret armament projects, since in general they had a priority in the assignment of underground facilities.

* * *

Quite recently, I have found a very interesting article, from 1946, which presents the "influx" of German technology in a new light (Harpes Magazine [or Harper's Magazine?—I.W.], October 1946). It also contains some intriguing descriptions of various German inventions. Perhaps not their character is the most important here, but the perception of the breakthroughs by the Allies. I have quoted only selected examples:

> Someone wrote to Wright Field recently, saying he understood this country had got together quite a collection of enemy war secrets, that many were now on public sale, and could he, please, be sent everything on German jet engines. The Air Documents Division of the Army Air Forces answered: "Sorry, but that would be fifty tons." Moreover, that fifty tons was just a small portion of what is today undoubtedly the biggest collection of captured enemy war secrets ever assembled. If you always thought of war secrets' as who hasn't?—as coming in sixes and sevens, as a few items of information readily handed on to the properly interested authorities, it may interest you to learn that the war secrets in this collection run into

Few people realise how a colossal progress the Germans have achieved in electronics. One of the less known subjects is the miniaturisation of electron valves. At the beginning of the war the Gema and Telefunken companies have started the production of so-called metal valves (later also ceramic), which were by an order of magnitude smaller than the existing glass valves. What's more important, however, is that they required even thousands times lower power. This valve, of the DL-11 type made by Telefunken (pentode), consumed merely 60 mW (milliwatts). It's worth noticing that it was comparable with the parameters of the transistors, which were introduced after the war. It made possible to manufacture quite a sophisticated electronic systems, which, for instance, could be installed inside rockets and could be powered from a battery. They were also mechanically resistant.

> the thousands, that the mass of documents in mountainous, and that there has never before been anything quite comparable to it. (…)
> What did we find? You'd like some outstanding examples from the war secrets collection?
> The head of the communications unit of Technical Industrial Intelligence Branch opened his desk drawer and took out the tiniest vacuum tube I had ever seen. It was about half thumb-size. Notice it is heavy porcelain-not glass-and thus virtually indestructible. It is a thousand watt-one-tenth the size of similar American tubes. Today our manufacturers know the secret of making it!… And here's something:
> He pulled some brown, papery-looking ribbon off a spool. It was a quarter-inch wide, with a dull and a shiny side. That's Magnetophone tape, he said. It's plastic, metallized on one side with iron oxide. In Germany that supplanted phonograph recordings. A day's radio program can be magnetized on one reel. You can demagnetize it, wipe it off, and put a new program on at any time. No needle; so absolutely no noise or record wear.
> An hour-long reel costs fifty cents.
> He showed me then what had been two of the most closely-guarded technical secrets of the war: the infra-red device which the Germans invented for seeing at night, and the remarkable diminutive generator which operated it. German cars could drive at any speed in a total black-out, seeing objects clear as day two hundred meters ahead.

Tanks with this device could spot targets two miles away. As a sniperscope it enabled German riflemen to pick off a man in total blackness. There was a sighting tube, and a selenium screen out front. The screen caught the incoming infra-red light, which drove electrons from the selenium along the tube to another screen which was electrically charged and fluorescent. A visible image appeared on this screen. Its clearness and its accuracy for aiming purposes were phenomenal.

Inside the tube, distortion of the stream of electrons by the earth's magnetism was even allowed for! The diminutive generator-five inches across-stepped up current from an ordinary flashlight battery to 15,000 volts. It had a walnut-sized motor which spun a rotor at 10,000 rpm-so fast that originally it had destroyed all lubricants with the great amount of ozone it produced. The Germans had developed a new grease; chlorinated paraffin oil.

The generator then ran 3,000 hours!

A canvas bag on the snipers back housed the device. His rifle had two triggers. He pressed one for a few seconds to operate the generator and the scope. Then the other to kill his man in the dark. "That captured secret," my guide declared, "we first used at Okinawa" to the bewilderment of the Japs."

We got, in addition, among these prize secrets, the technique and the machine for making the world's most remarkable electric condenser. Millions of condensers are essential to the radio and radar industry. Our condensers were always made of metal foil. This one is made of paper, coated with 1/250,000 of an inch of vaporized zinc. Forty per cent smaller, twenty per cent cheaper than our condensers, it is also self-healing. That is, if a breakdown occurs (like a fuse blowing out), the zinc film evaporates, the paper immediately insulates, and the condenser is right again. It keeps on working through multiple breakdowns at fifty per cent higher voltage than our condensers! To most American radio experts this is magic, double-distilled!

Mica was another thing. None is mined in Germany, so during the war our Signal Corps was mystified. Where was Germany getting it?

One day a certain piece of mica was handed to one of our experts in the U.S. Bureau of Mines for analysis and opinion. Natural mica, he reported, and no impurities. But the mica was synthetic. The Kaiser Wilhelm Institute for Silicate Research had discovered how to make it and something which had always eluded scientists-in large sheets.

We know now, thanks to FIAT teams, that ingredients of natural mica were melted in crucibles of carbon capable of taking 2,350 degrees of heat, and then—this was the real secret—cooked in a special way. Complete absence of vibration was the first essential. Then two forces directly perpendicular to each other were applied. One, vertically, was a controlled gradient of temperature in the cooling. At right angles to this, horizontally, was introduced a magnetic field. This forced the formation of the crystals in large laminated sheets on that plane.

"You see this"... the head of Communications Unit, TIIB, said to me. It was metal, and looked like a complicated doll's house with the roof off. It is the chassis, or frame, for a radio. To make the same thing, Americans would machine cut, hollow, shape, fit-a dozen different processes. This is done on a press in one operation. It is called the cold extrusion process. We do it some with soft, splattery metals. But by this process the Germans do it with cold steel! Thousands of parts now made as castings or drop forgings or from malleable iron can now be made this way. The production speed increase is a little matter of one thousand per cent. This one war secret alone, many American steel men believe, will revolutionize dozens of our metal fabrication industries. In textiles the war secrets collection has produced so many revelations that American textile men are a little dizzy. (...) But of all the industrial secrets, perhaps the biggest windfall came from the laboratories and plants of the great German cartel, I. G. Farbenindustrie. Never before, it is claimed, was there such a storehouse of secret information. It covers liquid and solid fuels, metallurgy, synthetic rubber, textiles, chemicals, plastics, drugs, dyes. One American dye authority declares: „It includes the production know-how and the secret formulas for over fifty thousand dyes. Many of them are faster and better than ours. Many are colors we were never able to make. The American dye industry will be advanced at least ten years." In matters of food, medicine, and branches of the military art the finds of the search teams were no less impressive. And in aeronautics and guided missiles they proved to be downright alarming. (...)

At a plant in Kiel, British searchers of the Joint Intelligence Objectives Committee found that cheese was be-

One of the examples of the achievements made possible by miniaturisation is this pocket tape recorder, produced by the Siemens und Halske company in the second half of the war. It recorded sound with a surprisingly high quality, being powered from a very small battery. Above one can see a shadow of a glass for cognac, which gives some idea about the size. Both the tape, the reels and the casing were made of plastics.

ing made-good quality Hollander and Tilitser by a new method at unheard of speed. Eighty minutes from the rennecing to the hooping of the curd, report the investigators. The cheese industry around the world had never been able to equal that.

Butter (in a creamery near Hamburg) was being produced by something long wished for by American butter makers; a continuous butter making machine. An invention of diary equipment manufacturers in Stuttgart, it took up less space than American churns and turned out fifteen hundred pounds an hour. The machine was promptly shipped to this country to be tested by the American Butter Institute. Among other food innovations was a German way of making yeast in almost limitless quantities. The waste sulphite liquor from the beechwood used to manufacture cellulose was treated with an organism known to bacteriologists as candida arborea at temperatures higher than ever used in yeast manufacture before. The finished product served as both animal and human food. Its caloric value is four times that of lean meat, and it contains twice as much protein.

The Germans also had developed new methods of preserving food by plastics and new, advanced refrigeration techniques. Refrigeration and air-conditioning on German U-boats had become so efficient that the submarines could travel from Germany to the Pacific, operate there for two months, and then return to Germany without having to take on fresh water for the crew. A secret plastics mixture (among its ingredients were polyvinyl acetate, chalk & talc) was used to coat bread and cheese. A loaf fresh from the oven was dipped, dried, redipped, then heated half an hour at 285 degrees. It would be unspoiled and good to eat eight months later.

As for medical secrets in this collection,_ one Army surgeon has remarked, some of them will save American medicine years of research; some of them are revolutionarylike, for instance, The German technique for treatment after prolonged and usually fatal exposure to cold. (…)

German medical researchers had discovered a way to produce synthetic blood plasma. Called capain, it was made on a commercial scale and equaled natural plasma in results. Another discovery was periston, a substitute for the blood liquid. An oxidation production of adrenalin (adrenichrome) was produced in quantity successfully only by the Germans and was used with good results in combating high blood pressure (of which 750,000 persons die annually in the United States). Today we have the secret of manufacture and considerable of the supply. (…)

But of highest significance for the future were the German secrets in aviation and in various types of missiles. The V-2 rocket which bombed London, an Army Air Force publication reports, was just a toy compared to what the Germans had up their sleeve. When the war ended, we now know, they had 138 types of guided missiles in various stages of production or development [!!!—I. W.], using every known kind of remote control and fuse: radio, radar, wire, continuous wave, acoustics, infra-red, light beams, and magnetics, to name some; and for power, all methods of jet propulsion for either subsonic or supersonic speeds. (…)

A long range rocket-motored bomber which, the war documents indicate, was never completed merely because of the wars quick ending, would have been capable of flight from Germany to New York in forty minutes. Pilot-guided from a pressurized cabin, it would have flown at an altitude of 154 miles. Launching was to be by catapult at 500 miles an hour, and the ship would rise to its maximum altitude in as short a time as four minutes. There, fuel exhausted, it would glide through the outer atmosphere, bearing down on its target. With one hundred bombers of this type the Germans hoped to destroy any city on earth in a few days operations.

Little wonder, then, that today Army Air Force experts declare publicly that in rocket power and guided missiles the Germans were ahead of us by at least ten years. The Germans even had devices ready which would take care of pilots forced to leave supersonic planes in flight. Normally a pilot who stuck his head out at such speeds would have it shorn off. His parachute on opening would burst in space. To prevent these calamitous happenings, an ejector seat had been invented which flung the pilot clear instantaneously. His latticed ribbons which checked his fall only after the down-drag of his weight began to close its holes. (…)

All such revelations naturally raise the question: was Germany so far advanced in air, rocket, and missile research that, given a little more time, she might have won the war? German secrets, as now disclosed, would seem to indicate that possibility. And the Deputy Commanding General of Army Air Force Intelligence, Air Technical Service Command, has told the Society of Aeronautical Engineers within the past few months: "The Germans were preparing rocket surprises for the whole world in general and England in particular which would have, it is believed, changed the course of the war if the invasion had been postponed for so short a time as half a year."

These were examples, which are impressive, although they rather increase the quantity of what we know about the German model of developing science and technology. But the words like "the greatest single source of this type of material in the world, the first orderly exploitation of an entire country's brainpower" and "a whole new glossary of German-English terms has had to be compiled on new technical and scientific items" clearly demonstrate that it was also a completely new quality. A very valuable insight into the mechanisms of the development of civilization. Because our civilization is in the state of general, systemic crisis, I believe that we do need such a knowledge—badly.

PART THREE

"KRIEGSENTSCHEIDEND"

"The most beautiful feeling, which we can experience, is the feeling of mystery. This is the source of authentic art-of true science. He, who has never felt this emotion, who has never tasted it, is for me like a dead. His eyes are closed."

—Albert Einstein

THE BELL

The third part of this book differs from the previous ones. It is not only a technical specification of a certain research project, but also a personal story—almost a diary of the toilsome research carried out by myself. This was carried out on three continents and took me four and a half years to complete. However I am convinced that this affair was worth it. I hope that with respect to at least the unusual nature of the project itself, this affair will also interest others. For this is something entirely new, and the facts introduced below have never been presented as part of a unified picture (with the exception of fragmentary specifications in my previous books). Of course there is no shortage of new elements, but to place them within the appropriate context, I must start from the very beginning…

Everything began in August 1997. I was visited by a certain man well-informed (even extremely) about various aspects of World War II. Since he had asked for his name not to be revealed, I will refer to him as an "anonymous historian." Once, in the 1980s, he had had access to many interesting documents of an intelligence nature and relating to the Third Reich. He had come into contact with them while analysing files of the so-called special military cell at the National Council at President Bierut's office (President of Poland until 1952).

During the first meeting in the summer of the aforementioned year he simply asked me a few questions. He had been intrigued by a piece of information from one of my books and was curious, whether I would be able to amplify several not entirely explained issues. Among other things he asked me if I had ever come into contact with a device developed by the Germans, which was code-named "the Bell," and made a sketch of it. On a circular base was some kind of bell jar, cylindrical in shape with a semicircular cap and hook, or some other clamping device at the top. The bell jar was supposed to be made of a ceramic material, resembling a high voltage insulator. Two metal cylinders or drums were located inside.

This description conveyed nothing to me and in principle I would have forgotten about the whole affair within a few weeks. But obviously this didn't happen, for several reasons, although at the time I still hadn't realised that the explanation of this whole story would become my life's ambition.

Firstly I was impressed by the level of knowledge of the person in conversation with me. This was no amateur living in a dream world. Of that I was sure.

Secondly he described the quite simply unearthly effects of this device's operation, arousing in me associations with the final scene from Spielberg's "Raiders of the Lost Ark," defining them as "absolutely shocking." He did this with authentic conviction, which sowed a seed of true restlessness in me. I could not overlook this, and was inwardly convinced that this was no mystification. I still didn't know it, but the seed had already started to germinate…

Thirdly he asked me the outright disarming yet seemingly trivial question: if I was able to state with full responsibility that the "Wunderwaffe," that "wonder weapon," was the V-1 or V-2, as was often mentioned. If in any German documents or in any original sources in general, I had come across information unravelling what the "Wunderwaffe" was. He stated that after all it could not have referred to the V-1 or V-2, since firstly these weapons had been from a military point of view not very effective (and therefore not "wonder") and secondly that the term "Wunderwaffe" had begun to appear in earnest already after the "V" weapons had been deployed in combat. This was indeed intriguing. Later from the point of view of this, I looked over various volumes from my library and in actual fact it appeared that some kind of unusual weapon had existed, practically unknown till this day. As far as I remember Goebbels' propaganda had promoted the "Wunderwaffe" even after the air raid on Dresden in February 1945. Some statement on this subject had been uttered by Goebbels during a speech made after the air raid.

In the biography of the Minister of Propaganda I found for example the following sentence concerning Goebbels' wife:[200] "Magda told her sister-in-law that Joseph **had seen** a new weapon, so visionary that it would undoubtedly bring the wonderful victory, which Hitler had promised to the Germans…" [bold by I.W.] This sentence was to have been uttered sometime around Christmas of 1944. So refer-

Joseph Goebbels

SS-Sturmbannführer Otto Skorzeny, because he was in command of the special "Jagdverbände" units, he was nicknamed the "first commando of the Third Reich." What is less known, however, is that in 1944 Skorzeny was also Hitler's plenipotentiary in charge of all matters related to the use of strategic weapons, ie. against the United States.

ence had been made to something that had physically existed at the end of that year, creating a "visionary" impression through its appearance alone—therefore it must have been something completely different to the weapons known up until then.

Later, in documents brought over from the American NARA archive at College Park near Washington, I found among other things a report referring to the interrogation of one of Otto Skorzeny's commandos.[201] He was Skorzeny's aide-de-camp, SS-Sturmbannführer Karl Radl, at the same time Chief of the VI-S/2 Division at the General Office for Reich's Security (RSHA Reichssicherheitshauptamt). Radl stated—would you believe it—that since the beginning of 1944 Skorzeny simply had no head to organise acts of sabotage on the enemy's rear areas, since he had come into contact with

An excerpt of the article from November 1947, cited in the text

the "wonder weapon" and as a result "been possessed" by the idea of "Sonderkampf" ("Special Warfare"), regarding the use of this weapon, to such an extent, that he considered it the only sure way to win the war. In light of the "Wunderwaffe" other matters became insignificant.

This issue reappeared again after the war. Skorzeny was to be found in Spain, where he had taken his secrets. The Spanish press, followed soon after by the American press announced, that he had tried to sell the secrets of the "wonder weapon." Some statements present in these reports were too shocking to "take their word" for it although they had been written in normal newspapers and not in the tabloids, in pursuit of the sensational. As far as I know, the American press had for the first time included information about the nature of the "Wunderwaffe" (?) in November 1947. From this annotation it followed that the article had arisen on the basis of intelligence agency information. It mentioned that among other things what was involved was some unusual flying object, with "electromagnetic" propulsion, which simultaneously had been "responsible for a wave of flying saucer observations over North America that summer."[202]

Really???

If this had been at all true, it would have signified after all, that the technology had already been sold.

A year and a half later the press published some even stranger information-declaration:[203]

The USAF knows what the flying saucers are and where they come from. (…) They are new flying machines based in Spain, whose flight principle is based on gyroscope [what an interesting comparison!—note I.W.] and were built by German scientists and technicians, who escaped from Germany."

Although for the time being we may pass over in silence any information about "flying saucers," as can be seen from the reports of the 1940s it does not at all unequivocally follow that the term "Wunderwaffe" could be attributed to the V-1 or V-2. In this context, the term began to appear much later in popular literature and without references to specific sources from the period of the war.

For the time being my reflection on this subject, caused by a "trivial" question, had led me to establish that something which had created such a "visionary" impression on Goebbels and Skorzeny must have been truly unearthly and that the "Wunderwaffe" remains something unknown. Therefore I arranged further meetings with my informant and tried to find out something more about the mysterious project: where and who had realised it, where did the information come from and the like. During these meetings I gained the following picture:

That mysterious device, "the Bell" ("die Glocke"), seemed at first glance relatively simple, although the unusual effects of its operation contradicted this. The description was admittedly incomplete and non-scientific, since it had originated from military personnel, who had not had access to all of the data, but even then it included many valuable details. The main part of "the bell" was made up of two massive cylinders-drums around one metre in diameter, which during the experiment span in opposite directions at tremendous speeds. The drums were made of a silvery metal and rotated around a common axis. The axis was formed by quite an unusual core, with a diameter of a dozen or so, to twenty centimetres, with its lower end fixed to "the Bell's" massive pedestal. It was made of a heavy, hard metal. Before each trial some kind of ceramic, oblong container was placed in the core (it was defined as a "vacuum flask"—?), surrounded by a layer of lead approx. 3 cm thick. It was approx. 1-1.5 m long and filled with a strange, metallic substance, with a violet-gold hue and preserving at room temperature the consistency of "slightly coagulated jelly." From the produced information it followed, that this substance was code-named "IRR XERUM-525" or "IRR SERUM-525" and contained among other constituents the thorium oxide and beryllium oxide [beryllia]. The name "Xeron" also appeared in the documentation. It was some kind of amalgam of mercury, probably containing various heavy isotopes.

Mercury, this time already in pure form, was also present inside the spinning cylinders. Before the start of each experiment, and perhaps also for its duration, the mercury was intensively cooled. Since information appeared about the use of large quantities of liquid gas, nitrogen and oxygen, it appeared that it was precisely these that were the cooling medium. The entire device, i.e. the cylinders and core was covered with the aforementioned ceramic housing, of a bell-like shape—a cylinder rounded at the top crowned with some kind of hook, or fastening. The entire device was about 1.5 m in diameter and about 2.5 m high. A very thick electrical cable led to the "cap." At the bottom on the other hand was situated a round and very solid (made of heavy metal) pedestal or base, with a diameter slightly larger than the ceramic housing.

This was about all that I had managed to establish during the aforementioned conversations about the device's construction. But it was more than enough to state, that this description did not match anything that we know about the Third Reich's secret weapons. The person in conversation with me had in any case emphasised that **not once had the term "weapon" been uttered in relation to the described device**. It was just a fragment of something greater, not in itself being any kind of weapon, despite having a very destructive effect on its environment.

Much more information than in the issue of construction remained about the course of the experiments themselves.

Each such experiment was carried out in a specially prepared chamber—a pool. In most cases it was located underground. Its surface was covered with ceramic tiles and the floor also with heavy rubber mats. The mats were destroyed after each test (!), while the tiles were washed—deactivated with a pink liquid resembling brine. In the case of tests conducted inside the chamber of an inactive mine, in 1945, such a chamber was always destroyed (blown up) after two-three tests. One of the individuals, a primary source of information, testified moreover that a special "set" had existed for conducting trials on open terrain. It was mounted on three railway carriages marked with large Red Cross symbols and consisted first and foremost of a power supply installation, connected to a high voltage line available at a given location. This individual was a certain Joachimm Ibrom, an employee of Deutsche Reichsbahn (German Railways), in the Opole (Oppeln) district. These railway carriages were later struck off the Opole Railway Headquarters stock, formally as the result of an air raid. They were burned with flame throwers and the remaining metal elements cut with acetylene blowpipes and dispersed over the site. An undeniably odd procedure. Later however I was to become convinced that all the information which I had gradually managed to find in the course of my private investigation was absolutely unprecedented and constituted not only "something new," but in general through a series of facts, gave a picture of a project fundamentally different from all that has been written about German research from the time of World War II. Besides I would never have sacrificed several years for something, which did not constitute an evident and concrete challenge. My aforementioned informer strongly emphasised that what was involved was **a uniquely classified project, the most secret research project ever realised in the Third Reich!** Therefore it is surely clear that regardless of the scale of difficulty it was worth verifying such a statement…

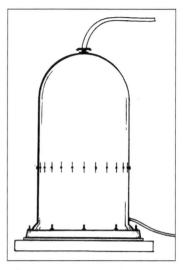

The approximate external appearance of "the bell"

The gate of Gross-Rosen concentration camp near Wrocław/Breslau. This was a completely different camp than for instance Auschwitz. Its primary aim was to supply the workforce to the numerous armament enterprises and underground construction sites in Lower Silesia. The vast majority of prisoners were dispersed in sub-camps.

However, let us return for the time being to the description of the aforementioned experiments.

First of all "the bell" itself was prepared along with the considerable power supply installation accompanying it. A whole set of cameras, movie cameras and probably also some measuring devices were placed on a special rack in the research chamber itself. Then a series of samples or objects were placed nearby, on which the effect of the emitted energy was tested. These were animal organisms (live lizards, rats, frogs, insects, snails and in all probability also... people—prisoners from KL Gross-Rosen)—plants (mosses, ferns, horsetails, fungi, moulds) as well as a whole series of substances of organic origin such as: white of an egg, blood, meat, milk and liquid fats. These preparations were of course carried out by scientists and technicians—fortunately their names are known, I will mention about this further on—as well as by a commando of prisoners from Gross-Rosen concentration camp, specially assigned to this work. It numbered up to 100 people and was code-named RWS-1. Just before the experiment the entire personnel was removed to a distance of 150-200 m at the same time employing individual, rubber protective suits and helmets or hard hats distinguished by large red visors. It then took some time for the drums inside "the bell" to get going. After reached the required velocity, the primary part of the test commenced, during which—which was emphasised—the device was connected to a high voltage and high intensity current. Probably because of this the whole device had to be efficiently cooled. This phase lasted from tens of seconds up to about 1.5 minutes. "The bell" revealed its operation in two ways: developing short-term and long-term effects—and perhaps also those about which we do not know. The former became immediately perceptible after the power had been switched on. These were: a characteristic sound, which could be described as something extremely similar to the humming of bees sealed in a bottle (hence the unofficial name "The Hive"—"Bienenstock" was also used in relation to "the Bell") as well as a series of electromagnetic effects. These consisted of the following: surges in surrounding 220 V electrical installations (bulbs "blowing") observed in the case of ground tests at distances exceeding 100 m, a bluish phosphorescence (blue glow) around "the bell"—obviously a result of the emission of ionising radiation, as well as a very strong magnetic field mentioned in the statements. In addition participants of the experiments felt disturbances of the nervous system's operation, such as formication ("pins and needles"), headaches and a metallic taste in the mouths.

After some time "long-term" effects appeared. At first some of the employees suffered disturbances of sleep, balance, problems with memory, muscle cramps and various types of ulceration. Later they succeeded in radically limiting these unfavourable effects.

The most shocking and at first totally inexplicable phenomena were observed in reference to the aforementioned organisms and substances subjected to tests in the research chamber itself. They suffered various types of damage, dominating being the disintegration of tissues structures, gelation and the stratification of liquids (among others blood) into distinctly divided fractions and others. From information made available it followed, that during the first phase of tests (May-June 1944) these kinds of side-effects caused the death of five out of seven scientists engaged in them. As a result, the whole first research team was dissolved. I have written "side-effects," since from the information which survived the war it clearly followed, that one of the main aims of the research was to limit them.

The most unusual changes were observed in the case of green plants. During the first phase, spanning about five hours after the test had been completed, the plants paled or became grey, suggesting chemical decomposition or the decay of chlorophyll. Extraordinary is that despite this, such a plant lived normally, by all appearances, for a further period—the order of a week. This was followed by immediate, almost rapid or cascade (8-14 hours) decomposition to a greasy substance, "with the consistency of rancid fat, resembling the mazout," enveloping the entire plant. This decomposition was devoid of all features characteristic of bacterial decomposition—among other things of smell. Besides, it was too rapid, giving the impression that all structure had decayed.

At the same time the formation of undefined crystalline structures was observed in the liquid organic substances, or something which resembled them. At that time these changes referred to the majority of samples.

On about January 10, 1945, they had managed, in an unknown way, to limit the number of damaged samples to about 12-15%. On March 25, a further clear drop in this number to 2-3% was recorded. Apart from this, another "side effect" emerged. After a certain number of experiments it was observed that mysterious "gas bubbles" were forming in the "the bell's" metal foundation...

SS-Obergruppenführer Oswald Pohl was in charge of the WVHA until the end of the war. He was a frequent guest at Gross-Rosen's chancellory. On the photograph he is standing sideways, to the left of Himmler.

I realise that this information, devoid of any kind of interpretation creates an unintelligible impression, that it lacks any sense, or is simply non-scientific. But this is not the case. Due to the unremitting work of myself and many specialists, we finally managed to unite it into a single, compact and intelligible whole, documenting many facts. For the time being however I am forced to present the information in such form, in how it became a starting point to further research. I ask therefore for continued patience…

At this "starting point" I also had at my disposal quite ample information concerning the organisational side of the whole project as well as a certain theoretical base.

I will begin with this second matter, as it is more directly related to the aforementioned phenomena. This "theoretical base" was only a set of scientific terms most often used in reference to various aspects of the described device's operation. From the start they were also quite mysterious. To such a degree that nobody before me had managed either to unite them or attribute them to specific contemporary work (and such attempts have been made—at least concerning the aspect of the analyses, about which I know). Two notions were obviously treated as fundamental, being:

"**Vortex compression**" and "**Separation of magnetic fields**."

Does this convey something to anybody? Obviously a rhetorical question.

Within the context of one of the people the problem of "a simulation of damping of vibrations towards the centre of spherical objects" appeared. In this case it concerned Dr. Elizabeth Adler, a mathematician from Königsberg University (this name appeared only once). In descriptions of "the bell's" effect on living organisms on the other hand the notion of "ambrosism" ("Ambrosismus") occurred. This was perhaps invented to honour one of the scientists, who admittedly was not a member of the research team, but was in some sense connected with the whole project. It concerned Dr. Otto Ambros, then chairman of the so-called "S" committee, responsible for chemical warfare preparations in Speer's Armament Ministry.

I must admit that from the beginning the plot connected to Ambros was totally belittled by myself, as not matching the whole picture. As it was to become evident a few years later, this was a big mistake—although there was never any doubt that chemical weapons were not responsible for "the bell's" operation, or any kind of chemical agent. Only in 2001 did I again turn my attention to the first note related to this affair from 1997, in which Ambros had appeared among two key names. As a result of this "omission," when information arrived about the importance of this person from another source, I had already managed to forget that his name existed in the original materials.

Some years ago I had made light of yet another fact on a similar basis, which did not match anything, partially because at that time it had already been designated as second-rate. From the present perspective however I can see that it was one of the hidden clues to the whole affair.

This was a reference about taking into consideration the process of transforming mercury into gold. I had ignored this as I came to the conclusion that whatever hadn't gone on inside "the bell," this process would have been economically unprofitable anyway and so of secondary importance, not describing the principle of operation of the whole "invention." The first conclusion actually turned out to be true, but the second part, not longer. For the aforementioned phenomenon can only occur in conditions characterised by quite narrow limitations, thus taking into account this or similar information allows one to exclude many incorrect explanations.

This is however only an added digression, but which indicates the great technical complexity of the whole issue. The first sketch of "the bell" which I had come into contact with gave me the impression of being something strikingly simple, although ultimately the whole device would turn out to be complex and technically sophisticated.

We will return again later to the technical issues, but now let us move on to the "organisational and personal" aspect.

The entire research project as such was created in January 1942 under the code-name "Tor" ("Gate"), which functioned until August 1943. After this it was renamed or rather divided into two "sub-projects." The code-name "Tor" was replaced by the code-names: "Chronos" and "Laternenträger." Both referred to "the bell," but the project had been divided into physical and medical-biological aspects. It was not established which code-name corresponded to a given aspect of the work. The system powering "the bell" probably received the code-name "Charite-Anlage."

The meaning of the Greek word "Chronos" is I suppose obvious, the German word "Laternenträger" looks somewhat less certain. From a literal point of view it corresponds to a man carrying a lantern (who in bygone days lighted street gas lamps). But one can look at this from yet another angle. It could be, as it was suggested to me, a not too literal translation of a certain ancient name—the name of "Lucifer," i.e. "he, who carries the light." Anyway, code-names cannot be treated too literally. They cannot reflect the true nature of a given issue, since it would then lose its point (unless with respect to an issue's particular unusualness there would be no risk of anyone guessing its true meaning anyway).

The entire work was supervised by the SS, at the same time their position in organisational structures was rather untypical. In general this made it considerably easier to keep the affair secret, among others because the SS was de facto an economic empire and so could alone ensure its own workforce and self-financing, without the necessity of making comprehensive data available to other institutions. Present-day "black" arms projects are being much the same realised, among others in the USA. On the other hand the leading role of the SS enabled the employment of a "special procedure" for classified protection on site. This relied e.g. on the possibility of guarding the site where work was being carried out by even two-three cordons of soldiers selected according to special criteria (e.g. from a foreign SS unit, not able to communicate with the locals, like in Fürstenstein) and on the other hand liquidating without hesi-

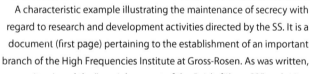

Emil Mazuw

A characteristic example illustrating the maintenance of secrecy with regard to research and development activities directed by the SS. It is a document (first page) pertaining to the establishment of an important branch of the High Frequencies Institute at Gross-Rosen. As was written, it enjoyed the "special support of the Reichsführer-SS," exploiting among others 150-200 prisoners, so something important must have been involved. However it is unknown to this day what this branch was specifically engaged in.
(AAN, Alexandrian Microfilms: T-175/ files of the Pers. Stab RFSS).

tation all persons who knew anything ("Geheimniträgern"—literally: "bearers of a secret") and were no longer considered necessary. This even referred to German citizens, if the matter's standing justified it. At any rate it was endeavoured in this way to keep secret till the end also projects "Chronos" and "Laternenträger." In the last days of April 1945, a convoy was formed from some medium and low ranking German cadre and the remnants of RWS-1 commando (in total 62 people). In consideration of the danger that they could fall into enemy hands, the order was issued to physically liquidate them, which was carried out on April 28 in an underground weapons factory near the town of Pattag-Neissebrück near Goerlitz (we will come to the locations where work was carried out later).

Let us return for the time being to the organisational issues. The whole project was co-ordinated by a special cell co-operating with the SS armament office, subordinate to the Waffen-SS. This cell was designated the "FEP," which was an abbreviation of the words "Forschungen, Entwicklungen, Patente" (research, developmental work, patents). The chief of this "FEP" cell was a certain Admiral Rhein, while the described project was co-ordinated by a quite mysterious individual, namely SS-Obergruppenführer (Four Star General) Emil Mazuw. Why mysterious? Simply because despite possessing one of the highest general's ranks in the SS, practically nothing is known about him. I got hold of his dossier in the USA in 1999, but through this he became in my eyes an even more obscure figure. It followed both from his dossier as well as cards from the course of his service, that Mazuw had been at the very top of the SS elite.[204] He was promoted to the rank of SS-Obergruppenführer on April 20, 1942, in other words he had possessed **the highest possible** SS rank at that time (in 1944 the SS-Oberst-Gruppenführer rank was further established, four people being promoted to it). He was awarded with the Honorary sabre of Reichsführer SS (Ehrendegen des RFSS) and honorary SS ring with skull and cross-bones (SS-Totenkopfring). Such a ring was given by Himmler for special service to the organisation. Their bearers constituted the highest caste of SS-men, given admittance to the greatest secrets. Each ring was personally dedicated by Himmler. In the event of the owner's death it had to be returned and was then exhibited as some kind of "relic" in a mystical sanctuary of the SS at Vogelsang castle. Mazuw already had it in 1936. He was therefore one of the powers behind the throne of the Third Reich, almost unknown to this day.

In 1936, he took part in a mysterious expedition to Iceland, organised by Himmler—the mystic, to search for the purest remnants of the original Aryan race. One can therefore assume that he was a member of the so-called "inner circle" of the Reichsführer SS. In likeness to King Arthur's knights he

The complex system of the SS ranks

SS	Equivalent in the Wehrmacht	US ranks
Privates		
SS-Mann	Grenadier	Private
SS-Oberschütze	Obergrenadier	–
SS-Sturmmann	Gefreiter	Private 1st class
NCOs		
SS-Rottenführer	Obergefreiter	Corporal
SS-Unterscharführer	Unteroffizier	Sergeant
SS-Scharführer	Unterfeldwebel	Staff sergeant
SS-Oberscharführer	Feldwebel	Technical sergeant
SS-Hauptscharführer	Oberfeldwebel	Master sergeant
SS-Stabsscharführer	Hauptfeldwebel	Sergeant major
SS-Sturmscharführer	Stabsfeldwebel	Sergeant major
Officers		
SS-Untersturmführer	Leutnant	2nd Lieutenant
SS-Obersturmführer	Oberleutnant	1st Lieutenant
SS-Hauptsturmführer	Hauptmann	Captain
SS-Sturmbannführer	Major	Major
SS-Obersturmbannführer	Oberstleutnant	Lieutenant colonel
SS-Standartenführer	Oberst	Colonel
SS-Oberführer	Oberst	Brigadier general
SS-Brigadeführer	Generalmajor	Major general
SS-Gruppenführer	Generalleutnant	Lieutenant general
SS-Obergruppenführer	General	General
SS-Oberst-Gruppenführer	Generaloberst	...

In practise, both military and SS ranks were used, simultaneously in the Waffen-SS and the police, e.g.: "SS-Brigadeführer und Generalmajor der Waffen-SS."

united twelve of the most faithful Obergruppenführers, regularly conducting meetings shrouded in deep secrecy in the crypt nicknamed "Walhalla" at Wewelsburg castle.

Most extraordinary however was that despite such a high position Mazuw not only never graduated from any military academy, but in general completed his education in the first (yes, first!) class of primary school (Volksschule)! This was clearly written in his dossier.[204] Mazuw didn't die until 1987, but I never managed to gain access to official records from his interrogations.

However, I was mostly interested in the scientific aspect of this enterprise. The individual responsible for the aspect relating to physics was Professor Walther Gerlach. At first I knew only that he had been one of the most outstanding scientists of the Third Reich. Soon it was to become evident that he had also been some kind of "power behind the throne."

After my informant had handed me his information I simply tried to find anything on the subject of Gerlach's scientific work from the time of the war. It became evident that such

A council of SS generals.
In the middle of the first row: Prof. Dr. Ernst Robert Grawitz.

information was so scant and fragmentary that the whole problem practically remained a "blank area." All the same Gerlach had directed the Reich's Scientific Research Council (Reichsforschungsrat, from December 2, 1943), so he must have been a key person from the point of view of scientific research important with respect to the war effort. After all during this period one would not expect that the management of German science would be turned over to a scientific researcher of butterflies or primitive people's folklore... The role of science during the total war was after all quite specific.

Admittedly Gerlach is often associated with the nuclear program, but he was never de facto directly involved in this matter and **never published any scientific work concerning nuclear weapons, or the construction of a nuclear reactor**. He constituted therefore a further part of the riddle, requiring an explanation practically from scratch... During my years of research, I managed to get hold of some scant, but extremely important information related to the war-time research work

Professor Walther Gerlach. On the second photograph he is inspecting the smouldering ruins left of his Institute of Physics, after an air raid.

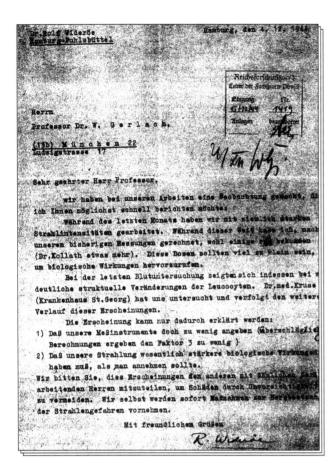

A vague letter to Gerlach, seized by the US "Alsos" mission from the archives of the Reich's Scientific Research Council. One of the scientists reports in it about the "structural changes of leucocytes," which occurred on a far larger scale, than theory predicted. It is not exactly known to what project this letter refers to.
(NARA/RG-319: Reports and messages 1946-51/Alsos Mission)

Professor Walther Gerlach during the war

of Gerlach, but we will come to that later. For now, I am all the time presenting the starting point of my research—the information, which I had acquired in 1997.

As I mentioned, the entire project was divided into two segments: physical and medical-biological. As far as the latter is concerned, SS-Gruppenführer Prof. Dr. Ernst Grawitz was Gerlach's equivalent, that is to say scientific co-manager of the project. He was at the same time chief of the medical service of the SS and police, chief of the Institute of Hygiene of the Waffen-SS as well as president of the German Red Cross. Nothing is known about his specific "contribution" to the realisation of the "Chronos"/"Laternenträger" project. Anyway I was never hot on his trail, as I considered it would be more promising to concentrate on the role of other people.

As far as the SS generals were concerned, yet another name surfaced: SS-Brigadeführer Heinrich Gärtner. He was responsible for ensuring electric energy supplies and probably for logistical issues in general. He was chief of the "research and development group" at the armament office of the SS (how it is possible, that such a person is practically unknown???). I was however mostly interested in physicists, since the most crucial and probably most difficult problem was the interpretation of the device itself (of "the bell"). The names of the physicists could be the thread leading to the solution of this riddle. Fortunately the personal data about the key people from this group was known. Apart from Prof. Gerlach and the aforementioned Dr. Elizabeth Adler, the name of a physicist from Darmstadt was also brought into light, who was responsible for the analysis and utilization of the initially mysterious phenomenon of "separation of magnetic fields," and for development of the high voltage generator to power "the bell." This was Dr. Kurt Debus. In 1997, I did not know anything else about him. It was however obvious that shedding light on his role would be the primary task. Two other individuals appeared out of "the background:" Dr. Edward Tholen [Tohlen?—I.W.] and Dr. Herbert Jensen. The former was to have worked at Peenemünde and was the designer of some kind of super-resistant alloy, used in the V-2 rocket structure. So he could have designed a fundamental part of "the bell," in which, considering the project's standing, materials "from the top shelf" had undoubtedly been used. This is only an assumption, but Tholen could also have worked on something, of which "the bell" was only a part.

I never managed to get hold of any information on the subject of Tholen, if in fact this had been his real name. I assumed that because of its mechanical and heat resistance coefficient/specific gravity, titanium or its alloy would have been

"We don't need Einsteins." A German poster from 1933, on which the author of the theory of relativity has been swept off the roof of an astronomical observatory bearing his name. Contrary to appearances this was not only antipathy of an ideological basis.

An excerpt of an intriguing report by the Polish Home Army Intelligence (resistance movement) (sygn. MM 3/44, from March 1944), probably regarding Tholen's invention. It describes the mysterious activities carried out in the underground "Mittelwerk" plant (among others the "thol" alloy). Regardless of the technical data given in the report it cannot refer to the V-2 for the simple reason that there is reference to "the end of 1943," while by December 5 of that year only the first technical drawings of the V-2 had reached "Mittelwerk." Therefore it must refer to previously unknown issues. (source: "Polish Home Army Intelligence"... see bibliography referring to part 1)

the most suitable material. Contrary to appearances and the suggestions of various people, the Germans were already using titanium at that tima very small scale. I managed to obtain (unfortunately incomplete) a British intelligence report about German research and development work on titanium, which was carried out among others by AEG—known to have been engaged in the "Chronos" project. In my copy of the report however I did not find any names.

The next interesting aspect, which loomed from data presented to me, was the issue of the companies engaged in this work and where it was carried out. This information referred mainly to the "final" stage of work—i.e. from the summer of 1944. Till November of that year the situation stood as follows:

The main laboratories were located in Lower Silesia (now in south-west Poland), in the town of Neumarkt (Środa Śląska), near the town of Leubus (Lubiąż) and were disguised as the facility "Schlesische Werkstätten Dr. Fürstenau and Co. GmbH (Schlewerk)." In reality what was involved was a site of secret research conducted under cover by AEG, Siemens and Bosch. As far as AEG is concerned, the name of Engineer Hellmann was brought into light. Among others the following establishments co-operated with this "facility":

- "Heeres Versuchsanstalt No.10." These were Wehrmacht laboratories, to which on November 1, 1943, an SS research team was co-opted, with the code designation BII.
- An underground facility on the grounds of Breslau-Stabelwitz (Wrocław Stabłowice) military airfield.
- Probably also the factories and laboratories connected with the nuclear program, located in the towns of Torgau, Dessau and Joachimstal. This assumption comes from the fact that after the war many shipping receipts and access passes were found in the adjacent buildings of Lubiąż, on which the names of these towns appeared.

Due to the threat of a new Soviet offensive and other factors, described further on, in November 1944, the main complex

Niederschlesien (Lower Silesia/Dolny Śląsk)

Lower Silesia was a special region in the Reich, whose significance rose systematically in due measure of the war's passage. It was distinguished by the richness of raw materials (from coal to uranium) as well as modern industry, often dispersed in the form of single plants, even in small towns in the mountains. The perfectly developed infrastructure favoured this, including a road and rail network. Its central location in German occupied Europe caused this region to be located far from the front lines and hardly bombed, to which the diversification of industry was instrumental. The mountainous regions has not been encircled by Wartime operations at all. From the point of view of war-time production Lower Silesia seemed to be therefore almost an oasis of fortune, situated in a diametrically different situation than e.g. the Ruhr basin, or great industrial cities such as Hamburg or Munich.

It was recognised as the "Reich's anti-aircraft shelter" ("Reichsluftschutzraum"), where in particular key branches of the armament industry were expanded and after the start of carpet bombing, industrial plants were evacuated here from other parts of the Reich. By the end of 1944 as many as 323 evacuated plants could be found on the grounds of the local Armament Inspection VIIIa. The construction of new plants was commenced, including at least 20 underground. Factories which were to produce aerial armaments "enjoyed" particular success. At the turn of 1945 and 1946 were to be completed among others two great underground complexes of facilities, associated with these weapons— "Riesa" (on the line of the towns of Waldenburg, Ludwigsdorf, commenced in the spring of 1943) and "Concordia" (in the triangle of the towns of Aslau/Zittau/Landershut, commenced in 1944).

Mazuw's personal card, from his files

The personal card of Rudolf Schuster

of laboratories was relocated to Fürstenstein (presently Książ) near Waldenburg (now Wałbrzych). They were specifically moved to a small underground facility lying not far from the so-called "old castle" (Altburg), in the vicinity of a great castle-palace, until 1930s in the possession of the von Pless Dukes. This place was located in the mountains, far away from the possible front line.

As it was to prove, these redeployments were only temporary. The research & development team was waiting for the preparation of a final research and production infrastructure. They did not wait for long. The last removal took place on about December 18, 1944. In the source materials which my informant had relied on, it was only mentioned that it involved "an inactive coal mine, adopted for military purposes—in the vicinity of Waldenburg." There are not many mines near Wałbrzych (especially "in the vicinity" and not "in"—where there are many) and so determining which was exploited for purposes other than that of mining coal, was in actual fact one of the simplest tasks. However we will return to this in a further section of the book...

In order to recapitulate and conclude the initial phase of my research, i.e. the level of knowledge from 1997, I will move on to the final issues—to the sources of both this information and information about the evacuation of the project at the end of April 1945 (the Sudety region in question was captured by the Russians no earlier than May 7-10). It so happens that both these issues are connected to each other, since the source of information were the officers responsible for evacuation.

As I have already mentioned, the last link in the whole

Karl Hanke,
Gauleiter of the Lower Silesia

chain of information sources (referring to the 1940s) was the military cell at the National Council (KRN), and specifically a certain Major Walczak, who without any records, according to a special procedure, drew up reports for President Bierut concerning matters of the highest importance. He co-operated with the Soviet counter-intelligence service (NKVD, "Smiersh"), and specifically with a special cell code-named "MIP."

A few years later Walczak followed in the footsteps of his professional colleagues and died in unexplained circumstances, but the documents remained. The first information connected with the affair that interests us arrived from the Polish Military Mission in Berlin, as far as I remember, in 1946. At the same time the intelligence officers co-operating with the mission seized in Germany a certain wanted German officer, called Schuster. He had false documents, declaring that he was Jewish.

I do not know if this had been a "shot in the dark," or not, at any rate his interrogations supplied some extremely interesting information. He was interrogated by Col. Władysław Szymański (the spelling Władymyr Szymanskyj also appeared —i.e. Ukrainian), using for this purpose the garages of the former Reich's Chancellory at Voss Strasse. He personally recorded the protocol—without a typist and interpreter. As it was later revealed, he had also co-operated with the same Soviet cell "Smiersh," which resulted in Russian activity aimed at taking over the matter and "hushing it up" on the Polish side, which incurred not only the loss of Col. Szymański, but also many other people.

As a result of the interrogations it turned out that the arrested "Jew" was in fact SS-Sturmbannführer (Major) Rudolf Schuster, son of Johann. An officer from Amt III of the RSHA. Since June 4, 1944, he had been responsible for transportation in the special evacuation commando SS "ELF," created by the Gauleiter of Lower Silesia, Karl Hanke. This commando had been carrying out transports as the company "Agricultural Fertilizers-Oskar Schwartz and Son."

Later, in the archives of the Berlin Document Center, I even discovered a photocopy of this man's service history sheet, from which it followed moreover that he had also been an officer of the "special duty office" at the General Administration and Economy Office of the SS (SS-WVHA, Amt A-V z.b.V). Schuster was the source of a significant part of the information presented on the previous pages. He also testified

The Ju 390 and its loading

that at the end of April 1945 a special aircraft from KG-200 group, a Junkers Ju 390, had taken materials connected with the "Chronos"/"Laternenträger" project and adopted a course for Bodo airbase in Norway. The aircraft was pale-blue and bore Swedish Air Force markings, and before its flight had been closely guarded and concealed with canvas cover. As far as I remember, the airfield in question was located near Świdnica (Schweidnitz), i.e. a dozen or so kilometres from Książ (Fürstenstein).

In Norway supervision of the transport was taken over by

SS-Obergruppenführer Hans Jüttner, the chief of the SS-Führungshauptamt (of which the Armament Office of the SS and the FEP were parts). Quite recently I have found a document stating that on the 17th of November 1941 an agreement was signed between Jüttner and the chief of the Wehrmacht's research office, marshall Wilhelm von Leeb, which paved the way for joint (Wehrmacht-SS) research projects. Strangely, however, it would be very hard to find any examples of such a joint undertakings – except for the project described here, which was coincidentally „born" shortly after this date.

SS-Obergruppenführer Jacob Sporrenberg, to whom we will return later…

Schuster also supplied information about his direct superior, the chief of the "Special Evacuation Commando SS-ELF." Despite intensive efforts undertaken after the war, he was never captured. He was:

SS-Obersturmbannführer (LtCol) Otto Neumann, son of Karl. An official of the Amt III of the RSHA, from June 28, 1944 chief of the ELF Commando with abode in Wrocław (Breslau). It was established that after the war he went into hiding under various names: Hans Hildebrant, Hans Erlich, Jacob van Ness (a Dutch citizen). In 1954 he was seen in Rhodesia; in 1964, in Switzerland.

Undoubtedly Neumann would have been an invaluable source of information and not only with respect to the project described in this book. He was one of the key individuals who executed the so-called operation "Regentröpfchen," launched by Bormann with the objective of evacuating the Reich's crucial resources, so that they could be later used to rebuild Germany's potential.

However Sporrenberg remained…

From June 28, 1944, he performed among others the function of "Commander for Special Evacuations" in the "North" District, i.e. in Norway. This was the so-called northern evacuation route leading through the Scandinavian countries, the southern one led through Italy and Spain. On May 11, Sporrenberg was arrested by the British and interrogated, among other things for being suspected of directing operation "Elster"—from March of that year, based on an attempt to bombard New York with V-1 missiles launched from U-boats, in all probability carrying biological warheads. However, considerably more serious accusations were to weigh heavily on Sporrenberg—involvement in crimes of genocide committed earlier in Poland. So he was soon deported to Poland. Preparations for the trial and the trial itself took several years. In view of a very specific threat of the death penalty, Sporrenberg talked willingly and a lot, calculating in this way that it would extend his life. Owing to Schuster's earlier testimonies it was known in advance on which issues to place the greatest pressure. A huge amount of information was generated, the majority of which however never became a part of the trial documentation and was treated as "top secret—materials for special purpose." In December 1952, the death sentence was announced, which was officially soon carried out. In reality however the day before

A translated evacuation document found by R. Schuster: lists of passports

The "Old Castle" (Altburg) in Fuerstenstein/ Fuerstein, 19th century drawing

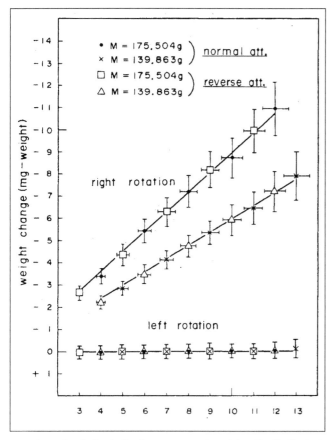

A graph referring to the Japanese experiment described in the text, on which weight reductions have been marked (in milligrams, on the vertical axis) as a function of rotational speed (in thousands of rpm, on the horizontal axis). The horizontal and vertical lines represent so-called standard deviation—possible measurement error. (source: see text)

the planned execution, by the so-called "operational measures" all persons obliged to be present during the execution have been changed, and instead of Sporrenberg a prisoner foreman (Kapo) from Stutthof concentration camp was executed, while Sporrenberg himself was flown to the USSR. In all probability he did not live there for very long—if only for the reason that the possibility of escape had to be excluded with 100% certainty. Sporrenberg however must have expected the deception and informed his family through his lawyer, since even in the 1960s petitions were sent for his release (to KC PZPR, central committee of the main party).

In 1947 Rudolf Schuster died suddenly in obscure circumstances (if he really died). Col. Szymański, who had interrogated him was killed in an air crash along with a group of witnesses, soon after completing the investigation. His superior and chief of the Polish Military Mission in Berlin General Jakub Prawin died in 1950 (as far as I remember, he capsized on his boat and drowned). Major Walczak was killed in a car crash. President Bierut went on an official visit to Moscow, where he suddenly became ill, died and returned in a coffin.

The Soviets took over the whole affair and cut off all links, in Poland only traces remained. Removing everybody was neither possible nor necessary, but in the existing situation further progress didn't come into play. Perhaps Poland's neighbours from the east had a moderately comprehensive picture of the issue at their disposal, however in Poland nobody was able to define either "the bell's" principle of operation, or why this device was considered so state-of-the-art.

Kommando ELF

The evacuation of materials associated with the "Chronos" and "Laternenträger" programmes was initially not supposed to be the aim of existence of the "Kommando of Special SS Evacuations, ELF." Polish and Soviet intelligence became interested in it for a completely different reason, which was also the guiding principle for its establishment. It had a connection with the creation of complex of eight underground depository shelters of the Reich's Main Security Office (RSHA) to the north-west of Wrocław (Breslau), in the town triangle of Parchwitz, Maltsch, Wohlau, which were to serve the central institutions of the Third Reich. The securing of transports associated with this was realised at the turn of 1944 and 1945, yet after the start of the Soviet January offensive.

These depository shelters…or underground bunkers were numbered among the most important such objects in the whole Reich and received the highest priority applied in this regard (AA). They had specific designations—I/AAX-SS/01 up to…08. It may sound strange, but their content has never been fully uncovered. Interrogations of German officers gave reason to believe that at least a large portion of the transported loads consisted of state bonds emitted for arms-producing companies, partially nationalised through their transfer to state-owned underground facilities. Apart from the officers themselves, numerous encrypted documents were intercepted, but despite persistent efforts, they have never been deciphered. The only outcome was the suggestion that they had been "translated" from a so-called synthetic language, consisting of "normal" words, but having different meanings.

Interrogations of the officers from this Kommando brought to light among other things an odd description of the following incident, which took place at the time given above: A local road near the town of Rogów (Polish). Three vehicles are travelling in the direction of the town of Praukau (Prawików)—a tank for transporting milk, an ambulance and a lorry. On the horizon appears a single Soviet aircraft, which drops a small bomb and flies away. The lorry is damaged, the vehicles stand still. Several SS officers are standing on the road. The one who got out of the lorry's cab begins to despair due to the "great loss of the German nation"—works of art from the Reich's Chancellory had suffered damage. The one who got out of the ambulance interrupts him sharply and travestying the words uttered by Christ during the "last supper," states: "This is a loss, which from the perspective of time will turn out to be insignificant, whereas for that which I am transporting, I answer only before history, for it is the blood and flesh of the Führer!" During the conversation it became evident that what was involved were the tissue samples of Adolf Hitler and twelve leading SS generals, "who along with the Führer were creating the Third Reich." The samples were to be found in a shining, metal container—a flat cylinder with low temperature and diameter approx. 2 m. It came from the Wewelsburg castle, whereas it was supposed to find its way to a facility (place) code-named "Wolfskranz." After a short stop the remaining vehicles continued on their way...

The Gauleiter

This was the party's representative in the largest party-territorial unit, called the Gau (Gau is more or less the equivalent of the present-day "Land"). The Gauleiter's authority was very great and rose in due measure of the war's passage, at the same time one should single out two turning points (of this rise):
1. On 16.XI.1942 they were appointed the "Commissioners of the Reich's Defence," which made them responsible for the "unified management of the economy" over their territories, for the entire civilian sector of the state, related to the war effort. In practise the opportunity of almost unlimited interference in the functioning of industrial plants had the greatest importance. A supreme instance in relation to the Gauleiters was Reichsleiter Martin Bormann.
2. On 25.IX.1944 Hitler signed a decree on the establishment of the "Volkssturm"...the territorial army, consisting of all men not yet incorporated into service, aged from 16 to 60. The Gauleiter had on principles of exclusivity the duty to create this formation as well as a primary influence on its command (he nominated commanders). The authorisations resulting from this were very significant -the manpower of the "Volkssturm" exceeded 5 million men, i.e. was in all probability even greater than of the regular armed forces. The Gauleiters were the most important representatives of state authority "in the field" (except for the front-line). Apart from this they often performed yet another function, e.g. they also had, as a general rule, honorary ranks of the SS-Gruppenführers.
In order to properly understand the position of a Gauleiter, one has to take into account yet one more fact—the so-called "Führerprinzip"—or "Führer principle" (incidentally, the word "Führer" doesn't have an exact English equivalent. It is not the same as the "leader," in my opinion "warlord," in a tribal sense, would be much closer to the truth). The "Führerprinzip" was one of the fundamentals of the Nazi political and administrative system, and consisted in granting a specific individual the attributes of an institution. According to this principle there wasn't, for instance, the institution of a "Higher SS and Police Command"—in a certain district, but the commander was the institution itself (Der Höhere SS und Polizei Fuhrer / HSSPF), likewise was often the case of a commandant (see today: the institution of a rector or bishop). The aim of such a system was mainly to simplify the decision-making process and mete out responsibilities, so that for instance decisions could be enforced immediately, also to underline the authority of the state. The Gauleiter, as one of the most important links in the Nazi structure of power, also functioned according to these rules—a man-institution in a "state of emergency state."

The RSHA and other central offices of the SS

The Reich's Main Security Office (RSHA... Reichssicherheitshauptamt) was one of the most important institutions of the country. It was the office of the SS and directly subordinate to the Reichsführer SS, Heinrich Himmler. It was established in September 1939 and operated until the end of the war. Apart from the inspectorates (among others—schools) and other detached cells, it consisted of six, and from the middle of 1940, of seven of the following offices:
I. Administration and Law,
II. This was engaged mainly in the control of publications, among others with the confines of the so-called "white intelligence." In 1940 it was re-named to Office VII, number II acquired a newly created office occupied with budget planning and economic issues.
III. Counter-Intelligence of the Security Service (SD), it embraced all domains of social, economic, cultural and scientific life in the Reich.
IV. The Gestapo—the secret political police, operating in the territory of the Reich and occupied territories, also in command of an organised body of agents inspecting German societies in neutral countries.
V. Office of the Reich's Criminal Police, broadly co-operated with SD and the Gestapo.
VI. SS Foreign Intelligence. In May 1944, it absorbed the foreign intelligence of the armed forces, i.e. the Wehrmacht (Abwehr—Ausland) and became the sole intelligence institution of the Reich. It was supported by Office VII. employed in analytical work relating to selected issues, mostly scientific.

Apart from the RSHA there existed 11 Central Offices of the SS, the most important being:
1. The Personal Staff of the Reichsführer SS—Himmler (Hauptamt Persönlicher Stab RFSS)—aide-de-camps, matters of protocol, distinctions, Ahnenerbe, Lebensborn, the realisation of the four-year plan plus matters particularly interesting Himmler.
2. The Main Office of the SS—the SS-Hauptamt (SS-HA)—recruitment and enlistment to the SS, chiefly matters associated with the formation of Waffen-SS combat units.
3. The Main Command Office of the SS—the SS-Führungshauptamt (SS-FHA). This performed the function of the headquarters of the Waffen-SS, unless these units were not subordinated to the front-line command of the Wehrmacht.
4. The Main Administrative-Economical Office of the SS

(SS-Wirtschafts- und Verwaltungshauptamt: SS-WVHA). This supervised the economic enterprises of the SS as well as economic exploitation of concentration camp prisoners.

Operation "Regentröpfchen"

Contrary to official propaganda the leaders of the Third Reich realised quite early on the inevitability of defeat. On 18 June 1944 in the hotel "Maison Rouge" in Strasburg the first of a whole series of councils took place, whose aim was to work out a method of transferring capital abroad and at the same time create the financial base for other evacuation measures. The initiator of the council was Reichsleiter Martin Bormann—chief of the NSDAP party machine, Hitler's personal secretary, SS-Obergruppenfuhrer and the Reichsminister all in one individual. Present among others were Fritz Thyssen, the Krupp family and leading officials of the I.G. Farben consortium. The "architects" of these measures were: Hjalmar Schacht—president of the Reichsbank and Hermann Josef Abs—a financial advisor to the Reich's government from Deutsche Bank and a financial advisor to the Vatican. Bormann's goal was to set in motion, based on the creation of financial principles, a widely planned evacuation operation, with the objective of securing the state's potential for the post-war period. It acquired the code-name "Regentröpfchen." "Target points" were: Argentina and other neutral countries, among others Turkey and Ireland. Its most important component part was operation "Eichhörnchen" regarding material goods. Others were among others:
1. "Aktion 1"
Viking
This was set up on 2 April and concerned the transfer of the scientific, political and military elite to so-called safe havens. The technical side of it was prepared by a member of Bormann's staff… Maximilian Erth (from September to October 1944). He was accompanied in his preparations by:
a) Philip Bouhler—chief of Adolf Hitler's private chancellory,
b) SS-Standartenführer Dr. Constantin Canaris (a nephew of admiral),
c) SS-Gruppenführer Dr. Gerhard Klopfer and in all probability:
d) Karl Hanke—Gauleiter of Lower Silesia (Niederschlesien).
2. "Aktion 2"
Läufer
Evacuation of the financial elite, linked with operation "Eichhörnchen," including a significant section of representatives of the aristocracy.
Like "Action 1," it was prepared almost exclusively by members of the SS.
Individuals responsible for the planning and co-ordination of measures were:
a) Franz Xaver Schwartz, treasurer of the NSDAP;
b) Franz Hofer, Gauleiter of the Tirol;
c) Ernst Hohenlohe-Langenburg. He was responsible among other things for negotiations regarding the transfer of resources from a special fund of A. Hitler called "Spende der deutschen Wirtschaft." This was a fund administering a special tax aimed at building up the economy. With this individual, co-worked among others: Otto Neumann and:
d) Bernhardt Krüger, chief of the VI F-4a Referat, employed among other things in the production of counterfeit money and forged documents, whose management was located in the town of Friedenthal. Bank-notes as well as identity documents were perfectly falsified, the almost unlimited access to them considerably facilitated the realisation of the described measures. B. Krüger survived the war, operating among other places in Argentina (along with his collaborator Dr. Herbert Scholz responsible for the transfer of foreign currencies to this country). The greatest role during the realisation of "Action 2" was played by the following individuals:
e) Grand Duke Christof von Hessen,
f) Wilhelm von Hessen,
g) Grand Duke Josias zu Waldeck und Pyrmont (Higher SS and Police Führer),
h) Prince Waldeck,
i) Prince Hohenzollern—Sigmaringen,
j) Prince Colonel Joachim von Auersberg—air attaché in Stockholm.

I realised that if the matter was to be treated seriously, I would have to prove its existence… on the basis of fully independent sources, since presenting the evidence in the form of Polish special service documents from 1945-1952 did not come, as it had become evident, at all into play. After the wave of archives destruction from the end of the 1980s, it was not even certain whether the documents still existed. Sisyphean labours therefore awaited me, estimated for years…

I could not count on making any progress in the problem of "documental verification" within the first few months. Therefore during this period I resolved to concentrate on an attempt to understand and explain the nature of this strange device—a generator of mysterious radiation.

This seemed simpler and as a result could deliver certain clues, as far as further directions of research were concerned. I resolved to assemble as large a group of consultants from the field of physics, as it would be possible.

"The bell" possessed so many characteristic features that finding some kind of unequivocal explanation seemed to be attainable, i.e. there was a basis for verifying different hypotheses. These features were chiefly:
- the employment of very high voltages;
- an emphasis of the phenomenon of "magnetic fields separation;"
- the occurrence of "vortex compression;"
- the fact that the device generated very powerful magnetic fields;
- the spinning of masses / bulky elements as a means to

achieve the above effects (directly or indirectly);
- as a result: the generation of powerful radiation;
- the continuous character of "the bell's" operation—i.e. non-pulse;
- the reference about transforming mercury into gold.

So there was no shortage of information, the hitherto existing barrier being the issue of understanding them. But it was difficult to find any device, which even approximately would be reconcilable with the above features.

Hypotheses so promising from appearances no longer came into consideration, like:
- A centrifuge for isotope separation. Only the spinning matched this concept, all other features excluded it.
- It was difficult to assume that "the bell" was to be simply a source of powerful ionising radiation, some kind of great X-ray tube transformed into a weapon. The emphasis to limit these effects contradicted this as well as information that "the bell" was not a weapon in itself. For this it would not be necessary to spin cylinders in opposite directions.
- Suspecting some form of turbine or electric motor did not convince me at all. A turbine does not generate magnetic fields destroying the surrounding electrical system, and besides is usually noisy. No electric motor can be supplied by high voltage since its winding would not withstand this.

In this situation I attempted to concentrate on another plot. Mercury is after all a metal with a very high specific gravity —20% higher than in the case of lead. Therefore perhaps the clue was the entry "spinning of masses"?

All the time I was making contacts with various consultants from the field of physics, asking in addition about this issue. One of them—Dr. Krzysztof Godwod from the Institute of Physics at the Polish Academy of Science advised me to contact Prof. Marek Demiański from the Institute of Theoretical Physics at the University of Warsaw, who was extremely well-informed as far as the phenomena caused by the very fast spinning of masses was concerned and would undoubtedly help me in solving this riddle…

In the meantime I received yet another signal referring to this trail. I was contacted by one of the scientists, to whom I had sent the description of the German research, with a request for an expert opinion. He was an employee of one of the Kraków institutes of the Polish Academy of Science: Dr. Mariusz Paszkowski—a man with a phenomenal research instinct and equally outstanding knowledge of physics. He stated that the description, I had sent him strongly matched work on the technical utilisation of effects associated with the Theory of Relativity, and that the clue to solving the riddle was the spinning of masses and occurrence of strong magnetic fields. What it concerned was that the adequately fast spinning of some object curves the space-time continuum, which in this case would signify the generation of a repulsive gravitational field! Paszkowski turned my attention to the fact that it had been attempted to exploit this branch of study before World War II

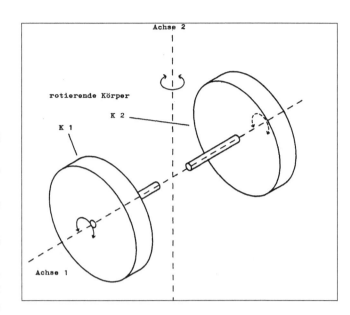

A schematic diagram of a simple, anti-gravity generating device. Drawing from patent documentation: DE 4017474A1.

in different countries, among others in Germany and the USA. As far as the latter country was concerned, the main establishment was located in Virginia (Virginia Polytechnic). The work carried out there was based on the spinning of balls made from a special steel alloy, suspended in the magnetic field. A velocity reaching 18 million revolutions per minute was achieved, when the spheres exploded under the influence of centrifugal forces. The goal was precisely to generate "anti-gravitation."

The Germans were also hot on the trail, although they had reserve for Einstein's theory. This came about not only because it was an example of "Jewish science," but also for purely scientific reasons. For Germany was the fatherland of quantum physics and the local physicists had serious difficulties with accepting the Theory of Relativity, which completely broke away from the quantum understanding of interactions in nature. Today we already know that this reserve was justified and that this theory—although confirmed by numerous astronomical

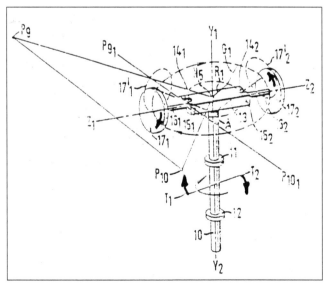

A diagram of a similar device… from Prof. Laithwaite's patent

observations, is incomplete and one day should be replaced by the quantum theory of gravitation.[205] But at that time it was a novelty, which perhaps gave the scientists in Hitler's service a kind of advantage... A sign of this school of "gravity generation" was among other things the work of O.C. Hilgenberg from 1931, treating this force as the resultant of the rotational motion of atoms (atomic spin)—i.e. as the "shadow effect" of atomic interactions.[206]

Such were the clues of Dr. Paszkowski. In any case they tallied to a large degree with the suggestions of Dr. Godwod as well as my own assumptions and led me to the same person—Prof. Marek Demiański—an expert on gravitation. Soon I met with him. The person in conversation with me showed a genuine interest in the matter, was clearly intrigued and also stated that ...vortex motion could be in this case the key to generating gravitation. Causing my amazement he even said that mercury would best suit this purpose "as a substance of high density and simultaneously as a liquid...i.e. a material characterised by a low viscosity." (The lower the viscosity, the greater the freedom of atomic spin). Later, already in front of the video camera lens, Prof. Demiański hypothesised further his considerations: "If indeed they had succeeded in aligning the axes of nuclei rotation in one direction, with the aid of a strong magnetic field..."

For me all of this began to look strangely familiar!

For the first time the different elements of this scientific jigsaw puzzle began to fit each other—not yet all, but everything in time. Prof. Demiański recalled that some years ago an article was published in one of the professional periodicals, in which measurements of "rotational" gravitation had been described, generated by the rotors of gyroscopes. This was the result of a wave of interest from the 1970s, when it was observed that a rotating gyroscope slightly but observably weakens the Earth's gravitational attraction.

We went to the Institutes' library, where after about 10 minutes we managed to recover this article. It had been published in 1989 in the prestigious magazine "Physical Review Letters"[207] and described the work of a group of Japanese scientists from Tohoku University. They were rather modest, i.e. they hadn't exceeded the normal speed ranges of industrial gyroscopes. Nevertheless they delivered shocking results.

Serial gyroscopes of quite small dimensions were tested, by weighing them on a very precise laboratory balance. A suitable weight was placed on one of the scale pans and on the other a vacuumed container with the gyroscope, to which ultra-thin wires led, supplying the gyroscope's motor. The container was made of glass so that through its panels and panels of the balance-cover it was possible to measure with the aid of a stroboscopic tachometer the rotational speed. In the next series of tests three gyroscopes were tested with rotors of masses: approx. 140 g (rotor median—52 mm), approx. 175 g (58 mm) and approx. 175.5 g (58 mm). Their rotors were accelerated to relatively low speeds of the order of 13,000 rpm., at the same time the measurements were carried out in both a state of rest and from a speed of 3,000 rpm upwards. Obviously rigorous care was taken to avoid any measurement errors and erroneous interpretations. Many other scientists (including from the editorial staff of Physical Review Letters) verified the course and results of the experiments. Even the influence of possible chemical transformations on the measurement results and influence of vestigial atmospheric convection [caused by temperatures differences] on the aerodynamic forces acting on the scale pans, were analysed. It took one and half years! These experiments delivered two unusual observations:

The anti-gravitational interaction (i.e. the magnitude of the gyroscope's decrease in weight) turned out to be far stronger than existing theories had predicted (based among others on the Einstein-Cartan theory). For the gyroscopes with a rotor weight of the order of 175 g it was admittedly only approx. 11 mg at 12,000 rpm, but the fact of there being a discrepancy with the predictions remains a fact. In the final part of the article it was written:

"If these theories are applied to describe our experiment, they give an extremely weak effect, but a giant reduction in weight (...) cannot be explained with them..."

Since the magnitude of the generated force was directly proportional to the rotational speed, it was possible on this basis to easily predict when the weight would fall to zero. It was calculated that this would occur with the two types of tested gyroscopes being accelerated to a speed of 3.27 MHz and 3.95 MHz (million rotations per second) respectively.

These speeds appeared to be simply astronomical and practically unattainable. I however acknowledged that since our understanding of these phenomena is limited anyway, perhaps it would be possible to increase this effect—which later proved to be true... By way of digression, it was precisely due to the "non-quantum" nature of Einstein's theory, that the results of the Japanese measurements didn't tally with the theoretical predictions. This theory simply fails, when it is necessary to calculate the relationship between gravitation and interactions on the atomic scale (such as atomic spin), even though it is undeniably a relationship of a fundamental nature.

The Theory of Relativity solution, which has in this case an application—the so-called Kerr metric and Cartan metric, was created to analyse astronomical phenomena and specifically phenomena occurring in the vicinity of a quickly rotating "black hole." Their usefulness for the description of gravitational effects on an atomic scale (rotating atomic nuclei) is thus limited. However there is no doubt that on this scale gravitational effects are significant. The space-time continuum in the immediate environment of atomic nuclei is often described not as a homogeneous medium, but rather as a kind of foam (gravity itself is the same as the bending of the space-time continuum). The problem is that on a large scale these effects neutralize each other, if only because the atoms rotate chaotically—usually in various directions...[205]

Despite this, the "classic" variant, using only the rotation of a mass, already finds application in various technical devices. There is already a large number of such patents, among others the German patent DE 4017474A1, British patent

WO 86/05852 or patents of Professor Laithwaite from Imperial College in London. During an official presentation in the mid-1970s, his device apparently lost several dozen percent of its weight![208] What is interesting, in both this as well as in the majority of similar cases, not one element is rotated, but two —in opposite directions. Again a familiar element?

Could the key lie here?

In scientific journals even analyses appeared indicating the possibility of a significant increase in the gravitational interaction as a result of "combining" the rotations with a strong electric charge.[209,210] This in turn recalls the reference to "very high voltages," used in the case of "the bell." In spite of this, all these elements connected current work on the generation of gravitation with that from the time of the war only indirectly. The key ideas in this last issue: "separation of magnetic fields" and "vortex compression" didn't seem to have a contemporary equivalent. But this was to be only for so long…

The precious clue proved to be hidden in the results of other contemporary research—research which a dozen or so years ago had wide repercussions in the world of science and is undeniably one of the precursors of a breakthrough taking place. It concerns work carried out under the direction of a Russian physicist, Yevgeniy Podkletnov[208,211] (in English literature the spelling Eugene Podkletnov is also met). Here is an excerpt from the description of the first revolutionary experiment, which was published by the British press:[211]

Finnish scientists are on the eve of revealing the details of the first anti-gravity device in the world. Measuring around 12 inches [30 cm] the device may significantly reduce the weight of a given object suspended above it. This property, which has been subjected to rigorous testing by scientists and whose description is to appear in journals devoted to physics next month, may prove to be the ignition spark of a technological revolution…
The Sunday Telegraph has learned that NASA is treating the issue seriously and will finance research with the aim of interpreting in what way can the anti-gravity device be used to propel flying vehicles.
Scientists from the Technological University in Tampere in Finland, who made the discovery, state that their device may become the heart of a new source of energy, in which it will serve to displace liquid through the turbines of current generators.
According to the declaration of Eugene Podkletnov, who conducted the research, the discovery was accidental. It occurred during the routine research of "superconductivity," i.e. the ability of some metals to lose their electrical resistance at very low temperatures.
The research team conducted experiments with a fast spinning super conducting ceramic [disc], suspended in a magnetic field generated by three electrical solenoids, at the same time the whole assembly was sealed in a container called a cryostat designed to maintain the low temperature.
"Our colleague entered the room in which the experiment was being carried out—recounts Dr. Podkletnov—who smoking a cigarette released a little smoke over the cryostat and noticed that the smoke kept on drifting upwards. This was very strange and we could not explain it."
The research revealed that objects placed above the device lost a small part of their weight, as if the device shielded the given object from the influence of [the Earth's] gravity—an impossible effect, in the opinion of most scientists. We presumed that perhaps some mistake had crept in—recounts Dr. Podkletnov—and we did everything we could to eliminate it. It became evident however that the effect didn't disappear. The team discovered that even the air pressure immediately above the device fell slightly, at the same time this effect was visible on every floor of the laboratory.
In recent years many so-called "anti-gravity" devices developed by amateurs as well as professional scientists have been rejected by the scientific establishment. That what distinguishes the last discovery from previous ones is that it defied the research of a sceptically disposed team of independent specialists and has been accepted as worthy of publication in such an important scientific journal like the "Journal of Physics D: Applied Physics."

Podkletnov conducted this experiment in 1996, achieving a weakening of the Earth's gravitation by 2-3 % (in 2000, already working for the Japanese Toshiba consortium he stated that he had cancelled gravity completely).[212] Most important is the idea of his invention itself—in all cases it involved the spinning of a superconducting disc in a spinning magnetic field. In 1996, this disc was made from a composition of yttrium, barium and copper oxides, achieving the "modest" speed of 5,000 rpm. I must confess that this issue absorbed my thoughts for a long time and no association ever crossed my mind that would allow this effect to be placed within the context of generally accepted physical knowledge. The main limitation resulted simply from the non-quantum nature of the Theory of Relativity, which makes it impossible to link for example gravity with an electromagnetic interaction. This caused that the associations sought after by myself in principle do not exist in academic physics. At the same time this added a certain aura of mystery to the work of Podkletnov.

I started to reflect on what was so inherently unusual in superconductors… After some time a kind of revelation occurred to me. A superconductor is after all not only a material characterised by a zero electrical resistance. It is also an ideal diamagnetic material. This means that it "does not let in" any magnetic field whatsoever. In the case of Podkletnov's devices it involved therefore something that should be called "separation of magnetic fields." Despite many previous attempts, this was the first event which had imparted some kind of practical meaning to the basic idea applied by the Germans. It was the second clue after "Kerr's metric" linking "the bell" with gravitation, or with relativistic physics in general. The only thing still lacking was linking this with the high voltages and strange radiation.

In any case I had succeeded in taking another step forward.

I spoke a couple of times with my British co-worker, Nick Cook, an aviation expert from "Jane's Defence Weekly," and who knew Podkletnov personally; on the subject of various current events from Podkletnov's "experimental plot." I heard from him among other things, that Podkletnov had specially studied material engineering so as to be able to make a highly resistant superconductor, capable of spinning at a very high speed. In 1999, the director of Toshiba, for whom the anti-gravity project had become "the apple of his eye," stood on such a plate 30 cm in diameter and was very proud that through this it was not damaged.

Seeing that the Russian had commenced his university education from the point of view of this work, it was obvious to me that he had been handed this "innovative" knowledge on a silver platter. I asked Nick to ask Podkletnov at the nearest opportunity what had made him think of the by no means obvious concept of "separation of magnetic fields"—"the clue." I was not at all surprised by the answer—it had come from access to the results of German work from the time of the war! Podkletnov officially admitted this and Nick Cook has mentioned this in his latest book.[208] Certainly it was awkward for Podkletnov to admit that he had had access to secret intelligence materials, so he later explained that he had derived his inspiration rather from Schauberger's work—but this "researcher" never used, or even contemplated using magnetic field sources, neither he had any idea about the superconductors.

Besides Podkletnov, Nick Cook had contact with another outstanding physicist engaged in work on anti-gravity. He was Dr. Hal Puthoff, the author of many revolutionary scientific works, and linked to aerospace consortiums such as General Electric (engines), Sperry, and also an employee of the CIA and NSA. Presently he holds the position of a director in the Institute for Advanced Studies in Texas. I asked Nick to convey to him my request for an interpretation of the "German" notion of "magnetic fields separation." Puthoff was to state that the links with relativistic physics are obvious…

So there was a further element to the puzzle, which—as usually happens in such cases, generated the next questions. Now I was "tormented" by the question of linking the "magnetic fields separation" with the high voltage occurring in the German specifications. Did the displacement of a magnetic field from superconductors (called the "Meissner-Ochsenfeld Effect") have any equivalent in the physics of high voltages?

I reflected upon the whole issue and quite naturally concluded that a high voltage current (as it later proved—over a million Volts) could not supply any engine or winding and the like. It must have led to a discharge, and consequently—it must have been a question of plasma physics. As it was to prove, this was the next step forward. At the same time I was reminded by the remark of Professor Demiański, who had stated that it is very important for the "active substance" to be characterised by a low viscosity. He had said this within the context of mercury, since by definition mercury has a lower viscosity than a solid body. If it indeed involved a key factor, it is possible to "extend" this reasoning and draw the conclusion that gas would have a lower viscosity than liquid, and lower than gas—plasma.

Therefore I began to look for a plasma equivalent of the Meissner-Ochsenfeld effect and… I soon found it…after a couple of days:

It became evident that under certain conditions plasma created through the flow of electric current, creates in turn a special kind of vortex. Such a vortex is called a plasmoid and is a stable or quasi-stable creation, being in a way a closed structure, capable of "living" for a certain time even after the disconnection of the power supply. After all it can even be considered proven that ball lightnings are plasmoids.[213,214] They are quite special formations and bring the next clue in solving our puzzle. There is no shortage of descriptions of ball lightnings, even coming from fully credible groups of witnesses, where there is mention for example of a ball lightning passing through a window pane or other obstacle, without its destruction.[215,217] Many such cases have been described among others in the very well documented work of Dr. Andrzej Marks,[216] published in Poland many years ago. Here there was even the description of an event in which a ball lightning entered through the cockpit's windscreen of a large passenger aircraft and flew along the central aisle practically through the whole aircraft—in full view of the crew and passengers. This was yet another fact, which turned my attention to plasma vortices. For this kind of penetration to be possible, relativistic physics and hence also gravity physics would have to be involved—this object must have simply curved the space-time continuum. I still did not know how this was possible, but familiarising myself with the scientific literature on this subject threw light on the issue.

It became evident that such types of plasma vortices are credited with a certain, unique feature—namely the lines of magnetic field force are almost completely closed. This is defined in the literature as a "magnetically closed system." Only due to this is a plasmoid of this type extremely stable for a plasma vortex—it is simply almost isolated from its environment. As far as the Theory of Relativity is concerned, the importance of this phenomenon (separation of magnetic fields) is interpreted like this: since the fields are "coupled" with the space-time continuum, the isolation of a field (in this case magnetic), or speaking the language of physics: ensuring the field's locality gives in effect a certain locality of the space-time continuum. In other words: separation of field(s) is the key to control gravity, because bending of space-time equals the generation of gravity and locality of space-time equals the screening of gravity. So we not only have "separation of magnetic fields" but in general a large similarity even to Podkletnov's experiments. It is after all a plasma vortex as well as a spinning magnetic field—moreover very strong and spinning very quickly. Very quickly—since the magnetic field very strongly compresses the plasma. The compression is so strong that it is even compared by some to the conditions prevailing during a nuclear explosion.[218] Exactly! Was this phenomenon, professionally called "pinch," not that sought after "vortex compression"? It had been managed to fit into place the next element of the puzzle…

Podkletnov's problems

The problems which limited the efficiency of the anti-gravitational devices of Podkletnov and his copiers (NASA, Boeing…) result directly from the limited mechanical strength of modern superconductors. Overcoming them is however possible even without interfering with the issues associated with the physics itself of this phenomenon. One can create a material which will resist much greater rotational speeds—and above all, which will be able to perform the function of a "fields separator" in the very construction of the flying craft. Such a possibility is given by the so-called fullerens.

They are peculiar crystalline structures, built from carbon atoms—in such a way that a layer, one-atom thick, creates a spatial figure with a regular crystalline structure—approximate to a sphere or tube. At that time they are empty inside. Very interesting quantum effects reveal themselves particularly in the tubes, which is associated among other things with the fact that they constitute "traps" for various particles and ions. The ions of certain metals, trapped in fulleren fibres cause e.g. that the entire material acquires the characteristics of a superconductor, at the same time it has a colossal superiority over Podkletnov's ceramic compounds—with regard to tearing resistance it is comparable to diamond. It will be possible (this is a question of a few years) to make a composite from fulleren fibres, able to resist a rotational velocity of the order of hundreds of thousands of rpm. Perhaps this will completely solve "Podkletnov's problems," all the more that it will be possible to make from a superconductor not only a rotating disc, but even a complete flying object.

Plasma

We know four states of matter aggregation: solid, liquid, gas and plasma…ionised gas. They differ in the degree of matter ordering, at the same time in plasma the atoms have the greatest freedom of movement (also rotational) and usually travel faster.

We know two fundamental types of plasma: "high-temperature" with a temperature the order of millions of K and "low-temperature" (thousands of degrees Kelvin), occurring among others in electrical discharge in a laboratory. In so far as the first type is a material practically completely ionised, in the second case we are dealing with a mixture of ions, free radicals, electrons and excited as well as non-ionised atoms. In plasma conducting an electric current magnetic fields may occur with a colossal concentration of energy, e.g. in rotating plasma the magnetic fields compress the plasma and compensates with a centrifugal force. In this case this phenomenon is called "toroidal pinch."

Familiar elements…a thick rubber mat, movable walls and other elements shielding against radiation

The "plasma focus" from the Institute of Plasma Physics and Laser Microfusion. The device is visible from the side…

…from the front…

…and from the rear, along with the author of this book.

These facts led to some very encouraging conclusions. It seemed that plasma physics was able to ensure a magnetic field strength (a charged plasma vortex itself generates a magnetic field on the basis of the so-called dynamo effect) and speed of rotation far higher than that of any mechanical system. It would obviously follow from this, that the German "bell" was some kind of "trap for plasma vortex."

At this stage of my quest I resolved to contact an expert in this field, and specifically someone specializing in the rotation of plasma.

I made my way first of all to the Institute of Nuclear Research in Świerk near Warsaw, and contacted Professor Marek Sadowski, who stated however that plasma vortices itself lay somewhat off the beaten track of his field and directed me to the Institute of Plasma Physics and Laser Microfusion in Warsaw's Bemowo district. Thanks to the help of the extremely kind Dr. Zagórski I finally (in 2001) made direct contact with the most appropriate person, who turned out to be

The cables reached as far as the second floor…

Dr. Marek Scholz, chief of a department in this institute. I made an appointment with him at his study. After a few days, full of tension and uncertainty, I made my way on a beautiful summer's day to the Bemowo institute. It so happened that to reach Dr. Scholz's study I had to cross a great hall full of various research devices (at its entrance I was greeted by the sign: "Warning! High voltages!"). I crossed its threshold and… was greatly shocked…

Before me stood "the bell"…or its contemporary counterpart. It appeared to embody a copy of the device from the descriptions of German prisoners of war, as if all the details had been imitated exactly with special care. Its housing was admittedly steel and the whole device had been rotated by 90°, but apart from this even the overall dimensions exactly matched that of the German device. Other elements familiar from the war-time description were visible as well, such as huge feeder cables as thick as arms, and above all…powerful anti-radiation shields, including the heavy rubber mats described earlier.

It became evident that ions accelerated by the flow of high voltage current reach such horrendous speeds (in some devices even 50-100 km per second) that during collisions a thermonuclear fusion reaction takes place. This causes strong X-ray and neutron radiation to be emitted during the course of the experiment. This time several further elements of the puzzle had been explained at one stroke: the anti-radiation shields and the radiation itself, the transformation of mercury into gold (undoubtedly on a small scale…thermonuclear fusion), the gas bubbles forming around the base of the German "bell" (without doubt the influence of neutrons), and even the "humming" sound.

One could state that we were "home and dry," if not for one "small" detail…the device from Bemowo, i.e. the so-called "plasma focus" is not, in spite of everything, a revolutionary achievement of humanity, something worth the lives of its own scientists. Something here was still missing…

It became evident that it was simply the spinning that was missing. Yes, plasma sometimes creates a kind of vortex, but this is usually a side effect. Nobody yet, nobody after the war… has built a "plasma focus" device chiefly for the fast spinning of heavy ions (I am obviously taking no account of top secret work)…the internal construction of every "plasma-focus" is purely static. The conception of rotating or counterrotating cylinders remains unknown. Nobody has struck upon the idea of doing this!

So there is still an opportunity to display one's talents in this field.

Due to the virgin nature of this field it was necessary to reflect on what specific way could it be possible to "spin" plasma…i.e. the ions of mercury.

This was a hard nut to crack. After many unsuccessful ideas and sleepless nights the following solution struck me:

I imagined a large, metal drum, in which a small amount of mercury was present. The drum would then be accelerated to a speed of say tens of thousands of revolutions per minute. Under the influence of the centrifugal force the mercury, as a liquid, would cover the walls of the drum creating a thin layer. After achieving the target speed a high voltage electrical discharge would be created between the circumference of the drum (the mercury layer)—and its axis—the core. Theoretically this would accelerate the ions of mercury towards the core, with a speed of many kilometres per second. But since the mercury would already possess a certain torque, in due measure of approaching the core its angular velocity would increase on a similar basis as in the case of a skater, who during a pirouette brings the arms close to the torso thus developing an increase in the rotational speed. In the case of the drum with mercury this would lead to an overlapping of the two speeds—created by a preservation of the torque and a result of the flow of electric current. From my approximate calculations it followed that by this means it would be possible to achieve a speed of the ultimate "compressed" vortex of the order of even hundreds of thousands of revolutions per second (if a linear velocity of 50 km/s is otherwise attainable, then with a vortex diameter of 6 cm this would signify a rotational speed of 180 thousand revolutions per second). Obviously this is only an estimated value, however one can see that by this means it is possible to achieve much more than Podkletnov's ceramic disc was able to withstand (approx. 100 revolutions/second). Of course the problem of the harmful, to put it mildly, radiation remains unresolved. Is this precisely why Podkletnov did not pursue the easy route? I recalled from somewhere else the remark made by a certain colonel, who "had heard that American pilots were falling apart" (in the 1960's) as well as a program broadcast by the Discovery Channel, dedicated to the famous "Area 51," titled "A billion dollar secret." Reference was made to the flights of strange objects over this base being terminated, when the employees of this super-secret centre "unmasked" it en masse, by lodging summonses in court due to the occurrence of various diseases and ulcerations, body fragments falling apart etc (i.e. the symptoms already familiar to us). Couldn't even the Americans overcome this problem?

In any case I introduced my concept of accelerating mercury ions through the spinning of drums to Dr. Marek Scholz from the Institute of Plasma Physics. He seemed to be extremely intrigued by it and acknowledged it to be "interesting." He said that he would try to prepare a complex analysis of this issue and assess the possibility of practically testing such a solution. He warned however that it was a complex task, which would take some time.

Before bidding farewell I asked however "for the sake of peace" if in his opinion, at first glance, such a device had any sense at all? He answered: "it must have sense, if something like that was made!" This constructive approach sounded encouraging…

We will yet return to this purely physical aspect of the "Chronos"/"Laternenträger" project. But this is only one of many aspects, which I have attempted to unravel, and I will now turn to the description of another: medical-biological. The issue of what kind of effect "the bell" exerted on living organisms and organic substances constituted one of the more serious challenges.

However, in this case I also managed to get in touch with a specialist, to whom… something incredible—everything was at once very clear. To such an extent that when I mentioned that plants were "dissolved" to the form of "a substance resembling rancid fat, but without any smell," this person showed me from memory some specific chemical formulas corresponding to this process.

This individual turned out to be the very pleasant Professor Alina Kacperska from the Department of Biology at Warsaw University. I found out from her that the whitening of plants was caused by so-called oxidative stress. This is based on the "photo-oxidation" of chlorophyll, i.e. its oxidation under the influence of photons. The oxidizing action is indicated in this instance by the lack of smell. In the next stage an enzyme called lipoxidase is released, which oxidises the fatty acids to the form of the aforementioned greasy substance (I had earlier assumed that it involved so-called "autolysis," i.e. the decomposition of a cell under the influence of its own enzymes, so I was not overly mistaken). Obviously chlorophyll and tissue are not oxidised by photons alone. If their energy is large enough they shatter the molecules of oxygen or chlorine into free radicals i.e. into single ions, whose reactivity is so high that they oxidise practically everything. This can be the effect of the interaction of highly energetic radiation of high intensity—for example close to the source.

The above sentences are a summary of my conversation with Professor Kacperska. At our farewell I received a photocopy of a book's excerpt, touching upon these "exotic" issues.[219] I learnt from it that in reality far more complex processes were involved than it would appear from this short summary. But only the chemical aspect was described, without any connection to the effects of particular types of radiation, so the description concludes little to solve our problems.

Despite the complex explanations of the Professor, I was struck by a certain divergence. The nature of the transformations caused by "the bell" appeared admittedly to be known to science, but their intensity still turned out to be unusually high when compared to phenomena researched by contemporary biologists.

A purely intuitive suspicion remained in my mind that the "oxidative stress" was not the only mechanism responsible for these transformations. I suspected that the effect of generated energy on the structure as such was also involved—on the degree of matter ordering. I recalled the work of the Russian scientist Genadiy Shipov and works of the German physicist, Professor Burkhard Heim (working during the war at the Goettingen University).[220,221,222] In all their works there was reference to changes being created in the structure of materials by artificially generated gravitational waves, also referring to metals. I am afraid however that a final settlement of this dilemma is not possible at the present stage of research…

For the time being let us leave it at that, as far as the physical and biological interpretation of the descriptions conveyed by the Germans is concerned.

As I have already mentioned, one of the key tasks, that I had set myself was proving, on the grounds of independent sources, the existence of this super secret and presently unknown research-armament project of the Third Reich. Even by definition this was to some extent an almost unrealisable task—if it was easy we would have known all about this a long time ago. This issue has taken up a great deal of my time and from the very beginning constituted a "No. 1 priority," but probably that's why I managed to gather particular evidence.

First and foremost this involved a "verification" of the wartime career of Professor Walther Gerlach. This was intriguing to the extent, that as I have already mentioned, despite being labelled as a nuclear physicist, he never directly took part in research linked to the construction of a nuclear reactor or weapon. So if in spite of this he was appointed as chief of the Reich's Scientific Research Council (Reichsforschungsrat), then, as I mentioned, this must have reflected in some way the importance of his scientific activity. Generally accessible sources proved however to be extremely scant, as far as any description of his activity was concerned, and at any rate scant in relation to the function performed by Gerlach…superior in relation to the whole of German physics. I never managed to get hold of any of his wartime scientific work. Whereas after the war Gerlach in general "switched over" to universal subjects. His main publications from this period were "The Physics of Everyday Life" (1956) and "Humanity and Scientific Research" (1962). He apparently wanted to cut himself off from the wartime period of his life. It is difficult to ascertain, if he felt obliged to maintain the secret…rather not. It is more likely that as a universally recognised authority he was afraid of his past. The prospect of questions being put to him like: "were you a fanatic of National Socialism?" or "did you take part in experiments on people?" or "did you want to poison half of Europe?" must have seemed horrifying (I will come back to this later).

A certain picture can however be created on the basis of his pre-war work.

It is no secret, that Gerlach was first and foremost a physicist engaged in magnetism, electricity and quantum electrodynamics. His most important publications from this period were entitled: "The Experimental Foundations of Quantum Theory"…1921, "Matter, Electricity, Energy"…1923 and "Magnetism"…1931.

He analysed the relationships between magnetism and atomic structure, calculated atomic spins and studied the spin polarisation of ions in a magnetic field[223]…simplifying one can ascertain that spin is the rotary motion of an atomic nucleus while spin polarisation is the ability of a group of nuclei to spin in the same direction. Gerlach was the co-author of the pioneering "Stern-Gerlach experiment," which brought him world-wide fame, and concerned precisely the behaviour of atomic nuclei in a magnetic field. In one of the most important publications referring to German wartime scientific work, this was defined as the realisation (still during the war) of research "with no practical significance," at the expense of the nuclear program plunged into a state of stagnation.

Was this so?

Could he have afforded this?

At any rate it is a fact, that Gerlach was not especially interested in the nuclear program, which he formally supervised on account of directing the Reich's Scientific Research Council. The Council had to regularly demand reports from this work, for whose writing he had neither the time, nor the desire. On one such report from the turn of 1944-45, which was out-of-date anyway, Gerlach changed the date to two months later before sending it, although in the end he didn't send it anyway.[223] The truth is that he was simply too busy with other issues, which he considered much more important.

As one can see, even a superficial analysis of his scientific work shows, that this was a person ideally suited to the research program described in this part of the book. A more discerning analysis only confirms this.

In 1924 for example, he wrote an article for the "Frankfurter Zeitung" referring to the possibility of transforming mercury into gold. In this way he had shown his attitude to the work of Professor Adolf Miethe from the Institute of Physics in Berlin…Charlottenburg.[225] It involved mercury being subjected to an electrical discharge. Gerlach stated that such a process was of course possible, but unprofitable. He wrote that if only the cost of electricity was to be taken into account, a gram of gold would cost one hundred thousand Marks.

Other work proves that Gerlach was faultlessly well-informed in these issues. In 1929, he wrote an article devoted to the fluorescence of mercury ions in a strong magnetic field, in other words referring to the behaviour of mercuric plasma.[226] He had obviously been engaged in these matters for a long time, because as far back as January 1925 he wrote to Arnold Sommerfeld about research on the spin (rotary motion of atoms) of ionised mercury…of mercuric plasma.[227,224] An exception concerning the post-war period is that in 1954…at a conference organised by AEG and his home scientific establishment…Munich University (of which he became rector after the war), he made a speech about research in the field of plasma physics, mentioning the creation of new elements and employing a voltage of half a million Volts to accelerate ions.[228]

A very interesting element, although of a somewhat different nature is embodied by the pre-war correspondence between Gerlach and Peter Kapica…the great Soviet physicist, later a Nobel prize winner.[224] It is interesting that, as it appears, a common attribute of both scientists was an interest in the nature of superconductivity. Of course the question comes to light…how could Gerlach have approached this phenomena. The assumption appears to be logical, that he could have approached this only from the aspect of the Meissner-Ochsenfeld effect—he was after all a "magnetician." If this is combined with plasma spin and with mercury, we find ourselves much closer to the "ultimate" phenomena. Especially if we realise that at the same time Gerlach became interested in the nature of ball lightnings. He left behind an article published in "Die Naturwissenschaften," in which he emphasised "the extremely strong induction activity of a flying ball."[229] So he must have been aware that a plasma vortex was involved, distinguished by a very strong magnetic field (spinning fast), of a special kind.

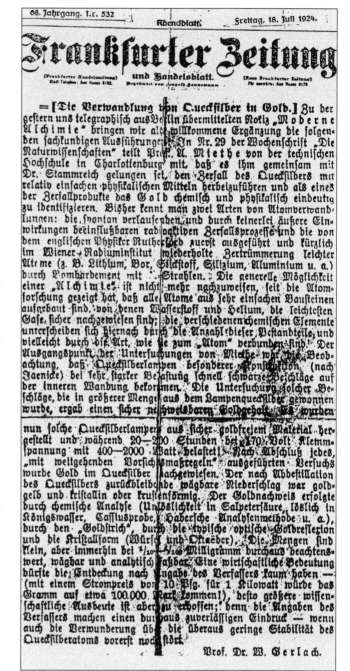

Gerlach's article referring to the "transformation of mercury into gold"

All of this however does not automatically prove that Professor Walther Gerlach was engaged in the realisation of a super secret, promising research project during the war. Of course it couldn't have proved this, as reference was made to pre-war work.

However such evidence exists.

At first it seemed to me that one of the simplest ways to acquire such evidence would be to gain access to recordings from the listening in of conversations held by leading German physicists after the end of the war. The British interned ten of the most outstanding scientists and during the period of June-December 1945 detained them (together) in the luxury estate of Farm Hall near Cambridge. Simultaneously an intelligence operation was carried out under the code-name "Epsilon," aimed at recording all conversations, which the interned scientists carried on with each other. After several decades this fact was finally revealed and the contents of the conversations were published in the form of a book.[230] I hoped to find in it the sought after details referring to Gerlach's activity, who had obviously been present in this group. I was however sorely disappointed. The revealed text was the equivalent of only 200-300 pages of typescript, i.e. roughly a single full day of conversations, held by a group of people. It is completely out of the question, that each of them uttered only a few sentences daily. They were after all worried by the arisen situation and in most cases had not seen each other for a long time. The recordings rendered accessible give the impression of being fragmentary, for example there are questions without answers or answers without questions. Issues are amplified, with no knowledge of where they came from. Here is an example of a not entirely clear exchange of opinions, including at the same time some intriguing elements [bold by I.W.]:[230]

> *Otto Hahn: Surely you are not in favour of such an inhuman weapon as the uranium bomb?*
> *Gerlach: No.* **We never worked on a bomb.** *I didn't believe that it would proceed so quickly. But I did think that we should do everything to construct sources of energy and exploit the possibilities for the future. After the first results, when the "cube method" had significantly increased the concentration, I spoke to Speer's right-hand man—Colonel Geist, as Speer was not available at the time, and later Sauckel from Weimar asked me: "What do you want to do with these things?" I replied: "In my opinion, a politician who is in possession of such an engine can achieve anything that he wants."*

Nowhere is there any explanation of what should be understood by the word "engine." The possible assumption that it involved a nuclear reactor is not at all obvious, since the Germans never called it an "engine." Irrespective of this even the statement itself: "we never worked on a bomb" may be shocking. There are a multitude of such insinuations in the recordings rendered accessible from Farm Hall. Here is another interesting curiosity, a quotation from p. 77 of the publication included in the bibliography (the previous one was from p. 80), forcing one to take a more discerning look at Gerlach's wartime role:[230]

> *Harteck: If we had worked on an even larger scale, we would have been killed by the Secret Service. Let's be glad that we are still alive. Let us celebrate this evening in this spirit.*
> *Diebner: Professor Gerlach would have been an Obergruppenführer and would now be sitting in Luxembourg as a war criminal.*
> *Korsching: If one hasn't got the courage, it is better he surrender at once.*
> *Gerlach: Please don't make such aggressive remarks!"*
> *[he leaves the room]*

Exactly! Perhaps Gerlach had been a high-ranking SS officer, on a similar basis as the other important scientists working for this organisation—for example von Braun or Prof. Grawitz? We will probably never get to know… From the materials in question it is difficult in any case to obtain any concise knowledge on the subject of Gerlach's work. The only interesting thing in my opinion, which still emerges, is evidence of the outright extreme Nazi fanaticism of this scientist, totally out of character even with the attitudes of the other interned men. When he learned about the dropping of the first atomic bomb he fell into a depression, started to cry and according to the British officers, intended to commit suicide.[233] But not because he was perturbed by the fate of the victims. As Professor Mark Walker explained:[233]

> *Gerlach affirmed that he had never supported the idea of constructing inhuman weapons, such as an atomic bomb. (…) He was however dejected, because the Americans had demonstrated their scientific superiority.*

Gerlach wanted to commit suicide, simply because he had once again experienced the defeat of the Third Reich! His associate… Werner Heisenberg, also interpreted it in this way.

Fortunately the way in which the materials from Farm Hall were classified was not 100% effective. For not only were documents or tape recordings left over from the affair, but also living people…

I tried to determine if there were any surviving witnesses, or best of all: historians' works based on more extensive factual materials, before "declassifying" (read: "abridging") of the original records. It turned out that something like that existed and furthermore…existed in Poland. Herbert Lipiński, a Polish historian, unfortunately no longer alive…and former employee of the Provincial Administration in Zielona Góra, had been engaged in a discerning analysis of this issue in the 1970s. The results of his work were published in the form of a series of articles…they referred of course to the German wartime scientific effort. He had had access to extensive source materials connected with Farm Hall. From his description something completely different followed, than from the "declassified" official version. Although the latter does not contain any specific references to the "Chronos"/"Laternenträger" project, at a time when practically nobody had yet heard about it, Lipiński wrote in reference to Gerlach that:[231] *the topics of conversations were* **most often:** *"atomic nuclei," "extraterrestial space," "magnetic fields" and "the Earth's gravitation."* [bold by I.W.]

For somebody who approaches German physics from the point of view of the nuclear programme, these ideas don't match each other at all and in connection with this appear to be devoid of any sense. But we know that they match…

They could not match any better!

If in the materials in question, we would like to discover evidence that the Germans had worked on combining magnetism with atomic physics to overcome the earth's gravity and enter outer space, then this evidence would not look any different than it does…

However this is not all.

There is also evidence that Gerlach's project had a truly special standing. Another researcher, this time British...Philip Henshall, was analysing Martin Bormann's (Hitler's secretary) diaries and noticed the fact that in his correspondence with Gerlach the subject of the "Wunderwaffe" appeared.[232,234]

Henshall wrote:

*At the end of 1944 Gerlach wrote to Bormann (...), that the project on which they were working might be **decisive for the war**." [bold by I.W.] As usual Gerlach was playing the role of a cautious scientist and did not want to state unequivocally that he had a "miracle weapon," which would end the war.*

Martin Bormann, on the photograph wearing the uniform of Obergruppenführer. He was the confidant of Hitler's greatest secrets and their scrupulous executioner.

This quotation demands a few remarks. First and foremost the question springs up of whether it referred to a nuclear weapon... this was however impossible. The Germans were simply not constructing a nuclear weapon, this was still very far off. Werner Heisenberg stated unequivocally after the war, that: "the Germans were interested in a nuclear reactor, but not a bomb."[233]

At the same time deserving reflection is the fact that in the decadent and at any rate practically hopeless stage of the war, Gerlach dared to write about a weapon, which would decisively ensure victory. After all by this alone he had assumed great responsibility. One did not hurl such slogans at the Führer without backup. In any case we have confirmation that **Gerlach was engaged in the realisation of a project "decisive for the war"**... this term is worth remembering, because it is going to reappear.

The term itself is also worth reflection. For it is, against all appearances, something completely new...as far as the designation of an arms programme is concerned. In 1944, when "total war effort" ("totaler Kriegseinsatz") was proclaimed, the term "important for the war" ("Kriegswichtig") appeared and began to function. It meant the lifting of other administrative restrictions. If for example in city "X" some work "important for the war" was being carried out, which demanded bronze, with the absence of another alternative, fences, monuments and fixtures were dismantled overnight and the next morning the bronze was at one's disposal. The term "Kriegswichtig" was the key that opened previously locked doors. The term "decisive for the war" ("Kriegsentscheidend") occurring (as it was to prove) as an official classification, was however something unusual and not only at the end of 1944. I personally analysed in depth cubic meters of German documentation referring to technology and never came across this term in a different context...as an official designation of any other research project or activity. So far I have not met anybody who was familiar with this term. As I have mentioned, we will return again to this issue. It is difficult however not to have the impression that what was to decisively change the outcome of the war, even in its final months or weeks, must have been equally shrouded in an unusual secrecy. This generates the next question: did Hitler know about all this? In the light of available information one should however answer this positively...and that unequivocally.

This is indicated simply by the presence of the Führer's personal secretary in the whole affair. Besides: if Goebbels and Skorzeny had known about the "Wunderwaffe," then Hitler would have known all the more. He undoubtedly had set his hopes on the project no less than Gerlach himself.

I presented above some attempts to explain Gerlach's role in the described project, however there is also some new information about the remaining scientists. It has become evident that several of them were brought over to United States within the confines of operation "Paperclip"—aimed at gaining outstanding German specialists.

Their personal files have survived, initiated in connection with this project, which I had the opportunity to acquaint myself with in the NARA archive at College Park. Admittedly many documents are still excluded from the files and kept in the CIA archive (which in itself constitutes emphatic evidence of their importance), but most are accessible.

The most important among those persons was undeniably Dr. Kurt Debus...responsible for the "separation of magnetic fields" issue as well as the power supply of "the bell." His personal file was also markedly thicker than those of the remaining scientists.[235]

His parent institution was the Institute of High Voltages at Darmstadt Polytechnic (Technische Hochschule). In 1942, he was transferred to the research institute of the AEG consortium in Berlin...Reinickendorf, in addition he also co-operated with the centre in Peenemünde. He was the author of several publications and patents regarding high voltage measurement technology. He developed among other things instruments for high pressure measurement and high voltage discharge parameters measurement. At the AEG research institute he constructed a power supply unit, supplying over 1 million volts current and took part in the equipping of a supersonic wind tunnel. He also took part in the development of measurement instruments for the V-2 test launch pads. In light of Gerlach's character there is little wonder that Debus was an exceptional fanatic, linked to the Nazi movement from the very beginning. He was a member of the SA from May 1, 1933, and of the SS from February 1, 1939. He was awarded with the cross of merit (Kriegsverdienstkreuz). Some interesting information found

> A search has been made of the records of the 7708 War Crimes Group and no record of Subject is on file in that office.
>
> Records of the 7970 CIC Group Central Registry indicate that DEBUS, Kurt denounced a former associate as a political defeatist in 1942, and caused him to appear before the Gestapo Headquarters. The former associate, Richard CRAEMER, was sentenced to two years imprisonment. CRAEMER eventually avoided actual imprisonment through the intervention of the German Electric Company (Berlin) which claimed his services were indispensible.
>
> **RESTRICTED**

An excerpt from Debus's files, referring to the "Debus-Crämer" case. It's worth to confront this with the short, official biography of Debus published by NASA (and available as a PDF), where he essentially was presented as an anti-Nazi. In fact an exceptionally ardent SS officer!

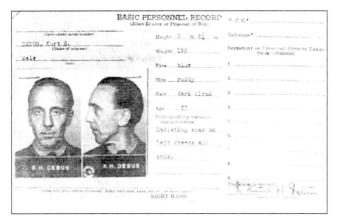

An excerpt of the American personnel record on Debus (Debus's file, see bibliography)

its way into his personal file to a large degree by chance.[235]

Everything began from the controversy surrounding his fanaticism. The Americans suspected that he was an SS officer. Debus declared on the other hand that he "only" had the rank of an NCO: SS-Staffelunterscharführer. Through the so-called de-Nazification court in occupied Germany all materials, that could be gathered, were assembled—especially court records. On this occasion an interesting issue was brought into light. In 1942, Debus had informed against one of his fellow co-workers from the AEG Research Institute…Richard Crämer, to the Gestapo. It all began very innocently. One day both met at work in the morning and Debus asked Crämer if he had slept well. Could a criminal affair arise out of something like this? It turned out, that yes. Crämer retorted: "I slept well, if not for that air-raid warning." This in turn caused Debus to remark: "well, the English shouldn't have started this war." Crämer smiled ironically and asked: "the English started it?" This smile was sufficient that Crämer had to make an appearance at Gestapo headquarters. He was sentenced to two years in prison. So for the second time we can see what is the meaning of an "exceptional fanatic." Such a mess and loss at the heart of the most secret research project! The leadership of AEG was horrified and felt obliged to explain why Crämer's detention was absolutely out of the question. **This explanation has survived**… I probably do not have to add that in a normal situation any attempt to likewise question the legitimacy of the Gestapo's activities, would mean balancing on the edge of life and death. Here however, the situation was far from normal.

The consortium headquarters and research institute in Berlin sent the adequate letters. I decided to translate almost the entire letter originating from the last institution, in my opinion the most substantial one. In addition it has been reproduced in the book. It was written by the chief of the AEG Research Institute, Prof. Dr. Carl Wilhelm Ramsauer. From the text of the document it follows…after collation with general knowledge on wartime scientific work—that the "Kriegsentscheidend" was de facto the highest degree of secrecy employed in the Third Reich—reserved for a single, exceptional case. Here are its contents (bold by myself):[235]

> *Certificate*
> *Mr. Engineer R. Crämer from the AEG transformers factory in Oberschöneweide is developing together with the AEG Research Institute a project concerning high voltages [Hochspannungsprojekt], which was contracted to AEG by the Ground Forces Armament Office [Heereswaffenamt] and is being realised under the code-name "Charite-Anlage," as a secret device important for the war. The realisation of this project is in half dependent on Mr. Crämer, who as the sole employee of AEG possesses necessary qualifications, concerning this special field of electricity. Without the co-operation of Mr. Crämer further realisation of this project is not possible. The research and development work must be carried out with full energy, at least until the end of war.*
> *The "important for the war" or "decisive for the war" importance of this project results from the following issues:*
> *1. The project is realised under special priority SS/1940, which is only granted in such special cases.*
> *2. Mr. Ministerial Director Prof. Dr. E. Schumann, director of the Research Division of the Ground Forces Armament Office has granted this project **the highest level** of urgency [the highest priority—I.W.], which has been described as **"decisive for the war"** ["kriegsentscheidend"] (compare the protocol from the briefing of 21.07.42, which may be submitted upon request).*
> *3. "The Plenipotentiary of the Marshal of the Reich for Nuclear Physics," Councillor of State Prof. Dr. A. Esau, President of the Physical-Technical Reich's Office [Physikalisch-Technische Reichsanstalt] (…) has explained the signing below, that in addition he will prove the necessity of carrying out this work in the interest of the war. A written confirmation will be submitted.*
> *[seal and signature]*

For the Gestapo and court these were apparently sufficient arguments, to immediately forget about the whole affair. Crämer did not go to prison. It is unknown what happened to him after the war. It is known however that Debus was used in various rocket and space projects of the US Armed Forces. He was probably acknowledged as one of the most important German

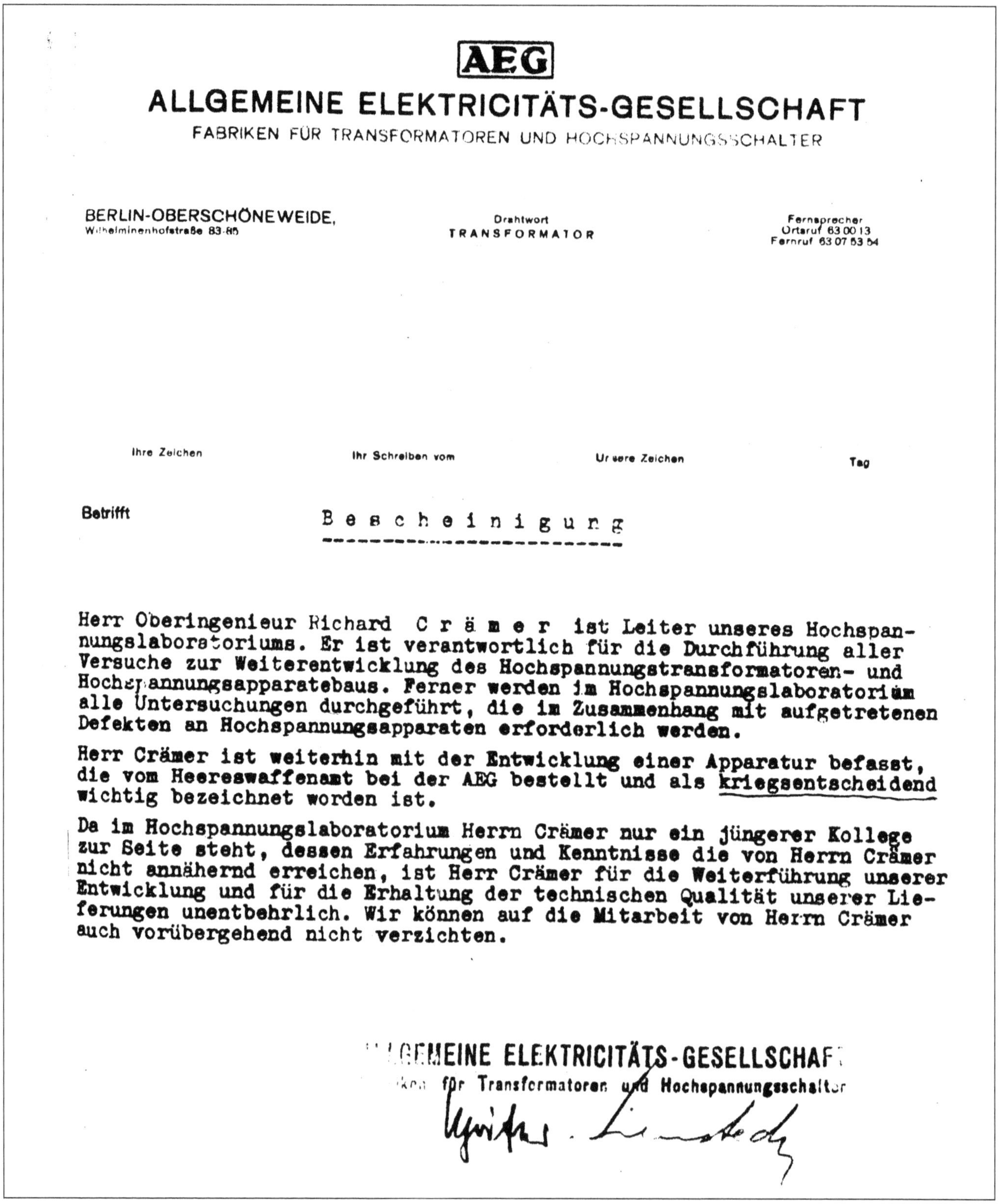

AEG

ALLGEMEINE ELEKTRICITÄTS-GESELLSCHAFT
FABRIKEN FÜR TRANSFORMATOREN UND HOCHSPANNUNGSSCHALTER

BERLIN-OBERSCHÖNEWEIDE, Wilhelminenhofstraße 83-85

Drahtwort: TRANSFORMATOR

Fernsprecher
Ortsruf 63 00 13
Fernruf 63 07 53 54

Ihre Zeichen   Ihr Schreiben vom   Unsere Zeichen   Tag

Betrifft: B e s c h e i n i g u n g

Herr Oberingenieur Richard Crämer ist Leiter unseres Hochspannungslaboratoriums. Er ist verantwortlich für die Durchführung aller Versuche zur Weiterentwicklung des Hochspannungstransformatoren- und Hochspannungsapparatebaus. Ferner werden im Hochspannungslaboratorium alle Untersuchungen durchgeführt, die im Zusammenhang mit aufgetretenen Defekten an Hochspannungsapparaten erforderlich werden.

Herr Crämer ist weiterhin mit der Entwicklung einer Apparatur befasst, die vom Heereswaffenamt bei der AEG bestellt und als kriegsentscheidend wichtig bezeichnet worden ist.

Da im Hochspannungslaboratorium Herrn Crämer nur ein jüngerer Kollege zur Seite steht, dessen Erfahrungen und Kenntnisse die von Herrn Crämer nicht annähernd erreichen, ist Herr Crämer für die Weiterführung unserer Entwicklung und für die Erhaltung der technischen Qualität unserer Lieferungen unentbehrlich. Wir können auf die Mitarbeit von Herrn Crämer auch vorübergehend nicht verzichten.

ALLGEMEINE ELEKTRICITÄTS-GESELLSCHAFT
für Transformatoren und Hochspannungsschalter

Above and on the next page: Documents from the Research Institute of the AEG confirming the existence of a research project related to high voltages and in terms of secrecy rated beyond any known classification. They don't say literally about plasma, or about gravitational propulsion, but quite recently even such a confirmation has been found!

scientists, since in 1963 he was appointed a director of the J.F. Kennedy Space Centre at Cape Canaveral. It must prompt reflection that one of the leading governmental facilities of the USA was directed by a fanatical SS-man. The arguments which stood behind this must have been really strong… He died in 1983 at the age of 75, leaving behind two daughters, who according to SS tradition were given old Germanic names: Ute Irmgard and Sigrid Monika. Perhaps they know something?

In the files of the "Paperclip" project I also found the personal files of several other interesting scientists, but there was

> **Prof. Dr. C. Ramsauer,**
> **AEG FORSCHUNGS-INSTITUT**
> FERNRUF • 492101
>
> Berlin-Reinickendorf/Ost 1, den 22. April 1943.
> Holländerstraße 31-34
>
> <u>Bescheinigung.</u>
>
> Herr Oberingenieur R. C r ä m e r von der AEG - Transformatorenfabrik in Oberschöneweide bearbeitet gemeinsam mit dem AEG - Forschungs-Institut ein Hochspannungsprojekt, das vom Heereswaffenamt bei der AEG bestellt ist und unter dem Decknamen <u>"Charité-Anlage" als geheimzuhaltende kriegswichtige Anlage</u> läuft. Die Durchführung dieses Projektes ist in ihrer einen Hälfte an die Person des Herrn Crämer geknüpft, der als einziger AEG-Angestellter die notwendigen Erfahrungen auf diesem elektrischen Sondergebiet besitzt. Ohne die Mitarbeit des Herrn Crämer ist daher die Durchführung dieses Projektes nicht möglich. Die Entwicklung muß mindestens bis zum Ende des Krieges mit aller Energie durchgeführt werden.
>
> Die kriegswichtige bezw. kriegsentscheidende Bedeutung dieses Projektes geht aus folgendem hervor :
>
> 1) Das Projekt läuft unter der Sonderstufe SS/1940, die nur in solchen Sonderfällen gewährt wird.
> 2) Herr Ministerialdirektor Professor Dr. E. Schumann, der Leiter der Abteilung Forschung im Heereswaffenamt, Charlottenburg, Hardenbergstr. 10, hat diesem Projekt die höchste <u>Dringlichkeitsstufe zugestanden, indem er es als "mit kriegsentscheidend" bezeichnet hat (vgl. Protokoll der Sitzung vom 21.7.42, das auf Wunsch vorgelegt werden kann).</u>
> 3) "Der Bevollmächtigte des Reichsmarschalls für Kernphysik", Herr Staatsrat Professor Dr. A. Esau, Präsident der Physikalisch-Technischen Reichsanstalt, Charlottenburg, Werner Siemensstr. 8-12, hat dem Unterzeichneten erklärt, daß er ebenfalls die Notwendigkeit, diese Entwicklung im Kriegsinteresse durchzuführen, bescheinigen wird. Die schriftliche Bescheinigung wird nachgereicht.
>
> **ALLGEMEINE ELEKTRIZITÄTS GESELLSCHAFT**
> **FORSCHUNGS INSTITUT**
> Der Direktor:
> *C. Ramsauer*

no information of any particular value. I began from Professor Hermann Oberth. I have not mentioned him until now, because reference to him appeared to some extent in the background of the project in question. Immediately after the war a document was discovered in some buildings in Środa Śląska, on which his name appeared. It referred to a delegation of several scientists, who had arrived from Prague, stopped on the way in Wrocław /Breslau and Środa Śląska /Neumarkt (3 days) and then made their way to Torgau. The business trip took place between September 15 and September 25, 1944, and consisted of the following individuals: Professor Hermann Oberth, Herbert Jensen, Dr. Edward Tholen, Dr. Elizabeth Adler and two others, whose names are illegible. The importance of this information lies in the fact that Hermann Oberth was the most outstanding specialist in the world engaged in space flight theory, with a far superior authority than that of the young Von Braun (at that time 32). In short he represented a potential, which is undoubtedly not wasted, particularly when a rocket programme was being put into effect, projecting several decades into the future. Like earlier in the case of Professor Gerlach this information reveals to us a certain unusual and significant fact…significant for the work being carried out. Namely that in principle it is unknown what Prof. Oberth was engaged in during the war. One could have the impression that this is some kind of light at the end of the tunnel, which until now has been cloaked in the darkness of night. After all it is known for sure that Oberth was not connected with the centre in Peenemünde, since in this case he would have undoubtedly held at least one of the positions of command, in other words the fact of his engagement would have been known (thousands of specialists employed there worked after the war in other countries, from the USA and USSR to even Egypt and so is out of the question that a possible secret of this kind could not be kept hidden). So it seems that some kind of alternative program had existed, being carried out for a long time, and quite a serious one at that. This is indicated even by fact alone of there being a lack of information on the wartime work of Oberth. The most important information appearing from his American personal file is that he was in general brought to the USA. A certain curiosity is represented

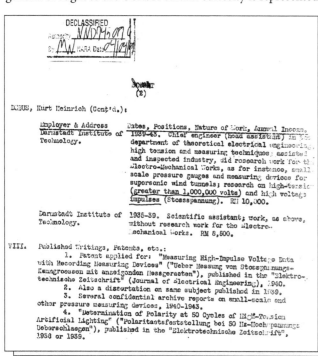

A document from Oberth's files, referring to the necessity of locating and recruiting Herbert Jensen. (Oberth's files)

An excerpt from K. Debus's American files, from which it appears among other things that he had worked on the generation of "artificial lightning" exceeding 1 million volts. Could something like this have had any connection with a weapon "decisive for the war"?

by the fact that he was interested in… occultism. Information more worthy of keeping in mind is included in another document. It was signed by a US intelligence colonel and contains the "categorical demand" to establish the identity and recruit a German scientist, Herbert P. Jensen, within the confines of project "Paperclip." This name is already familiar to us. It seems that the Americans were not interested in Oberth due to his purely theoretical pre-war analyses, but for altogether more specific reasons. It looks after all like it had been attempted to recreate the former research team—the information about Jensen probably came from Oberth, since it was found in his personal file.[236] In this context it is worth taking into account that after the end of his internment at Farm Hall, Gerlach was also transferred to the USA and intensively interrogated. He finally returned to Germany, but American secret services had taken possession of his "work diary." Almost nothing is known about it apart from the fact that as a valuable trophy it ended up in the CIA archive.[238]

When I was in the USA, the files of yet another person interested me, who the Americans depended on. It concerned Prof. Hubertus Strughold. His name had never appeared in documents referring to the "Chronos"/"Laternenträger" project, but in one of the interviews he confessed that in 1945 he

One of the wartime tunnels under the "new" castle

The "new" castle at Książ / Fürstenstein

A fragment of the "old" castle

Ludwikowice—one of two entries to the "mine" valley

The ravine of the Pełcznica stream (the stream is visible at the bottom) and the ruins of the "old" castle rising above it

had tested some kind of "space flight simulator" in the vaults of Książ (Fürstenstein).[239] Strughold was a pioneer in the field of space medicine. However, I only established that he found his way to the USA on 3 August 1947 and played a key role in preparations for the first American manned space flight…[237]

We end on this the investigation of the scientists' fates and activities. There is however one more aspect of the work, which was barely pointed out on the previous pages—reference to where the work was carried out and the specific facilities. I will remind one that it concerned in chronological order: the underground laboratory in Środa Śląska (Neumarkt) near Lubiąż (Leubus), the underground facility in the vicinity of the so-called old castle in Książ (Fürstenstein) and the "inactive coal mine near Wałbrzych (Waldenburg), which had been taken over for these purposes." Nothing substantial is known about the laboratory in Środa Śląska. The issue of Książ appears much better. A complex of two extremely picturesque castles…"old" and "new" is located there. It is no secret that during the war this area was closely guarded and was under the "guardianship" of the military…chiefly of the Luftwaffe and the SS. Two underground facilities were dug and tunnelled there on a large scale, on several levels. The one under the "new" castle is relatively well known, because part of the system has been researched and catalogued. The known section consists of tunnels with a total length of approx. 1 km. However in all

The access road to the "old" castle… and the remnants of a characteristic entry gate from the time of the war. An identical one appeared on the road leading to the "new" castle.

Ludwikowice—one of the buildings of the mining complex with a German name and swastika still visible under recent paint—this building no longer exists

A labyrinth of concrete roads around the mine lead to numerous bunkers scattered in the forest

One of the bunkers camouflaged with trees

The outlet of one of ventilation shafts, leading to a wartime level; similarly camouflaged

Here, throughout the entire levelled ground, once extended a large railway station, around 100 metres from the mine's main shaft

probability most of the facility is not yet known, because out of at least six former entrances currently only one is accessible and after the war the main elevator shaft connecting the castle with the underground levels was filled in. Work was probably commenced in 1943, exploiting among others a commando of prisoners from Gross-Rosen, Italian specialists and approx. 400 Soviet miners from the Donbas region.

The purpose of the facility has never been fully unravelled. In all probability the Führer's command post was to be established here, code-named "Rüdiger," although on the other hand it is known that during the final months of the war a Luftwaffe research station code-named "Wetterstelle" was transferred to these vaults. Among other things the electronic equipment and bombsights from shot down Allied aircraft were tested here. From the reports of some prisoners it appears that this complex was connected to an underground facility in the nearby Sowie Mountains (German: Eulengebirge), by a tunnel 16-18 km in length.[240,241] This facility or complex of facilities was code-named "Riese."

Much less is known about the facility under the "old" castle, distant from the "new" castle by only approx. 500 m. In spite of the short distance, both complexes of tunnels and halls have so far been treated separately, since the 60-80 m deep ravine of the Pełcznica stream divides them (German: Polsnitz). In spite of this the possibility of a connection existing between them is not excluded. Mr. Tadeusz Słowikowski among others has quoted such reports,...the local researcher, who has devoted half of his life on researching Książ and probably has the most documents on this subject. In addition the first professional and in general first geological reconnaissance of the underground facilities under the "new" castle ... conducted in 1960 by the District Mine Rescue Station in Wałbrzych (when the main shaft was still accessible) ... validated such suppositions. Four levels of tunnels were catalogued at that time ... and not the two widely known at present. The lowest was located "at the level of the stream."[242] Such a number of levels was also indicated by such a simple detail like the still existing until not long ago, lift distributor box.

The facility under the "old" castle has been completely inaccessible since 1948. Its existence however was confirmed by the two main researchers of Książ, the aforementioned Tadeusz Słowikowski as well as Sławomir Orłowski from Wrocław. The

Ludwikowice—the "fly trap." When viewed from inside, the ring gives the impression of a finished whole.

A post-war inscription on one of the pillars: "swimming in the pool is forbidden." Therefore it must have once been filled with water

A fragment of the pool's border

former stated moreover that the whole area around the castle was closely guarded during the war by at least two rings of Luftwaffe posts, and was also prepared to show the blown up entrance to the underground facility. Orłowski described in one of his articles (from 1977) the mysterious events connected with this facility:[243]

Among people researching the old plans of Książ the conviction prevails that the small castle had a well developed system of underground passages, which at the bottom of the ravine and Pełcznica brook were connected to the oldest castle corridors. (…) I heard the solution of this riddle in 1967. (…) I met Ivan Konkov in a group of Soviet tourists, combatants from the last war, who had fought in Lower Silesia. (…) Just then Konkov recalled that he had been stationed at Księżno castle… as Książ was called at the time. (…) Seeing my interest in this subject, Konkov asked if I knew why the small castle at the summit had been destroyed. I asked for his account of the event. And here is his story:

"*One day some soldiers spotted two people moving about in a grove of evergreen rhododendrons. Being called to halt, they fled into the undergrowth, returning fire. (…)*

The escaping men dodged and wove, but when the ring tightened, they disappeared on the grounds of "the old castle." Several soldiers reached the castle yard, but here they were caught by strong machine gun fire. (…) Just then a powerful explosion occurred. The walls of the gothic ruins shook and crashed onto the yard, which undulated and collapsed from the force of an underground explosion."

Vegetation has covered part of the "fly trap's" structure from Ludwikowice. On this photograph a fragment of the "pool's" border is visible (1) and located much closer to the centre, one of the pillars (2).

A modern "fly trap" for testing helicopters at a plant of PZL

The installation channel/duct, leading from the power plant (approx. 50 m away) up to the "fly trap," the ring of which is visible behind the trees. The lower part of the photograph shows a typical high voltage cable, pulled out of the duct.

The remnants of a warning notice..."caution poison," on one of the bunkers near Ludwikowice

Even today the smell escaping from the ventilation funnels betrays the presence of chemical weapons

In this way, in 1948, the guards from Werwolf sealed the secret of these underground vaults. Little information has survived, which would allow it to be unravelled. The reason for this was the close guard of the whole region and the fact that almost all prisoners who worked here during the final months of the war, were slaughtered. Some unusual information however has appeared. I heard from Mr. Słowikowski, not asking him at all about such matters, that at the turn of 1944-45 one of the inhabitants had seen some objects vertically landing and taking off, which he called "flying barrels." On June 23, 1999, the Polish television network TVN presented the report of an anonymous witness—a woman, whose husband was taken as a prisoner of war after the Warsaw uprising, and worked and lived at Książ castle until the last days of the war. He survived the execution by firing squad of the prisoners in April 1945 thanks to a steel corset, protecting his damaged spine. He stated that he had seen there some rounded, experimental flying vehicles. Jerzy Rostkowski, an extremely reliable researcher and the author of a film about the history of Książ, in turn got hold of a written account of a Gross-Rosen prisoner, working at Książ. One day he had overheard a conversation between SS-men, concerning a kind of "super weapon" that was being created here. He remembered the following words: "Only those in the forests or high in the mountains will survive, because no shelter will protect them from death."

The above accounts are in principle a compilation of already known information related to Książ. But finally I managed to

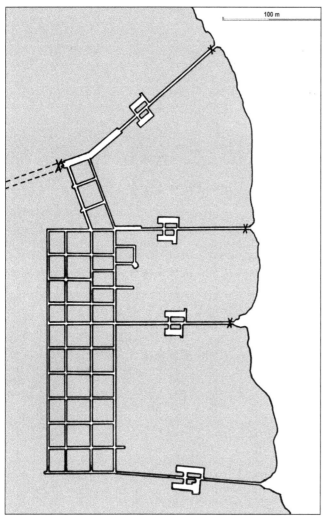

Plans of some sections of the "Riese" complex...the "Włodarz" facility

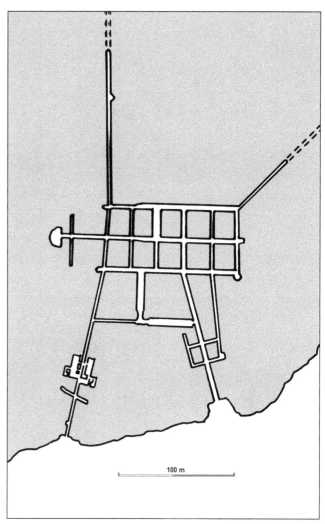

The "Osówka" facility

Cooling towers (CH) once stood next to the power plant. They looked completely unlike the wartime "fly trap."

get hold of some new facts, which put everything in a completely new light. Before this happened however, I concentrated on unravelling the role of yet another mysterious place—that undefined "former coal mine near Wałbrzych (Waldenburg)."

For this objective I contacted a scientist, an employee of the Polish Academy of Sciences, who was engaged in mine research, and especially in the history of mines in this area—Michał Banaś from Cracow. I had the impression that my question gave him some kind of relief, since shortly before he had "discovered" a rather strange mine, meeting the above criteria and he was curious what it had been used for during the war. Simultaneously he did not know of any other facility, which would match this intrigue…and this one matched to perfection. This mine was located in Ludwikowice Kłodzkie (then: Ludwigsdorf) and until 1945 bore the name "Wenzeslaus." It is situated about 20 km from Książ and practically adjoins the earlier mentioned underground facilities complex ("Riese") in the Sowie Mountains. Michał Banaś managed to gather a great deal of documentation, from which it appeared, at least concerning the ground section, that great changes had taken place here during the war. It was enough to compare pre-war, wartime and post-war aerial photographs. Banaś established that in 1931 a big accident had occurred here, which caused the mine to become bankrupt and be taken over by the state. From initially available information it appeared that during the war an explosives factory was established here, and at the same time some chemical plants of the Dynamit A.G. consortium.[244]

"Riese"—some of the underground tunnels

One of the ventilation collectors, evidently from the time of war, in Ludwikowice. It is a comouflaged bunker, which was protecting the shaft and housed the fans. Above it are four concrete ventilation chimneys, each having internal cross-section of 1 m². I have to admit, that I wrote a book about Hitler's underground factories, but this is the largest such an object, that I have ever seen. Larger may only be the elevator shafts. It looks, that the giant mine was treated as just a starting point, just the first element of a larger complex. The second one would be the level(s), to which the wartime ventilation shafts are leading (top level of the mine is about 100 m deeper). The third one forms infrastructure on the surface. The fourth one would consist of the nearby "Riesa" or "Riese" complex…after finishing of the connecting tunnel, mentioned by Prof. Mołdawa.

"Riese"…one of many piles of fossilized cement bags, that the Germans never managed to use

One of the entrances to the underground facility

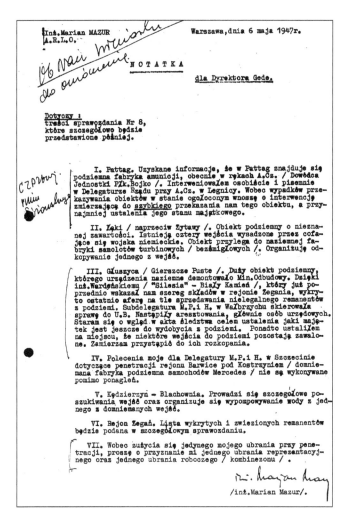

A document from the so-called Operation to Disarm the Oder Line from 1947, in which there is reference to the existence of a large underground complex near the town of Głuszyca (Wüstegiersdorf). Machines/engines were plundered from it and the entrances probably later blown up. Now only legends circulate about Głuszyca's huge underground facility. On the margin is a note to send the document to the Minister of Public Security, Różański.

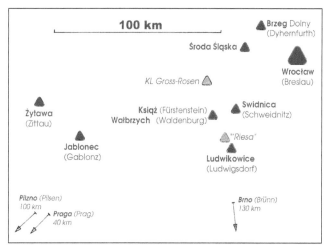

The location of the places/facilities described in this part of the book, in relation to each other

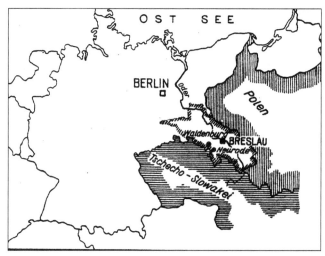

A simple map showing the location of Lower Silesia in the Reich… according to the borders from 1938

"Riese"…a stone slab with an engraved German eagle… a memento of the war years

Since from Banaś's accounts it appeared that this was generally an interesting area, we made our way to this place…

The whole area, in the centre of which was located the main lift shaft, proved to be the interior of a deep valley, which was accessible only through two "mountain passes." Since the remnants of watch-towers could be seen in them, it was obvious that the whole area had been closely guarded, and its configuration caused that in this way the whole valley was physically cut off the outside world.

Even first glance at this once "prohibited zone" was sufficient to be seized by the irrefutable impression that a lot had gone on here during the war, and truly important things at that. First and foremost was one struck by the unusually developed infrastructure, typical for underground facilities of a military nature, and not for mines—kilometres of perfect concrete roads in forests, a mass of bunkers, the remnants of gates and fences within the sealed zone and the like. The other unusual detail, even for military facilities, was the painstaking camouflage. All the reinforced concrete buildings and bunkers possessed

Cross-section of the "Fly trap." It is standing in a "water tank" or "pool," that has significantly larger diameter, than the visible upper ring. Presently the tank is filled with ground and slag. Possible connection with the installation duct has not been explored. The "entablements" in the centre of the tank has been described by a witness...presently are not visible.

either so-called trough roofs, on which normal sized trees were growing or were covered with earth—for just the same effect, to which camouflage paint was added.

One structure especially intrigued us, one of the concrete roads leading us to it. On an area of flat ground was a dodecagon-shaped pool, with a diameter of almost 40 m. Its edges were painted with a still visible green camouflage paint. Inside the pool towered a pretty unusual structure. It was formed by 12 massive pillars about 10-12 metres high, connected at the top by a dodecagon-shaped, reinforced concrete ring with a diameter of around 30-31 m. It was strange, since it did not have any roof and between the pillars there were no walls, or even the slightest trace of them. The structure gave the impression of being a complete whole, plastered and painted with green paint. Only on the circumference of the ring at the top were some kind of metal fixtures visible. At first, before we had taken a more thorough look, it appeared to be the scaffolding of an incomplete cooling tower or cooling stack, belonging to the nearby, pre-war power plant. However Michał Banaś produced a photograph from 1934 and said: "the power plant had its own cooling towers, they stood here..., this does not resemble any cooling tower." In his opinion it was first of all too low in relation to its diameter and secondly the pillars would have had to have been connected by walls. Besides, this explanation was contradicted first and foremost by the pool... larger than the structure itself.

The whole structure gave a pretty eerie impression, but simultaneously reminded me of something. At the time I still did not know of what, but after returning home I retrieved a photograph of something very similar. It was called a "fly trap" and was designed to test objects vertically taking off and landing...

Later Nick Cook from "Jane's," who arrived from Great Britain among other things to see "it," showed me yet another photograph, I daresay resembling even more the structure from Ludwikowice. It bore the caption: "AVRO's test-rig for the Mach 4 Project 1794 Saucer"...designed for testing disc-shaped flying objects, built in the Canadian factory AVRO. What a provoking explanation... Of course I am not sure if it is real, although the photograph of the aforementioned, mysterious structure has been reproduced a few years ago in several thousand books and so far I have not met any clear-cut alternative explanation. The "fly trap" has also been shown several times on television, as something of an unknown purpose... also with no response.

Pretty unusual in all of this is the aforementioned pool. It matches the description of the surface covered with ceramic tiles and washed with brine. There is yet another detail—a coupling link exists between the "fly trap" and the power plant standing nearby. This is a concrete installation duct (made from prefabricated elements), through which cables were led under the central part of the pool. I write this, because remnants of them are still there... such a cable... as thick as a man's arm ...can be seen on one of the photographs. As Michał Banaś explained, high voltage current is not used in mines, and in those with a gas hazard—like this one, the electrical power supply is replaced by a pneumatic one. So as to avoid any misunderstanding, an explanation is necessary: a gas hazard only exists during the mining of a deposit, there is no problem when the underground space serves other purposes.

In order to obtain additional information, we decided to talk with the inhabitants of the buildings lying in the valley. We wanted to discover somebody who had been settled here relatively early (the Germans were displaced) and had seen more traces of wartime activity. We made contact with such a person, Mr. Henryk Lasak, who had found his way here in 1947. However he knew little, although he remembered that in the area of the power plant and the fly trap some notices prohibiting entry with the "SS" insignia had been present. We heard from others, that a huge amount of chemical weapons had been produced here, containers of which are still sealed in bunkers and that from time to time sappers find chemical ammunition. We found such bunkers in the forest with the danger notices: "caution, poison!"

Thanks to the help of Mr. Piotr Kałuża we managed to make contact with somebody who had worked in the buildings of the former mine in 1948: Mr. Frank Szczogel from Lądek Zdrój. He was to have discovered at that time the personal files of workers employed there during the war. It appeared

A modern aerial photograph of the "fly trap's" surroundings (from the collection of Jerzy Cera)

Left: a similar photograph from 1954. It is almost identical to the Allied aerial photograph from the autumn of 1944 on which, for the first time, the mysterious ring-like structure appeared.

Balls of Fire Stalk U. S. Fighters In Night Assaults Over Germany

By The Associated Press.

AMERICAN NIGHT FIGHTER BASE, France, Jan. 1—The Germans have thrown something new into the night skies over Germany —the weird, mysterious "foo-fighter," balls of fire that race alongside the wings of American Beaufighters flying intruder missions over the Reich.

American pilots have been encountering the eerie "foo-fighter" for more than a month in their night flights. No one apparently knows exactly what this sky weapon is.

The balls of fire appear suddenly and accompany the planes for miles. They appear to be radio-controlled from the ground and keep up with planes flying at an hour

Donald Meiers of Chicago said. "One is red balls of fire which appear off our wing tips and fly along with us; the second is a vertical row of three balls of fire which fly in front of us, and the third is a group of about fifteen lights which appear off in the distance—like a Christmas tree up in the air—and flicker on and off."

The pilots of this night-fighter squadron—in operation since September, 1943—find these fiery balls the weirdest thing that they have yet encountered. They are convinced that the "foo-fighter" is designed to be a psychological as well as a military weapon, although it is not the nature of the

One of the press articles from 1944, describing the unusual objects observed over Germany

Above: Ludwikowice. One of the wartime roads in the forest—sections of a former entry gate are visible—remnants of a second security ring within the closed area.

Left: Ludwikowice. One of the bunkers. The method of providing camouflage is clearly visibile.

Right: Part of the "Fly trap". Remnants of green camouflage paint are still visible

The "fly trap" from Ludwikowice

A document from the archives of the Reich's Scientific Research Council, referring to an assessment of the "Foo-fighters" phenomenon by the Allies. It had been assessed as a very serious danger.
(AAN/ Alexandrian Microfilms, files of the Reichsforschungsrat).

from the files that a large number of electrical engineers and mechanics had been employed…therefore not only explosives and chemical weapons had been involved. A former female prisoner discovered by Mr. Kałuża stated in turn, that those working underground had to take some kind of tablets several times daily and that the staff walked around in white lab-coats. Michał Banaś then established, that the mine had been a true underground city. Excavations were found on many levels, up to 610 m down and had occupied an area of 9 x 16 km horizontally. From the south-east they approached a distance of only 200 m to the mysterious "Riese" complex, built during the war by tens of thousands of prisoners from Gross-Rosen (according to Dr. Jacek Wilczur from the Main Commission for the Research of Nazi War Crimes by over 70,000,[246] according to other sources, at least 40,000[247]). So the whole issue began to appear on a totally different scale! The only problem, that remained was that since the end of the war the mine has been completely flooded with water and there is no access to the underground section.

It is made clear by the existing documentation, that even before the war the combined area of the tunnels and halls of "Wenzeslaus" amounted to thirty million square metres (!)… it was one of the largest coal mines in Europe.[268] This means about 50 times more, than in the case of the giant Mittelwerk complex near Nordhausen.

Since I was unable on the basis of existing sources to unequivocally determine the connection between the sites in question where work was carried out, I turned to an authority in this field—**the only living person who had had direct access to the German documents and other wartime sources**. This individual was Prof. Mieczysław Mołdawa, who as a prisoner of Gross-Rosen and expert, worked in the so-called technical chancellery of the camp, making various plans (for example plans of nuclear shelters for German cities), but generally concerning "camp" construction projects.

Taking this opportunity, Mołdawa stated that he had designed cooling towers (he was a Professor of building engineering), but authoritatively pronounced that the construction from Ludwikowice was without fail not one. Prof. Mołdawa had direct contact with Hans Kammler and Oswald Pohl. I turned to him, since in his book concerning Gross-Rosen there was some information which could be some kind of link:[244]

> *Kommando "Fürstenstein" in Książ near Wałbrzychu, on the grounds of a* **Luftwaffe headquarters with an aerial weapons study centre** *and special inspection (Sonderinspektion) of underground factories construction in the Sowie Mountains massif. A small Kommando, established in 1944 and* **administratively connected to the nearby Kommando "Wüstegiersdorf,"** *building an arms production complex, working on the construction of camouflaged chambers, shelters and store houses* **for military research posts** *(p. 192).*

Wüstegiersdorf is present day Głuszyca—the official German location of the "Riese" complex.

First and foremost the questions arose of what was the mysterious "aerial weapons study centre," and what in the aforementioned mountains was to be produced… Only the element linking Książ with the "Wenzeslaus" mine was missing.

Armed with a video camera and backup of tape I headed, after prior appointment, for Prof. Mołdawa's apartment.

A photograph of the "shining balls" in the skies over Germany, published in the 1940s

> ### Hans Kammler
>
> Dr. Kammler was, alongside Karl Otto Saur and Xavier Dorsch from Speer's Ministry, one of several individuals responsible for the most important armament undertakings. His position rose systematically in due measure of the war's passage, proportionally to the growing influence of the SS on the production of arms and their deployment. He was born in 1901. After completing technical university studies he began work in the Air Ministry, and after the creation of the SS passed along with a group of Luftwaffe specialists to work at its headquarters (SS-Hauptamt). Initially he performed one of the managerial functions in the department responsible for finances and construction enterprises. In February 1942 he became chief of "group C" in the Main Economic-Administrative Office of the SS (WVHA). Driving the expansion of SS concentration camps and building engineering, he contributed to the tremendous rise in its income as well as political and economic standing. From the beginning of 1943 Kammler acquired moreover a series of "special tasks" to carry out, connected with particular armament projects. According to Speer's words he displayed during their execution "an unusual energy, freshness and ruthlessness." After Himmler as the SS-Reichsfuhrer had taken control over the V-1 and V-2 programmes he became responsible for their production and military deployment. At the turn of 1944 and 1945 he directed not only programmes directly subordinate to the SS but entire branches of the armament industry... among others the production of jet fighters and guided surface-to-air rockets. He was also responsible for construction of the underground armaments facilities associated with them. Officially he was commander of an army corps for special missions (Armeekorps z.b.V.). In March 1945 he had the good fortune to inspect the Czechoslovakian stretch of the front. After this event nobody knew what became of him. Perhaps he died, though it is unlikely that this would never been recorded.

Frankly speaking I had no hope that he would be able to tell me anything more than he had written in his book. However he was able to—and delivered some outright crucial information.

He began with the issue of the so-called "Jägerstab." Therefore before I begin to give an account of the conversation, I will explain what this was. The year 1943 was a period of systematic loss of the Luftwaffe's control over the skies of Germany. Air raids were sowing the seeds of devastation in the German armament industry. Counteracting this was only possible with the aid of extraordinary measures. As a result, on 1 March 1944 the so-called "Jägerprogramm" was brought to life and the institution which was to put it into effect...the "Jägerstab." It involved accelerating the production of new fighters and other revolutionary aerial weapons. The basis for it was to be formed by underground factories, resistant against air raids. They were initially to be under the authority of Speer's Armament Ministry and the Luftwaffe, but during 1944 the SS and specifically the "Rüstungsstab" of SS-Gruppenführer Hans Kammler seized ever greater control over these projects. By the end of 1944, he had already seized control over both Luftwaffe research work and the construction of underground factories.

At the instant that the "Jägerprogram" was called into being it became a key armament programme. Few researchers fully realise this, but the leaders of the Third Reich assumed, that the new weapons would cause a complete change in the situation in the skies over Germany and the Luftwaffe to regain air superiority. The mass use of the revolutionary armament was to be synchronised with the counteroffensive in the Ardennes.²⁶⁵

We return to the account of Prof. Mołdawa.

He delivered the first, very important piece of information: the "Jägerstab" was based in Żytawa (Zittau), whereas in Książ (Fürstenstein)—"under the control" of the Luftwaffe—was situated its department responsible for the development of new weapons. It was precisely the most revolutionary of them that was to be produced in the Sowie Mountains. A special commando of prisoners has been created specifically for this purpose... The unusual care about the prisoners (non-Jewish) assigned to realise this task testified to the great hopes that the Luftwaffe command had for this work. Apart from standard camp rations they were guaranteed a pretty solid military diet including milk (!). The airmen, knowing the brutality of the SS-men didn't even allow them to escort the prisoners. This obviously gives an idea of the high priority granted to what they were working on.

According to Mołdawa, this "forgotten" Kommando, controlled by the Luftwaffe was large and operated in complete abstraction from the rest of the Kommandos, administered by this concentration camp. It appears that none of its members survived the war...

As a supplement to this information, Prof. Mołdawa pointed out, that not very far from Książ (about 85 km to the West), in Zelezny Brod a "planning centre for operations with the use of strategic, radio-guided weapons" was constructed for the Luftwaffe. The person in conversation with me emphasized that it was hard not to notice that a series of key facilities had

Hans Kammler on the left. A photograph from August 1944.

Osówka—the known part of the "Riese" complex

been concentrated in such a narrow region. When treated as a whole, they give a picture of a defined armament program.

I asked of course: what was this weapon?

According to him, it was to be a weapon "decisive for the war" (!), composed so to say of two components. The first was a flying object or combination of such objects with revolutionary new capabilities. It was to reach "safely" all major cities of the enemies, Prof. Mołdawa mentioning "Moscow, London and New York." The person in conversation with me did not know what kind of object it would have been. He knows that various variants had been considered, he saw for example the

Part of the former abatises—"Riese"

sketch of a heavy bomber transporting an unmanned rocket-propelled aircraft. But this referred to the initial phase, when various concepts were being considered. He does not know what was chosen.

The second component on the other hand, was to be the most lethal weapon of the Third Reich and of World War II… the latest generation of chemical weapons (tabun, sarin, soman). After all they were also produced mainly in Lower Silesia, and… in Ludwikowice.

In connection with this I asked Prof. Mołdawa about the role of the mine already familiar to us. He replied that it had been part of the whole production complex. After the war he talked with an engineer who designed the connection between the mine's underground facilities and the "Riesa" complex (exactly! "Riesa" and not "Riese,"—he stated that he remembers this precisely, so I will use this code-name from now on). It is not known if this connection was actually made. Only part of the "Riesa" complex is presently known… approx. 90,000 m³, while just a comparison of the consumption of materials and size of the labour force with similar facilities suggests at least 250,000 m³. In the unanimous opinions of Prof. Mołdawa and myself, such a connection (a mine plus a horizontally constructed complex) resulted simply from the fact that the mine had admittedly a huge cubic capacity at its disposal, but had a very serious and irremovable shortcoming. The main transportation lines… two large shafts…were characterised by a series of limitations. First and foremost large elements…finished products could not be

Reference to German work on a "flying saucer" in the NARA Archive
(see: text)

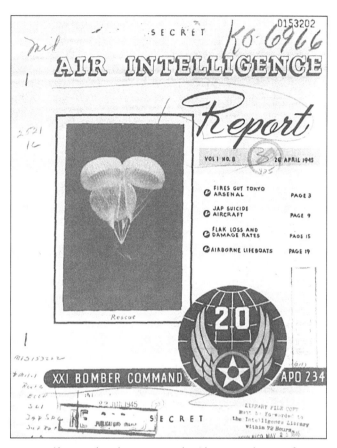

Above and on the next page: the title page and excerpt from an American intelligence summary referring to "Foo-fighters"

moved through them, although the mine was perfectly suited for e.g. the production of chemical weapons...as transportation was ensured by pipelines. In a given case however a separate final assembly line was needed...precisely "Riesa," characterised by huge halls and horizontal transportation lines leading to the entrances.

"Riesa" or "Riese" is another outstanding "object" on our journey. It is a complex of underground facilities, evidently unfinished, constructed on the circumference of a mountain massif, at the same level. Probably these facilities were to be connected near the centre of the massif, forming one underground structure over 3 km wide and about 5 km long (from east to west). It is very likely, that some part(s) of "Riese"/"Riesa" has not been "rediscovered" after the war, because there are large discrepancies between the German data and what can be seen today...e.g. there were large concentrations of the workforce in places where nothing has been found. According to probably the most reliable publication on this subject, written by Prof. Seidler ("Unvollendete Anlagen"), the complex was to be finished by August 1945 and was about to consume 359,100 cubic metres of concrete, which is e.g. significantly more, than the quantity of concrete used in the Empire State Building (about half of the entire volume of the building). Obviously, this doesn't include the "Wenzeslaus" mine and other materials used for the construction. Professor Seidler wrote, that part of "Riesa" was about to be comprised of various headquarters... for Hitler, military staffs and certain central institutions, while the remaining part was intended for industrial purposes—for the production of a new weapon.

It looks then like the true equivalent of the purely mythical "alpine redoubt," which existed mostly on paper. A kind of underground capital of the state, connected with an arsenal of the last resort???

In this way, mostly thanks to the information from Prof. Mołdawa, all the puzzle's elements had been credibly fitted into place, including elements on the map. They occupy after all a large portion of the map. If one is to believe the accounts about the tunnel connecting the Sowie Mountains with Książ, we obtain an underground city 25 km in length!

The information (technical) presented above raises the fundamental question: could the described system have been "decisive for the war"?

Decidedly yes!

A carrier of weapons of mass destruction (WMD), out of reach of enemy fighters (manoeuvrability, speed) and with a long range, would have been the materialization of a technological jump of incalculable consequences. According to official data the Germans produced 10,000 tonnes of phosphoro-organic chemical weapons. This data however refers to only two factories, while in reality there were at least four of them. This amounts to around 100 billion doses resulting in immediate death!!! The possibility of them being safely transported to the enemy's main population centres would have signified a doubling of the number of war victims in the course of a week! It would have been enough to possess a "fleet" of around 50 large objects, which would have corresponded to approx. one tenth of a percent of the number of aircraft produced in the Third Reich during 1944. Germany could therefore have won the war even at the turn of 1944-45—as it followed in a letter from Gerlach to Bormann. In light of this it becomes clear why in Germany there had been no equivalent to the American "Manhattan" project—which engaged well over 200,000 people, while the bomb dropped on Hiroshima killed directly "only" 78,000 people (more or less the same as the largest conventional air raid on Dresden). In the conditions of that time a nuclear weapon was simply devoid of any military justification (at least in the Reich, the Americans had no "chemical alternative"). Apart from all of this the question obviously arises of whether the Germans managed to bring into production and employ a revolutionary type of propulsion. This question seems rhetori-

Has this development any relationship to our combat report of "balls of fire"? While it becomes readily apparent that all of these reports have not stemmed from the same causative source, it does not appear beyond the realm of possibility that some of the so-called "balls of fire" may have been generated by the rocket motor of the "Viper". In one form or another as many as 302 sightings by 140 crews which may be classified under the heading "balls of fire" have been reported. While a large percentage of these sightings have continued for at most two-three minutes, some have persisted for as long as 15 minutes. A few are quoted below.

1.

"A yellow ball of fire about six inches in diameter observed moving upward at an angle of about 75 degrees. Fire burned out at 8,000 feet. Six white balls of fire seen to come up from Nagoya. A greyish ball of fire about the size of a soccer ball passing below the aircraft in the target area. A red ball of fire was seen coming up from below."

2.

"A red ball of fire was seen dropping slowly from 1,000 feet to 500 feet below and ahead of one B-29. Then it dropped like a bullet and exploded on the ground. One crew reported seeing some object going away from them just before they reached the target. It was believed to have been a fighter. Flames seemed to be coming from it and as it turned two balls of fire came up toward the B-29 from behind but did not get close to it."

3.

"Unidentified enemy aircraft launching what appeared to be a fireball on the withdrawal course."

4.

"In one instance three balls of fire appeared to be launched from enemy aircraft."

5.

"Four balls of fire about size of a fighter plane, flared at level height then drifted to earth flaring up or burning."

6.

"Ball of fire first seen at five o'clock level about 300 yards behind B-29. Near as can be determined ball of fire was about the size of a basketball. When evasive action taken in form of turns, ball of fire turned inside B-29 and kept following. Appeared that each time B-29 made a turn ball of fire fell behind but on straightaway it would make up lost distance. B-29 lost altitude, going down to 6,000 feet in order to gain air speed and finally turned back toward coast. It was estimated that ball of fire followed for about five or six minutes.

7.

"One crew member thought he saw short streamer behind ball of fire, which was faint and not steady. This light appeared to fade when ball of fire was making turn, then increased after resuming straight course. Streamer of light was seen for about one minute after ball of fire headed back toward coast, when it faded abruptly. Blister gunner of this crew thought he saw wing in connection with ball of fire; and that wing had navigation light burning on left wing tip."

8.

"Just after leaving secondary target on course of 120 degrees and at 7,000 feet, observed what appeared to be ball of fire following at about four o'clock. B-29 immediately took evasive action gaining and losing 500 feet and

SECRET

cal, but it is not because… it is known that they did.

This can be proved (although nobody before has succeeded in doing this). The documents which confirm this, rather unequivocally, are among other places in the NARA archive, and their copies—in my hands. As one can surmise, in this case it concerns unmanned objects—the reported problems with radiation "side effects" probably excluded the participation of a pilot, still at the beginning of 1945.

The documents in question refer to something which in the Allied mass media were labelled "Foo fighters." In the last weeks of 1944, and more precisely in the second half of November, Allied pilots started to come into contact with a new phenomenon. Luminous, rounded flying objects were sighted, which sometimes only shadowed the approaching aircraft and sometimes performed strange aerobatics within their formations. Anti-aircraft weapons were totally ineffective against them, while radar sets ceased to work. The objects emitted strong electromagnetic energy, damaging some devices on board the aircraft. They were shining with a very strong light, most often white, red, orange or amber. They appeared above all in the region of the Rhine river line, although worth taking note of is the fact, that a smaller number of observations were also recorded over Japan.[248] Reports mentioned on the whole objects generally smaller than aircraft. Soon the Allied press started to write about this. A good deal of confusion was caused e.g. by an account printed in "The New York Times."[249]

> *Yesterday, during a night air raid on Hamburg, a mysterious, luminous ball appeared near an Allied bomber squadron, which despite many attacks of escorting fighters, appeared to be indestructible. This mysterious and undoubtedly Hitler's new weapon, very effectively jammed all radio-communication. None of our experts managed to explain as previously, what the principle of operation of these "luminous balls" was based on and through which at tremendous speeds* **demonstrated manoeuvrability at variance with the laws of aerodynamics!** *[bold by I.W].*

This is by no means an isolated example. The day before a similar description had appeared in the British daily "South Wales Argus":

> *The Germans have produced for Christmas their new, secret weapon. It is clearly an anti-aircraft weapon and resembles the glass balls, with which one decorates Christmas trees at Christmas. They were observed over German territory, sometimes in groups. They are silver (…).*

The culmination of reports on these objects came at the turn of 1944-45. A very interesting article devoted to these observations was published on January 2, 1945, in the important American newspaper "New York Herald Tribune":

> *BALLS OF FIRE STALK U.S. FIGHTERS IN NIGHT ASSAULTS OVER GERMANY.*
> *By The Associated Press.*
> AMERICAN NIGHT FIGHTER BASE, France, Jan. 1
> *The Germans have thrown something new into the night skies over Germany—the weird, mysterious "foo-fighter," balls of fire that race alongside the wings of American Beaufighters flying intruder missions over the Reich. American pilots have been encountering the eerie "foo-fighter" for more than a month in their night flights. No one apparently knows exactly what this sky weapon is. The balls of fire appear suddenly and accompany the planes for miles. They appear to be radio-controlled from the ground and keep up with planes flying (…)*
> *Donald Meiers of Chicago said:*
> *"One is red balls of fire which appear off our wing tips and fly along with us; the second is a vertical row of three balls of fire which fly in front of us, and the third is a group of about fifteen lights which appear off in the distance—like a Christmas tree up in the air—and flicker on and off." The pilots of this night-fighter squadron—in operation since September, 1943—find these fiery balls the weirdest thing that they have yet encountered. They are convinced that the "foo-fighter" is designed to be a psychological as well as a military weapon (…)."*

According to some reports the "Foo-fighters," or "Kugelblitz"—as they were supposed to be called by the Germans—sometimes also shot down the approaching bombers, creating explosions. From other accounts it followed that the objects were huge, rotating and emitted a large amount of heat.[248] The most contradictory to common sense—in the opinion of the pilots—seemed to be however their ability for tremendous accelerations. But all of this was "only" press reports, although when I was once reviewing the microfilmed archive of the Reich's Scientific Research Council I was taken aback by a report on those accounts.[250] The "Council" was obviously very interested in the Allied reaction… However I managed to get hold of a considerably more important source than the press reports.

In the archive at College Park (NARA) I managed to retrieve a US Air Force intelligence report from April 1945, describing this phenomenon extensively.[251] It is a really shocking document. Above all it says that as many as 302 observations of such objects were recorded (!), made by 140 aircrews. So we are talking about a mass phenomenon. What more—it can be clearly seen that it was a Luftwaffe weapon. New elements appearing in the report, were among others:

- Observations referring to Germany and Japan to an equal degree.
- Several incidents, in which there was mention of the "Kugelblitzs" (?) **being carried by enemy fighters**.
- Observation of several objects crashing into the ground, which were accompanied by explosions—so they were in all probability material creations of technology.
- The conclusion of the report's authors, that it is difficult to

explain the whole phenomenon by referring to jet or rocket propulsion. The expression: "a revolutionary method of defence" was mentioned.

Is an intelligence analysis, referring to 140 sources reliable material?

It is proof.

The Germans and Japanese, probably together, created flying objects having absolutely revolutionary characteristics, although they managed to employ in practise only the unmanned version.

In the NARA archive I found something else…

The huge majority of archive materials are unfortunately not precisely catalogued, however the main collection of documents handed over from the US military intelligence service (over 10,000 boxes) is provided with some kind of catalogue. In it is a considerable section referring to German research and development work from the field of aviation. It obviously aroused my interest. I found there something amazing —referring to work on a "flying saucer." Unfortunately in the main collection itself the document was not present—only a reference that it had been excluded from the files, so I have reproduced only the catalogue sheet.[252] Such work, irrespective of other reasons, can no longer be solely the subject of speculation.

In the intelligence report[251] mentioned a little earlier a totally new element appeared, which raised the question about the possible co-operation between the Reich and Japan in the field in question. Was this possible?

It is known from other sources that a certain exchange of technical information had existed, in the confines of which technical documentation of the Me 163 and Me 262 among other things had been given to the Japanese. Here however probably something greater was involved. I did not exactly check the Japanese plot, but there is some information suggesting the possibility of joint work on mercuric propulsion.

Since both Allies were separated by huge distances and enemy armies, almost exclusively the submarines of both sides were used for the transportation of materials within the confines of the technical exchange, which ferried like shuttles. The strange thing in all of this is that at the end of the war very large amounts of … mercury started to "appear" on their decks. So large that this had undoubtedly taken place at something's cost, after all the submarines' internal capacity was very limited and fuel was "worth its weight in gold." **Mercury was evidently a strategic resource for both sides**, which is not possible to explain in any "conventional" way.

The first such information had intrigued me long before now, and I had chanced upon it in a book about German polar research.[253] There was reference in it to a U-859 submarine, which was supposed to have departed from Germany in April 1944 and soon after was sunk in the Malakka strait near Indonesia. Its cargo was retrieved in 1972, when it turned out that it contained 33 tonnes of mercury(!). In October 2001 the British Discovery Channel broadcast a documentary devoted to the great Japanese I-52 submarine, which in the summer of 1944 was sunk in the Atlantic, near Gibraltar, by an American bomber. It had taken from Germany plans and strategic materials, including a large amount of mercury. The cargo had been transferred at open sea from a U-530 U-boat, incidentally, the same, which later reached Argentina in an evacuation mission.

I even found in a book devoted to Hitler's gold information that in one of the hidden deposit storehouses a "huge amount of mercury" had been found, apart from gold ingots.[254]

These are all incidents not uncovered until after the war and in reality we do not know what truth hides behind them. However considerably more is known about one such cargo, since it was captured in entirety (the U-boat plus its crew) by the Americans—even in May 1945, which the Americans have disclosed. The vessel—a Type X-B U-234—left Norwegian fjords on April 16, 1945. It was to sail to the West, then to the Southern Atlantic and finally to Japan. Apart from the crew, on board were: three Japanese—two officers and a submarine design specialist—seven German military specialists from various fields, a military judge, who was to take over the Richard Sorge case in Tokyo, the new German military attaché as well as two experts from Messerschmitt.[232]

Very interesting and symptomatic is the fact, that although the captain was aware of the capitulation of the Reich on May 8, he recognised his mission as valid and sailed on. Only when the submarine's radio watch notified him that Japan had recognised all agreements signed with the Third Reich as invalid, did Captain Johann Fehler decide to surrender to the Americans. On May 19, the U-234 stood moored at Portsmouth wharf, in the American State of New Hampshire. Its very interesting cargo was precisely catalogued within the next few days.[255] The total cargo mass was 95 tonnes, if not counting the lead ingots placed in the keel to balance the vessel. There was all manner of "goods," among others: new medicines, various electronic devices for aircraft, planospheric lenses, 56 kg of uranium oxide ["10 boxes, 56 kg," which some authors interpret as 560 kg], electrical coils, various munitions, a kind of

The Škoda (SS) plant in Pilsen was considered so important, that the real factory was camouflaged, while 5 km away a wooden mock-up of the entire plant was built (on the photo). It fulfilled its role at least once.

Japanese giant submarines in Tokyo's harbour. Their displacement reached up to 6,500 tonnes.

fire control system, various Junkers plans, plans of the Me 323 and many smaller shipments.

First and foremost however, the U-234 was carrying mercury—in total 24,112 kg! As far as its mass was concerned, this was an absolutely dominating position. Of course the crew knew nothing about its purpose.[232, 255]

Isn't this information indicating a previously unknown aspect of World War II? No particular professional knowledge is needed to come to the conclusion that at the end of the war they strived to launch, at any price, the production of some weapon, to which mercury was the key. We are talking after all about typically industrial amounts...

This issue once again beckons the question of what happened to the project in question after the war. I have to confess that I initially considered Argentina to be the most probable country of destination. Everything after all was evacuated within official confines, probably within the confines of Martin Bormann's "Generalplan 1945." Part of it was the so-called "multi-plane" operation of strategic evacuations code-named "Regentröpfchen." Its main objective was to secure the capital and cadre for the future re-building of the state's potential. Argentina was the principle target link of these evacuations, several U-boats arriving there, undoubtedly including the U-977, U-530, and in all probability the U-650, which at the beginning of July 1945 "vanished" near the coast of the country in question. The trace, which had been the flight of the Ju 390 from Lower Silesia was severed after all near the sole, still moderately safe U-boat base of Trondheim. Apart from this, in the testimonies of one of the Germans the name of the evacuation's target location appeared (only once), "Ebores," which was supposed to lie somewhere "far to the south." It was never managed to unravel this question, in all probability it refers to a code-name and not a geographical name. A series of state-of-the-art aviation projects were continued in Argentina, and near the town of San Carlos de Bariloche (or shorter: Bariloche) a team of German scientists realised a not fully clear project related to plasma and high voltages. A large research centre, located on an island, was built especially for them and one of the key persons on the team was Engineer Hellmann from the AEG consortium. In any case I have described these issues extensively in a separate book.[256] When I was there I spent several hours in conversation with one of the directors of a research centre in the mountains near Bariloche— with a scientist well-informed in the "post-German" archives (P. Florido). I showed him the characteristic features of "the bell," but he was not familiar with any of them. Faced with a lack of specific clues, I was forced to abandon the Argentine trail.

It has been suggested to me many times, without providing the specific sources of these suggestions, that the Germans headed in the direction of one of the polar regions. However nothing confirms this and I do not think that such a possibility was particularly likely. The main problems were the extremely difficult geographical conditions, including navigational difficulties and the lack of infrastructure. The only plot from this group which I consider worth attention, is that which was initiated by the pronouncement of the U-boat Fleet Commander, Karl Dönitz from 1938, which was reproduced in the German press of that time:[257]

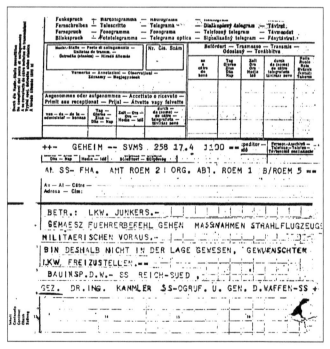

The last sign of life, that Kammler left behind. It was a secret cable, sent on April 17, 1945, to the Waffen-SS command (SS-Führungshauptamt), in which Kammler refused to "make the Junkers transport vehicle available (return it)." (source: "Teufel oder..." see bibliography)

The first page of a statement referring to the research and development work conducted by the Waffen-SS at the Škoda plant in Brno/Bruenn (from 1942).
(AAN, Alexandrian Microfilms: T-175/ files of the Pers. Stab RFSS)

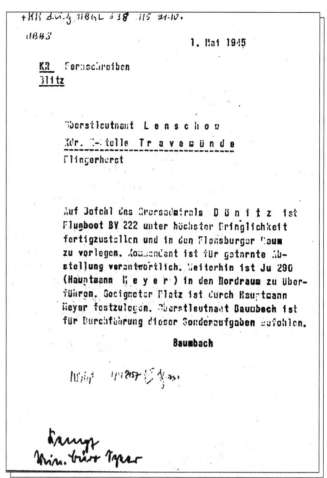

A teletype from the General Inspector of the Luftwaffe's bomber air force, Werner Baumbach, ordering the urgent redeployment of the giant BV-222 hydroplane and Ju-290 to the Flensburg region. It was sent on May 1, 1945, and, as it follows from the memoirs of one of the KG-200's pilots, was related to the plan to evacuate the Third Reich's highest authorities to Greenland.

Die deutsche Kriegsmarine ist stoltz. Sie baute für ihren Führer und Reichskanzler Adolf Hitler einen absolut uneinnehmbaren Versteck, wo er vor allen seinen Feinden sicher sein wird.

The English translation reads as follows:

The German navy may be proud. It has built for our Führer and Chancellor of the Reich Adolf Hitler an absolutely impenetrable hiding-place, where he will be safe from all enemies.

Everything indicates that it referred to an underground U-boat base located in Greenland bearing the code-name "Biberdam" (in all probability presently on the grounds of the American base of "Thule"). This variant was confirmed by a former pilot from KG-200 group.[258] He wrote in his memoirs about "a hiding-place on the coast of Greenland." Even on May 1, 1945, Dönitz gave the order for the leaders of the Third Reich to be transferred there. But in the end this never came about, and all the more nothing indicates that "our" project found its way there.

It turns out that the most important traces point to… the United States. Even if the press reports from 1947, describing Skorzeny's contacts with US intelligence representatives and attempts to sell the "super weapon" had been false, it still remains a fact that he "advertised" this weapon in the press. But above all it is the fates of the scientists which point to the North American plot. All key persons found themselves or were to find themselves after the war in the United States (H. Jensen was intensively sought after by US intelligence).

An interesting and not fully unravelled clue here is the fate of SS-Gruppenführer Kammler—at the end of the war and afterwards. He was without fail a primary person in the organizational hierarchy associated with the Lower-Silesian "Chronos"/"Laternenträger" project, being after all a frequent guest at the Gross-Rosen head-office. Independently of the "Jägerstab's" department responsible for research and development work, Kammler's "Rüstungsstab" had an analogous department in the former plants of Skoda in Pilzno (Pilsen)

The last known photograph of the Ju 390 aircraft, taken in mid-April 1945 at an airfield in Prague. At that time it was the only existing specimen of this type.

and Brno (Brünn)—160 km to the south of Książ/Fürstenstein. The Waffen-SS liaison staff, which co-ordinated activity in that region, was in turn situated in Prague. All of these locations indicate as well the last known region of Kammler's abode. This is interesting in so far as all knowledge about him ends at the moment in which the last known element appears in his war-time CV—the flight of the Ju 390. In the period in question only one specimen of this type existed and it seems that Kammler simply flew away in it, together with "his" project.

The last known document, which was left after him, was a cable from April 17, 1945, referring to the use of a "Junkers transport vehicle" ("LKW. Junkers").[259,208] Simultaneously from this period, from mid-April 1945, comes the last known photograph of the Ju 390—taken in Prague, during preparations for flight.[258] Then the trace comes to an end—there was no aircraft, and no Kammler. Although he was intensively hunted after the war, he completely "vanished." The opinion prevails that only one of the superpowers could have so effectively hidden a high-ranking SS general. It is known from other sources that the Americans (Patton's army) drove a wedge deep into Czechoslovakia among other reasons to reach the employees and documents of Kammler's "department" and lengthily interrogated all of them. These materials are to this day kept top secret.[259,208] This is very intriguing and once again points to the "American trail." There are many more similar corroborations. I am going to present only a chosen example…

The Argentine Trace

In the course of my research I finally considered the possibility of an evacuation to South America, and specifically to Argentina, as unlikely. One cannot however completely exclude this, although the location itself of Bodo base, in the north of Norway, rather suggests a Japanese direction. The Ju 390 could, as the sole aircraft, overcome this route without a stop-over—flying over the Barents Sea, and Northern Scandinavia is the part of Europe closest to Japan. But why is it not possible to completely reject the Argentine trace? Apart from the elements mentioned in the primary section of the text, there exist the following:

—in 2001 one of the persons co-operating with me in the gathering of information came to see me, reporting a quite strange affair—his acquaintance, the son of a Polish diplomat residing in Uruguay during the war, had shown him a photograph of an aircraft already known to us. It was the Ju 390 (only one specimen of which existed in 1945), standing on a field airstrip, hacked out of the jungle, in Uruguay, near the border with Argentina. It differed from the known version only in the lack of any kind of markings and the presence of a small, transparent cupola for astronavigation, placed in the fuselage. This photograph was supposedly taken precisely in 1945. I personally haven't seen it, and it hasn't been examined, but I have great trust in my co-worker and treat the whole affair as a valuable signal.

—Could such an event have ever taken place? Certainly it cannot be excluded. The Ju 390 prototype was one of the first aircraft adapted for mid-air refuelling and in favourable circumstances would have been able to fly non-stop even so far. This was the region in South America closest to Europe, where the Germans felt relatively safe. Not far away was located the Nazi enclave of "Isla Paraiso" and further to the south-west, the next (among others "Heide" near the border with Paraguay, "Santa Rosa" near Cordoba as well as "Cordier" and "Aldea" near Bariloche…).

—The possibility at least of the existence of a plan of continuation of research work in Argentina still matches the profile of "post-German" research carried out in a nuclear centre near Bariloche…built specially for the Germans. Indeed officially researched was the behaviour of plasma under high voltage conditions in connection with a thermonuclear project, but other explanations were not even denied by an experienced employee of the centre, who to my amazement stated that not all the materials from this period have been declassified to this very day.

—Throughout the whole of the war and directly after it, the Germans operated in Argentina quite confidently, with the tacit assent of local intelligence. Rare raw materials and a vast amount of grain were bought… through neutral countries. It even came to this that they single-handedly worked mines. In Patagonia operated a coal mine called "La Tungar." It was closed before the war, since the coal was contaminated. As it became evident, one of these contaminants was the ore of vanadium, a metal very precious in the armament industry. A German company officially operating in Argentina (like hundreds of others) contracted over 200 miners, bought up the whole area and exploited the ore until it was exhausted.

—The context of the operations of evacuational commando "ELF" also points to an Argentine plot. It was engaged among others in the realisation of the so-called "Operation Eichhörnchen"… financial evacuations, leading to the aforementioned country. Apart from this, after the war a German messenger was arrested who had tried to make contact with people remaining in the region of the Środa Śląska / Lubiąż. He possessed ID papers issued in Uruguay.

In 1981 in the Polish magazine "Przekrój" an article was published, written by Mr. Arnold Mostowicz, devoted to the alleged crash of some disc-shaped flying object near Laredo in the USA (1948). Mostowicz presented it as a UFO crash. In response to this article the editorial staff received a rather unusual letter, signed "Robert Allan Kolitzky," excerpts of which I will take the liberty to quote below:[260]

In my first sentence I wrote that it was fortunate that the article had found its way to me, so perhaps I will explain in short why I think so: well, as far back as 1948-49 I had the possibility to speak my mind about the Laredo crash. I was authorised to make these statements by the fact that I had lived in 1948 in Laredo, and on 7 July, around 18.00, along with my brother Miron and co-worker Dewey Reynolds, was already at the scene of the crash. The first search party reached the site, after receiving our radio report, at around 23.00, and scientists from the military laboratory, just like Mr Mostowicz mentioned, no earlier than around 2.00 a.m., already on 8 July. And so, along with the two people accompanying me, we were the first people who saw the still burning object and remnants of its passenger. Our recorded, written down and confirmed testimonies are present in archives and in their time have been made available to interested persons. Nobody ever demanded from me to keep all the details of this issue secret, so I never felt obliged to keep silent about it. (…)
b) the creature, which died in the crash, was a laboratory rhesus monkey. I can even provide much more detailed information about him, or rather her, since this monkey was female. She was called Imu, was born in a Boston breeding institute and as an adult specimen was sold to the army, along with a group of 23 other specimens. This was in May 1947. Imu was not 86.3 cm tall, and certainly not 135 cm, as some sources have mentioned. Admittedly the body was quite seriously injured, and its overall dimensions ranged between 67 and 72 cm. Citing the reliable report of Dr. D.C. Hagen, I will also give additional information about the animal's height before the flight and crash: 96.5 cm.
The uncanny details about the body's appearance, provided for the first time by Spencer and Haskins, are only their inventions or arose in the imaginations of people who had sold information to them. It is true that the disfigurement, caused by trauma and fire, considerably altered the external and partially internal appearance of Imu. Despite this it was enough to examine more precisely the dead body, to reject as groundless all suspicions of the creature's extraterrestrial origin. (…)
d) one can assume that none of the superpowers in 1948 had at their command rockets with a range of 1,600 km. I am not going to deliberate whether a radar in Washington registered the cross-country flight of the object that interests us , or whether the object took off (or as some prefer… was launched) from White Sands. I do not want to discuss something, of which I am not sure. I do not want to suggest anything. I can however ensure Mr. Mostowicz about one thing: that, what fell, exploded and burned at Laredo, was neither a spacecraft from another world, nor the V-2 rocket. The trail leading to weapons formerly belonging to the Germans is correct. However no rocket was ever involved here. If in the case of Imu I could provide more precise information to those interested, then regarding the object, of which she had been a passenger, I know much less. For obvious reasons I was not informed about all details. I can state however that the dimensions of the flying object and remarks about both its circular shape and lack of any traces of conventional propulsion more or less agree with the facts. From talks with fellow officers I can conclude that one of the devices brought over from Germany and known under the joint name of "Kugelblitz" was destroyed in the accident. These devices had a completely different design to known rockets or aircraft. I have good reason to believe that some of them possessed ionic-mercuric propulsion. The Callowey report contains enough information to figure out that our army possessed at that time many V-2 rockets, several prototypes of the lesser known V-3 and V-4 rockets and about thirty Kugelblitz of various types, fired combat missiles (also remote-controlled), as well as piloted machines.

During the years 1948/49/50 the first more important experiments with the devices in question were carried out. The device which fell in the vicinity of Laredo had a differently marked out flight route and just like all the previous ones should have only flown over uninhabited areas. However soon after take-off ground services lost control of it, and the last record of the device's operation came an hour before its impact. (...)

f) as I have already mentioned the monkey-pilot's body no longer exists. It is possible however that certain preserved organs have survived. The rest was burned on 16.07.1948. I do not know if any elements of the destroyed device have remained. Many of them were collected from the crash site. I saw some of these parts in a laboratory in the year after the crash. Whatever happened, it probably does not matter, since it is known that many complete specimens of the Kugelblitz are present in US Army collections. Also there are prototypes of machines, built on the basis of their design.

That is all that I have to say on this subject (...)
ROBERT A. KOLITZKY

The possibility of this technology being seized by the Americans in effect forms the next question...why do we still not know about this officially?

I do not know if my explanation will be correct, however it is possible to explain it. If such revolutionary technology is involved, as the Germans assessed it, then common military practice orders this type of weapon to be kept top secret, and we will not find out anything about it until the outbreak of a truly major war. There is no doubt that work in this field was and is still being carried out under the protectorate of the US government. Proof of this is a book by Cook—a recognised analyst in the field of modern aeronautical engineering.[208] I myself found very specific references, for example when searching for scientific work describing the rotation of plasma. Many scientific articles contained annotations about the financing of work by the US government. To be more specific...in a publication devoted to the possibilities of generating plasma vortices, rotating in opposite directions and interacting with a magnetic field, the symbols of grants and contracts of the following institutions appeared: "Air Force Office of Scientific Research," "Air Force Cambridge Research Laboratories, Office of Aerospace Research" and "NASA."[261] In 1998 reports appeared about a new USAF reconnaissance aircraft, propelled by "rotating, superconductive mercuric plasma at a pressure of 250,000 atmospheres." It is supposed to be designated the TR-3B. And even on the eastern side of the iron curtain quite serious reports about analogous Soviet work appeared, as early as the 1950s. It was even acknowledged in specialist military magazine that the Russians were building prototypes...[262] So there is a chance that the present day will finish writing the sequel to the war-time story. As a result perhaps we will find out for example what was the unusual jelly-like metallic substance filling the core of "the bell," or what happened to the giant Ju 390...

Anti-gravitation in the Universe

The year 1999 turned out to be a turning point in the history of science. A number of scientists and astronomers from such renowned institutions as Cornell University, the Space Telescope Institute or Lawrence Livermore Laboratory, after examining the latest data came (independently) to the conclusion that the Universe in which we live is expanding at a greater and greater speed. In the society of physicists and in specialist articles the once fashionable term "anti-gravitation" once again began to function—this time as a dominating force in the Universe, but only visible at large distances. It is it which repulses huge galaxies away from each other. After a previous discovery that stars in galaxies travelled along completely different paths than was expected, (the question of "dark matter") once again it became evident that our knowledge about the nature of gravitation was only fragmentary and moreover at least to a large extent erroneous. For the described discoveries contradict the predictions of theorists. It became clear that in the scientific description of the world, the chapter referring to gravitation consisted mostly of blank spots, and moreover certain pages were placed in it back to front. In spite of this, surely many years must yet pass before the wall of the mental barrier finally crumbles and funds for researching anti-gravity in laboratories become significantly increased. Many scientists simply do not see the link between the possibility of this force existing in the Universe and the possibility of researching it on the Earth.

This data has been recently confirmed by NASA to have "a one-percent margin of error"—thanks to the latest observations. In its edition from June 2003, the American magazine "Astronomy" quoted that: "Only 4 percent of the Universe is made up of ordinary matter. Some 23 percent is cold, dark matter and 73 percent [!!!—I.W.] is thought to be dark energy... an anti-gravitational force, that is accelerating the rate at which the universe expands." According to one of the scientists involved in the research, "this is the beginning of a new stage in our study of the universe"—this is something that doesn't fit very well with Einstein's theory.

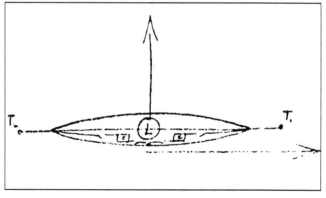

A sketch of an antigravity craft from a 1950s military publication (WPL 7/58... see bibliography). Why a disc?

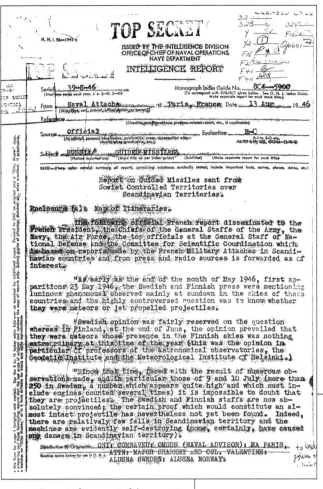

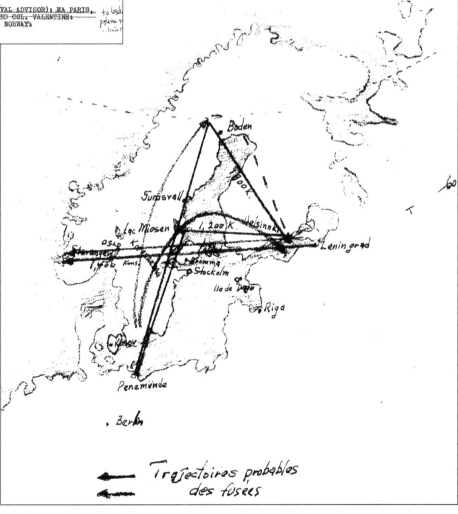

Excerpts from a set of documents referring to the objects observed over Scandinavia, including a description of their extraordinary flight characteristics, as well as a map prepared by French secret services, showing the most typical trajectories. (source: see bibliography)

As far as the Russians are concerned, it is also very probable that they seized and developed the German achievements in this field. Even Podkletnov's declaration proves this. Some time ago I received a letter from a reader, whose grandfather had served immediately after the war in the area of the former German rocket range at Ustka (on the Baltic coastline, which was strewn all over with similar ranges). I will not cite the letter since it follows from the author's suggestion that he wouldn't have wished this. His grandfather was supposed to have been a witness to the crash of a strange flying object in June 1945, which in the opinion of Soviet officers had been concealed in bunkers not yet explored at that time. In the shattered wreckage found on sand dunes, it is rumoured that the bodies of two pilots were found, "with attached Nazi emblems." I would have surely belittled the above account, if not for two odd looking details: according to the Soviet officers the Germans had managed to evacuate some objects to Spain, and initial examination of the propulsion revealed the presence of "an unknown, jelly-like substance."

I could have left all of this unmentioned, however I have added a few of my own opinions on this subject after writing the main part of the book, because I managed to get hold of some documents, I dare to say, confirming this story. During my last archival query in the USA I acquired some intelligence documents, until recently top secret, referring to the so-called Scandinavian wave of strange flying object sightings from the summer of 1946.[264] Over 300 cross-country flights were observed over Norway, Sweden and Finland. This issue has already been described in the press, but the intelligence reports revealed its origin. It turns out that the V-1 and V-2 missiles couldn't possibly have been involved, because the range of the "guests" as a general rule greatly exceeded 1,000 km; apart from this their trajectories were rather flat and often involved ceilings of the order of only several hundred metres.

On the grounds of these observations the objects were divided into two groups: "cylindrical" and "brilliant fiery balls." They could perform sudden manoeuvres and reach a speed of "up to 2,800 km/h"! In short, here we have typical descriptions of the "Kugelblitz." What more, none of the crashed "balls" was ever found, because, as it was ascertained, the objects were equipped with self-destructive mechanisms. This also appeared in the case of the "Kugelblitz"!

> *The Scandinavian Epilogue*
>
> The passing flights of strange objects from June-September 1946 were decidedly the "event of the year" throughout the whole of Scandinavia. As practically all objects approached from the Baltic, the conviction dominated that this way the Soviets wanted to "support" their diplomatic operations towards this part of Europe… namely enforce on it neutrality in the new, cold war. For obvious reasons this threat was treated seriously and was accurately recognised… above all by the Swedes. Relatively close to the Polish coastline, approx. 250 km to the north of Leba… on the island of Gotland, was located the telemetric station of their radio-intelligence and radio-counterintelligence (FRÅ). As local authors mention (E. Svahn, A. Liljegren, L. Gross), it was noticed that the passing flights of the "guests" were preceded by strange radio emissions, arriving from directions coinciding with the localisation of the post-German bases of Peenemünde, Ustka and Łeba. On several occasions a Swedish reconnaissance plane tried to approach them, but Soviet fighters always appeared in the skies at the right moment. In spite of this, the connection of the Russians with these events was never proved… first and foremost because debris was never found of a size enabling their identification. For these objects always exploded on hitting the ground or surface of the water. They also flew too fast and often too low, so as to be able to photograph them close up, e.g. from the deck of a fighter. An analysis of all information connected with this was engaged in by the so-called Jacobsson Committee, appointed by the Swedish Ministry of Defence. As far as I know, these analyses are still classified.
>
> The Americans were particularly interested in the progress of the Jacobsson Committee's work, among other things on July 23 Admiral James Forrestal held a conference with its members.
>
> An intercepted French intelligence report had been attached to the set of American reports, prepared for the President of France (hence the very serious nature of the issue involved). From its text and attached maps it evidently appears that the objects were mainly flying from the area of Peenemünde! According to the French, the fact that the trials were conducted on foreign territory simply resulted from such courses being imposed by the location of the former German installations on the Baltic coastline.[264] Materials of this kind, of which there is no shortage, unequivocally call on us to treat the existence of the "Wunderwaffe" as a fact…

However even in this case yet one more issue remains…how did it happen that scientists from the 1940s understood exactly where they were heading? They had applied after all ideas from XXI century physics. How is it, that they were conscious of bringing about a major turning-point? What arguments did they lay down (before the launch of work) that caused them to win the race for funds with the great and influential armament consortiums? And they won decisively…

It seems that the whole issue is something more than just a technical problem. The unusualness of all this is summed up by the fact, that descriptions of mercuric propulsion had appeared as long ago as in ancient times—in alchemy and old Hindu books—one can easily check this. The "Samarangana-sutradhara," a book at least 2,000 years old, said for example: "By means of the power latent in the mercury which sets the **driving whirlwind** in motion [bold by I.W.], a man sitting inside may travel a great distance in the sky in a most marvellous manner."[263]

It may prove that an explanation of all the technical questions related to work from the time of the war, will reveal a far greater mystery…

The whole affair is probably best summed up by the words of Albert Speer: "I sympathised with anyone who tried to unravel any kind of undertaking taken up by Himmler… the ingeniously weaved swindles and dreadful ideas, which this uncanny individual was capable of, seemed to come from a different world."[266]

MORE ABOUT THE PHYSICS OF THE "BELL"

After certain consideration I decided to add a kind of summary at the end of this edition, in which I could clarify misunderstandings (circulating mostly on the Internet), present latest findings, as well as comment on various hypotheses and conclusions of other authors analyzing the aforementioned German project. To my amazement, several other authors, American and British, have immersed themselves in the obscure reality of the Third Reich's most secret research project. What they contributed may generally be divided into two categories:

1. Speculations as to the "Bell's" purpose (which is still uncertain).
2. Hypotheses pertaining to its "modus operandi," or the physical principle/effect which the "Bell" exploited.

Even if some of the information contained in these books is inaccurate (an insignificant percentage, in fact) or represents "dead ends," in my view it is nevertheless invaluable. It represents a kind of free market of ideas, which may eventually enrich the final picture, especially in connection with the fact that I never managed to clarify all issues in "The Truth about the Wunderwaffe." The first edition of this book, from 2003, which these authors referred to, also lacked important information gathered during the following six years.

The list of books touching on or developing this topic is quite extensive:

1. First and foremost there are numerous publications by Joseph P. Farrell (at least four), the most significant being: "The SS Brotherhood of the Bell" and his recent work titled "Secrets of the Unified Field" (AUP, published respectively in 2006 and 2008).
2. "Hitler's Suppressed and Still-Secret Weapons, Science and Technology" by Henry Stevens (AUP 2007).
3. "The Rise of the Fourth Reich" by Jim Marrs, from 2008 (Harper Collins Publ.).
4. "Hitler's Terror Weapons: from V-1 to Vimana" by Geoffrey Brooks (published in the UK by Pen & Sword, in 2002).

I will start from Farrell's books and from his hypotheses leading him to believe that the "Bell" was supposed to be a kind of weapon itself. He writes in the "Secrets of the Unified Field":

Among the hodgepodge of bizarre research projects there were scattered clues that the Nazis were indeed deliberately researching the possibilities of weapons based on electromagnetic energy. And scattered among these there were further clues that the Nazi regime was researching the possibilities of a weapon based on torsion physics and the tearing of the space-time continuum, the same kind of physics that led to the astonishing and unanticipated results of the Philadelphia Experiment. The Nazis were in effect after a kind of "torsion bomb." The clues, however, were not only scattered throughout the hodgepodge of projects the Allied intelligence teams encountered as they entered the Reich, but the projects themselves were similarly spread across the landscape of Nazi Germany. The Allied intelligence teams can therefore hardly be blamed for having missed the clues, for who would have reasonably thought that the Nazis were after a weapon that was potentially far more destructive than mere atom bombs? But there was a final reason the clues were missed, besides their dispersion across the catalogue of projects and the landscape of Germany. They were missed because, like so many other cases of classified research in Nazi Germany, they were most likely coordinated by the super-secret think tank of SS-Obergruppenführer Dr. Ing. Hans Kammler, the "Kammlerstab."

The author touches on two important issues in this quotation. The first—a "torsion bomb"—is not particularly convincing. We don't even know whether such a thing could exist at all. Perhaps his assumptions are accurate, but in my opinion, what they initiated had to be clear and capable of being proven (here I mean demonstrating the feasibility and physical sense) and its potential had to be obvious. Otherwise the highest priority wouldn't have been granted. Such a priority meant a great burden of responsibility for the scientists engaged—they were treated as potential saviors of the Third Reich. They had to do this and there was no place for pure speculations when the decision to initiate the project was taken. Obviously this does not eliminate the possibility that during research some interesting side effects emerged which they attempted to explore, but the main direction had to be, in my opinion, rather straightforward. Perhaps Joseph's approach was similar, depicting one of the possibilities, and therefore cannot be treated as an obvious error. Summarizing: I do not agree that the "Bell's" main purpose was to be a "torsion bomb," but we should bear in mind that one of the objectives could have been to carry out basic research in torsion physics as well. They had certain grounds to expect a specific outcome, but couldn't have been in a position to predict everything (the deaths of some members of the German personnel in the project's initial phase testifies to this).

My own reservations also refer to speculations that the "Bell" was supposed to be some kind of time-machine, which has been described by Joseph P. Farrell and Jim Marrs, as well as Henry Stevens in "Hitler's Suppressed and Still-Secret Weapons..." We also do not know if it would be at all possible in this case. I am also concerned that they had greater preoccupations at that time... I am aware that at least in the case of Stevens' book this was based on the testimony of a witness, but when he stated that the experiments had something to do with the nature of time, he could equally have meant that this only involved relativistic physics, not necessarily that they were constructing a time-machine. I am not trying to "debunk" it, all I would like to say is that I have certain, perhaps irrational reservations as to this interpretation. It also would not have justified the top priority—by not saving the Third Reich.

The above quotation from the "Secrets of the Unified Field" contains another important matter that I will take the liberty to comment on. It concerns the reasons why the Allied intelligence services "missed the clues." Farrell himself develops this notion further on (p. TK):

The fact that it had not been penetrated during the war by any Allied or Soviet intelligence operation led both Igor Witkowski, in his book "The truth about the Wunderwaffe," and I in my books "Reich of the Black Sun" and "The SS Brotherhood of the Bell," to speculate that U.S. General Patton's Third Army thrusts into Bohemia, Austria, and south central Germany and the Harz Mountains region were too precisely coordinated with some of Kammler's most secret facilities to be coincidental. Rather, they seem to have been steered to their objectives by someone highly placed in the US Intelligence.

Well, I have to admit that this seemed the case, and would be stupid not to follow this thread and attempt an investigation. When I did so, however, it ceased to be so clear. I couldn't find any information confirming that Allied intelligence officers, at least those detached to the Third Army, were aware of the issue of "Kammler's secrets" at all—in particular those in question. I described this to a large extent in "our" (Polish) second volume of "The Truth about the Wunderwaffe," the most significant parts being incorporated into the extended, newer American edition of this book. My personal impression is that at that time—in the Spring of 1945, before the end of the war—very few in the Allied armies even suspected the scope of the technological breakthroughs realized in the Third Reich. Please note the "gradual" character of Operation "Paperclip" (designed to "import" the best rocket and aviation experts). Only after interrogating the first wave did the American officers realize what they had missed and only then decide to greatly expand the operation. But such knowledge wasn't common in the spring of 1945. The worst was precisely the case with the secret activities of the SS. To this day very few people (and I mean really few...) are aware of the fundamental fact that the SS had its own armament office (*Mil. Amt*, the equivalent of the Wehrmacht's *Heereswaffenamt*), even fewer realize that within it existed a secret research and development section (FEP) that never even mentioned the name of its chief—Otto Schwab, in spite of his high rank of general—for Kammler himself wasn't personally responsible for the science as such. The situation may change after the publication of this book, but in 2007, when I was writing this about this (specifically SS activities in occupied Czechoslovakia and Schwab), I could not find literally anything about him even on the Internet.

It does not seem likely therefore that Allied intelligence "grasped" this field to a significant degree, certainly they did not have the overall picture. I was more amazed that Soviet expertise represented a markedly higher level. In Czechoslovakia (nicknamed the "SS model state" by Hitler himself) they had, if not the overall picture, then at least significant parts of it. They acted quicker and had greater resources at their disposal. The example of Jachymov is telling in this respect—when they realized they had access to probably the most important nuclear research facility of the Third Reich, they never dispatched some team in a 4WD car, but whole "battalions"—without asking for permission and immediately after receiving the information (a battalion usually numbers up to 500 soldiers). That is how it worked... Perhaps it can be explained by the different "role" of technological know-how in the USSR. While many Westerners treated the German experts with suspicion, the Soviet Union was simply addicted to Western technological innovations. The great industrial revolution of the 1930's simply would not have taken place, if not for the influx of Western technology (chiefly American). Regardless of the reasons, they were efficient in this business, and although it would be hard to imagine a more "ideologized" state than the USSR just after the war, their approach in this case had less to do with ideology than the Americans', for example. I remember (and I will probably never forget) a conversation with a former serviceman, who worked in the Academy of General Staff in Poland, who some 12 or more years ago talked about captured German documents referring to "Riese" and the "special weapons" that were supposed to be manufactured there. Among others he mentioned the name of the scientific chief or consultant of the Soviet intelligence team tasked with interpreting the reconnaissance in this case—Professor Artsimovich. He was a plasma and thermonuclear physicist. Then I wasn't yet aware of the connection with plasma physics, but it means that already in 1945, probably even before entering the area of these mountains, the Soviets knew which field of physics was crucial in this respect. When we take into account how pioneering and difficult a field it was, this was quite an achievement—and commands my respect. Of course we should also bear in mind that the SS was in a unique position, thanks to which they were quite effective in keeping something secret. The unique combination of state-of-the-art technology, capitalism, socialism and slavery, with a mixture of ruthlessness, gave them both the potential and the means to isolate from the outside world even tens of thousands of workers and Jewish specialists etc,. who could be eliminated afterwards, almost without leaving a trace. The most likely source of leaks—the workforce—could at least be greatly reduced. This recalls, by the way, the issue of what happened to the project after the war. So far the most likely hypothesis, in my opinion, is that it was evacuated to Argentina—this thread was originally presented by Geoffrey Brooks in "Hitler's Terror Weapons." Such a possibility was "introduced" by one of the original sources, a German courier who was arrested and interrogated in Poland shortly after the war. It appears that there were two transports associated with this project (I don't know what specifically was evacuated, whether it involved the "Bell" itself, or just the documentation…). One by plane to Uruguay, near the town of Gualeguay, where a landing strip was prepared in the jungle, and the other to a place that in all probability pertained to the codename "Ebores," mentioned by the courier. It was identified with a place on the island of Tierra del Fuego, within or near the Bay of Thetis, in the extreme south of Argentina. I have no definitive proof, but I have been there several times attempting to investigate these facts, and an evacuation to this country generally seems quite likely. It is also possible that there was an attempt to "revive" this project in the Nuclear Research Center near Bariloche, built essentially for the German scientists evacuated from Europe. Officially its purpose was to pursue a certain way of building a thermonuclear reactor, under the supervision of scientist Ronald Richter, where a thermonuclear reaction was to be initiated by the discharge of a high voltage electric current. In a book written by a leading contemporary Argentine nuclear physicist, Professor Mario Mariscotti ("El secreto atomico de Huemul"), he wrote among other things that this project lacked scientific grounds and wasn't really scientific, while something else was carried out there as well, in secrecy. Namely that a team from the former laboratory of Manfred von Ardenne, from Berlin, was assembled there and simply continued the research started in Germany. It was the research of "ball lightnings," or we should rather interpret this as the research of ball lightning-like vortices in plasma—plasma solitons in other words. It may be interpreted as something really important, plus a cover up for the press… The latter being the project that "lacked scientific grounds." Back in 1997 when I was at that nuclear center (Centro Atomico Bariloche), I asked one of the scientists whether the description of the "Bell" was in any way familiar to him. His answer was negative, although he added after a while that even he didn't have access to all documentation from that period—quite a strange declaration heard from one of the directors of the Center. It seems even more likely, therefore, that some elements of this project ended up there, "in the far south," but still there is no proof.

Farrell's books contain numerous interesting clues or facts, but let me reflect for a moment on a significant detail that requires correcting. He wrote in the "Secrets of the Unified Field" that (pages 268-269):

Inside the Bell were placed two counter-rotating drums, into which was poured its "fuel," a mysterious compound code-named "IRR Xerum 525," a heavy liquid-gooey substance of violet reddish hue. (…) Once inserted into the Bell, this mysterious compound 525 was then rotated in the counter-rotating drums through high electrical potentials. When this happened, the Bell glowed a pale blue glow and in some accounts that saw it tested outside, levitated. While in operation the Bell buzzed, earned it its nickname among the Germans, "die Bienenstock," or the Beehive.

Well, this mysterious "Serum" or "Xerum," according to the witnesses mentioned earlier, was not placed inside the spinning cylinders or disks. It was a dense, jelly-like substance, some amalgam of mercury and other heavy elements (I wouldn't say that it was a compound…) that was kept inside the axis of the device. It was a cylindrical container shielded by a layer of lead. It wasn't subjected therefore to a high voltage electrical discharge. The source of heavy, mercuric ions that were accelerated during the experiment, was just mercury (presumably pure). A certain amount, probably rather small, was placed inside the drums or disks. It was then heated, ionized and accelerated by the high voltage discharge, creating two counter-spinning plasma vortices. The disks were placed one over the other—not one within the other, as is described in some other publications. Perhaps it was my fault that I hadn't specified this clearly enough.

Generally the "Bell" operated like a classical plasma accelerator (which as such was officially invented shortly after the war), but with few significant exceptions which have to be described. The ions were accelerated from the circumference, which was the positive electrode, toward the axis, the negative electrode. The closer they were to the axis, the higher their angular momentum and angular velocity, and the stronger the magnetic field. Theoretically they should have reached an equilibrium near the axis, creating donut-like vortices, gradually compressing under influence of the magnetic fields—with diameter and thickness decreasing while gaining speed—with at least some percentage of the ions inevitably hitting the axis. Because of their very high kinetic energies (very heavy nuclei, when compared to electrons for example), they underwent nuclear reactions during collisions, certainly emitting large amounts of neutrons and in effect causing nuclear transformations within the "Serum/Xerum" which filled the container within the axis. It is still unclear whether this material was a kind of "waste," a reservoir for the harmful by-products of the reactions, or perhaps its production was the main objective. I will return to this question, for Joseph P. Farrell has proposed quite an interesting hypothesis. For the time being, let us continue with the description of the "Bell's" operation…

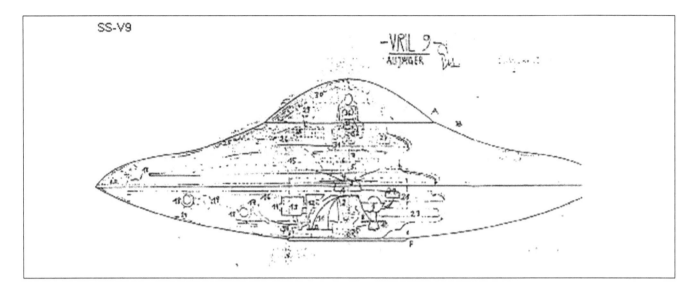

One of the examples of technical drawings showing the Bell in a "wider context." The Bell seems to be placed under the dome, in the axis of the craft. It doesn't "end" however where the pedestal was in the test device, but it looks like it was connected to some flat, wide structure occupying the lower level. One may have the impression that it was the "core" (containing the "Serum 525") that connected it with the rest. Perhaps the main purpose of this device was to produce some ORME-like substance (a mix of high-spin isomers) which then worked like a superconductor, shielding the fields along the bottom of the craft. Perhaps that's why it had to be more or less flat, to provide sufficient lifting power? It's of course only a speculation, because the authenticity of this drawing isn't certain. It's worth emphasizing that spin-isomers were discovered and made by Werner Heisenberg, the description of the separation of fields in superconductors was the merit of Walther Meissner and Robert Ochsenfeld (also Germans), while the first comprehensive description of the phenomena occuring inside the superconductors was published by the brothers Fritz and Heinz London—all that in the 1930's. Our understanding of these phenomena hasn't changed significantly since then, if not the opposite…

This device differed from a "conventional" plasma accelerator also by the fact that it constituted a kind of "tandem" of two accelerators, with vortices accelerated in opposite directions and with their magnetic fields directed oppositely. Generally speaking a plasma accelerator with a perpendicular constituent of the magnetic field is known, described after the war as being designed in a beehive configuration—at least in the German nomenclature. It is unclear to me whether this has anything to do with the sound produced, resembling the humming of a large transformer station (and of bees), or if it refers to the shape of the vortex in such a configuration, which to some extent resembles an oval beehive. It seems simple, but so far I haven't found any confirmation that the "Bell's" configuration itself was ever tested after the war. It is also very rare for an accelerator to be specifically designed to spin heavy ions and I have never heard of any accelerator using mercury…

One may ask at this stage: why? Does it not make sense? If not, then why was it so important for the Germans? Or perhaps something is wrong with our reasoning?

I am not sure about this, but from the classical point of view today the "Bell" might be generally perceived by scientists as a potential source of trouble, for in such a configuration one may expect massive radiation caused by collisions of the ions—although not massive enough to justify turning it directly into a weapon, and the cost/effect coefficient would be rather discouraging. It could make sense only under the condition that contrary to a conventional accelerator its purpose wouldn't be just to accelerate ions, but to create specific, strong fields, not attainable in any other way (which at the present time is rarely the case, unless we have in mind a *tokamak*, but there would be no point in building tandem *tokamaks*). The "Bell's" purpose differed from that of any accelerator known to me in that it was probably designed to create soliton-type vortices, in which the lines of magnetic fields are isolated, or as the Germans defined: "separated" from external fields. Such an interpretation is justified by the appearance of the term "separation of magnetic fields" in the original materials, which clearly points toward solitons and also has relativistic connotations—i.e. a link to gravity. This is precisely torsion physics and what apparently most interested them was the mutual interaction of two such solitons (or their magnetic fields, extremely strong—we shouldn't forget about this detail). I have some suspicions as to the sense of such counter-spinning, but have to admit that I do not fully understand this. Nevertheless it has to make sense, since this element appears in several independent, contemporary approaches to antigravity…

Summarizing: the "Bell" should have generated antigravity, but could have been useful not just from the point of view of propulsion. Incidentally: I am not aware of any original source saying that it levitated, nor did it have any hieroglyphic inscriptions on its circumference, as was suggested by animation presented in a certain TV documentary and later circulating on the Internet ("UFO Hunters" from the History Channel, the computer animation of which is largely inaccurate). The gravitational or rather antigravitational field was probably capable of something else, no less interesting, and in my view a hypothesis pertaining to this aspect is Farrell's most significant contribution. He has associated the substance that filled the cylindrical container inside the "Bell's" core—the mysterious Xerum or Serum—with certain by-products of nuclear reactions known as isomers. As he wrote in "The SS Brotherhood of the Bell":

> Discovered in 1921, isomers are simply metastable or "extremely stable" forms of atoms that are brought about by a state of excitation of protons or neutrons in their nucleus, such that they require the change in their spin, before they can release their pent-up energy. We are now a step closer to understanding the Bell, since it now appears that the device, due to the presence of "Xerum 525" was much more than a high-voltage, counter-rotating plasma trap, as conjectured by Witkowski. It certainly was that, but it was also a kind of reactor as well.

These substances, isomers, do not differ from normal atomic nuclei by the number of protons or neutrons, as isotopes do, yet their energy is different because unlike their "regular" equivalents, the nuclides (protons and neutrons) are spinning more or less in the same direction (generally they are always spinning, only that usually their angular momentums cancel each other). As a physicist would say: their spin is polarized. This makes them a much better "raw material" for an antigravity generator. I would even say that this is the key to a future revolution in this field. It coincides with the amazing discovery that various elements and isotopes, regardless of their mass, generate different antigravitational forces—see the article from Physics Letters B from 16 February 1989, p. 137 (the difference observable in the case of various elements alone, not isomers, even if they have the same mass, is the order of +/- 50%). Incidentally: isomers were discovered by Werner Heisenberg, a close co-worker of Professor Walther Gerlach! As I described earlier, during the war he worked in occupied Czechoslovakia, in the town of Aussig (the German name of the town Ústí nad Labem, not the code-name of some facility as I thought previously). Very interesting information on the extraordinary significance of isomers has also been presented by Jim Marrs in his "Rise of the Fourth Reich":

> Hudson himself obtained eleven worldwide patents on his "Orbitally Rearranged Monoatomic Elements (ORME). Hudson found that the nuclei of such monoatomic matter acted in an unusual manner. Under certain circumstances, they began spinning and creating strangely deformed shapes. Oddly, as these nuclei spun, they began to come apart on their own. It was found, for example, that in the element rhodium 103, the nucleus became deformed in a ration of two to one, which made it twice as long as it is wide, and entered a high-spin rate. When all electrons are brought under the control of the nucleus of an atom, the nucleus attains a "highward," or high-spin, rate. When reaching a state of reciprocal relationship, the electrons turn to pure white light and the individual atoms fall apart, producing a white monoatomic powder. (...) By the early 1990's scientific papers were being published by the Niels Bohr Institute and Argonne National and Oak Ridge National Laboratories, substantiating the existence of these high-spin, monoatomic elements and their power as superconductors. (...)
> Puthoff [Dr. Hal Puthoff, director of the Institute for Advanced Studies in Austin, Texas—I.W.] concluded that the powder was "exotic matter" capable of bending space and time. The material's antigravitational properties were confirmed when it was shown, that a weighing pan weighed less when the powder was placed in it than it did empty. The matter had passed its antigravitational properties to the pan. Adding to their amazement, it was found that when the white powder was heated to a certain degree, not only did its weight disappear, but the powder itself vanished from sight. When a spatula was used to stir around in the pan, there apparently was nothing there. Yet, as the material cooled, it reappeared in its original configuration. The material had not simply disappeared, it apparently had moved into another dimension.
> This hypothesis leads to the conclusion that the "Bell" might have been used to manufacture a certain substance—perhaps a byproduct of "irradiation," such as being subjected to extremely powerful magnetic fields. Perhaps this was the main purpose, we may never know.

Joseph P. Farrell noticed one more interesting fact, this time referring to one of the German scientists—Dr. Elizabeth Adler, a mathematician. He wrote in the "Secrets of the Unified Field":

> While there is no direct evidence, there is a very significant bit of indirect evidence, and it has already been mentioned, one of those obvious facts that one might overlook unless attention were drawn to it: Elizabeth Adler was a mathematician from the University of Königsberg. And of course, the University of Konigsberg was home to the very first Unified Field Theory, the "higher-dimensional" unified theory of Theodor Franz Eduard Kaluza, the first theory successfully to unite mathematically the gravitational equations of General Relativity with Maxwell's electromagnetic field equations. It was the very theory that led to the whole Unified Field Theory craze of the 1920's and 1930's in the first place. Dr. Kaluza was, of course, a Privatdozent or a kind of "adjunct" in the University's mathematic department, and, after his paper has been published, was certainly a high-profile figure. Dr. Elizabeth Adler could hardly have been present at the University in the same department without having heard of his name. (...) we have an indicator that perhaps the Nazis were experimenting with a modified Kaluza-Klein theory, one which incorporated the torsion tensor itself.

This association makes sense to me. It seems more likely that they based their work on the theory of gravitation developed during the war by Professor Pascual Jordan. To a large extent this resembled the Kaluza-Klein theory, only that it took the spinning of an electromagnetic field into account.

Also quite impressive is the book by Henry Stevens. He managed to gain access to numerous documents, among others a previously unknown set of documents pertaining to Ronald Richter, the chemist who headed the plasma research project in Argentina just after the war. In one of them Richter himself stated that he was interested not just in thermonuclear reactions, but also in:

> *Excitation of space structure by pulsation-controlled plasma explosion, testing the limitations of quantum mechanics and quantum dynamics, experimental approach to the unified field theory and to the velocity of propagation of gravity a.s.o.*

It's worth noting that his approach to gravitation was "experimental," when he was only active in the plasma field, therefore we may draw the conclusion that he tested gravitational effects, presumably associated with spinning. When we put all the pieces of information referring to Argentina together (around 20 German physicists worked there), then it is hard not to get the impression that the entire "Chronos"/"Laternenträger" project was indeed evacuated there. No less significant is the fact, quoted by Stevens, that during the war Richter was an employee of the AEG Research Institute in Berlin as well…

In the aforementioned book we also find an independent confirmation of the "Bell's" existence, although the information refers mostly to the so-called "Fly trap" (as the circular test rig is called in Poland), known also as "The henge." In this case I will allow myself to quote a slightly longer passage:

> *Witkowski's research shows that the Bell was always operated underground. This leaves us to wonder at the purpose of the Stonehenge-like structure with the hooks at its top. (…) We have physical evidence that something was going on. We have the mysterious Stonehenge-like structure. We have Igor Witkowski whose peek into the NKVD secret files [files of our military counterintelligence, to be exact—I.W.] led him to this unexplained place.*
>
> *This is where I stood in the Winter of 2002 when a chance conversation occurred with a friend of mine. His name is Greg Rowe and I have corresponded with him for several years as the result of a chance contact. Greg is a trained engineer, as was his father. Once Greg told me that his father had worked for NASA at the Huntsville, Alabama facility. Knowing that some German Paperclip scientists worked there, I asked Greg if any of these worked with his father. Greg replied with a list of German scientists with whom his father has worked and a few words about each one of them. One of these, Otto Cerny, seemed particularly interesting from Greg's description so I pressed him for more information. What follows is a compilation of several e-mails on the subject of Otto Cerny. Cerny was Greg's father's boss. Greg knew the whole family and went to school with Cerny's son, whom Greg called by name. Greg also referred to Cerny's wife by her first name and gave other small details of their family life. When he was somewhere between 12 and 14 years old, Greg and his family were invited to dinner at the Cerny's house. This would have been between the years 1960-62. Greg sat and listened as the older men talked.*
>
> *Otto Cerny was an engineer and physicist. He had worked at Peenemünde on a variety of projects. That was why he was in the United States, to work on rockets, and why he was a Paperclip scientist to begin with. But it is his work prior to coming to Peenemünde, which was the subject of discussion that night. Cerny said that he had worked near Breslau [now Wrocław in Poland—I.W.] in the early years of the war. It was there that he met his wife, who worked in Breslau at a hospital, where she was employed doing physical rehabilitation work. Cerny continued describing this work that night in Alabama, dismissing it at first as "weird experiments on the nature of time." Greg's father must have picked up on this comment, because the two men quickly became involved, according to Greg, in a deep discussion concerning the nature of time. Greg told me that it seemed to him now that at all times Cerny was a little vague in his statements, choosing his words and being careful of what he said, almost as if he were under some sort of hidden duress. Greg listened to the two men attentively, but did not enter into the conversation himself. At that time and place in polite adult company, as Greg explains, a child did not speak unless spoken to. In considering what follows and what was relayed to me, it should be remembered that Greg had not yet read Nick Cook's book, or even heard of the work of Igor Witkowski. Greg remembers that Cerny drew a circle of stones, which Greg said was "like a Greek Temple," with a ring around the top. Then Greg added a feature not mentioned by Cook or Witkowski: "and some sort of ring inside of that." This second ring was like a hoop of metal, from which something hung—like an oscilloscope or a TV screen. Greg went on to mention the atomic symbol as a means of description. The atomic symbol has a nucleus, around which orbit electrons. The electrons have two orbits, one within the other and are moving independently from each other. From this description it follows that this structure contained two movable and independently adjustable fixtures, from which something hung, perhaps as with a gyroscope. (…) Greg reports that, after a pause, Cerny cut more pastry and poured another cup of coffee, then changed the subject to a jet engine, which had been built at Peenemünde, for an unmanned rocket other than the V-1.*

Stevens made an attempt to ascertain the credibility of his source, generally with positive results. A broader description is of course in his book.

That is about all I wanted to comment on in reference to the other, aforementioned publications. Allow me now to summarize the most significant physical aspects of the "Bell."

After several years of "digging" in various archives I realized that contrary to common opinion, there are entire segments of WWII history that remain unclear to this day. Some interpretations were accepted and persisted simply because they were well publicized and elements that did not fit the accepted scheme were usually omitted. Some, on the other hand, were just fragmentary compilations and their authors often didn't realize that what they saw was just half of the picture. This was the case with the German nuclear programs, for example. There was no unified nuclear program, although the researchers tacitly assumed that there was some equivalent of the American Manhattan Project. In fact, the most important and most interesting activities took place in the latter eastern zone, beyond the reach of the Allied intelligence services and consequently of post-war historians as well. In fact the Third Reich's effort in this field consisted of numerous independent projects, in certain cases exploring very different notions or ideas (serving different aims in some cases), which then were not as confined to well established theoretical canon as today. During my research I occasionally encountered information indicating quite clearly that an offshoot of one of these programs had a clear connection to gravity research, but in a way quite different from Einstein's theory—and not just for ideological reasons.

I realize that any mention of an alternative (while equally valuable) approach to gravitation sounds at the very least suspicious to a scientist. At last we have a well verified theory that so far explains everything very well. Or does it? I propose starting considerations from this fundamental question.

Previously the dominating conviction was that although the General Theory of Relativity (GTR) was not very useful in a laboratory on Earth, because it could not really lead to the construction of any technical devices utilizing it, it worked almost perfectly in the Universe. This notion was so strong that it effectively suppressed any competitive explanations for many decades. But in due measure of the exponential influx of new data, especially from astronomy, its credibility has been undermined. In fact, the General Theory of Relativity has not changed for almost 100 years not because it explains everything perfectly, but for the simple reason that it cannot be developed any further to encompass newly emerging fields, despite divergence from new data. Recent discoveries show that there is some new form or "incarnation" of gravity—dark energy, which manifests itself in the form of a repulsive gravitational force, repelling galaxy clusters away from each other at steadily increasing rate. In fact it represents some three quarters of all the energy equivalent of the Universe. Although Einstein introduced a constant, known as the cosmological constant, which is supposed to explain this, there is nothing which could indicate that he himself understood its nature. In other words it is quite hard to accept the GTR, while accepting also that the dominating form of gravity has a different nature (unknown!)—and is negative. Einstein did not accept that gravity may be negative; such an equation, as a derivative, was formulated later, and not by himself (we will move onto this later). The comparison between "normal gravity" and "dark energy" in the Universe is such that Einstein's theory really describes only around 10% of known gravity.

I know—there still exist mainstream scientists who will say that it all just concerns the cosmological constant. I can accept that, but the aforementioned discovery signals something more important as well, which cannot be ignored so lightly. Namely it appears that this negative gravity is not directly associated with… mass, because it is literally everywhere. The vacuum filling the gigantic voids between galaxy clusters is a source of it to the same extent as the space around us (that is why here on Earth it is insignificant, because the space time continuum around us is relatively small, while we are close to large masses). The only credible and sensible explanation of this fact is that the only possible source of such an interaction are the quantum fluctuations of space time itself (spin of these quanta?). This is something completely different than what Einstein's equation describes. It implies in turn that we will not achieve any progress in our understanding unless we resort to quantum physics. Incidentally, in my modest opinion astronomical discoveries also tell us that antigravity is not some exotic interaction, visible only in the vicinity of a massive, spinning black hole, as physicists imagine. It seems that it is a manifestation of some fundamental aspect of nature, and should not be as difficult to generate as we generally think. We do not really know what it is—and this is the problem.

This is not the only new fact that casts a dubious light on Einstein's theory. According to the theory's predictions, gravitational waves should exist. Numerous research projects have been initiated lasting for over a decade, employing various sophisticated technologies, but not even the slightest trace of such waves was detected. Nothing at all, which suggests they probably do not exist. Therefore the general picture is not as clear as textbooks say.

The above introduction served the purpose of opening a sort of window, through which we may take a "new" look at the very fact that approaches to gravitation may be different, and should also take into account that the description of the entire remaining physical world is different—a quantum one. It is obviously an introduction to the approach worked out by certain scientists from the Third Reich. This may have been a strange state, where ideology played far too important a role. However—in this case the prejudice towards Einstein actually opened up new avenues. And we should not forget that the centers in Munich, Heidelberg and Göttingen were all considered of worldwide importance. After all, Germany was the Motherland of Quantum Physics.

I started my research with Professor Walther Gerlach, who became Chief of the Reich's [Scientific] Research Council

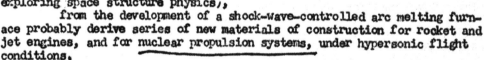

One of the greatest achievements of the recent years, in shedding a new light on the "Bell." is the merit of an American researcher, Henry Stevens. He has "discovered" a file on Ronald Richter in the NARA archive, who was the head of the German/Argentine nuclear research project (Project Huemul), but earlier he was a colleague of Kurt Debus from the Research Institute of the AEG. The files describes both periods, shedding therefore a light also on the high voltage project of the AEG, described earlier. What's crucial here, is that it clearly says that it was a plasma research, aimed at generating gravity. The area has been described as "nuclear propulsion systems," i.e. in the same way as in the case of Pilsen, described further!

(Reichsforschungsrat) during the war, although the opinion prevailing in literature is such that his expertise had nothing to do with the war effort (this contradiction is also illustrated by the fact that despite such a high position, at present he is almost unknown). To me his alleged lack of significance to the war effort seemed virtually impossible and did not make any sense. As certain information was passed to me suggesting quite the opposite, I ventured to start taking a closer look at this individual.

Before World War II, Gerlach was known mostly due to his experiments with heavy atomic nuclei, which were accelerated. He was co-author of the famous Stern-Gerlach experiment, in 1921, which proved the quantisation of atomic spin and magnetic moment. His approach to nuclear physics was mostly "magnetic," being the author of numerous publications on electromagnetism. At some point, before the war, he expressed interest in certain phenomena which could shed a new "brighter" light on the nature of magnetism. In the biography obtained at his university in Munich (where he was a rector), I found certain references that are generally omitted in other works. He was interested in magnetic effects in plasma. For example: as early as 1925 he wrote a letter to Arnold Sommerfeld about an experiment in which the spin of mercury ions (mercuric plasma) was measured. Knowing that Gerlach was a "magnetician," we may assume that the main area of interest were the magnetic effects associated with this. A couple of years later, in 1929, he partially described this research on mercury in an article published in the February edition of "Helvetica Physica Acta." Very few people know that this was not just a theoretical interest; after the war, in 1954, at a conference organized by AEG (crucial during the war in this respect) he revealed that during the war plasma was generated using an electric current with a voltage of half a million Volts, therefore some kind of plasma accelerator must have been used. This is strange, because officially such devices only appeared a decade later, in the 1950s. One may ask: well, what's so special about that? After all it's only history…

Not exactly, for not long after the first experiments with mercuric plasma in the 1930s, Gerlach started exchanging correspondence with a Soviet physicist, the Noble Prize winner Piotr Kapica, or Kapitsa if you like (incidentally: he held Soviet citizenship, but his parents were Polish) and this correspondence has revealed the much less conventional purpose of this only apparently trivial research. The next step of Gerlach's plasma research was on ball lightnings. He exchanged ideas with Kapica, who shared his interest in this field. I have not read these letters myself and do not know all the details mentioned in them, but it does not seem hard to figure out what direction they represented in Gerlach's approach. Kapica won the Nobel Prize for his research on … superfluidity, including his theory pertaining to it. How does this match Gerlach's area of expertise? Does it match at all? At first glance it doesn't, if we look at it from a proper perspective however, it matches perfectly, only that it reveals the different nature of the work of one of the best German physicists (even his plasma-related research finds no reflection in textbooks). In all probability ball lightnings, which so fascinated both Gerlach and Kapica, clearly indicate torsion physics and can be identified as so-called plasma solitons—quite peculiar, magnetically isolated systems—vortices containing high energy plasma. Such an assumption is confirmed by modern experiments and is the only way to sensibly explain why these vortices are so stable—by five or six orders of magnitude more stable than "common" vortices in plasma. Therefore their nature must be very different from common "sparks," which last for so short a time that their existence in a plasma accelerator or a plasma-focus device may be recorded only on ultra fast photographs. Why do ball lightnings not dissipate energy in the same way? Can this discovery contribute anything to our considerations?

Such solitons are isolated systems just as vortices of electric current are in superconductors. Certain vortices (in a superconductor they are represented by spinning atoms) display something very strange—they completely isolate themselves from the external environment—in the electromagnetic sense. They may exist in certain kind of liquids, solids and gases as well as in plasma. The analogy between plasma and superconductors is all the more relevant, as plasma is generally also a source of strong magnetic field. Another kind of soliton are the vortices in … superfluids. Another kind, but they all share the abovementioned, fundamental property. For example: a vortex in liquid helium will never stop, if only a sufficiently low temperature is maintained—there is no friction, therefore no heat is emitted and consequently the vortex does not lose energy by emitting electromagnetic energy. One may venture a description and analogy that the electromagnetic field in such a vortex resembles a closed bubble. Kapica's description of them was the first successful attempt to physically describe this

The mysterious FEP research structure of the Waffen-SS was headed by SS-Gruppenführer Otto Schwab and the deeply concealed FEP cell itself was probably located in the small town of Glau, in the south-western outskirts of Berlin. It was detached from the armament office of the Waffen-SS, headed by SS-Obergruppenführer Jüttner, mentioned earlier. In one of the recently uncovered documents it was referred to as "SS-Führungshauptamt, Amtsgruppe A, Technisches Amt VIII, FEP," which at the same time illustrates the organisational scheme. This photograph shows Schwab in 1934, in SA uniform (probably, since there is no other photograph of him to compare with). He was quite a bright doctor of physics, specializing in nuclear physics and carried out his own research as well.

phenomenon of solitons as such, i.e. solitons in plasma—the aforementioned ball lightnings, which Gerlach tried to generate. Why is this apparently interesting? Simply because from today's perspective we can see the clear relativistic context of this direction of research (related to gravity). Why do vortices supposedly have anything to do with gravity?

Shortly after World War One a French physicist, Elie Cartan, formulated a derivative of Einstein's theory, including an equation which was later accepted as the Einstein—Cartan equation. He introduced a term with a negative sign, meaning that a spinning mass may be the source of a negative component of gravity (or antigravity). Later on, during World War II, the German physicist Professor Pascual Jordan in turn developed a theory, according to which a spinning magnetic field may also be a source of negative gravity—i.e. a rotating mass which emits a magnetic field would be a more effective source of negative gravity than a mass without this field. This theory has nothing to do with so-called frame dragging, a description developed by Joseph Lense and Hans Thirring before World War II, although Thirring was a co-worker of Gerlach during the war. All these works were based on the notion that a gravitational field is "coupled" with electromagnetic fields, which means if some kind of separation of spinning magnetic fields can be achieved, this should cause some kind of separation of reference frames in a relativistic sense. It means, in other words, that the rapid spinning of strong and "separated" magnetic

Professor Pascual Jordan. He was the author of a forgotten, German quantum approach to gravity. He would have received a Nobel Prize for it shortly after the war, if it wasn't revealed that he worked in Peenemunde. Even in the 1950's and 60's he was recognized as one of the very few outstanding physicists dealing with gravitation in the world.

fields may be a much more effective source of induced gravity than the spinning of masses only.

How useful can it really be?

Most scientists consider that spinning may create significant gravitational forces only in extreme circumstances, such as in the vicinity of fast spinning black holes or neutron stars—where both the intensity of fields and masses are very large, and certainly not possible in a laboratory. This is however based on false assumptions. Experiments carried out in laboratories in the past (it was a particularly fashionable field in the 1970s…), involving gyroscope rotors spun with "modest" speeds the order of 10,000 rpm, demonstrated that although these effects amounted to only milligrams of weight loss, it is worth emphasizing that this was several orders of magnitude greater than the Einstein—Cartan equation predicted. We must bear this in mind because these cases indicate clearly that although we can say the effect as such is real, or will be real, a total "debunking" of these phenomena on the basis of such calculations is rather out of place. We can only imagine that in this case the spinning rotor is in fact a "set" of rotating atoms and particles, which have their own spins, and therefore a "true" description should take into account the quantum character of these constituents. Incidentally: the main property of solitons has an entirely quantum nature.

The discrepancies are even larger when we rotate not just mass, but also its magnetic field. The outstanding German physicist Pascual Jordan, working at the time with Wolfgang Pauli, demonstrated as early as 1928 that "quantum fields could be made to behave in the way predicted by special relativity" during, as he called it, "coordinate transformations." The latter phrase means a situation in which fields are separated from the external environment. A simple model of such a situation is a spinning magnet plus a superconductor, which should completely separate the magnetic fields (thanks to the Meissner—Ochsenfeld effect, discovered before WWII). So-called high temperature superconductors currently available have severe limitations, being ceramic discs which cannot withstand very strong centrifugal forces, nor strong magnetic fields. In spite of this experiments carried out by Podkletnov in Finland, later in the Japanese company Toshiba and successfully repeated among others by NASA in recent years, have demonstrated that a 1% reduction in weight is relatively easy—when one takes into account that the concentration of energy in this case is in fact very modest. Here the contradiction with the derivatives of Einstein's theory is even stronger—in due measure of the greater tie with quantum effects.

1% or 5% is not much when we think about some technical application, there is however an open window to this newly emerging field—opened in fact by Gerlach, Jordan and other scientists working before and during World War II. It consists of the following facts:

1. In plasma, thanks to a phenomenon known as pinch, we may achieve magnetic fields stronger by several orders of magnitude than what a "poor ceramic superconductor," or even a piece of metal, will ever be able to withstand.
2. The same refers to velocities—plasma vortices compress themselves under influence of the magnetic field, compensating in this way the centrifugal force. Recent data states about 5 or 10% of the speed of light, achieved by ions (Sandia National Laboratory in New Mexico, project "Shiva Star"), thus completely out of this world! As we recall, superconductors were rotated with velocities the order of only 10,000 rpm…
3. The same refers to the "solitonic" properties of plasma vortices, which do not depend directly on the intensity of magnetic field—superconductors lose their ability to separate magnetic fields above a certain, relatively small field intensity.
4. Various materials may have the same mass, but as "sources" of induced gravity in plasma their effectiveness may be different, regardless of their mass, because of the different "isotopic spins" of various ions (the isomers). This means that not all elements are suitable for this purpose, but mercury—chosen by Gerlach—is certainly one of the best, among other factors because it is characterized by a very low viscosity.

We can say therefore that this is certainly a promising direction, especially when we realize that it was not continued after the war, at least not openly. Even when we only extrapolate the results achieved with superconductors linearly—i.e. multiplying the intensity of magnetic fields and angular velocities (taking into account the difference in performances between mechanical systems and plasma), we should easily achieve numbers (percentages) worth investigating.

We should not be distracted however by theoretical predictions based on derivatives of the General Theory of Relativity. One of the few experiments to explore this newly emerging field was carried out in the UK in the 1980s and described in *Physics Letters B* (16th Feb. 1989), in which relativistic effects generated by heavy ions spun in an accelerator were measured. It is described in the last chapter. It would have been extremely difficult to weigh atomic nuclei moving at tremendous speeds, but it was quite easy to measure their inertia. Inertia equals gravitational force and of course depended on the rate of spinning. Its reduction amounted to as much as 50%! It represents the largest "antigravitational" effect known so far, at least to me, which confirms the hopes pinned on plasma, despite the lack of "separation of fields" in the aforementioned case. Once again it confirms that calculations based on Einstein's theory and applied to the quantum realm should not be treated literally.

There are a number of mathematical analyses available on "gravitational solitons" (such as a book bearing this title at amazon.co.uk), but they all have a certain fault. Their very existence suggests that the notion as such makes sense, but involve such complex mathematics that a friend of mine, a physicist for whom solitons are part of his professional subject of activity, said that the suspicion simply arose whether they were all relevant and useful. When I looked at these equations and matrixes, an association arose in my mind with the ptolemaeic theory of epicycles. It was a tremendously complex, one may say "Sisyphean" attempt to describe (on the basis of the GTR!) something that in fact is fundamentally different in nature from Einstein's non-quantum theory—a quantum world of atomic spins and particles moving at tremendous speeds. The Einstein—Cartan equation is very short and simple, while the discrepancy already amounts to several orders of magnitude… The description is complex, but in my modest opinion it pertains to something that is in fact very simple, only different from what we know. One of the basic manifestations of nature. Nevertheless it is certain that the properties of solitons should significantly increase the reduction in gravitational force—vide: Jordan's theory (known also as the Jordan—Thiry theory).

The "advantage" of such a situation—the fundamental differences between these two physical realms—is however that nobody has yet attempted to investigate the aforementioned relationships experimentally. 'No theory in the mind and the eyes do not see the facts'. I will allow myself to mention modestly at this point that breakthroughs rarely happen because nature

> Recently it was possible to at least partially explain the role of SS-Gruppenfuhrer Emil Mazuw, thanks to new, surprising findings. It looks that he was in some way associated with the Himmler's research organisation named "Das Ahnenerbe." It's not clear whether he was its member, but he undoubtedly took part in Ahnenerbe's various activities, and that rather important ones. In this light it seems justified to at least admit a possibility that Himmler's interest might have had something to do with the clues "acquired" by various Ahnenerbe expeditions to Asia, especially with the expedition from 1939 to Tibet, which has delivered very extensive collection of manuscripts containing information on mercuric propulsion. Information contained in these manuscripts—in the 108 volumes of "Kanjur," but in other sources as well—was meticulously deciphered and analysed in one of the Ahnenerbe's research posts, in the Mittersill castle in Austria. These efforts enjoyed quite a significant interest on the part of Himmler. His copy of Kanjur contained not just technical descriptions of the mercuric propulsion, but even a couple of interesting drawings. All these Tibetan manuscripts were in high demand even before the war, mostly because they refered to a popular, then mysterious subject of an unkown energy called "vril". In fact it was a Tibetan name for a force unifying electromagnetism with gravitation (there is no shortage of such a descriptions). This illustration presents the cover of a pre-war German publication titled "Vril: the cosmic, primordial force." If—lets imagine—Himmler would show the translations to one of his scientists familiar with the unification of forces, such as Jordan or Gerlach, they probably wouldn't have any problems with finding out what it's all about. Mazuw seems now to be a kind of missing link in this respect.

suddenly reveals some of its secrets. Usually it is a consequence of the very simple fact that people begin to see something that was around them all the time. We should also bear in mind that the connections between gravity and the rest of the physical, quantum world are certain, we should not treat them as something abstract. There are phenomena in which gravitational and electromagnetic effects are generated as manifestations of the same physical reality, at the same time. Why is it so difficult to come to the conclusion that we should start from them?

* * *

Recently, I tried to find any verification of the experiments carried out by Podkletnov with the superconductors, as they show that we should take a closer look at the Einstein's theory. Einstein himself didn't see any need to take into account the spinning of mass, let alone the spinning of electromagnetic field associated with mass, or influence of the separation of electromagnetic fields on gravity. Only his followers developed such approaches, although their descriptions are almost worthless if we would like to calculate generated forces. Even the simple experiment with spinning gyros demonstrates that discrepancies between experimentally measured forces and theoretical predictions amount to several orders of magnitude. Such a derivatives of Einstein's theory are significant only as much as they show that there exists a connection at all, but one just cannot "prove" that antigravitational effects are insignificantly small, as the scientists still believe, because these derivatives do not agree with experiments and therefore are not credible basis for such a claims. Therefore the Podkletnov's experiments seemed to be important in this respect, for they demonstrate that the phenomenon of "separation of magnetic fields" is something that shouldn't be marginalized (which in superconductors occurs thanks to the Meissner-Ochsenfeld effect, separating electromagnetic fields totally). In December of 2009, I have found an interesting article, published on a scientific news website, named "Science Daily," from March 26, 2006. It describes the confirmation of Podkletnov's discovery (the source is: http://www.sciencedaily.com/releases/2006/03/060325232140.htm). I will let myself to quote short fragments:

> *Scientists funded by the European Space Agency have measured the gravitational equivalent of a magnetic field for the first time in a laboratory. Under certain special conditions the effect is much larger than expected from general relativity and could help physicists to make a significant step towards the long-sought-after quantum theory of gravity. Just as a moving electrical charge creates a magnetic field, so a moving mass generates a gravitomagnetic field. According to Einstein's Theory of General Relativity, the effect is virtually negligible. However, Martin Tajmar, ARC Seibersdorf Research GmbH, Austria; Clovis de Matos, ESA-HQ, Paris; and colleagues have measured the effect in a laboratory. (…)*
> *It demonstrates that a superconductive gyroscope is capable of generating a powerful gravitomagnetic field, and is therefore the gravitational counterpart of the magnetic coil. Depending on further confirmation, this effect could form the basis for a new technological domain, which would have numerous applications in space and other high-tech sectors" says de Matos. Although just 100 millionths of the acceleration due to the Earth's gravitational field, the measured field is a surprising one hundred million trillion times larger than Einstein's General Relativity predicts [10^{21}???—Igor W.]. Initially, the researchers were reluctant to believe their own results.*
> *"We ran more than 250 experiments, improved the facility over 3 years and discussed the validity of the results for 8 months before making this announcement. Now we are confident about the measurement," says Tajmar. (…)*
> *"If confirmed, this would be a major breakthrough," says Tajmar, "it opens up a new means of investigating general relativity and its consequences in the quantum world."*
> *The results were presented at a one-day conference at ESA's European Space and Technology Research Centre (ESTEC), in the Netherlands, 21 March 2006.*

It seems that this article is an almost exact copy of the article from the European Space Agency's own website http://www.esa.int/esaMI/GSP/SEM0L6OVGJE_0.html.

The consequences of this breakthrough couldn't be greater… First and foremost it confirms that separation of magnetic fields, which was exploited in the "Bell" according to the German description itself, is the key to understanding and influencing gravity—of course also that gravity may be influenced, and that without tremendous energies. Therefore if we would manage to achieve much higher level of energies, velocities, and higher level of separation of fields (which should take place in gravitational solitons in plasma, as in the "Bell"), the effects may be amazing. It means: huge, when compared to the previous theoretical predictions. The key to connecting gravity with quantum phenomena is emerging before our very eyes. In fact it's the emergence of The New Physics—physics of the Third Millennium! Overcoming this theoretical obstacle means, from a historical perspective, opening of the doors to the Universe. It's just happening now, the rest is only a matter of time.

HOW HITLER PLANNED TO WIN THE WAR IN 1944-1945

After years of research I realized that what I had described regarding "The Bell" project was only a piece in a larger puzzle. Now, while I work on this updated edition, I realize it can be included in the broader context of scientific works. This and many other projects were part of the efforts of the SS to create a unbeatable strategic arsenal. This field has never been fully explored by historians since the war. For example: Western researchers were hardly in a position to explain what the SS was doing in territories located south of the border with Lower Silesia, nicknamed by Hitler himself the "SS Model State" (SS-Musterstaat).

1. For years I have been trying to investigate the truth behind certain undertakings carried out by the Third Reich in Lower Silesia—including monstrous facilities like the elusive to this day "Riese," research carried out by the SS in Ksiaz and other locations in the nearby mountains, as well as the connection between the "underground industrial complex" and Professor Strughold's space research in Szczawno Zdrój… I know two people who had access to both original information and documents—from the war or reconnaissance carried out directly afterwards. I particularly have in mind my aforementioned informant at the beginning of the chapter about the Bell as well as Professor Mieczysław Mołdawa, who spent the final years of the war in the Technical Office at camp Gross-Rosen—through its manpower "serving" many Lower Silesian armament undertakings. In numerous talks with these individuals the names of places on the other side of the mountains—in the present Czech Republic—were always mentioned, although I had never asked about this. After some time it must have become evident that certain connections existed which needed to be explained if one wanted to find the answer to questions concerning Lower Silesia itself.
2. It is no secret that the so-called Protectorate of Bohemia and Moravia had actually been in the sole hands of the SS. No "civil" government administration had existed, like that for example appointed in occupied Poland. The highest authority was an SS General. Obviously this scenario was developed further. It was not immediately clear how Himmler tried to turn this to his advantage, but was worth looking into in more detail. This was clearly emphasized by the fact that the trail left by Kammler also led there—and that his mission's objective and nature was so unclear.
3. In 2000, I met a researcher from Great Britain, Nick Cook, who was interested in certain work carried out by the Germans in Lower Silesia. Thus we ventured on a common "quest." I consider him a highly regarded journalist, an analyst from "Jane's" publishing group—publisher of numerous magazines and military yearbooks, considered worldwide in a league of its own. A few years later Cook published a book describing his detective attempts at explaining secrets associated with a series of state-of-the-art aeronautical projects—starting with those carried out by the SS during the war, and ending with the latest American research. What surprised me were the quite numerous references to all that occurred in the Protectorate, particularly at the end of the war—although based on sources already known. For not only numerous aircraft factories but also research posts had been located there, including those connected with the SS—even ramjet propelled fighters were designed at Skoda…

All of this confirmed my intuitive convictions that the key to many important issues including the most interesting projects connected with World War II resided precisely there—or rather in archives where materials on this subject can be found. Also encouraging was the fact that almost nobody up until now had seriously taken up this issue—by which I mean locating so-called "primary sources," particularly with respect to research projects. I came to the conclusion that due to the scale of these operations something must have survived, at least some shred of information which would shed light on something. Motivated by this I began a full-scale search that would prove (or refute) that:

1. More important and more interesting events had taken place in the Czech Republic than what had been duplicated so far in literature.
2. The SS had played a key role.

3. Secrets regarding "Lower Silesian" projects are closely connected to what happened south of the present-day state border, sometimes only a dozen or so kilometres away…

I will leave my readers to be the judge of this work.

My first move relied on examining the source Nick Cook used—a source which was to reveal a completely new reality. It was a book by the aforementioned Tom Agoston, based on his own reconnaissance from the period directly after the war (he wrote among others reports describing the Nuremburg process), but first and foremost on talks with Skoda's former director—Dr. Wilhelm Voss.[138] So let us begin with who this informer was…and what in reality were the former Skoda factories themselves. Let us begin with the latter issue…

Before the war Czechoslovakia had a relatively modern and developed armament industry. It was dominated by one group—Skoda itself, strongly tied to American capital. By virtue of the Munich treaty, as soon as Czechoslovakia was annexed Skoda was taken over by the Third Reich. It initially found itself in Krupp's sphere of influence, before finally becoming part of the Hermann Göring Werke group. The firm was merged with Ceska Zbrojovka (renamed Waffenwerke Brünn), so becoming a Waffen Union company.

Dr. Wilhelm Voss was director from 1938 to January 27, 1945—to the moment when Göring dismissed him for refusing to receive two of his special, permanent envoys. He was even barred from entering the factories, but enjoyed the huge respect of Czech personnel, so this never had much practical significance. He simultaneously fulfilled (formally) the role of a department head in Speer's ministry—although resided mostly in the Protectorate. He was known for caring about the plant's condition as well as worker satisfaction. He pushed through for example the transition of pay into the Czech Koruna, a loophole enabling wages to be significantly higher. By virtue of the role he fulfilled and private connections, Voss had access to many high ranking secrets. It follows from his report for example that:[138]

The careers of Kammler and Voss overlapped at Skoda, where they jointly set up and operated what was generally regarded by insiders as the Reich's most advanced high-technology military research centre. Working as a totally independent uncover operation for the SS, the centre was under the special auspices of Hitler and Himmler. Going outside the scope and field of Skoda's internationally coveted general research and development division, it worked closely with Krupp [reminiscent of Głuszyca in the context of "Riese"—footnote I.W.] and was mainly concerned with analysis of captured equipment, including aircraft [reminiscent of Ksiaz—I.W.], and copying or improving the latest technical features. In so doing the SS group was to go beyond the first generation of secret weapons.

Its purpose was to pave the way for building nuclear-powered aircraft, working on the application of nuclear energy for propelling missiles and aircraft; laser beams, then still referred to as "death-rays"; a variety of homing rockets; and to seek other potential areas for high-technology breakthrough. In modern high-tech jargon, the operation would probably be referred to as an "SS research think tank." Some work on second-generation secret weapons, including the application of nuclear propulsion for aircraft and missiles—was already well advanced.

The above profile follows—one should emphasize—directly from Dr. Voss's account!

Propulsion for guided missiles? For such a non-existent strategic weapon? It is hard not to associate this with the chemical weapons thread so present in Czechoslovakia, including Pilsen itself…

This matter requires some reflection. The first say superficial interpretation of the above account is an association with something resembling present-day atomic powered submarines—where the system is powered by a normal reactor as in a power station. It is obvious that such an interpretation must be rejected. The Germans had great difficulty in building any working nuclear reactor whatsoever, so the variant that reactors would have been mass produced—in some kind of miniaturized form to fit into a projectile is simply out of the question. It wouldn't be possible to achieve this even today, despite great advances in the field of nuclear energy. Besides even in ten or thirty years such a hypothetical system would not be competitive against a standard rocket engine. It is worth drawing attention to the fact that Voss mentioned "well advanced" work. In any case Kammler wouldn't have carried out the aforementioned mission to Pilsen a month before the war's end (where the centre was located), if the prospect of utilizing the results of his work to repel enemy armies was not actually within reach, and if it had not been an entirely specific arsenal…

There obviously exists a more realistic and simpler solution than a reactor-thruster system. Something which propels and is based on phenomena from the field of nuclear physics, but need not have any connection with a standard reactor—along with fuel rods, lead shielding and the like. Is this not the same as that described in the chapter about the Bell? Perhaps, but we still cannot be sure…

There is yet another question which arises while on this subject.

As readers of my previous books are well aware, one of the more interesting threads concerning work carried out by the Germans in Lower Silesia, towards the end of the war, was the research of Professor Hubert Strughold in Szczawno Zdrój. His research was carried out in an underground facility subordinate to the Luftwaffe, near Ksiaz, which he described in an interview with a Polish journalist, Kakolewski. In the interview he described a device in which full scale "space flights were simulated," for the cockpit was manned. Control problems supposedly emerged in connection with vibrations. This constitutes significant information that can be interpreted as such:

1. That a supersonic wind tunnel was situated underground for e.g. rocket or capsule testing on a 1:1 scale, though this would have been such a huge installation (the order of half a football pitch), that constructing it underground can more or less be ruled out. First and foremost building a tunnel on a 1:1 scale would have been most irrational.
2. Or one could of course assume that the simulator was a centrifuge, although a centrifuge is not controlled and no vibrations arise from its operation.
3. A third interpretation is that a structure was tested not so much with regard to moving air, but with active thrust on a test rig. But a space rocket with a working engine would never be tested underground! Unless, according to Voss, the Germans were counting on entirely different propulsion for their strategic weapon. Perhaps this was the connection? Needless to say we do not know—these are only unbinding thoughts...

Voss's reports along with the materials which Agoston himself found include much more interesting information:[138]

> *The SS research operation at Skoda had been set up without the knowledge of Göring, Speer, or the German research centres. The builders of the V-1 and the V-2 were likewise kept out of the picture. The undercover SS research operation fitted in with Himmler's dream that, as the Rheingold of Nibelungs, if shaped into a ring, would give its possessor mastery of the world, so would the SS team give the Greater-Greater Reich mastery of much of the world.*
>
> *A study of intelligence reports shows that blueprints, drawings, calculations, and other relevant documentation or materials were protected by a triple ring of SS counter-intelligence specialists Himmler had assigned to Pilsen to prevent security leaks and sabotage in the research divisions and the plant in general. The SS team was internally referred to as the Kammler Group. (...)*
> *The funding for the Pilsen SS operation was channelled through Voss, who thus was able to remain in the picture, as he recalled when describing the set-up to me in Frankfurt in 1949. In the course of several extended interviews in Frankfurt and at his home in Bavaria, Voss spoke of his past activities with unique frankness. Skoda's overtly operating research and development division, working closely with the SS group on some projects, had provided a convenient cover for the Kammler Group specialists, culled in great secrecy from Germany's research institutes to supplement the in-house experts. All were picked for their know-how and not their party records, Voss said. All had to have the ability to tackle visionary projects. A number were Czech nationals. Some had worked in the United States before the war.*
> *Working for the Kammler project had provided new opportunities for the experts. (...) Many scientists, anxious to see their work in print, even if it was kept top secret, prepared papers for a central office of scientific reports, which circulated them to specific recipients. Some of these reports were used as the basis for selecting candidates for employment at Skoda.*
> *Himmler put top priority on routing all Waffen-SS research and weapons development contracts to Skoda and regarded smooth cooperation between the Waffen-SS armaments office and Skoda of utmost importance [Note I.W.: "Technisches Amt VIII" at SS-Führungshauptamt crops up again!]. This is reflected in correspondence between Voss and Himmler. To ensure this cooperation, Himmler set up a Waffen-SS liaison unit at Skoda, putting Voss in charge. Voss reported directly to Himmler."*

The information presented in earlier chapters depicts a concentration of SS activity in the Protectorate of Bohemia and Moravia. It indicates areas of interest specific for this territory, in some way the reason for the existence of the "SS-Musterstaat." This concerns first and foremost chemical weapons and strategic delivery systems for an arsenal of mass destruction. But the question arises of why this was not confined to the Protectorate, namely why were certain elements of the emerging greater plan located nevertheless in Lower Silesia?

We don't know the answer, but I believe this matter can be explained quite convincingly. Czechoslovakia had the advantage that SS services responsible for the development of new weapons controlled extensive industrial and research infrastructure there. Despite Speer's boasting, SS control over many facilities was preserved throughout the war—for the simple reason that the Armament Ministry had no possibility of even determining what work certain research posts connected to the SS-Führungshauptamt were engaged in. In spite of these favourable circumstances, the Protectorate did have certain faults which could be avoided by situating specific elements of the program to the north of the Sudeten Mountains.

Lower Silesia had the main advantage in that it was inhabited solely by a German population. It was possible to execute full-scale undertakings here. In the Protectorate this would presumably have been significantly more difficult and risky. Another reason could have been the greater availability of manpower, provided by Gross-Rosen concentration camp. This was created chiefly to build armament facilities—approximately 20 km to the north of Wałbrzych. Lower Silesia also had a better developed transport infrastructure. But first and foremost it was a safer region.

Overall: the facts known so far as well as new information published in this book give a relatively coherent picture of preparations for the "decisive" phase of the war, which in all probability was embodied by the German term "Sonderkampf" (see the chapter titled "The Bell"). This picture is comprised of the following elements:

1. Pilsen—the main SS research centre, coordinating work on a new propulsion system, based on phenomena from the field of nuclear physics.[138,139]

The Skoda works in Pilsen were one of the largest in the occupied Czechoslovakia. They wouldn't be unusual however, if not for the fact that their director was an SS officer and generally the SS had the most to say in Czechoslovakia (then: "the Protectorate"). These were the reasons for which on the grounds of these works the SS has placed one of the most secret research post in the whole Reich, cooperating with the top secret FEP structure, still practically unknown. General Kammler was nominated by Hitler in April of 1945 (!) the "Reich's plenipotentiary for the reconstruction of the research center in Pilsen regarding nuclear technology for the propulsion of guided weapons and aircraft" (the center was damaged in an earlier air raid). Here we can see SS officers receiving the delegation of Japanese armament specialists on the grounds of these works, which is another indication of German-Japanese cooperation in this field. By the way: the officer with SS hat, having SS insignia on the collar, but without SS "runen" on the other side, is someone who was an officer of another armed service, only detached to the SS and for some reason subordinate to its command.

2. Sroda Slaska [Neumarkt]—probably an auxiliary SS scientific research team ("co-opted" to Wehrmacht laboratories). The source here is "merely" an individual who in the past analysed materials from the interrogations of evacuation commando members [known by the author].

3. A number of facilities directly connected with chemical weapons, constituting the "SS arsenal." They can be divided into those employed in the production of combat poison agents themselves (Brzeg Dolny, Decin, Kolin, Pardubice),[132,135] "weapon systems"—such as warheads, aerial installations (again Pilsen, Tanwald, Brno, Cakovice),[135] as well as production in derivative areas—gas masks (Zamky)[135] and portable dispersion devices (Zlin).[135] The materials presented in the chapter on chemical warfare preparations clearly portray the huge scale of this aspect of armaments in the Third Reich and its close connections with the SS. There is one more detail included in this area: if one is to believe Keith Sander's report on the "chemical" version of the "Feuerlilie" rocket from one of the previous chapters (I personally consider him reliable), then one must draw attention to the fact that work on this missile, including preparations for its production, were carried out right up until January 1945 by the Ardelt Werke plants in Wrocław [Breslau]—thus within the same, specific area.

4. The Research and Development Department of the Jägerstab (Fighter Staff) in Ksiaz near Wałbrzych. The information source here is a former prisoner of Gross-Rosen, Professor M. Mołdawa, who worked in the camp's Technical (Design) Office, through which he had access to otherwise unobtainable information (officially and unofficially: he heard numerous conversations between the Germans, admittedly had "occasional" though personal contact with Kammler—in 1944 he even received a bar of chocolate from him, as a thank-you for the perfectly crafted brass plate on his office door…). In his book devoted to Gross-Rosen—later supplemented with statements I recorded—Mołdawa wrote about the connection between the research post in Ksiaz and the "Riese" complex in the understanding that a strategic weapon was to have been manufactured there designed (among others) in Ksiaz itself. It was to transport a chemical arsenal and work carried out by the SS in cooperation with the Luftwaffe.[144] One can therefore venture the assumption that in so far as Pilsen fulfilled the leading role as the research centre working on the propulsion itself, then Ksiaz was analogously the centre which coordinated work on the weapon delivery system as such, along with testing and so forth.

5. Ksiaz—the SS outpost connected with high frequency research. As written previously, various work was carried out here. We by no means know all that this involved (see: Speer's report), so one cannot exclude it having some kind of role in fulfilling the plans described here in a territorial context. It could for example have worked on guidance or navigation systems. One of the sources includes documents from the Personal Staff of the SS-Reichsführer.[145]

6. Szczawno Zdrój [Bad Salzbrunn] (lying just between Wałbrzych and Ksiaz, both Ksiaz and Szczawno constitute de facto the northern suburbs of Wałbrzych). An underground "space research" outpost was located in the area of one of the present-day sanatoriums, led by Professor Hubertus Strughold, formally subordinated to the Luftwaffe Institute of Flight Medicine. It was described earlier and I consider its activity requires no special commentary. The information source here is Strughold himself—and the interview he gave to Kakolewski. The facilities mentioned in points 4-6 are located virtually in the same area, approximately 2 km from each other. Difficult to consider a coincidence…

7. Experimental facilities. One can assume that these were in all likelihood located in Lower Silesia (Ksiaz—the "Old Castle," Ludwikowice), in the direct vicinity of "Riese." The source here is Professor Mołdawa and a former analyst, mentioned in point 2. Allow me to quote in addition the report of a former prisoner, Jadwiga Debiec:[149] *There also existed a camp in Ludwigsdorf [Ludwikowice] that almost nobody knew about. The people there were led in columns and each column was different. Their skin was*

8. Munich (BMW)—according to intelligence sources the only facility directly employed in work on a strategic weapon for transporting a chemical arsenal, which was located beyond the described region. This weapon system was designed to carry out an attack on the USA.[135] In an intelligence report cited in one of the preceding chapters—there was mention of "an unknown type of poison gas" (phosphor-organic agents were unknown to the Allies).

9. Production—"Riese" in the Sowie Mountains. Its role in the program is testified by Professor Mołdawa, including a direct connection to the research department in Ksiaz;[144] a report from Dr. Jacek Wilczur from the Main Commission for Investigating Nazi War Crimes presented further on (based on military counter-intelligence materials and witness reports from Germany), as well as a report from Jerzy Cera. The latter was one of the first researchers of the complex in the Sowie Mountains. In his book he described how when leading a group of cadets he had to put up with the company of Soviet "journalist," who later turned out to be an educated officer at...the Baikonur Space Centre. This was at the beginning of the 1970s.[146] Mołdawa unequivocally connected the planned production of a strategic weapon with a chemical arsenal to this site. Allow me in turn to quote an excerpt of a statement from Dr. Wilczur on this subject:[147] *We have the authority to claim today based on research, searches and investigations carried out in the last dozen or so years, that making us believe the complex in the Sowie Mountains was designed for Hitler's future headquarters is a lie. We obviously do not rule out that a certain segment or part of the building was designed for Hitler's headquarters. But all of this could not have encompassed tens of kilometres of underground tunnels. This is nonsense which would appear obvious not only to an expert from the field of military construction or fortifications, but even somebody with no specialist knowledge. (...) According to our knowledge it was designed to be a huge armaments complex, in which the production of a special purpose weapon was planned, including weapons of mass destruction. (...) Using obscene forms of pressure and compulsion on local Germans who had stayed for some time after the war in the vicinity of "Riese," military counter-intelligence obtained information that stands by my opinion as to the purpose of the facilities in the Sowie Mountains.* It cannot be ruled out that within the framework of the aforementioned SS and Luftwaffe plans, the employment of a nuclear arsenal was also planned—as warheads or bombs (probably in the future when the production of war chemical agents had already been mastered on an industrial scale). It is worth remembering in this context that large facilities (two) were located in the area described here, which according to intelligence were not so much "engaged in work on a nuclear weapon," but "were building" or "manufacturing" it! This refers to the facility "Aussig" near Jachymov in the Czech Republic[135] as well as "Sagan" (Zagan) in Lower Silesia.[148] "Aussig" was, as in

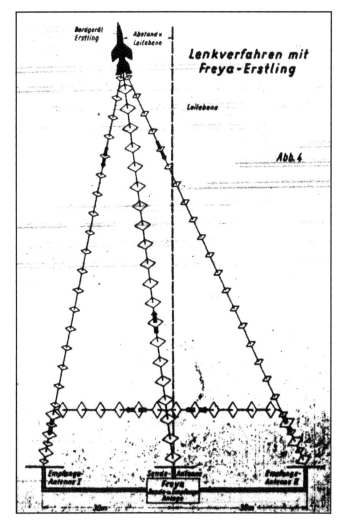

One of the crucial elements of the prepared strategic arsenal in general were the methods of providing accuracy at distances of hundreds, if not thousands of kilometers. The main solution was coupling of the missile with an interferometric radar system, which measured deviation of the flightpath and transmitted coded, correcting signals. The first such a sophisticated system was developed for the A-4b missile (range around 650 km) and entangled the Freya radars plus a transmitter and receiver of the data, code-named Erstling. Hence it was named Freya-Erstling.

coloured, as a matter of fact their bodies were soaked in colours. The yellow and blue columns most transfixed me [chemicals production?—I.W.]. They were closely guarded and you could see how exhausted they were, not walking but simply crawling. I knew they worked in the underground rooms of a power plant and were treated inhumanely. A group of Jews arrived, who after some time were apparently killed and replaced. Red Cross documents exist proving that two labour camps (Gross-Rosen sub-camps) in Ludwikowice were organisationally subordinated to AL "Wüstegiersdorf" (Głuszyca)—i.e. connected to "Riese."[149] It follows after all from Professor Mołdawa's account in his book on Gross-Rosen that "Commando Wüstegiersdorf" was subordinated to the Luftwaffe research department in Ksiaz working on "new aerial weapons," suggesting an analogous connection in the case of Ludwikowice.

For the purpose of correcting the navigational systems of the weapons meant to be used against USA (the A9/A10 missile in particular), the Germans have created the Wassermann-Erstling system, which contained three Wassermann radars connected in an interferometric grid. They had the range of only around 400 km, but were needed only until the missile leaved Earth's atmosphere. The See-Elefant radar, visible standing on the higher masts, was also planned to be used, although it was not active in the guiding role, but it had the advantage that in good conditions it could detect echo even from the eastern shore of Canada. It was the first operationally used radar based on the temperature inversion in the ionosphere. The first such a system was built on the Danish island of Röm, second one was under construction in 1945 in Norway, code-named "V-Stelle Gaustadt."

many cases described here, a classified underground plant. "Aussig" seems even to be the key to understanding the German approach to nuclear weapons, for this team was directed by the "star" of German nuclear physics, Werner Heisenberg, who worked there in complete secrecy. This team was previously completely unknown! This picture is supplemented by post-war documents stating that after the war the Soviets recovered specific German nuclear weapon components from Jachymov.[151] The following is in turn a quotation from the aforementioned American report:[135] University of Dresden. Said to be heart of German development of the secret weapons. At the University of Dresden a group of chemists, engineers and professors are working under the direction of Heisenberg. Associates of Heisenberg are working on a by-product of radium at the Schicht mines in Aussig near radium mines at Jachymov in the Protectorate. (Cable, M/A Ankara, #99, 11 Apr 44, MIS Journal 298, 12 Apr 44. S.) It seems likely that this research team had something to do with a mysterious "radium-mercury amalgam" that was shipped to Japan, we know among other facts that the Japanese submarine I-29 transported such cargo.[150] New docu-

ments (which I discovered) contribute of course to the opinion described in an earlier chapter—that there was no unified German nuclear program. The nuclear theme also appears in several other places. They include Kowary as well as Dr. Wilczur's report from a military penetration of a section of "Riese" at the beginning of the 1960s, in which he mentioned the discovery of uranium ore remains in underground rooms.[130] The former power engineer of the complex—Anton Dalmus—also mentioned this. In addition: it is worth pointing out that the connection of work described in the chapter about the Bell with Jachymov was previously mentioned by my informer. (It has been mentioned by him under its German name: St. Joachimstal, see page 247). Information about a ship transporting graphite blocks on the Odra River should also be included in the same group of reports. It was bombed and sunk on August 17, 1944, in the vicinity of Szczecin [Stettin]. New details have recently been published about this ship:[152] it was called the "Artushof" and 38 graphite blocks among others measuring 2.5 m by 0.6 m were recovered from its cargo hold after the war. The similarity of this case to an order in 1941 is striking, where the Siemens "Pla-

nia Werke" plant in Racibórz manufactured 100 blocks measuring 3 m by 0.6 m (designed for the first reactor with a graphite moderator, which "didn't work" due to graphite contamination—to this day the location of which is unknown). The 1944 episode may be connected with a graphite reactor in Gottow, but would the blocks have been transported via the Odra River in this case? In any case the reactor had a stack in the shape of a sphere, so the moderator's blocks must also have been shaped to fit it. This may testify to the existence of another reactor of this type, perhaps in Lower Silesia.
10. Železny Brod—approximately 40 km to the south-west of Kowary in the Czech Republic. A command/staff facility was built here at the end of the war connected with the planned use of "long-range guided weapons." It was subordinated to the Luftwaffe.[144]
11. Ksiaz—this location appears in the current list for the fourth time. It pretty much represented the certain location of Hitler's headquarters (an underground command post of central level). In a straight line Ksiaz was located only a dozen or so kilometres from "Riese" and according to a report from one of the prisoners an underground connection had already been built through which an electric railway was to run...

All in all these preparations employed, along with prisoners, well over 50,000 people, which gives a picture of something like the equivalent of the American "Manhattan" project—managed by the SS and in "cooperation" with the Luftwaffe (although here the nuclear arsenal was in the background). Was this program not implemented due to technological problems connected with delivery systems, delay in building an industrial base or perhaps because somebody in the Third Reich was too afraid of the irreversible rise in influence of the SS (which also cannot be ruled out)—this we will probably never know.

A remark in passing (italic):

The scenario of undertakings approved for execution brings to mind certain associations with the Japanese program to develop a weapon of mass destruction—as if the same model had been copied. Different teams and centres also operated completely independently, in certain cases unaware of each other's existence. This is why for a long time researchers only concentrated for example on the activity of the Tokyo Riken Institute, which dealt with isotope separation, through which a fundamental element evaded them: That what Lower Silesia and the Sudeten Mountains were for the Reich (a safe, mountainous rear area with infrastructure), Korea was for Japan, particularly its mountainous north. Analogous underground facilities were built there, likewise exploiting the "benefit" of almost unlimited, unpaid manpower (POWs, Korean peasants, as well as...Japanese children). This region also had a large concentration of industry. A huge complex was built there marked with the Latin (!) letters "NZ," where led by the future Noble Prize winner Professor Hideki Yukawa, a program to manufacture a nuclear weapon was carried out deep in the mountains, which the Americans were totally unaware of.

This program was carried out on behalf of the Japanese Navy. The equipment of "NZ" was also removed after the war by the Russians... Programs connected with chemical and biological weapons were subjected to analogous "compartmentalization" [source: R. Wilcox—"Japan's Secret War"].

One can obviously reflect on why the Czech-Lower Silesian "puzzle" is still incomplete, why over sixty years was needed to spot certain connections, why there are almost no original documents on these undertakings that were "decisive for the war." One must realize though that in the case of such classified projects few documented traces usually remained. After all we do not even know the names of people who worked for the SS in Pilsen, and no documentation exists from Otto Schwab's "Design Office"—particularly the "FEP" cell! A book by Karlsch among others devoted to atomic research illustrates this phenomenon:[140]

As far as the strict maintenance of secrecy by the Germans themselves is concerned, Hitler gave the respective order in this matter as far back as June 1940. In due measure of the worsening situation in the war for Germany, Hitler became ever more distrustful. His personal pilot Hans Baur reported that Hitler suspected there to be a spy in his inner circle—"at the end of 1944 at atmosphere of suspicion dominated at Hitler's headquarters." Hitler would usually conduct particularly important talks face to face, including atomic research. (...) In larger company the "Wunderwaffe" were spoken of in very general terms; when the conversation became more specific, it was forbidden to take minutes.

Talks on important topics were also conducted face to face in the case of individuals responsible for nuclear research, and important information was not recorded in duty calendars. This is shown by the example of Walther Gerlach (...). His secretary Gisela Guderian kept an up-to-date diary and wrote all of his correspondence by typewriter. When anybody visited Gerlach to show him the latest research report, she had to leave the room. It was forbidden to record very important meetings. (...) Shortly before the end of the war many secret documents were destroyed in a more or less systematic way. Research centre files from the Navy, Air Force, SS and Reich Post Office show particularly large gaps.

It is impossible to remove all traces and block all leakages with 100 per cent effectiveness (hence the intelligence reports shown in this book); but in the existing situation not much more can be done to complete the picture—unless one managed to find in Russia the results of reconnaissance carried out by "Smiersz" ("MIF") ... "MIF" was the scientific and facility reconnaissance section of the Red Army's counter-intelligence, which gathered information on such facilities and scientific research.

Certain elements of the picture have already been presented in the chapter about the Bell. Allow me to supplement it with yet another example of this still secretive episode, part of a project

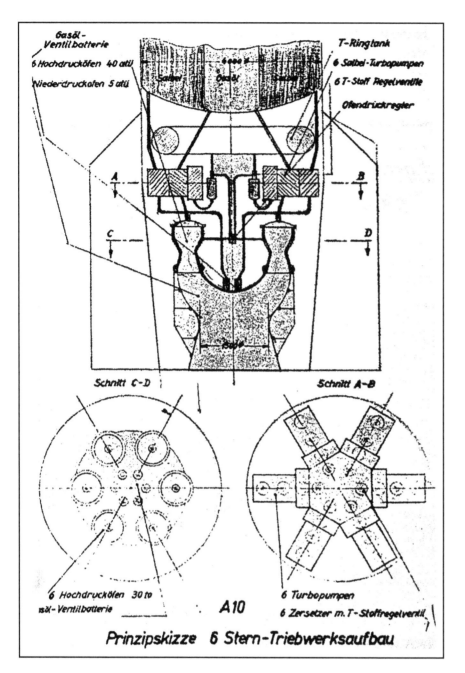

One of the little know facts is that the giant engine for the A9/A10 intercontinental missile was designed almost two years before the end of the war, although it was modified yet. The original drawing presented here shows the accepted configuration. It consisted of six modified engines from the V-2 plus a common chamber and a common nozzle, which performed the function similar to an afterburner in the jet engine. Only the fuel, without oxidizer, was to be injected there, and this fuel was… oil. Thanks to this it was possible to achieve thrust of 200 tonnes, instead of 156 tonnes with only six improved V-2 engines. The most important advantage was however the simplicity, thanks to which such an engine could be built within a couple of months.

that does not match known patterns in any way—merely as an example of the fact that in certain cases we only know shreds of information. Here is a further excerpt from Karlsch's book:[140]

The Lindemayer Group researched materials used in the production of rocket parts. In addition technicians working in this team were given a special task. They tested the stability of aluminium spheres of varying size. These experiments were commenced in Anklam, as Ingeborge Brandt herself recalls, who aged sixteen was given a duty assignment to work in the office of the Lindemayer team: "Some kind of terrace was situated in front of our room. It was called the control stand. Some kind of sphere was placed there (…). I estimated its circumference to be approximately 1.8 m. An account given by Irene König is more precise, who worked in Anklam as a telegraphist and repeatedly saw experiments being carried out with the sphere: "There were two spheres of differing size, one placed inside the other, and these spheres evaporated. At first I thought they were boiling water inside them, but never dared ask, as everything was so secretive. Following an air raid we were transferred to Friedland. Lindemayer travelled to Nordhausen [perhaps this is connected with the "strange" work described earlier at "Mittelwerk"?— I.W.] and Johann Grüner became manager of the team. In Friedland the spheres spun in a boiler at high speed. Sometimes powerful roaring and rumbling could be heard, and the engineers told us they were carrying out experiments with pressure regulators. It should be mentioned that these types of large aluminium sphere were not used in any rocket designs. The observed vapour could have originated from dry ice that was used to cool an unknown material placed in the sphere."

One of the underestimated "constituents" of the German plans for a strategic war was the "stealthy" Type XXI submarine. It was considered so revolutionary, that according to German estimates (and Skorzeny's view…) around 300 of them would be enough to isolate the european theatre of operations from the United States and cut off the United Kingdom from shipments of raw materials. It wasn't such a distant perspective, for before capitulation 119 of them were already in service. The photograph shows the conning tower.

These accounts can be associated with a certain, equally unusual (though perhaps distorted) report of Polish Home Army Intelligence, from January 1944:[16]

> *Information from January from a German Sergeant. They have already prepared 6 million "Leuchtkugel" ["glowing spheres"—I.W.]. Further production in progress. Supposedly a caustic, scorching liquid that destroys everything, dispersed using all manner of projectiles.*

The information summarized at the beginning of the chapter on the SS armament program recalls again the issue of an alleged "Sudeten Redoubt." Such a presumption is justified in the sense that a conglomeration of a huge industrial base connected with a new weapon (supposedly to change the course of the war), with command posts at the highest level including even Hitler's headquarters, was built in a single area. This is also confirmed by the aforementioned document concerning plans to transfer SS central institutions to Czechoslovakia—prepared by Kammler. In this case the "redoubt" was certainly more real than the mythical "Alpine Redoubt." In his book Tom Agoston devoted a little space to this last question.[138] Allow me to quote some excerpts from his interesting account, for it enables us to compare the real undertakings from Lower Silesia and the Protectorate with the "Alpine myth"—which in all probability was to divert the Allies' attention away from what was actually being prepared…

> *The factor that most worried the Allied commands was a sudden flow, as of March 1945, of US intelligence reports from the Office of Strategic Services (OSS, forerunner of the Central Intelligence Agency, CIA). They reported that the Germans were preparing an "armed and defended Alpine National Redoubt" in "Southern Germany and Northern Austria," to enable Nazi leaders to "carry on the fight." According to OSS reports, the "Redoubt" was to be defended by "most sophisticated secret weapons and elite troops, trained to organize the Reich's resurrection and liberation of Germany from the Occupying Forces." The "Redoubt" was said to have its operational command near Salzkammergut in Austria and its supreme command at Obersalzberg, Hitler's retreat near Berchtesgaden [precisely why it was bombed in the final days of the war!—I.W.].*
>
> *The "National Redoubt" in fact never existed. It proved to be nothing more than a propaganda myth. Yet the Allies accepted the OSS reports at their face value. This misjudgement not only swept away the basic target of Overlord, it also created a political running sore and flashpoint at the most critical interface between East and West in Europe. Ultimately, it contributed to a basic change in the face of Europe.*
>
> *Today, (…) the record of events that April 1945 reveals an alarming degree of political naiveté. And the naifs were plentiful on both sides of the Atlantic.*
>
> *From March onward, the OSS report had taken hold to an extent that they began to influence Allied tactical thinking. By the spring of 1945, the myth of a "National Redoubt" had rapidly mushroomed among the military, despite the caveats from British and US military intelligence. (…)*
>
> *There appears to be no evidence that Allied military leaders had taken steps to probe the validity of the early OSS reports, despite their ponderous implications. The earlier OSS intelligence forecast, the first to mention the "Redoubt" complex, was published by the analysis branch of the OSS on 29 December 1944. The report, marked CLASSIFIED, was keynoted by the prediction that "the Germans would probably establish reduits [sic] when the fighting has stopped." It continued:*
>
> *The heart of the plan would probably be a reduit, a mountain strongpoint, fortified and provisioned in advance, defended by strong guerrilla garrisons, serving as a base for marauding guerrilla bands. From such retreats the Nazis would probably be able to prolong resistance for a*

considerable time... It would be unwise to rule out that from some top party Chieftains, probably too well known to disappear underground, the reduit would provide the stage for a spectacular last stand fight.

Subsequent events showed that the "Redoubt" had no more substance than the deception operations the Wehrmacht routinely used. They continuously circulated rumours about the pending arrival of fictitious reinforcements, rapidly changing dummy headquarters and "advance personnel," even putting inflatable rubber tanks in the field to deceive photo interpreters. These were the types of rumours that the OSS should have analysed and scotched, instead of disseminating them over a period of several months, thereby influencing military and political strategy to the benefit of the Soviet Union. Eisenhower's command decision to seize the "Redoubt" before the Nazis could organize it for defence was knocked into a cocked hat when Allied troops discovered that the "Redoubt" was a phantom. It was no more than a composite of unimplemented proposals submitted to Hitler by Nazi fanatics in Bavaria, and was subsequently shown to be a realistic ruse-de-guerre furthered by the disinformation experts of Propaganda Minster Josef Goebbels, who targeted leaks to neutral correspondents in Berlin.

The Goebbels-huckstered ruse had taken such deep root within the Allied Supreme Command that correspondents at Supreme Headquarters, Allied Expeditionary Force (SHAEF), including myself, were actually shown an Allied map purporting to mark Nazi military dispositions within the "Redoubt." Bradley, later chief of the US Joint Chiefs of Staff, had to concede shortly after the war that: "...not until the campaign ended were we to learn that this Redoubt existed largely in the imagination of a few fanatic Nazis. It grew into so exaggerated a scheme that I am astonished that we could have believed it as innocently as we did. But while it persisted, this legend of the Redoubt was too ominous a threat to ignore, and in consequence it shaped our tactical thinking during the closing weeks of the war."

With such intelligence and decision making as revealed by the "Alpine Redoubt Case," the Allies stood no chance in the technological-intelligence race with the Russians—in an area which concentrated the most valuable armaments projects of the Third Reich. Moreover, they also had little opportunity for the accurate reconnaissance of key targets in Czechoslovakia and Lower Silesia, particularly with two or three counter-intelligence rings. By the way, if only those "naive" analysts had known what surprise the Germans really had in store for them...

The existing system gave specific benefits to the Germans even during the war—as far as Allied planning of huge, entire operations on the basis of "information" that had no connection with reality was concerned. Because of this research outposts and production facilities connected with the "SS Plan" were never bombed (with the exception of Pilsen, perhaps as a result

Hans Kammler. In the last 6 months of the war he has become the second most important person in the Reich, after Hitler.

of being "coincidentally" situated on the premises of Skoda). The Germans therefore had full freedom of operation, right up until the last days of the war! Recently I managed to find one of Kammler's documents in the German Bundesarchiv, which supports the hypothesis that "Riese" and other industrial/research and command facilities in that area formed a kind of redoubt of last resort. In this document Kammler answered Himmler's request to construct new office and command buildings for the SS in Berlin. He rejected this order because the institution he directed was fully engaged in constructing the special S-3 undertaking. What was the S-3? According to my list of codenames for German underground facilities (in all over 800!) it was, among other details, something larger than "Riese." It encompassed, in addition to other facilities in that area—first and foremost Hitler's new headquarters beneath Ksiaz/Fürstenstein Castle. It was a mirror image of the S-4 complex under construction in Turingia. The document's context suggests that this was the largest SS project of all, dated 12.12.1944. The connection with new command centres makes it quite a good candidate for a redoubt![153]

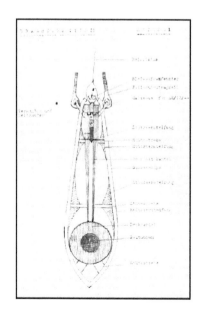

German cross-section of the plutonium bomb, declassified recently by the Russians

The Czech town of Zelezny Brod. According to a Gross Rosen prisoner, professor M. Moldawa, the only witness who worked in the concentration camp's chancellery and survived, the Luftwaffe and the SS were jointly constructing there an underground strategic command center "for long range, radio controlled weapons." I went there, to the city hall, where I asked the people responsible for land management and forestry, whether they know about any underground facility. They knew nothing, except for the fact that the main hotel in town was occupied in 1944 by Luftwaffe officers, although there was no airfield—it's in the mountains.

One of the mysteries of this period concerns something nobody has been able to explain reasonably until now. Namely the subject of secret mercury shipments to Japan—already described by myself in the chapter about the Bell. I described cases connected with several U-boats, although I must strongly emphasize that as they were cloaked in secrecy we know in essence very little about them, the cargo of which was revealed in accidental or otherwise ways (U-234, U-859 and U-530 in connection with Japanese I-52 and UIT-25). The world later found out about these cases only because the submarines were sunk and it was later possible to reach their cargo holds, or—as in the case of U-234—in the wake of surrender to the Americans. Despite the majority of their voyages remaining unknown, available data still paints a picture of a mass phenomenon. The scale of these undertakings seems to be typically industrial! Recently it has been possible to add one more case to this intriguing data...On January 10, 2007, an article was published in the reputable "International Herald Tribune" which described the discovery of the U-864—another submarine, sunk on February 9, 1945 directly off the coast of Norway. In this case the U-boat's entire cargo consisted of mercury, which excludes the pointless theory that this dangerous metal was ballast—the Third Reich had more important issues at hand than shipping ballast to Japan! Here is an excerpt from the article:[154]

I knew it was a submarine," said Karisen, a 73-year-old retired ship's pilot, gesturing across the choppy gray expanse of the North Sea toward the site of the blast on Feb. 9, 1945.
On that day, he said in an interview, he was out gathering peat with his grandmother on this island, then, like the rest of Norway, under Nazi occupation. "It was a big explosion but soon after it was calm and there was nothing to see," he said. "But in the German garrison there was a lot of activity."
That is hardly surprising. The plume of water and debris, rising 60 meters, or 200 feet, into the air, was the result of a torpedo fired from a prowling British submarine, the Venturer, that struck the German sub U-864 amidships at the very start of a clandestine voyage to Japan.
As the German vessel sank in two parts into more than 120 meters of water, it took with it not only the 73 men on board, but also 65 tons of mercury for the Japanese munitions industry and, some historical accounts say, a newly developed German jet-fighter engine [this is surely a distorted echo of "The Truth about the Wunderwaffe" English edition; whereas fulminating mercury was no longer used in ammunition production during the Second World War!—I.W.]—technology that was supposed to give the Axis powers an edge in the closing stages of the war. Much later there were rumours—and they remained just that—that the vessel was carrying fabled Nazi gold or even Hitler's last will and testament. (...)
The long saga of U-864, however, is far from over. Many of the canisters containing the liquid mercury are corroding. Small amounts of mercury have seeped out and Norwegian government tests around the wreck have detected slightly raised amounts of the metal in crabs and fish—the country's second biggest export after oil and gas. (...)
Officials in Norway believe that up to a third of the 1,857 flasks of mercury carefully stowed along the keel of the submarine now lie strewn over the seabed, many of them buried in mud, their condition unknown.

Kammler's document, which I found in the Berlin's Bundesarchiv. It states that the SS (SS-WVHA) was engaged in the construction of a "Special Construction Undertaking S-III" (Sonderbauvorhaben S-III) "with all its forces" and therefore was unable to execute other tasks. It follows from this document that it was somewhere close to the border between Lower Silesia (Niederschlesien) and the former Czechoslovakia (the Protectorate). From a list of German underground projects it follows that S-III encompassed Ksiaz, Riese and probably other facilities in that area. The main conclusion is that all these facilities, research posts etc. were part of something much larger, therefore that it was something serving one purpose and that it was the most important project of the SS. Nobody realised that before... It seems that it was reaching from research infrastructure, through the industry (Riese, Ludwikowice) up to the command infrastructure, which included Hitler's command post under the Ksiaz castle and a telecommunication center, probably code-named Rudiger.

I managed to find further information on this topic and other "mercury" cargo submarines in the aforementioned book on "Axis" operations in the Indian Ocean and Far East.[150] In reference to the U-864 it mentioned this as being the only case of two submerged submarines in combat during the war that ended in one being sunk—the commander of the British vessel ascertained his target's position by sonar and probably fired a homing "Fido" torpedo.

The book presents many new facts about these strange shipments. It also clearly depicts how infernally risky and costly this undertaking was. Only 42% of all shipments reached the Far East—thus these missions can be conveniently called semi-suicidal! Despite such great odds and unsatisfied deficits on both sides of the alliance, the fact that mercury made up 45% of all cargo despatched to Japan by submarine must cause reflection—it was decidedly "No. 1" (926.7 tons)! Perhaps this figure also includes mercury in the form of a radioactive amalgam by way of aforementioned submarine I-29. The huge significance of this metal enables mercury deliveries to be treated as the main aim of the "underwater exchanges"—a goal which nobody has been able to explain so far.

Only 43% of this metal reached the designated ports of arrival—396 tons. This data was based on generally accessible sources and does not encompass the secret flotilla revealed by the Americans during the intelligence operation codenamed "Lusty" (partly destroyed) whose existence the book's author apparently wasn't aware of. But he described many other U-boats that carried out "mercury" missions. Allow me to briefly list them for a fuller picture:

- In addition to the "Torelli" taken over from the Italians (UIT-25), other vessels from this squadron shipped mercury. In this context there was mention of submarines codenamed "Aquila II" and "Aquila III"—it is unknown if this concerns both or only one of them.

- On May 10, 1943, U-511 left Lorient, commanded by Fritz Schneewind, carrying "containers of mercury, a Daimler-Benz 3000 HP engine, Type IX U-boat plans and passengers." On July 15, it moored at Penang in Indonesia, and exactly one month later reached Kobe in Japan.
- Mercury transports clearly intensified in mid 1944. On August 22, two U-boats left Bordeaux: U-180 commanded by Rolf Reisen and U-195 by Friedrich Steinfeldt. The former contained "1,843 mercury cylinders," the latter: "54.4 tons of mercury"—and a number of other commodities, among others spare parts for U-boats based in the Far East. Only U-195 reached its destination on December 28. U-862 also moored in Penang on September 9, commanded by Heinrich Timm, supplying "mercury cylinders." A couple of weeks later U-861 appeared in Penang after five months at sea. In this case mercury comprised the main part of the cargo—as much as 120 tons. In general the majority of "mercury" submarine voyages described in the chapter about the Bell also fall into this period, among others U-859 (1,959 steel cylinders, sunk at the end of September 1944) and the huge Japanese I-52.
- Intensive attempts to supply their ally with this apparently crucial raw material were carried out uninterrupted, but the risk increased almost with every month. In spite of this on August 24, 1944, the huge U-boat U-219 was dispatched from Bordeaux, a special Type X freighter loaded with: "an entire transmitting radio station destined for Kobe, diesel spare parts, a torpedo balancing station for Penang, hospital medicaments, equipment for a German radio station in Singapore, spare parts for Arado 196 aircraft and boxes of aluminium ingots. Unpolished optical glass and steel cylinders of mercury were placed in the vessel's keel." The U-219 reached Jakarta (Batavia) on December 11, 1944, and was the final U-boat that made it to the Far East. The following three were sunk and the fourth—the aforementioned U-234—surrendered to the Americans.[150]

One more issue connected to the chapter about the Bell…

It involves finding the answer to the question of what exactly was the propulsion that the secret SS think tank in Pilsen and related research posts worked on. How can this be explained without going beyond the scope of nuclear physics on the one hand, but on the other—taking into account the criterion of technical feasibility in that period (though it is obvious an avant-garde concept was involved through and through). That which I intend to present in the following section of this chapter is a certain hypothesis, based on scientific fundamentals. It is the development of an interpretation of a German document published in the chapter about the Bell. It is however a separate issue and does not require previous familiarization with its content, not being in addition a simple duplication.

At first glance it would appear that propulsion fulfilling the above criteria was plainly something impossible to achieve, after all the monopoly of rocket propulsion has not been broken to this day. Perhaps it involved some pipe dream, a great technical-scientific mistake, some attempt to create a *perpetual motion machine* and the like? If however a great armaments program was based on this, and work was "well advanced," then the prospect of this breakthrough's practical application was real and imminent. The official and unprecedented classification of the project as "decisive for the war" must also have been based on concrete data. The task that I have set myself depends after all on finding a meaningful explanation in accordance with accounts of the device's operation included in the chapter about the Bell. Only if I am able to unambiguously demonstrate that it was impossible to achieve, can we then assume that some great misconception was involved.

We usually associate nuclear physics with a reactor or bomb. It need not have been so in this case. Nuclear physics is also the source of a number of phenomena of different kind. In a way this ensues from the fact that in the atomic world we can achieve much greater speed and energy density than we know from everyday experience. One type of such phenomena is Einstein's Theory of Relativity. In this way, at least according to the theory, one can generate a useful force from the point of view of propulsion. Let us first examine this widely known academic canon:

Einstein's theory has been developed at various times, so-called "solutions" being derived from it to describe various systems or situations, taking into account factors omitted by the author himself. One such derivative (obviously generally accepted by science) is the Einstein-Cartan Theory, formulated in 1918. It took into account not so much the existence of mass itself, but also its spinning motion. A term with a negative sign appeared in the equation, meaning that at a sufficiently high rotational speed a body begins to repel itself—negative gravity dominating over positive. In other words, antigravitation is involved. Conducted experiments demonstrate that just on the basis of the relations described in this equation, this should be the order of several million rotations a second.[155] I attempted to consult someone who had worked precisely on this derivative of Einstein's equation. It emerged that there were very few specialists of this kind in Poland, however in the summer of 2006, I contacted (i.e. telephoned) Professor Andrzej Trautman from the Institute of Theoretical Physics at the University of Warsaw—perhaps the best expert in Poland. I presented him with the following problem:

I obviously described the German project in a few words, adding: *"I decided to approach you, because a term suggesting a relativistic context appears in the account [i.e. concerning the Theory of Relativity]—the notion of "magnetic field separation," in connection with angular ions travelling at great speeds."* I attempted to further expand on this issue, substantiating my suspicions, but the Professor interrupted me: "I think, Sir, that this is impossible. We have calculated this [Cartan's equation] for "black holes" and it emerged that the antigravitational effect

A document found in the Bundesarchiv, probably from Himmler's field command post, describing the "urgent need" for chemists, expressed by Kammler, in February 1945! It may suggest that he was preparing a chemical offensive until the very end. As in the case of the numerous reports presented in the chapter regarding the chemical arsenal, it's a completely new reality, never described before as part of large scale, comprehensive preparations.

is completely insignificant. It wouldn't be possible to achieve this in a laboratory."

I again tried to add what I was specifically basing my assumptions on and what elements I had in mind, but Professor Trautman wasn't in the least interested. The specific reason I telephoned him—with the intention of arranging a longer talk in person—was proof itself that it was possible in a laboratory! For I had in front of me a photocopy of an article that had already been published a dozen or so years earlier in the journal "Physics Letters B." In theory one of the leading Polish scientists in this field should have known about this long ago—much more than myself. However quite the opposite happened—he did not want to hear a word of it—"it's impossible and that's the end of it!" (I had already grown accustomed to this). And I would have revealed to him not just theoretical predictions, but those resulting from a very interesting experiment…[156]

The article described the surprising results of experiments carried out using a heavy ion accelerator—at the Daresbury Laboratory in Great Britain. It was actually a system involving two accelerators (tandem type) that not only accelerated ions, but also caused them to spin. Frankly speaking, for years I had unsuccessfully tried to obtain the results of measurements of this kind. It was particularly difficult as typically the ions of light elements are accelerated, few are interested in the energy resulting from their spin and first and foremost—so far I had not come across any publication in which all this would have been analysed from the view of the Theory of Relativity (specifically antigravitation). Here however it was "all in one," moreover: the ions were spun at such tremendous speeds that deformations of the nuclei under the centrifugal force were the order of… 50%! Thus the limits of what was technically possible were reached, only the "benefits" of other phenomena were not exploited that could greatly increase the effects we are interested in, but which we will leave for now. For the time being we are looking at what refers directly to the Einstein-Carter Theory (a spinning mass as a source of energy).

Admittedly Mercury was not tested, mentioned in the context of the German "bell," but other very heavy elements were: Gadolinium, Terbium and Dysprosium. Despite this the analogy to the German account was quite large—apart from the absence of "magnetic field separation" and what this involved. What did the results of these experiments determine? Two surprising facts were revealed:[156]

1. Gravitational attraction or repulsion was not measured, since at these nuclei speeds and energies this would have been, to put it mildly, quite difficult. This effect can however be measured very accurately somewhat differently. It is known after all that there exists an equality sign between the force of gravitational interaction and inertia (neutralisation of gravitation would mean neutralisation of inertia and vice versa). Accordingly the inertia of moving nuclei was measured. It emerged that at high rotational speeds the fall in gravitation = inertia reached as much as 50%! I suspect this does not agree with the calculations of Professor Trautman and his colleagues!

2. The other new fact was that these results were slightly different for particular elements—at the same energies! The article stated: *"The figure shows that there are large differences in the behaviour (…) as a function of rotational frequency for different examples. For ^{150}Gd and ^{151}Tb the values of dynamic moments of inertia are very high at low frequencies [low spinning speeds—I.W.] and decrease rapidly with increasing frequency, while for ^{149}Gd they decrease less rapidly and finally for $^{151,\ 152}$Dy they are almost constant."* In this case so-called isotopic spin plays a role—the different resultant speed of spin of the protons and neutrons forming the atomic nucleus.

The latter fact—the varying "ability" to weaken gravitation and inertia by different elements and isotopes again recalls the question of why Professor Gerlach, the leading individual in the German project, chose Mercury. Mercury and its ionized form had become his main subject of interest even before the

of an already emerging revolution in physics. In this case the main stimulus comes from astronomy. It emerged that the movements of galaxies are entirely different from what can be concluded on the basis of the Theory of Relativity—they repel each other more than they attract each other! (which of course also contradicts Professor Trautman's position, indeed our entire past knowledge...) An astronomer described thus:[157]

A true breakthrough was achieved in cosmology at the end of the last century. It suddenly emerged that both our past idea of the makeup of the Universe as well as its history and future was not just incomplete, but wrong! The mysterious cosmological constant (or a constituent of the Universe simulating it), introduced by Albert Einstein, and later considered by himself a great mistake and forgotten, exerts a substantial influence on our Universe, acting as antigravitation and accelerating expansion, whilst the energy density associated with it constitutes approximately 70% of the Universe's entire density. (...) There, where the areas of applying quantum physics (small scale) and the theory of gravitation (large masses and energies) begin to overlap, should emerge New Physics. But so far we have no such theory at our disposal, though the apparently crazy ideas which regularly appear in scientific journals may give us a foretaste of a new vision of the Universe. (...)
Perhaps the problem lies in the very form of the General Theory of Relativity, and not in some additional constituent of the Universe?

An article published in the American popular science journal "Natural History" referring to experiments conducted in space on board an orbital laboratory stated:[158]

What about the possibility that Gravity Probe B will show the way to new physics? According to one Nobel laureate, the physicist Chen Ning Yang of Stony Brook University in New York, Einstein's general theory of relativity is likely to be amended in a way that somehow entangles spin and rotation.

In general the above discoveries demonstrate that antigravitation is in no way an "exotic" interaction and should be more easily generated than thought up until now. How?

Spinning motion itself is still inadequate, although some quite interesting pointers do exist.

Previously I mentioned a theory formulated directly after World War I by Josef Lense and Hans Thirring—a German and Austrian. They formulated a derivative of Einstein's theory, which also took into account spinning motion. This was not however their most important area, though Thirring remained a close associate of Professor Gerlach during the war. The point is that on the basis of their theory it wouldn't have been possible (as emerged in the meantime) to significantly increase the efficiency of a possible propulsion.

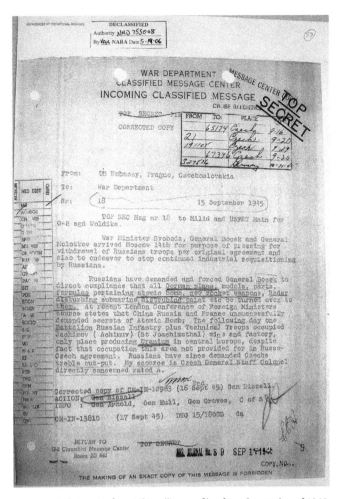

A document from US intelligence files, from September of 1945, describing the importance of Jachymov (St. Joachimstal) from the point of view of German nuclear research. It says, among other things, about "parts of atomic bomb." It also illustrates how the Soviet technical intelligence worked... Jachymov was mentioned already in the context of the Bell, but it wasn't the only really crucial nuclear facility in the occupied Czechoslovakia. In Pilsen was the research post (a team of "visionary" SS physicists) working on revolutionary nuclear propulsion. In Usti nad Labem (Aussig) there was, for example, Heisenberg's research team (producing among other things some "radium-mercury amalgam," for Japan!). All that was, for obvious reasons, beyond the sight of the Western historians. Nobody even suspected that Werner Heisenberg, the most prominent German nuclear scientist, had his own team working on the bomb—if it indeed was related to the bomb. As in the case of Gerlach, Oberth or Jordan, an assumption was tacitly adopted that they were not incorporated into the system. It's also because nobody just saw it as a system...

war. To obtain the answer to this question we must presumably go back to the 1920s and establish why Gerlach "bet on" spinning the ions of this metal—and at the same time engaged in research on the nature of ball lightnings. Did correspondence with Piotr Kapica influence his decision? Did his inspiration in this matter originate in the USSR? The answer in this case is probably hidden in Russian archives...

As I previously mentioned, what was proved during the described experiments is only one of several approaching signs

Probably the most important place in the Third Reich, and one of the most deeply classified, where all the threads linked to the most advanced research projects met—part of a larger SS compound in Glau, near Berlin. Officially it was the Waffen-SS Artillery School No 1, in fact it was a cover-up for the Schwab's nuclear research institute and for the FEP/Waffen-SS cell, which controlled research on revolutionary propulsion (regarding the Bell, but also Pilsen). One of the new sources in this respect, a German book by H. Meier titled "Forschung als Waffe, Band 2," says also that FEP/Waffen-SS coordinated projects referring to chemical weapons, run jointly with the Wehrmacht. Therefore one may say that Glau was the key to the planned new destructive offensive, against which the Allies probably wouldn't have too many means of defense.

However the right theory existed, at least in general outlines even during the war, only that its author was someone else. I mentioned in the chapter about the Bell that one of the basic terms used in the German account was the enigmatic from the outset "separation of magnetic fields." As I wrote, it was paramount to check whether this was not the area that could increase the "gravitational effect." The answer to this question could simultaneously confirm or negate the legitimacy of adopting the "gravitational hypothesis." In any event this was quite important information. It emerged that this indeed had an evident connection with gravitation, based not on the Lense-Thirring theory but the aforementioned "other" theory, formulated by Professor Pascual Jordan (a German with ancestors harking back to the Spanish aristocracy—hence his curiously sounding name).

Jordan was, like Gerlach, one of the most outstanding German scientists of the war period. His achievements were so exceptional that he was virtually a certain candidate for the 1954 Nobel Prize. However it was finally received by Max Born, when Jordan was effectively disqualified after his wartime collaboration with the Nazi Party and SS was brought to light. Based on research from the war period Jordan wrote a number of publications on both gravitation and quantum physics as well as common areas where the two fields met (which few are currently engaged in…). He determined among other things that the possibility of artificially affecting gravitation through the spinning of a powerful magnetic (electromagnetic) field existed. Precisely on the same principle that I described in the chapter about the Bell: a field (magnetic) cannot be treated in isolation from the space it is situated in and if it would be possible for us to accomplish complete separation of spinning fields, we would achieve some kind of gravitational shielding. Analyses indicate that this effect is only significant when a very high spinning speed, large energy field and degree of field "separation" the order of 99% is achieved. The final effect should be a product of these three factors.

The theory of field separation was "perfected" shortly after the war, connected to a version from the French physicist Thiry and is currently known as the "Jordan—Thiry theory." Before I move on to its significance in the context were are interested in, allow me to devote a few paragraphs to describing the individual himself, Pascual Jordan.[159, 160, 161]

He was born on 18 X 1902 in Hannover, hence in 1944 he was only 42. He graduated from the prestigious University of Göttingen, probably best in the world as far as the exact sciences were concerned—in the mid 1920s. He also began his research career there as the assistant of mathematician Richard Courant and physicist Max Born. In 1936 he received the degree of reader (assistant professor in the US) and became a lecturer at the university. Initially he was only interested in quantum field theory (electromagnetism), but shortly before the war gravitation and its ties with quantum theory became his main research area. His main works from this period were "The Notion of Quantum Theory" from 1936 ("Anschauliche Quantentheorie") and "20th Century Physics," also from 1936. In 1941 he also published an exception, being rather a projection of his personal interests—the book "Physics and the Secret of Organic Life." Werner Heisenberg was his close co-worker and co-author of numerous articles during this period, and he was in close contact with Professor Gerlach. Jordan had joined the Nazi party as far back as 1933, where he quickly became an active member. A year later he was also a member of the SA. This certainly made it easier for him to climb the ranks of his scientific career in the coming years, but party chiefs did not particularly trust him. Despite the fact that Jordan could be recognised as an ardent national socialist, he never renounced his admiration for Einstein and did not distance himself from outstanding Jewish scientists like Courant, Born and Wolfgang Pauli. Jordan's exceptional talent was rapidly appreciated however by the military and specifically by the Luftwaffe, who among others had sent him to Peenemünde as far back as 1936—in connection with the research base's expansion. Thus this was virtually on a convergent course with the respective stage in Debus's activity, although the latter was a member of the SS. However little is known about the Luftwaffe wartime projects which Jordan was engaged in. We know that up until 1944 he also held the civilian post of Professor of Theoretical Physics at Rostock University and in 1942 he was decorated by the highest authorities of the Third Reich with the Max Plank Medal for outstanding service. Only after the war could he publish his comprehensive work on the connections between quantum physics and gravitation: in 1952 the book "Gravitational Force and the Universe" ("Schwerkraft und Weltall") appeared along with "The Atom and the Universe" ("Atom und Weltall") four years later. In the 1960s he was among others a member of the "Gravity Research Foundation" in New Boston. Irrespective of this he was also a member of the Bundestag (West German Parliament), where he could be recognised as an ardent nationalist; in 1957 he launched a campaign to arm the Bundeswehr with tactical nuclear weapons and in 1965 protested against the recognition of the Oder-Neisse Line as the permanent border with Poland. It is a characteristic fact that identical fanaticism was also displayed by the remaining scientists active in this field (described in the chapter about the Bell): Gerlach and Debus.

What does Jordan's theory signify in practise when assessing a conception which in all probability became a revolutionary propulsion "connected with nuclear physics"? It means of course that it would be possible to turn the aforementioned "fifty percent" cancellation of gravity into more than 100%, if only the theory's criteria could be fulfilled.

As I have already stated, the only way to create a heavy ion vortex under conditions of field separation with sufficiently large field intensity is through plasma physics, specifically so-called plasma solitons—something that also interested Gerlach (ball lightnings in the context presented in letters to Mr Kapica). This question has recently been enjoying greater interest in science. Besides if it was not so interesting and promising, the very costly American project "Shiva Star" would not have been continued for so many years in secret in the area of a guarded airbase (incidentally the name "Shiva Star" refers to messages from Mahabharata).

The Germans had at their disposal a foundation not only in the form of Gerlach's know-how (he measured the spins of plasma ions) and Jordan's theory, but this also explains why Gerlach corresponded with Professor Kapica (a Soviet scientist, but with Polish parents, by the way). Kapica received the Nobel Peace Prize for the discovery and description of superfluidity. One should bear in mind that vortices in such liquids are solitons, and what Piotr Kapica did was provide some kind of supplement to Jordan's works, enabling a complete picture to be created.

It is clear, at the same time, that the main subject of interest on the part of Gerlach when corresponding with Kapica were ball lightnings—i.e. plasma solitons.

If one wonders, therefore, whether it was even possible for the Nazis to achieve something that would have taken us half a century longer—then we can really see what the burden of Einstein's theory means. It was simpler than it appeared, they just exploited a connection, an "entrance" to a new world, that is already observable in nature.

BIBLIOGRAPHY

Part I

BIOS = British Intelligence Objectives Sub-Committee
CIOS = Combined Intelligence Objectives Sub-Committee
(Intelligence summaries)

1. A. Speer, "Erinnerungen," Frankfurt am Main, 1969.
2. Records of the NAIC/Wright-Patterson AFB: "History of AAF participation in project Paperclip. March 1945-March 1947."
3. D. Hölsken, "V-missiles of the Third Reich," Monogram Aviation Publications, 1994.
4. J. B. King & J. Batchelor, "Deutsche Geheimwaffen," 1975.
5. F. Welczar, "Pestka wiśni," Przekrój, issue 1089/1966.
6. "Rockets and guided missiles" CIOS-report, Item No. 4, 6, File No. XXVI-II-56 (1945).
7. W. Kozakiewicz, J. Wiśniewski & S. Żukowski, "Broń rakietowa," Publ: Główny Instytut Mechaniki, 1951.
8. "Institutes of the Bevollmächtiger für Hochfrequenzforschung," CIOS-report, File No XXX-37, Item No 1, 7. (1946).
9. T. Bartkowiak, "Wunderwaffe zawiodła," Nadodrze, issue 14/1969.
10. J. Nowak-Jeziorański, "Prawda o Peenemünde," Rzeczpospolita, 4-5 XI 2000.
11. M. Wojewódzki, "Akcja V-1-V-2," Warszawa, 1972.
12. A. Glass, S. Kordaczuk & D. Stępniewska, "Wywiad Armii Krajowej w walce z V-1 i V-2," Publ: Mirage, 2000.
13. T. Belerski, "Polacy rozpracowali tajemnice niemieckie," Rzeczpospolita, 1-2 IX 2001.
14. T. Bazylko, "Wunderwaffe rozszyfrowana," Za Wolność i Lud, issue 1/1961.
15. M. Wojewódzki, "Jak uczeni polscy rozszyfrowali tajemnicę hitlerowskiej rakiety V-2," Stolica, issue 27/1963.
16. J. Sroka, "Poligon V-2 na Podlasiu," Za Wolność i Lud, issue 9/1967.
17. Z. Niepokój, "Przewoziłem największą tajemnicę wojny," Za Wolność i Lud, issue 20/1965.
18. F. Welczar, "Stonoga nie będzie strzelać," Przekrój, issue 1088/1966.
19. A. Marks, "Widziałem V-3," Przekrój, issue 1259/1969.
20. A. Turra, "Heeresversuchsstelle Hillersleben," Publ: Podzun-Pallas, 1998.
21. J. Miranda, P. Mercado, "Die geheimen Wunderwaffen des III Reiches," Publ: Flugzeug Publikations GmbH, 1995.
22. I. Bednarek, S. Sokołowski, "Fanfary i werble," Publ: Śląsk, 1966.
23. F. Hahn, "Waffen und Geheimwaffen..." Wetzar, 1995.
24. W. Dornberger, "V-2: der Schuss ins Weltall," Esslingen, 1952.
25. T. Burakowski, A. Sala "Rakiety i pociski kierowane," Publ: MON, 1960.
26. D. Masters, "German Jet Genesis," Publ: Jane's Publications, 1982.
27. K. Kens, H. Nowarra, "Die deutschen Flugzeuge 1933-1945," Publ: Lehmans, 1972.
28. R. Michulec, "Luftwaffe 1935-1945 cz.4," Publ: AJ-Press, 1997.
29. W. Bączkowski, "Samoloty odrzutowe," Publ: Iglica/Agencja Wydawnicza CB, 2000.
30. R. Ford, "German Secret Weapons in W.W. II," Publ: Amber Books, 2000.
31. J. Osuchowski, "Gusen: przedsionek piekła," Publ: MON, 1961.
32. K.W. Müller, W. Schilling, "Deckname Lachs," Publ: Heinrich Jung Verlagsgess, 1995.
33. "Messerschmitt bombproof assembly plant," CIOS-report, Item No. 25, File No. XXVI 44.
34. H.W. Wichert, "Decknamenverzeichnis deutscher unterirdischer Bauten, U-Bootbunker, ölanlagen, chemischer Anlagen und WIFO-Anlagen," Publ: Johann Schulte (year unknown).
35. K. Margry, "Nordhausen," After the Battle, issue 101 (1998).
36. J. Gałaś, S. Newiak, "Flossenbürg: nieznany obóz zagłady," Publ: Śląsk, 1975.
37. I. Witkowski, "Podziemne krolestwo Hitlera" ("Hitler's underground kingdom"), Warsaw, 2006.
38. NARA/Air Intelligence Summary No. 53 (United States Strategic Air Forces in Europe), November 12, 1944.
39. S. Fleischer, M. Ryś, "Ar-234 Blitz," Publ: AJ-Press, 1997.

40 M. Ryś, "Horten Ho-229," Nowa Technika Wojskowa, issue 7-8/2001.
41 H.P. Dabrowski, "Flying wings of the Horten brothers," Publ: Schiffer, 1995.
42 H.P. Dabrowski, "The Horten flying wing in World War II," Publ: Schiffer, 1991.
43 "Secret German aircraft projects of 1945," Publ: Toros Publications, 1997.
44 R. Stanley, "Der Beitrag deutscher Luftfahrtingenieure zur argentinischer Luftfahrtforschung und Entwicklung nach 1945: das Wirken der Gruppe Tank in Argentinien 1947-1955," In: "Nationalsozialismus und Argentinien," Publ: Peter Lang, 1995.
45 W. Wagner "Kurt Tank. Konstrukteur und Testpilot bei Focke-Wulf" ("Die deutsche Luftfahrt" vol. 1), Munich, 1980.
46 U. Goni, "Peron y los Alemanes," Publ: Editorial Sudamericana (Arg.), 1998.
47 M. Mariscotti, "El proyecto atomico de Huemul," Publ: Sigma (Arg.), 1996.
48 T. Bower, "The Paperclip conspiracy, the hunt for the Nazi scientists," Boston/Toronto, 1987.
49 L. Adamczewski, "Tajemnicza studnia w Lubaniu," Głos Wielkopolski, issue 9 Nov. 1998.
50 O. Skorzeny "La guerre inconnue," Publ: Albin Michel (Paris), 1975.
51 M. Korzun, "1000 słów o materiałach wybuchowych i wybuchu," Publ: MON, 1986.
52 NARA/RG-319, Entry 82A: "Reports and messages 1946-1951 (Alsos Mission)."
53 BIOS Final report No. 148, Item No. 1: "German Betatrons" (1946).
54 AAN/ Alexandrian Microfilms- records of "Persönlicher Stab Reichsführer-SS" (T-175). Folder 360114 (360/14?).
55 "Gesselschaft für Gerätebau," CIOS-report, Item No. 4, File No. XXI-59 (1946).
56 "German tank design trends," CIOS-report, Item Nos. 18, 19, File No. XXIX-58 (1945).
57 "Wojna pancerna," Gazety wojenne, issue No. 85.
58 A. Kiński, P. Żurkowski "Czołg superciężki E-100," Nowa Technika Wojskowa, issue 12/1994.
59 A. Kiński, "Jaki był IS-2?" Nowa Technika Wojskowa, issue 6/2001.
60 "German development of hydraulic couplings and torque converters: J. M. Voith, Heidenheim/Brenz," CIOS-report, Item No. 18, File No. XXIX-34 (1945).
61 "Report on German development of gas turbines for armoured fighting vehicles," BIOS Final report No. 98, Item Nos. 18, 26.
62 "The ZF electro-magnetic transmission, with a special application for the Panther tank," BIOS Final report No. 579, Item No. 18.
63 W. Trojca, "Pz. Kpfw. V Panther," Publ: AJ-Press, 1999.
64 "German infra-red driving and fire control equipment, Fallingböstel," CIOS-report, Item No. 9, File No. XXIV-7 (1945).
65 Z. Hak, "Kuriozni zbrojni projekty…," Publ: FORT-Print, 1995.
66 "Ferromagnetic materials for radar absorption," BIOS Final report No. 869, Item No.1.
67 "Work of Prof. Hütting on ferromagnetic substances for use in radar camouflage," BIOS Final report No. 871, Item No. 1.
68 "The Schornsteinfeger Project," CIOS-report, File No. XXVI-24.
69 "Production and further investigation of Wesch anti-radar material," BIOS Final report No. 132.
70 "Sound absorbent coatings for submarines," CIOS-report, File No. XXIV-8, Item No. 1.
71 "German plastic developments," CIOS-report, File No. XIII-6,7, Item No. 22.
72 W. Trojca, "U-Bootwaffe 1939-1945 cz.4," Publ: AJ-Press, 1999.
73 "Operation of the Type-XVII 2500 HP hydrogen peroxide turbine propulsion plant for submarines," CIOS-report, File No. XXX-110, Item No. 12.
74 "German naval closed cycle Diesel development for submerged propulsion," CIOS-report, File No. XXX-76, Item No. 12.
75 "Recoilless guns development of Rheinmetall-Borsig," CIOS-report, File No. XXVII-27, Item No. 2 (1946).
76 "Development of weapons by Rheinmetall-Borsig," CIOS-report File No. XXXI-63, Item No. 2 (1946).
77 S. Pataj, "Artyleria lądowa 1871-1970," Publ: MON, 1975.
78 "Airborne recoilless 88-mm gun," United States Strategic Air Forces in Europe/Air Intelligence Summary, No. 58 (17 XII 1944).
79 AAN/Alexandrian Microfilms-records of "Reichsforschungsrat."
80 Collective work: "Indywidualna broń strzelecka Drugiej Wojny Światowej," Publ: Lampart, 2000.
81 M. Bryja, "Piechota niemiecka vol. 3," Publ: Militaria, 2000.
82 "German infra-red devices and associated investigations," CIOS-report, File No. XXX-108, Item Nos. 1, 9 (1945).
83 "German Seehund apparatus," CIOS-report, File No. XI-8, Item No. 9. (1945).
84 "German infra-red devices and associated investigations-report No. 2," CIOS-report, File No. XXX-9, Item No. 9 (1946).
85 NARA/ RG-319 Entry-82A "Reports and messages/Alsos Mission," records of the Reichsforschungsrat.
86 T. Rajewska, "Nadzieja Kriegsmarine," Tygodnik Morski, issue 21/1971.
87 AAN /Alexandrian Microfilms-records of: "Persönlicher Stab Reichsführer-SS," (T-175/324).
88 J. Chalecki, "Lunety noktowizyjne," Wojskowy Przegląd Techniczny, issue 11/1984.
89 "German research on rectifiers and semiconductors," BIOS Final report No. 725, Item Nos. 1, 7, 9.
90 "German infra-red equipment in the Kiel area," CIOS-report, File No. XXX-3, Item No. 1.

91　M. Walker, "German national socialism and the quest for nuclear power," Publ: Cambridge University Press, 1989.
92　AAN/Alexandrian Microfilms-records of "Persönlicher Stab Reichsführer-SS" (T-175/208).
93　"The invention of Hans Coler, relating to an alleged new source of power," BIOS Final report No. 1043, Item No. 31 (1946).
94　E. Łokas, "Ciemna materia we Wszechświecie," Wiedza i Życie, issue 10/1998.
95　A.G. Riess, "Universal peekaboo," Nature, 16 Sept. 1999.
96　R.H. Bailey, "The Air War in Europe," Publ: Time-Life Books, 1981.
97　"Druga Wojna Światowa w powietrzu" (memoirs of the Allied pilots), Publ: Szramus, 2000.

Part II

100　F. Hahn, "Waffen und Geheimwaffen… 1933-1945."
101　T. Burakowski, A. Sala "Rakiety i pociski kierowane," Publ: MON, 1960.
102　W. Kozakiewicz et.al., "Broń rakietowa," Publ: Główny Instytut Mechaniki, 1951.
103　"Institutes of the Bevollmächtiger für Hochfrequenzforschung," CIOS-report, File No. XXX-37, Item Nos 1, 7 (1946).
104　"German infrared equipment in the Kiel area," CIOS-report, Item No. 1, File No XXX-3.
105　"Rockets and guided missiles," CIOS-report, ITEM No 4, 6. File No. XXVIII-56 (1945).
106　"Restricted summary of German controlled missiles," CIOS-report, Item No 4, 6. File No. XXIX - 55 (1945).
107　M. Ryś, "Rakiety i radary," Nowa Technika Wojskowa, issues 5, 7/1999.
108　M. Bornemann, "Geheimprojekt Mittelbau," Publ: Bernard und Graefe, 1994.
109　"Aerodynamics of rockets and ramjet research and development work at Luftfahrtforschungsanstalt Hermann Goering, Völkenrode," CIOS-report, Item No 4, 6. File No. XXVII-67 (1946).
110　M. Griehl, "Fla-Rakete Schmetterling," Flugzeug, issue 4, 5/1998.
111　"Description of the construction and performance of the anti-aircraft rocket Enzian E4," CIOS-report, Items No. 4, 6, File No. XXVII-66 (1946).
112　"German development of homing devices," CIOS-report Item No. 1. File No. XXVI-57 (1946).
113　"The I.T.T., Siemens and Robert Bosch organizations," CIOS-report, Item Nos 1, 7, 9, File No XXXI-38.
114　A. Olszewski, "Tajemnice poligonu Nord," Tygonik Demokr, issue 37/1964.
115　K. Nicpoń, "Bomby kierowane cz. I," Nowa Technika Wojskowa, issue 9/1997.
116　T. Gander, P. Chamberlain, "Small arms, artillery and special weapons of the Third Reich,"
117　"Proximity fuze development. Rheinmetall Borsig A.G. / Mülhausen," CIOS-report, Item No. 3, File No. XXVI-1 (1945).
118　"Survey of German ramjet developments," CIOS Final report, Item No. 6, File No XXX-81.
119　NARA / "United States Strategic Air Forces in Europe, Air Intelligence Summary No 74" (8.IV.1945).
120　NARA/RG-319 "Army Intelligence Document Files" Box 1 (documents pertaining to Hans J. Kaeppeler, interned in Camp Perry / Ohio, ISN 31G 2507960).
121　R. Fuks, "Prawdziwe zadanie prof. Kurta Blome: broń bakteriologiczna w arsenale III Rzeszy," Za Wolność i Lud, issues: 51, 52/1974, 1, 2/1975.
122　"Technical information on Tabun and Sarin, I.G. Farbenindustrie A.G., Frankfurt / Main," CIOS-report. Item No 22, File No. XXIII-24 (1945).
123　IPN / microfilm "Akta Bergeamt Waldenburg-Nord."
124　"Wykaz obiektów opuszczonych i niewłaściwie zagospodarowanych (stan na 13 lutego 1953 r.)," document reproduced in: "Przegląd Techniczny", June 11, 1995.
125　U.S. Air Force History Office/Bolling AFB, microfilm "Operation Lusty."
126　NARA/RG-38 "Intelligence Division top secret reports of Naval Attaches 1944-1947. Box 9."
127　Records of Wright-Patterson Air Force Base/National Air Intelligence Center: "History of AAF participation in project Paperclip may 1945-march 1947 / Study No 214."
128　I. Witkowski, "Hitler w Argentynie i IV Rzesza" ("Hitler in Argentina and the Fourth Reich"), Warsaw, 2009.
129　K. Mallory, A. Ottar, "The architecture of war," New York, 1973.
130　Interview with Dr. Jacek Wilczur, recorded by the author of this book. Copy in my possession.
131　NARA/RG-319 "Joint Intelligence Objectives Agency" (JIOA) report "German underground installations." Part 3: "Various installations of general interest"
132　D. Czerniewicz, "Gaz! Gaz! Gaz!," Odkrywca magazine (Polish), issue 4/2006.
133　A. Speer, "Der Sklavenstaat" (English edition titled "Infiltration"), Publ: MacMillan, 1981.
134　J. Lamparska, "Tajemnice ukrytych skarbów," Publ: ASIA Press, 1995.
135　NARA/RG-319 "P-Files": Intelligence Research Project #363: "German secret weapon manufacture," September 19, 1944.
136　F. Gibb, "U-Boat find revives nazi escape theory," The Times, May 2, 1983.
137　NARA/RG-319 BIOS final report No. 782, Item No. 8: "Interrogation of professor Ferdinand Flury and dr Wolfgang Wirth on the toxicology of chemical warfare agents".
138　T. Agoston, "Blunder! How the US gave away nazi super-secrets to Russia," Publ: W. Kimber, 1985.
139　A. Schulz, G. Wegmann, D. Zinke, "Die Generale der Waffen-SS und der Polizei," Publ: Biblio-Verlag, Bissendorf, 2005.

140 R. Karlsch, "Hitlers Bombe," Publ. Deutsche Verlags-Anstalt, 2005.
141 Bundesarchiv Koblenz, NS-19, "Akten des Personliches Stabes des RFSS," Folder 3021.
142 Bundesarchiv Koblenz, NS-19, "Akten des Personliches Stabes des RFSS," Folders 3546, 3800.
143 A. Glass, "Meldunki miesięczne wywiadu przemysłowego KG ZWZ/AK 1941-1944" (Monthly reports of the Home Army's Industrial Intelligence). Published as a facsimile by Naczelna Dyrekcja Archiwów Państwowych and the chancellery of the Prime Minister of Poland, Warsaw, 2000.
144 Interview with professor Mieczysław Mołdawa, recorded by the author of this book. Copy in my possession.
145 AAN/T-175 (Alexandrian Microfilms): Records of "Personalstab des RFSS," various documents referring to the works on high frequencies.
146 J. Cera, "Tajemnice Gór Sowich," Publ: Inter-Cera, 1998.
147 P. Maszkowski, "Olbrzym (-iej) zagadki ciąg dalszy," Odkrywca magazine (Polish), issue 11/2004.
148 NARA/RG-319 "P-Files," FIAT files (Field Information Agency Technical): Intelligence Research Project No. 876 "German underground factories producing military equipment," October 18, 1944.
149 Z. Dawidowicz, "Akcja Sowa 71," Part 3: "Ludwigsdorf," Odkrywca magazine (Polish), issue 2/2006.
150 L. Paterson, "Hitler's Grey Wolfes. U-Boats in the Indian Ocean," Polish edition, Publ: L&L, 2005.
151 NARA/RG-319, "Top secret incoming and outgoing cables 1942-1952," Czechoslovakia.
152 R. Owidzki, "Prace wodne PPT," Odkrywca magazine, issue 2/2007.
153 Bundesarchiv NS-19, folder 3346.
154 A. Cowell, W. Gibbs, "German sub manaces North Sea 61 years after sinking," International Herald Tribune, January 10, 2007.
155 W. Kopczyński, A. Trautman, "Czasoprzestrzeń i grawitacja," Publ: PWN, 1981.
156 P. Fallon et al., "Superdeformed bands...," Physics Letters B, February 16, 1989.
157 M. Frąckowiak, "Neutrina i ciemna energia: niezwykłe połączenie," Urania, issue 2/2005.
158 A. Fisher, "Testing Einstein (again)," Natural History, issue 3/2005.
159 "Who is who in nazi Germany," Publ: Wiederfield and Nicolsa (London), 1982.
160 Brockhaus Encyclopaedia, issue from the 1960's.
161 http://en.wikipedia.org.
162 V. Belinski, E. Verdaguer, "Gravitational solitons," Publ: Cambridge University Press, 2005.

Part III

200 R. Manvell, H. Fraenkel, "Goebbels," Publ: Czytelnik, 1972.
201 NARA /RG-319: "Security classified intelligence and investigative dossiers 1939-1976, Box 8."
202 L. Shapiro, "Spies bid for Franco's weapons," The Denver Post, November 7, 1947.
203 Washington Daily News, May 14, 1949.
204 NARA, personal records of the SS (microfilms of the Berlin Document Center).
205 M. Kaku, J. Trainer, "Beyond Einstein," 1987.
206 O.C. Hilgenberg, "Über Gravitation, Tromben und Wellen in bewegten Medien," Publ: Giesmann und Bartsch, 1931.
207 H. Hayasaka, S. Takeuchi, "Anomalous Weight Reduction on a Gyroscope's Right Rotations around Vertical Axis of the Earth," Physical Review Letters, issue 25/1989.
208 N. Cook, "The hunt for Zero Point," Publ: Century, 2001.
209 S. Deser, "Equivalence principle violation, antigravity and anyons...," Classical and Quantum Gravity Supplement/1992.
210 G.D. Rathod, T.M. Karade, "Advance in Perihelion due to Electrogravitational Field," Annalen der Physik, 7. Folge, Band 46, Helf 6/1989.
211 Matthews, I. Sample, "Breakthrough as scientists beat gravity," Sunday Telegraph, September 1, 1996.
212 N. Cook, "Warp drive: when?" Jane's Defence Weekly, July 26, 2000.
213 S.A. Majorow et al., "Metastable state of supercooled plasma," Physica Scripta, issue 4/1994.
214 J. Barry, "Ball Lightning and Bead Lightning..." Publ: Plenum Press, 1980.
215 T. Matsumoto, "Observation of tiny Ball Lightning...," Supplement to Fusion Technology from January 23, 1994.
216 A. Marks, "Pioruny kuliste...," book published in the 70's.
217 K. Shoulders, "Energy conversion using high charge density," US patent No. 5,123,039.
218 W. Bostick, "Recent experimental results of the plasma-focus group at Darmstadt...," Journal of fusion energy, Vol. 3, issue 1/1985.
219 B.D. Mc Kersie, Y.Y. Leshem, "Stress and stress coping in cultivated plants," Publ: Kluwer, 1994.
220 G. Shipov, "Theory of the physical vacuum," I used the summary available in the internet.
221 B. Heim, "Elementarstrukturen der Materie," Vol. 2, Publ: Resch Verlag (Innsbruck), 1984.
222 B. Heim, "Elementarstrukturen...," Vol. 1, Publ: Resch Verlag, 1989.
223 D. Irving, "Kryptonim Virushaus," Publ: Książka i Wiedza, 1971 (Polish edition).
224 R. Heinrich, H.R. Bachmann, "Walther Gerlach. Physiker, Lehrer, Organisator," München, 1989.
225 W. Gerlach, "Die Verwandlung von Quecksilber in Gold," Frankfurter Zeitung, July 18, 1924.

226 W. Gerlach, "Über Bandenfluoreszenz des Quecksilbers im Magnetfeld," Helvetica Physica Acta, 2/1929, pages 280-281.
227 Gerlach's letter to Arnold Sommerfeld from January 1, 1925.
228 W. Gerlach, "Experimente zur Kernphysik" in the materials from the conference "AEG Vortragswoche an der Technischen Hochschule München," May 3-7, 1954.
229 W. Gerlach, "Über die Beobachtung eines Kugelblitzes," Die Naturwissenschaften, issue 15/1927, page 552.
230 "Operation Epsilon: The Farm Hall transcripts," Publ: Institute of Physics, 1993.
231 H. Lipiński, "Kulisy atomowego wyścigu," Gazeta Lubuska, issue 178/1976 (magazine).
232 P. Henshall, "Vengeance..." Publ: Sutton, 1995.
233 M. Walker, "Nazi science. Myth, Truth...," 1995.
234 M. Bormann, "Bormann Vermerke. Hitler's Secret Conversations 1941-1944," Publ: Signet Books, 1961.
235 NARA/RG-330 "Foreign scientist case files," Box 28, folder: "K. Debus."
236 NARA/RG-330 "Foreign scientist case files," Box 120, folder: "H. Oberth."
237 NARA/RG-330 "Foreign scientist case files," Box 164, folder: "H. Strughold."
238 L. Groves, "Now it can be told", 1962.
239 K. Kąkolewski, "Co u pana słychać?" a book.
240 M. Dudziak, "Riese," Publ: JMK, 1996.
241 J. Lamparska, "Tajemnice ukrytych...," Publ: ASIA Press, 1995.
242 A. Szymura, "Zamkowe Podziemia," Światowid, issue 14/1961.
243 S. Orłowski, "Tajny obiekt Fürstenstein, pierwszy świadek," Życie Warszawy, September 17-18, 1977.
244 M. Mołdawa, "Gross Rosen," Publ: MON, 1990.
245 Interview with Prof. M. Mołdawa recorded on a video tape, in author's possession.
246 J. Wilczur, "Zagłada Wielkiej Sowy," Prawo i Życie, issue 17/1964.
247 J. Cera, "Tajemnice Gór Sowich," Publ: Inter-Cera, 1998.
248 R. Vesco, "Intercept but don't shoot," Publ: Grove Press, 1971.
249 "The New York Times," December 14, 1944.
250 AAN/Alexandrian Microfilms, records of "Reichsforschungsrat," Doc. 001484, report: J-9181.
251 NARA/Intelligence Publications: Air Intelligence Report. Vol. 1, No. 8, April 26, 1945. (XXI Bomber Command).
252 NARA/RG-319 "Index to the ID File," Box 117.
253 Ch. Friedrich, "Secret nazi polar expeditions," 1979.
254 J. Sayer, D. Botting, "Złoto III Rzeszy," Publ: Sensacje XX wieku, 1999 (Polish edition).
255 NARA /RG-38, Box 13. Document OP-20-3-G1-A.
256 I. Witkowski, "Code-name Regentröpfchen...," to be published in 2004.
257 Dönitz's speech "Der Stürmer," June 17, 1938.
258 P.W. Stahl, "Geheimgeschwader KG-200," Publ: Motorbuch Verlag, 1978.
259 T. Agoston, "Teufel oder Technokrat. Hitlers graue Eminenz," Publ: Nikol, 1993.
260 R.A. Kolitzky, "A jednak małpa," Przekrój, September 6, 1981.
261 W.H. Bostick et. al., "Pair Production of Plasma Vortices" Phys. of Fluids. Vol. 9 (1966), page 2078.
262 O. Gedymin, "Antygrawitacja i możliwości jej realizacji," Wojsk. Przegląd Lotniczy, issue 7/1958.
263 "Samaranganasutradhara," Fragment from D. Childress, "Vimana...," Publ: AUP, 1995.
264 NARA/RG-38 "Intelligence Division Top Secret reports of Naval Attaches 1944-1947," Box 11, report 35-S-46 and others.
265 W. Lüdde-Neurath, "Regierung Dönitz, die letzten Tage des Dritten Reiches," Göttingen, 1964.
266 H. Thomas, "SS-1, the unlikely death of Heinrich Himmler," 2001.
267 H. Lommel, "Geheimprojekte der DFS vom Höhenaufklärer bis zum Raumgleiter 1935-1945," Motorbuch Verlag, 2000.
268 Eberhard and Rita Völkel, "Ludwigsdorf im Eulengebirge", published by the authors.

ACKNOWLEDGMENTS

I wish to sincerely thank the individuals mentioned below for their valuable opinions and help in enabling me to acquire the source materials, requisite for the principle part of this book being made: Prof. M. Mołdawa, Prof. M. Demiański, Prof. M. Sadowski, Prof. A. Kacperska, Prof. J. Auleitner, Dr. K. Godwod, Dr. Hal Puthoff, Dr. M. Scholz, Dr. A. Marks, Dr. M. Paszkowski, Dr. R. Zagórski, N. Cook from "Jane's Defence Weekly," M. Banaś, R. Bernatowicz, G. Brooks, J. Cera, J.P. Farrell, A. Kotarski, A. Kuczyński, J. Lamparska, R. Leśniakiewicz, W. Łuszczewski, J. Marrs, H. Stevens, W. Wiktorowski, M. Wiśniewski, B. Wróbel, as well as the employees of the Institute for National Remembrance (Warsaw), the AAN, CAW and ADM Archives (Warsaw), the National Archives and Records Administration (College Park, USA), the National Air Intelligence Center (Dayton, USA) and Centro Atomico Bariloche (Argentina), and also all those whose names I have not mentioned.

The following images originate from the Bundesarchiv, Koblenz:
p. viii (above) Bild 146-1992-093-13A / Photo: Hanns Hubmann
p. viii (below) Bild 183-H28426
p. 3 Bild 146-1992-068-19A
p. 44 (below) Bild 183-L21649 / Photo: Lysiak
p. 112 (above) Plak 003-009-223

The author welcomes any correspondence related to the topics of this book. Please e-mail to:

contact@igorwitkowski.com

ABOUT THE AUTHOR

Igor Witkowski was born in 1963 in Warsaw, Poland. Since 1990 he has worked as a journalist dealing with military technology as well as the history of World War II and the development of science and technology. He was editor-in-chief of the magazines "Technika Wojskowa" ("Military Technology") and "World War II." So far he has also written over 50 books and around 250 articles on these subjects, as well as on the development of our civilization. Since 1997 he has dedicated himself mainly to the collection and analysis of materials — the sources for this book. He has worked in various archives on three continents. "The Truth about the Wunderwaffe" is the crowning achievement of this work.

CPSIA information can be obtained
at www.ICGtesting.com
Printed in the USA
BVHW011028030222
627755BV00005B/404